AF335214

Microelectronic Packaging

New Trends in Electrochemical Technology Series

Series Editors

M. Datta
Cooligy, Inc., California, USA

T. Osaka
Waseda University, Japan

Volume 1
Energy Storage Systems in Electronics
T. Osaka and M. Datta

Volume 2
Electrochemical Microsystem Technologies
J.W. Schultze, T. Osaka and M. Datta

Volume 3
Microelectronic Packaging
M. Datta, T. Osaka and J.W. Schultze

New Trends in Electrochemical Technology

Microelectronic Packaging

EDITED BY

M. Datta
T. Osaka
J.W. Schultze

CRC PRESS

Boca Raton   London   New York   Washington, D.C.

Includes bibliographical references and index.
ISBN 0-415-31190-X (alk. paper)
 1. Microelectronic packaging. I. Datta, Madhav. II. Osaka, Tetsuya, 1945- III. Schultze, J. W. (Joachim Walter) IV. Series.

TK7870.15.M53 2004
621.381'046--dc22   2004050337

Contents

Contributors

Haruo Akahoshi, Applied Electrochemistry Unit, Department of Materials and Devices Research, Materials Research Laboratory, Hitachi, Ltd., 7-1-1 Oomika-cho, Hitachi-shi, Ibaraki-ken 319-1292, Japan

Vadim Bogush, Department of Physical Electronics, Tel-Aviv University, Ramat-Aviv, 69978, Israel

Jeffery W. Butterbaugh, FSI International, 3455 Lyman Blvd., Chaska, MN 55318, USA

Akira Chinda, Research and Development Center, Hitachi Cable, Ltd., 3-1-1 Sukegawa, Hitachi City, Ibraki-pref., 317-0065, Japan

Nathan Croitoru, Department of Physical Electronics, Tel-Aviv University, Ramat-Aviv, 69978, Israel

Madhav Datta, Cooligy, Inc. 2370, Charleston Road, Mountain View, CA 94034, USA

Valery M. Dubin, Logic Technology Development, Intel Corporation, 2501 NW 229th Street, Hillsboro, OR 97124, USA

Moshe Eizenberg, Technion IIT, Haifa, 32000, Israel

Darrel R. Frear, Freescale Semiconductor, MD EL 725, 2100 East Elliot Road, Tempe, AZ 85284, USA

Dakin Fulton, Semitool Inc., 655 West Reserve Drive, Kalispell, Montana 59901, USA

Kazuhiro Ikuina, Infineon Technologies, Corporate Assembly and Test Division, Takanawa Park Tower 12F, 3-20-14 Higashi Gotanda, Shinagawa-ku, Tokyo 141-0022, Japan

Alexandra Inberg, Department of Physical Electronics, Tel-Aviv University, Ramat-Aviv, 69978, Israel

Saikumar Jayaraman, Assembly Technology Development, Intel Corporation, MS CH5-159, Chandler, AZ 85226, USA

Norio Kimura, Ebara Technologies, Inc., San Jose, CA 95134, USA

Amit Kohn, Technion IIT, Haifa, 32000, Israel

Paul Kohl, School of Chemical and Biomolecular Engineering, Georgia Institute of Technnology, 311 Ferst Drive, Atlanta, GA 30332-0100, USA

Sol Krongelb, T.J. Watson Research Center, IBM Corporation, P.O. Box 218, Yorktown Heights, NY 10598, USA

Sergey Lopatin, AMD, 1 AMD Place, Sunnyvale, CA 94008, USA

W.H. Lytle, Motorola Inc., MS MDEL 725, 1200 E. Elliot Rd., Tempe, AZ 85284, USA

Thomas Marieb, Logic Technology Development, Intel Corporation, 2501 NW 229th Street, Hillsboro, OR 97124, USA

Peter Moon, Logic Technology Development, Intel Corporation, 2501 NW 229th Street, Hillsboro, OR 97124, USA

Yutaka Okinaka, Advanced Research Institute for Science and Engineering, Waseda University, 3-4-1 Okubo, Shinjuku-ku, Tokyo 169-8555, Japan

Eric D. Perfecto, IBM Microelectronics, East Fishkill Facitlity, Route 52, Hopewell Junction, NY 12533, USA

Nick Petrov, Tel-Aviv University, Tel-Aviv, 69978, Israel

Tom Ritzdorf, Semitool Inc., 655 West Reserve Drive, Kalispell, MT 59901, USA

Lubomyr T. Romankiw, T.J. Watson Research Center, IBM Corporation, P.O. Box 218, Yorktown Heights, NY 10598, USA

Yosi Shacham-Diamand, Department of Physical Electronics, Tel-Aviv University, Ramat-Aviv, 69978, Israel

Sadasivan Shankar, Logic Technology Development, Intel Corporation, Santa Clara, CA 95052, USA

Yuzo Shimada, Functional Materials Research Laboratories, NEC Corporation, 1-1, Mayazaki 4-chome, Miyamae-ku, Kawasaki, Kanagawa, 216-8555, Japan

Harsono S. Simka, Logic Technology Development, Intel Corporation, Santa Clara, CA 95052, USA

Yelena Sverdlov, Department of Physical Electronics, Tel-Aviv University, Ramat-Aviv, 69978, Israel

John Tang, Assembly Technology Development, Intel Corporation, MS CH5-158, Chandler, AZ 85226, USA

Manabu Tsujimura, Ebara Technologies, Inc., San Jose, CA 95134, USA

Kazuaki Utsumi, NEC Lamilion Energy Ltd., 1-1, Mayazaki 4-chome, Miyamae-ku, Kawasaki, Kanagawa, 216-8555, Japan

Shinichi Wakabayashi, Shinko Electric Industries Co. Ltd., 36 Kita-Owaribe, Nagano-shi, 381-0014, Japan

Vijay S. Wakharkar, Assembly Technology Development, Intel Corporation, MS CH5-159, Chandler, AZ 85226, USA

David K. Watts, ACuTE, Inc., 761, Santa Ray Avenue, Oakland, CA 94610, USA

Keith K.H. Wong, IBM Microelectronics, East Fishkill Facility, Route 52, Hopewell Junction, NY 12533, USA

Osamu Yoshioka, Research and Development Center, Hitachi Cable, Ltd., Otemachi Building 1-6-1, Otemachi, Chiyoda-ku, Tokyo, 100-8166, Japan

Preface to the series

While electrochemical technology has been playing an important role in the manufacturing of chemicals, metals, batteries, fuel cells, and other electric utility industries, its contribution in the evolution of electronics and communication industries has been phenomenal. During the past several decades the electronics industry has been through a very rapid evolution from thick to thin films and to ever increasing miniaturization. Electrochemical technology played a decisive role in the direction of this evolution. With an increasing understanding of transport processes, current distribution problems, process monitoring and control issues, and the ability to develop environmentally friendly processes, electrochemists and electrochemical engineers have been able to face the challenges presented by the electronics industry. Several new and unique electrochemical processes and technologies have been introduced. They include advanced plating and etching technologies, cleaning and planarization technologies, and silicon processing technologies. These technologies have made a significant impact in storage, packaging, interconnects, and several other aspects of the microelectronics industry. The importance of smart energy storage systems in the sophistication of advanced portable electronic devices is becoming increasingly recognized. Microelectromechanical systems, sensors, and nanotechnology are some of the other evolving areas where electrochemical processing technologies are playing an increasingly important role. In the aerospace and automobile industries, machining and finishing of complicated shaped components of different materials including high strength alloys have been possible due to development and implementation of advanced electrochemical machining, polishing, anodization, and plating technologies.

The series is intended to report recent advances made in the development of electrochemical technologies as applied to microelectronics, information and other high-tech industries as well as to heavy industries. The emphasis will be on the application of newly developed concepts and materials. Internationally recognized experts from scientific and industrial communities will provide the current status and future trends of a technology area. The first two volumes were devoted to energy storage systems for electronics and electrochemical microsystem technologies. This volume covers the field of microelectronic packaging. Future volumes in the series will include topics such as nanotechnology, advanced chip metallization, electro-chemically fabricated tailored materials, and micro/nano electromechanical systems (MEMS, NEMS).

Tetsuya Osaka, Waseda University
Madhav Datta, Cooligy, Inc.

Preface

Spectacular scientific and engineering advances in microelectronics have brought about a pervasive and beneficial influence in our day-to-day living from all forms of communication to extremely delicate and sophisticated medical applications to supercomputers for defense and space exploration. All these started with the invention of the transistor in 1948 and the integrated circuit chips soon thereafter. Continued miniaturization trends in integrated circuits led to the migration of inter circuit wiring and connections from boards, cards, and modules to the chip itself.

In the microprocessor industry, developments in transistor technology continue to follow an exponential progress represented by Moore's law — doubling the number of devices per chip every 18 months. These developments have been so rapid that all other technologies to put these devices to work have been challenged. Meeting these challenges requires development of advanced packaging solutions. Indeed, solutions using advanced materials for microprocessor scaling, heat management systems, and improvements in package substrates continue to drive major packaging efforts while market constraints continue to exert significant cost pressures. In the current situation where semiconductor chips are becoming commodities, the microelectronics packaging aspects are becoming the value added part of the product. Accordingly, the microelectronics industry is placing a heavy emphasis on packaging technologies.

Phenomenal miniaturization and increased performance in microelectronics are results of continuous developments in electronic materials, processing technologies, and unique integration schemes. Along with lithography and vacuum technologies, electrochemical processing played a decisive role in micro- and nano-scale processing. Compared to competing vacuum processing, electrochemical processing technologies have emerged as more environmentally friendly and cost-effective micro- and nano-fabrication methods. Electrochemical processing has thus become an integral part of wafer processing fabrications and an enabling technology in many aspects of microelectronic packaging.

The present book is intended to capture the impact of electrochemical technology in various levels of microelectronic packaging. Traditionally, interconnections within a chip are considered outside the realm of packaging technologies. This book, on the other hand, emphasizes the importance of chip wiring as a key aspect of microelectronic packaging and focuses on electrochemical processing as an enabler of advanced chip metallization.

Packaging hierarchy and the importance of advanced processing tools have been taken into consideration in planning the scope and organization of the book. The book has been divided in five parts. The introductory chapter in Part I outlines some of the basic aspects of electrochemical processing, defines the microelectronic packaging hierarchy, and emphasizes the impact of electrochemical technology in different levels of packaging. Part II consists of three chapters on chip metallization covering topics from current technology status to research and development efforts in resolving electromigration issues through development of robust barrier layers and alternative chip metallization materials. Chapters in Part III, devoted to key aspects of chip–package interconnect technologies, deal with the development trends in tape carrier, area array chip package interconnection by flip-chip (C4) technology, wafer-level packaging with embedded air gap, and lead-free flip-chip technologies. Five chapters on packages, boards, and connectors are assembled together in Part IV covering aspects of materials development, technology trends in ceramic packages and multichip modules, bumping technology for advanced packages, plated through-hole technology for boards, and electroplated contact materials. Emphasizing the importance of processing tools in enabling technology development, Part V contains chapters on chemical–mechanical planarization, electroplating, and wet etching/cleaning tools.

As noted earlier, microelectronic packaging industry is evolving rapidly. Accordingly, many novel miniaturized packaging concepts are currently being developed which include technologies such as System In Package, Micro Chip Scale packaging, and three-dimensional chip stacking. These technologies which also use electrochemical processing for interconnects have not been covered separately in this book. This is consistent with the objective of the book which is to document the role of electrochemical processing in microelectronic packaging rather than describing the different packaging technology concepts. An important aspect of microelectronic packaging not covered in this book involves thermal management systems, which include electro-kinetic pumps and micro-heat exchangers. We expect to cover these two topics in another volume dedicated to sensors, actuators, micromechanics, and micro-electromechanical systems in general. Some other topics that are important but not included in this book include environmental aspects of electrochemical processing and electrochemical processing of semiconductors and compound semiconductors for device fabrication.

We sincerely thank the authors and referees for their help in the making of this book.

Madhav Datta
Tetsuya Osaka
J. Walter Schultze

Part I

Introduction

1 Electrochemical processing technologies and their impact in microelectronic packaging

Madhav Datta

1.1 Introduction

Increasing demand for high-performance microelectronic products is placing a heavy emphasis on the need for rapid advances in packaging technologies. In the semiconductor industry, developments in transistor technology continue to follow an exponential progress represented by Moore's law – doubling the number of devices per chip every 18 months. Successful research scale demonstration of a 20 nm transistor is a testimony to the advances made in this field.[1] These developments have been so rapid that all other technologies to put these devices to work have been challenged. In order to integrate such highly performing transistors, the industry is currently focusing on the development of advanced microelectronic packaging technologies. They include alternative chip metallization with much improved electromigration properties; alternative chip–package interconnect technologies and materials that are environmentally friendly and are compatible with material changes in chips and packages; and thinner and denser packages and printed circuit boards (PCBs). Electrochemical processing technologies are expected to make significant impact on all of these aspects of microelectronic packaging.

Advances in electrochemical technology have played a major role in the phenomenal growth of storage, chip interconnects, microelectronic packaging, microelectro-mechanical systems (MEMS), and many other microelectronic and micromechanical components. Figure 1.1 shows key technology and product developments in the electronics industry that are directly linked to the advances in electrochemical processing. The vertical arrows in Figure 1.1 indicate the approximate year at which a viable manufacturing technology for a given product was first demonstrated. Further chronological developments and advances in these technologies have not been captured in the figure. No attempt has been made to present an exhaustive list of the contributions of electrochemical technology in different electronic products. Rather, only some selected key items are highlighted that demonstrate the impact of electrochemical technology in the miniaturization and sophistication of electronic products.

Electrochemical processing entered the electronics industry in the 1940s to fabricate PCBs by chemical etching of patterned copper laminates.[2] The first commercial application of electroless copper was reported in the mid-1950s with the

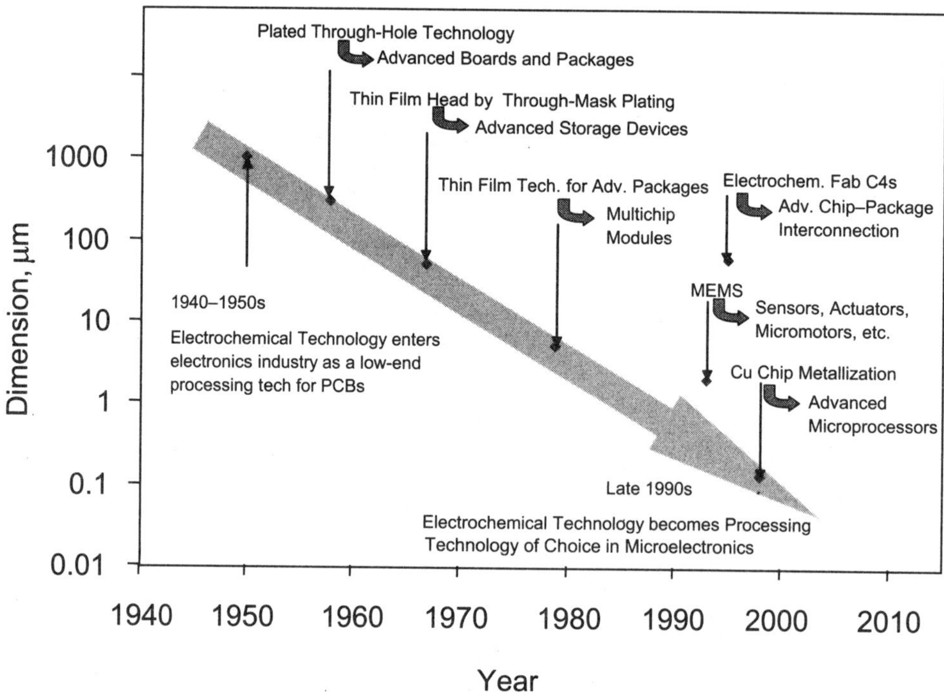

Figure 1.1 The impact of electrochemical technology in microelectronics and its miniaturization trends. Vertical arrows indicate the approximate year at which a viable manufacturing technology for a given component was first demonstrated.

development of plating solutions for plated through-hole (PTH) printed wiring boards.[3] Continued advances in high aspect ratio PTH technology, and in fine line wiring technology have contributed to the development of advanced boards and packages that are used today. An important breakthrough in electrochemical processing took place in conjunction with the developments in lithography in the late 1960s and early 1970s, when the application of through-mask plating for thin film heads was demonstrated and implemented in high-volume manufacturing.[4] Continued efforts on the development of novel magnetic materials and their precision processing have led to the advanced storage devices.[5] In the 1970s and 1980s, electrochemical processing technologies started to make a wider impact in different areas of microelectronics that included connectors and interconnects, metallization for multichip modules, and other microelectronic packages.[6,7] Phenomenal advances in electrochemical processing occurred in the 1990s when it enabled fabrication of MEMS and cost-performance flip-chip interconnects; and in 1997, it enabled a paradigm shift in chip making with the introduction of Cu metallization.[8–10] Indeed, the dual-Damascene process for copper chip metallization, and the flip-chip (C4) technology for high density, area array chip–package interconnection have placed electrochemical processing technologies among the most sophisticated processing technologies employed in the microelectronics industry today.[11] Due to its cost–performance advantages, electrochemical processing now has emerged as the

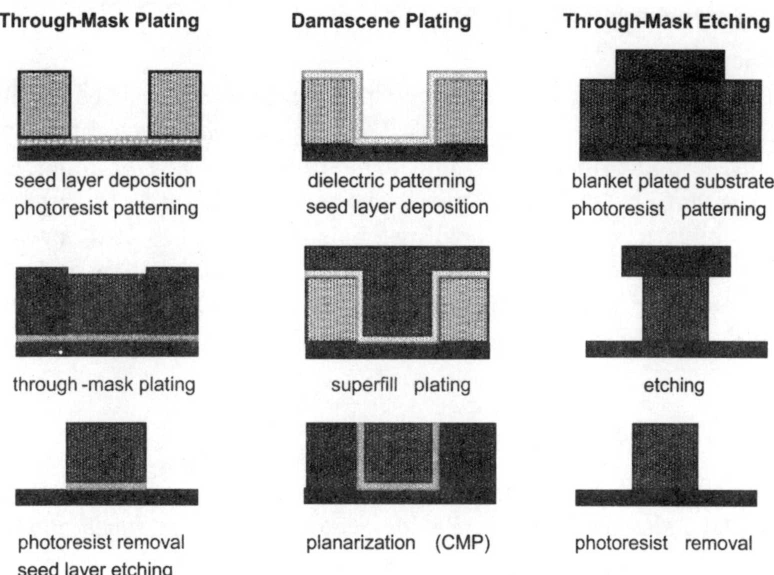

Figure 1.2   Three different types of electrochemical processing techniques applicable in the fabrication of a metallic structure.

technology of choice for the manufacturing of a variety of microelectronic components from low-end consumer products to advanced microprocessors.

1.2   Electrochemical processing technologies

Electrochemical processing technologies employ both electrolytic and electroless methods of metal deposition and dissolution. Depending on the application, processing may involve deposition/dissolution of blanket layers or it may involve localized through-mask fabrication of micro- or nanostructures. Processes that depend on redox reactions without the application of electric current include electroless plating and chemical etching. The electrolytic processes include cathodic processes such as electroplating and electroforming, and anodic processes such as electroetching and electropolishing. An essential requirement of electrolytic processing is the presence of a continuous metallic layer (seed layer), which conducts electric current from the contact at the sample edge to all points in the sample where deposition/dissolution is desired.

Three different types of electrochemical processing techniques applicable in the fabrication of a metallic structure are (1) through-mask plating, (2) Damascene plating, and (3) through-mask etching (Figure 1.2). Damascene plating is another version of patterned plating, which brought about a revolution in the fabrication of chip interconnect and in electrochemical processing. In Damascene plating, the patterned dielectric remains intact and forms a functional part of the structure. A continuous seed layer conformally covers the patterned dielectric. Plating occurs all over the surface thus creating challenges for void-free structure fabrication.[10] Chemical–mechanical polishing (CMP) is used for planarization and removal of excess "overburden" metal, and seed layers.

Advances in electrochemical processing technologies have been possible due to simultaneous progress in different areas. They include continuous improvements and innovations in photolithography, fundamental understanding of engineering principles that govern electrochemical micro- and nanofabrication processes, understanding of the ability to produce tailored materials and structures, and development of high yielding electrochemical processing tools that are compatible with ultra-clean semiconductor fabrications. In the following, the basic principles of electrochemical metal deposition and removal processes are described.

1.2.1   Metal deposition processes

Electrochemical metal deposition processes by electrolytic as well as electroless methods find industrial applications for decorative surface finish and functional purposes in electronics, micro-mechanics, automotive, heavy engineering, and consumer products. They include processes such as blanket plating, pattern plating, and electroforming.

1.2.1.1   Electrolytic plating

Electrolytic plating (also known as electroplating or electrodeposition) involves cathodic deposition of metals, alloys, and other conducting materials by the application of an external current. Electroplating is widely employed in a variety of applications ranging from blanket deposition for wear resistance and corrosion resistance to nanoscale feature fabrication for ultra-large-scale integration (ULSI). Depending on the application, the plated thickness may vary from few angstroms of uniformly deposited compact films to electroformed structures that are millimeters thick. In many applications, metals, alloys, and metal matrix composites are deposited for the fabrication of single or multilayered structures and coatings. Compared to vacuum deposition processes, electrolytic deposition is less expensive and it possesses some unique features that make it an enabling technology for applications such as chip metallization and flip-chip bumping. On the other hand, electroplating process control is sometimes complicated due to unknown proprietary additives used in the bath. Although some aspects of electroplating still remain empirical, the gap between fundamental understanding of electrodeposition processes and their manufacturing application is narrowing.[12]

Development of electroplating processes and tools for fabrication of micro- and nanostructures requires a thorough understanding of the underlying electrochemical engineering principles. This involves the understanding and application of the principles of mass transport, and current distribution. Mass transport limits the rate in electrochemical processing and influences the current distribution and the microstructure. While a detailed description of electrochemical mass transport and current distribution is beyond the scope of this chapter, their importance in influencing electrochemical processing will be highlighted in the following paragraphs.

Mass transport and current distribution are of critical importance for the optimization of electrochemical processes and tools.[13] They determine the thickness and uniformity of blanket deposited surfaces. In patterned processing, they also determine the shape evolution, leveling, and superfilling. Three different types of current distribution are generally distinguished.[14] When the electrode potential does not vary significantly with current density, the primary current distribution prevails, which depends solely on the cell geometry. In the presence of electrode polarization, secondary current distribution prevails, which is determined by the charge transfer kinetics. And finally, when the electrochemical processing is influenced by mass transport conditions, the tertiary current distribution prevails, which depends on both the potential distribution in the bulk and on the local rate of mass transport. For optimization of tools and process parameters, it is often sufficient to consider the limiting cases of primary current distribution or mass transport controlled tertiary current distribution.

In the electrochemical processing of patterned structures, additional factors influence the current distribution. In such cases, the current distribution is considered on three different scales: substrate scale, repeating pattern scale, and feature scale.[15] At the substrate level, the current distribution is governed by the overall cell geometry and the uniformity of current distribution is generally achieved by using auxiliary electrodes or current shielding concepts. On the pattern scale, the current distribution depends on the feature geometry and the spacing. Current density on a feature is higher when it is spaced farther away from a neighboring feature. On a feature scale, the current distribution evolves with time due to the continuously changing shape of the feature. In electroplating, the current distribution within the features is influenced by the use of suitable additives.[10,15,16]

The influence of additives on cathodic leveling during electrodeposition has been studied by many authors.[10,15–19] A leveling agent is a suitably chosen additive that generally acts as an inhibitor for the metal deposition reaction. It is consumed at the cathode and its reaction rate is mass transport controlled.[16–19] Since peaks are more accessible than valleys, they are more strongly inhibited by the additives, leading to preferential metal deposition into recess. These concepts have been used to develop understanding of superfilling during dual-Damascene plating and to develop electroplating baths for Cu interconnects.[10]

Electrodeposition offers the possibility of fabricating structures with tailored materials and properties. Small changes in process conditions can have an enormous effect on the microstructure and composition of the deposit and hence its properties. These small changes may be in the form of additives or complexing agents or in the deposition parameters. Precise control of the additives and process parameters is extremely critical to provide reproducibility and uniformity of deposition. Frequently, internal stress develops in electrodeposits, which can cause cracking or loss of adhesion. Mismatch between substrate and deposit, grain coalescence during growth, and incorporation of additives or hydrogen may contribute to the internal stress. It is well known that electrodeposition under limiting current conditions leads to dendritic or powdery deposits.[20,21] Therefore, the value of the operating current density with respect to the mass transport limited current density is a critical parameter for deposit morphology.

Copper and solder alloys are the most commonly electrodeposited materials in the microelectronics industry. An enormous amount of literature is available on the mechanism of copper deposition.[20–25] The role of inhibitors and of applied current density on the deposit morphology has been reported in the literature.[21,23–25] Increasing inhibition and/or increasing current density produces fine-grained structures. Kinetically limited growth tends to favor compact columnar or equiaxial growth in copper deposits while mass transport limited growth favors formation of loose dendritic deposits.

Commercial copper baths are generally acid sulfate solutions that contain small amounts of chloride ions in conjunction with two or more organic additives. These additives are essential for influencing the structure or roughness of the deposit.[20,21] During electroplating of micro-features, they are essential for leveling and super-filling.[10,26] Additives that act as inhibitors tend to promote formation of fine equiaxial grains, which may lead to increased internal stress. Electrodeposited copper may contain non-equilibrium grain structures, which spontaneously recrystallize even at room temperature.[27] As a consequence, structure-dependent properties such as sheet resistance, and internal stress of deposits may change slowly with time after deposition.

Pb-containing solders are the principal joining materials used in the micro-electronics industry today.[9] These solders are mainly high lead solders with high melting temperature, especially 97Pb3Sn and 95Pb5Sn, or the low melting temperature eutectic composition (63Sn37Pb). PbSn alloy codeposition is one of the simplest alloy electroplating processes due to the closeness of the standard cathodic reduction potentials; Pb and Sn from their ionic state in solution are separated by only 10 mV. Furthermore, since the reduction potential of hydrogen is slightly more noble (positive) than their potentials, and since both of them have high hydrogen over-potential, their alloy can be electrodeposited from acid solutions with high cathode efficiency close to 100%. Commercially available methane sulfonate based lead and tin proprietary electrolytes are commonly used for solder deposition.[9,28] The addition agents in electrolyte have been found to control both the deposition morphology and the alloy composition. Use of pulsating current has been found to produce fine-grained deposits and minimize the need for additives.[29]

1.2.1.2   *Electroless plating*

Electroless plating is an autocatalytic metal deposition process from solution. The process does not require the use of an external current. In electroless deposition, the electrons required for the metal reduction are supplied by the simultaneous oxidation of a reducing agent in the solution. For the process to take place heterogeneously, the metallizing surface must be energetically favorable for oxidation of the reducing agent and it must also be electronically conductive for the transfer of electrons to take place. Plating is initiated on a catalyzed surface and is sustained by the catalytic nature of the plated metal surface itself. For oxidation of formaldehyde, for example, copper is the best catalytic surface followed by gold, silver, platinum, palladium, nickel, and cobalt.

Compared to electroplating, electroless plating is a relatively slower process and requires elaborate bath control. However, it offers several advantages that make it a method of choice in certain applications. Electroless metallization plates uniformly over all surfaces, regardless of size and shape. It may be plated onto non-conductors or conductive surfaces that do not share electrical continuity. The ability to plate large racks of substrates simultaneously is an advantage in some applications. Because of these advantages, electroless processes are used for a variety of metallization applications that include Cu, Ni(P), Co(P), CoW(P), and Au for wiring, seed layer, diffusion barrier layer, and corrosion and wear protection layers.[30–34] Copper is the most widely metallization layer that is deposited by electroless plating. The most important commercial application of electroless plating is in the plated through-hole process for fabricating printed wiring boards.[7,34] Other applications include buildup of conducting lines in multichip modules, functional plating on plastics for electromagnetic interference (EMI) shielding of electronic components, composite connectors, and molded interconnect devices.[31,34,35] The use of electroless copper in chip metallization has also been demonstrated but it has yet to be commercially implemented.

The basic components of an electroless plating bath are the metal salt and a reducing agent. In an electroless copper bath, for example, the source for copper is a simple cupric salt such as copper sulfate, chloride, or nitrate. Various common reducing agents for electroless copper bath suggested in the literature include formaldehyde, dimethylamine borane, hypophosphite, and oxalic acid.[30,34] In practice, however, formaldehyde is the most popular reducing agent for electroless copper baths. In spite of considerable pressures exerted on the plating industry by environmental and regulatory agencies because of its toxicity, formaldehyde baths continue to be commercially popular because of their cost effectiveness and ease of control.

Formaldehyde based electroless copper baths employ high pH, above 12. Since simple copper salts are insoluble at pH 4 and above, a complexing or chelating agent is necessary. These agents may be one of the following: ethylenediamine-tetraacetic acid (EDTA), alkanol amines, or tartaric acid. At present most of the electroless copper baths are based on EDTA. Other components in the bath are the additives that are generally a proprietary portion of the formulation. They include: stabilizers, rate enhancers, and in some cases surfactants. Stabilizers are employed at low concentrations, typically 1–100 ppm, and constitute compounds such as thiourea, mercaptobenzothiazole, other sulfur compounds, cyanide or ferrocyanide salts, and heterocyclic nitrogen compounds. Rate promoters are present in the solution at concentrations 0.1 M or higher. The rate promoters are inorganic salts such as chlorides, chlorates, nitrates, ammonium salts, perchlorates, molybdates, and tungstates.[30,34]

Electroless copper baths are generally operated with air sparging to saturate the solution with oxygen. Dissolved oxygen is effective in stabilizing electroless systems from spontaneous decomposition since oxygen prevents disproportionation by further oxidation of cuprous ions in the bath to cupric ions.[36] The solubility of dissolved oxygen in the solution is proportional to the salt content. For formaldehyde based systems, this is governed by the accumulation of formate. According to the mixed potential theory, the magnitude of the oxygen reduction current can alter the potential of the depositing surface.

The conductivity and catalytic nature of the substrate dictate the initial growth behavior in electroless copper deposition. The microstructure of copper deposits follows the crystallographic characteristics of the substrate. However, controlled fine-grained deposits can be obtained by using specific additives that can strongly adsorb on the copper surface and can modify the natural grain development. Epitaxial growth is inherent with electroless copper deposition. This is due to high mobility of copper atoms and also due to the fact that copper recrystallizes easily. Furthermore, since copper surface is a good catalyst for formaldehyde oxidation and is good conductor for electrons, nucleation is expected to be uniform on all growth fronts, favoring epitaxial growth. This behavior allows one to tailor the grain structure of electroless copper films.

1.2.2   Metal removal processes

Electrochemical metal removal processes are commonly known as wet etching. Processes where the energy source for the metal dissolution reaction is derived from the etchant are known as chemical etching. Electroetching, on the other hand, relies on the passage of an external electric current for metal dissolution to take place at the workpiece, which is made an anode in an electrolytic cell. Compared to dry etching technologies, wet etching methods are used in microfabrication because of their selectivity, high etch rates, and relatively low capital investment.

1.2.2.1   Chemical etching

Wet chemical etching involves removal of unwanted material by the exposure of a workpiece to an etchant whereby the exposed material is oxidized by the reactivity of the etchant to produce reaction products that are carried away from the surface by the medium. Wet chemical etching involves conversion of a solid insoluble material to a soluble form. The extended lattice of metal atoms in the solid state is broken down so that these atoms can enter the solution as soluble compounds. This is accompanied by removal of electrons from the metal. These electrons are accepted by the etchant, which acts as an oxidizing agent. For example, in the chemical etching of copper by ferric chloride solution, the etchant is reduced to produce different complexes in solution. The complexation reaction is dependent on the amount of chloride ions and water molecules. The metal removal reaction typically involves several sequential steps. The dissolution kinetics is controlled by the chemical reactivity of the species involved (activation controlled). In many systems, the metal removal rate is determined by the speed at which the reaction product is removed from the surface and the fresh reactant is supplied to the surface (mass transport controlled). Temperature variations also profoundly influence the kinetics of metal removal reaction.

In through-mask wet etching, shape evolution is controlled by several phenomena such as dissolution kinetics, surface films, and mass transport.[13,37] Wet chemical etching is generally isotropic in nature (i.e., the material is etched both vertically and laterally at the same rate). The etch boundary, therefore, recedes at a 45 degree angle relative to the surface. A study of the effect of gravity field on chemical etching using an ultracentrifuge showed that a certain degree of anisotropy in chemical etching can be obtained if the acceleration field is such that the denser fluid, present near the wall, is drawn out of the cavity.[38] This leads to the development of eddies within the cavities thus resulting in an anisotropy in mass transport rate. Under optimal conditions the etch factor can thus be reduced, permitting fabrication of deeper cavities than under pure mass transport controlled conditions.[38]

Wet chemical etching is employed in the electronics industry to fabricate a variety of components. They range from removal of seed layers in chip making to two-sided through-mask etching of thick sheets of molybdenum and stainless steel for fabrication of evaporation and screening masks.[9,37] Chemical etching baths contain chemicals that are generally corrosive and toxic, thus posing safety and disposal issues. In many manufacturing processes involving chemical etching, waste treatment and disposal costs often surpass actual etching process costs.[37] Ever increasing cost of incineration and imposition of landfill restrictions are the main reasons behind the need for developing processes with particular emphasis on waste disposal issues.

1.2.2.2 Chemical–mechanical polishing

Chemical–mechanical polishing (CMP) involves intimate contact between a wafer surface and a pad that is charged with carefully formulated colloidal slurry. The relative motion between the wafer and the pad combined with applied pressure and chemical activity of the slurry results in polishing of the wafer surface. CMP has found application in various aspects of semiconductor manufacturing from transistors to chip interconnects and involve polishing/planarization of a variety of materials that include silicon, silicon dioxide, silicon nitride, tungsten, tungsten nitride, tantalum, tantalum nitride, titanium, titanium nitride, platinum, copper, and polymers.

As the name implies, material removal and planarization in CMP is a combination of chemical and mechanical action. The material removal rate is dependent on variables such as the type of pad, slurry composition, pad conditioning, and applied pressure. However, the effect of chemical action vs. mechanical action is not easily separable. Similar to electroplating, an understanding of the physics of CMP has been generated by modeling the process on wafer scale and feature scale.[39–42] The most common wafer scale model for material removal is the classical Preston's equation,[43] which relates material removal to the applied pressure and the linear velocity at the interface. Wafer scale modeling also took into account the tribological aspects of the fluid film between the wafer and the pad.[40] Feature scale modeling approach assumed that the material removal is related to sharp edges that are defined as slurry particles embedded in the fibers of the polishing pads. The elastic modulus of the slurry and the material to be polished are the critical factors in determining the CMP removal rate.[41] These models, however, do not directly include the contribution of chemical action from the slurry. Based on the development of CMP process for tungsten and copper, Kaufman et al.[44] suggested a model that includes the action of a metal etchant, a metal passivation agent, and an abrasive agent. According to the model, actions of these agents result in the removal of passivating films from the high spots while such films continue to protect the low spots. Continuous cycles of formation, removal, and reformation of the passivating layer continue during the CMP process. However, this model also fails to explain the exact role of the etchant. Indeed, the mechanisms involved in CMP are far from being understood.

The material removal and polishing at a given point in CMP is determined by the topography and the relative heights of the surrounding features. Coplanarity and dishing are the critical issues that form the basis of a CMP development process. A detailed description of these and other aspects in CMP is given in Chapter 15.

Development of CMP process for Cu was one of the milestones that made the implementation of Cu interconnect possible. An important aspect of the Cu CMP process development involved consideration of the interactions between Cu plating and Cu CMP. These interactions are at both local and global levels and create challenges in process integration. Overfill effect of dense lines during electroplating is an example of such interactions. Overfill effects are known to leave residual Cu in dense line during CMP, resulting in electrical failure due to shorting. On the other hand, overpolishing of wafers to clear the Cu residues leads to excessive dishing in wide Cu lines. Another example is the need for significant excess plating to fill up the wide lines for geometric leveling during plating. However, such "overburden copper" places heavy burden on the CMP process. Optimization of these processes therefore involves a closer interaction between CMP and electroplating teams.

1.2.2.3   *Electrochemical etching and electropolishing*

Electrochemical etching (or electrochemical micromachining) involves material removal from a workpiece by making it an anode in an electrolytic cell in which a nontoxic salt solution is used as an electrolyte and controlled metal removal takes place by the application of an external current.[37] Electrochemical machining (ECM) is practiced in aerospace, automobile, and other heavy industries for shaping, milling, deburring, and other finishing operations of generally large parts. Application of the principles of ECM to process thin films and microstructures is known as electroetching or electrochemical micromachining (EMM).[37]

Most of the thin films of metals and alloys that are of interest in the microelectronics industry can be anodically dissolved in commonly employed ECM electrolytes (such as neutral salt solution of sodium nitrate, sulfate, or chloride). The material removal rate in EMM depends on the specific electrochemical behavior of the metal electrolyte system and is determined by the applied current density according to Faraday law. With proper considerations of transport parameters and high current density for metal dissolution, desired metal removal rates can be obtained. A unique feature of EMM is its ability to provide controlled surface finish. Under conditions where the solubility limit of the anodically produced salt is exceeded, a salt film precipitates on the anode surface. This leads to a mass transport controlled anodic limiting current and surface brightening.[45-47] The high current densities needed to achieve surface brightening lead to electrolyte heating and issues related to the ability to make electrical contact with thin films. The use of pulsating current, with short pulses and high peak currents, permits one to exceed the solubility limit while keeping the average current density very low.[47] Pulsed current EMM has, indeed, been demonstrated to be extremely effective in the patterning of thin films and foils.[48]

Electropolishing involves anodic leveling and anodic brightening of a surface and is generally carried out in concentrated acids or organic solvent based solutions.[47,49] Anodic leveling is due to a non-uniform current distribution on protruding and receding parts of a rough surface. Theoretical models of anodic leveling based on current distribution considerations do not account for possible effects of crystallographic orientation on dissolution kinetics.[49] Such effects on a microscopic scale are responsible for the fact that anodic dissolution under activation control usually leads to an optically dull surface appearance due to crystallographic etching. Surface brightening is achieved only under conditions where the dissolution mechanism is independent of the crystallographic structure. This is achieved when the rate of anodic dissolution reaction is mass transport controlled.[47,49] Different transport mechanisms can limit the anodic dissolution reaction under electro-polishing conditions. As mentioned above, during high rate dissolution in neutral salt solutions, the transport of reaction products away from the anode is rate limiting and salt film precipitation occurs at the surface.[45-47] A salt film mechanism has also been found to be responsible for electropolishing of titanium and tantalum in methanol-based electrolytes.[50] On the other hand, during electropolishing of stain-less steel[51] and copper[52] in concentrated sulfuric acid and phosphoric acid based electrolytes, the transport of acceptor species (water molecules) has been proposed as the rate-limiting mechanism.

Development of EMM and electropolishing processes for applications in elec-tronics requires an understanding of the challenges associated with thin film metal removal processes. The most important of these is the issue related to the loss of electrical contact.[53,54] This problem may arise at the point of contact where the passage of high current on the thin film may lead to excessive heating or if the contact point is exposed to the electrolyte, preferential dissolution at these points will take place. Under both conditions, global electrical contact to the sample will be lost. Another aspect of the loss of electrical contact may arise due to different photoresist feature sizes on the sample being etched, where different current distribution conditions exist within the cavities of small vs. large features. Small features clear away, while an island of material remains within the large patterned features, leading to a premature stoppage of the etching process.[53] Other challenges include the ability to provide straight and smooth walls and minimized undercutting. Understanding of some of these aspects led to the development of precision EMM processes and tools for wafer processing.[54]

1.2.3 *Precision electrochemical processing tools*

Electrochemical tools include electroplating, electroless plating, chemical etching, electroetching, cleaning, and CMP. These tools differ in configuration, complexity, and sophistication depending on their application. For example, a typical line used for copper plating/etching for PCBs consists of plating, etching, and rinse tanks that can handle large panels of boards. The automation of these systems consists of large hoists on linear tanks to move the tracks between various tanks for treatment, electroplating, electroless plating, etching, post-treatment, and various rinsing steps associated with the process.

On the other hand, plating, etching, and cleaning tools for semiconductor wafer processing must meet some basic requirements that are essential in a fab processing environment.[55] Automated wafer processing equipment must provide facilities to unload dry wafers from a cassette or a front opening unified pod (FOUP), and deliver dry, processed wafers back to the same carrier. These requirements necessitate the integration of multiple processing cells in a single set of automated process equipment. Robust hardware and processes must ensure the high yield that is required in semiconductor manufacturing.

Providing desired fluid agitation at the reacting surface is one of the main considerations in the design of a precision tool. Different agitation systems that are applicable include paddle action, fountain flow, channel flow, electrolytic jet, slotted jet, and multi-nozzle systems.[46,48,54–56] Sample orientation, electrolyte heating, steady-state bath control, and provisions for filtration are some of the other important engineering aspects that need to be taken into account in the design and fabrication of precision electrochemical processing systems.

In case of electrolytic processing, electrical contact to the wafer, terminal effect, edge exclusion zone management, backside contamination requirements, and thickness uniformity are some of the key issues that dictate the design of a tool for semiconductor wafer processing. The reactor design takes into account all of these issues and is specific for a given set of process chemistry and conditions. For electroplating, two types of wafer processing tools are commercially available: paddle cells and fountain cells.[55] For chemical etching, spray units are the most commonly used commercially available systems. Electroetching is a newly emerging area;[54] industry standard tools are, therefore, not yet available.

1.3 Microelectronic packaging

Packaging is the science of establishing interconnections with electrical components such as transistors, diodes, capacitors, and resistors to form circuits. The individual circuits must be interconnected to form functional entities. A chip communicates with other chips in the circuit through an input/output (I/O) system of interconnects, and the chip and its circuitry are dependent on the package for support and protection. Indeed, major functions of the package are to provide a path for the electrical current that powers the circuit on the chip, distribute the signals to and from the chip, dissipate the heat generated by the circuit, and provide mechanical support and environmental protection to the components and interconnections. The objective of microelectronic packaging is to ensure that the chips and interconnections are packaged efficiently and reliably.

Traditionally, interconnections within a chip have been considered outside the realm of packaging technologies.[57] However, with increasing integration, a steadily

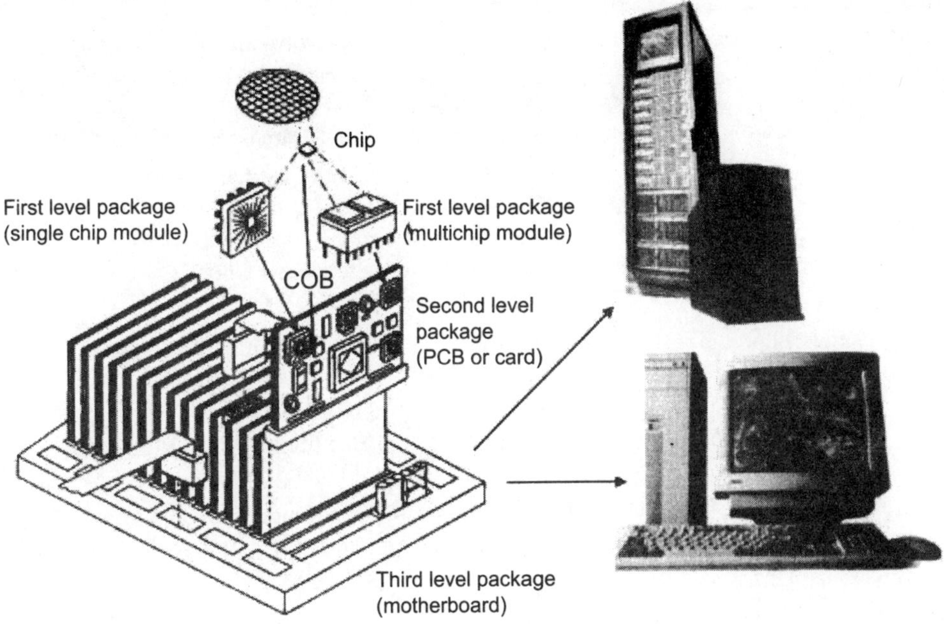

Figure 1.3　　Typical packaging hierarchy showing different levels of packaging.[57,58] Chip level (zero level) packaging includes chip metallization, and provisions for chip–package interconnection. The first level packaging is the assembly of chip and substrate (organic or ceramic package) to form a single or a multiple chip module. In some cases chips may be directly attached to the card (chip-on-board, COB). The second level packaging is the assembly of chip modules and other components on PCBs. The third level packaging differs; depending on the sophistication of the system it may involve several PCBs plugged into a motherboard.

increasing percentage of wiring is migrating into the chip, thus making the semiconductor thin film wiring a very important aspect of microelectronic packaging. Figure 1.3 shows a typical packaging hierarchy.[57,58] The chip level or wafer-level packaging generally focuses on solder bumps for flip-chip technology and is also known as zero level packaging.[57,58] In order to emphasize the changing trends in chip-level packaging, in this chapter the chip-level packaging will include interconnect metallization, and provisions for chip–package interconnection such as flip-chip bumping, wire bonding, and tape automated bonding (TAB). The first level of packaging involves joining of chip(s) to a substrate, which may form a single chip module (SCM) or a multichip module (MCM). Depending on the thermal cycle environment, the substrates for SCM/MCM may be either organic or ceramic packages. In some cases, the chips (without a substrate) are directly attached to boards (direct chip attach) known as chip-on-board (COB). In the second level of packaging, packaged SCMs, MCMs, and other components are assembled on a PCB or card. PCBs are generally copper-clad sheets of epoxy–glass laminates with plated through-hole interconnections. The third level of packaging may vary depending on the system. In a desktop, several PCBs are plugged into a motherboard, while in a

Table 1.1 Electrochemical processing technologies in microelectronic packaging.

Packaging level	Package item/type	Function	Electrochemical process for fabrication
Chip level	Cu metallization	Chip interconnect	Dual-Damascene plating, CMP
	Seed layer	Electrical continuity for Cu plating	Electroless Cu
	Diffusion barrier	Improve Cu electromigration	Electroless Co(P), CoW(P)
Chip–package interconnection	Flip-chip (C4)	Area array interconnection	Electroplating of solders, chemical/electroetching of seed/BLM layers
	Wire bonding	Peripheral interconnection	Electroless (Ni/Au) finishing of bond pads
	TAB	Peripheral interconnection	Electroplating of chip bumps (Au/solder), electroplating/etching of leads
First level packages	Plastic packages	Chip and PCB interconnection, thermal path for heat dissipation from chip	Electroplating, electroless plating, etching for wiring; PTH for multilayer packages
	Ceramic packages		Electroplating, electroless plating, etching, CMP for multilayer thin film modules
Second/third level packages	PCBs, cards	Global wiring and mechanical platform for modules and components	Electroplating Cu foils, etching for file lines; pattern electrolytic, electroless plating of lines; PTH for multilayer boards
Miscellaneous	Heat sink	Heat dissipation from chip	Electroless Ni/Au on Cu matrix
	Connectors	Repeated separation and reconnection of electrical pathway	Electrolytic or electroless plating of Au, Pd, Pd alloys

handheld calculator the outer shell is the third level of packaging. On the other hand, a workstation or a mainframe uses several motherboards within an enclosed box. Electrochemical processing technologies have made significant impact on the evolution of all packaging levels described above. These processing technologies may vary from low-end electroless gold coating of the copper heat sink that are attached to the back of the chip, to the fabrication of extremely precise nanoscale features of chip interconnect metallization (Table 1.1). Accordingly, the degree of sophistication of tools and processes vary based on whether they are applied in package fabrication or in wafer fabrication.

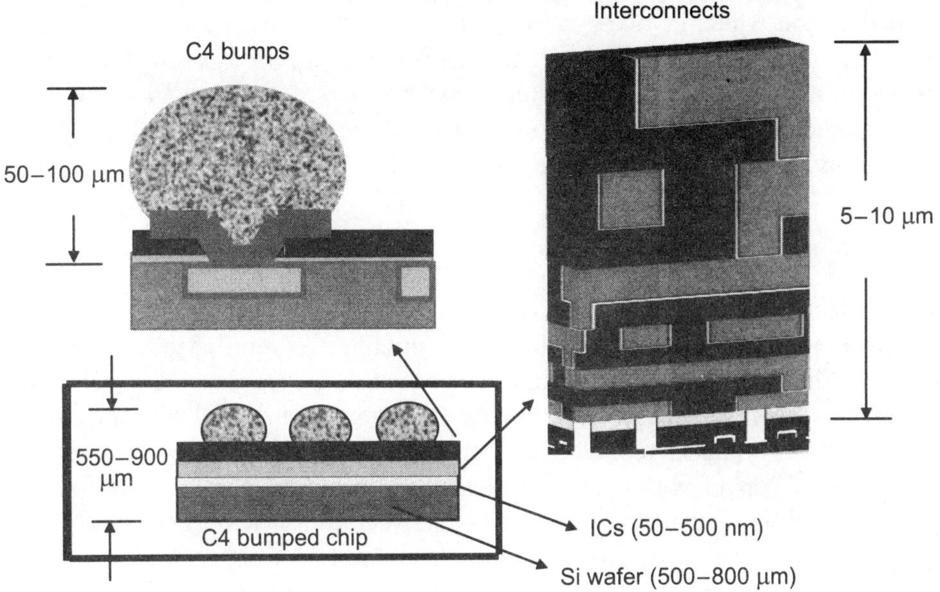

Figure 1.4   Different components of a chip fabricated on a silicon wafer.[11] Transistors and other devices are first formed on silicon, followed by fabrication of interconnect metallization. Passivation layer and C4 processing provide hermetic sealing and further connection to the chip. Also shown is the typical thickness range of each component.

1.4   Impact of electrochemical processing technologies in different levels of microelectronic packaging

1.4.1   Chip level interconnections

Figure 1.4 shows a simplified picture of the components of a chip fabricated on silicon wafers.[11] Silicon wafers are generally 500–800 µm thick. Leading chip manufacturers currently use 300 mm wafers that can accommodate in excess of 500 chips. Transistors and other devices are fabricated on these Si wafers using the so-called "front-end" operations. A single chip (approximately 1 × 1 cm) these days contains hundreds of millions of such devices that are 50–500 nm thick. Indeed, Intel's latest 90 nm technology will pack 330 million transistors in an area measuring only 109 mm^3.[59] These devices are connected by an efficient wiring layout consisting of a stack of six to seven metallization layers embedded in dielectric materials. These interconnect wirings are fabricated by the dual-Damascene electroplating process. The metallization layers consist of lines that are laid perpendicular to each other in the adjoining levels, and are connected by vias. The total height of interconnects vary between 5 and 10 µm. The next step is to hermetically seal the chip while keeping provisions for further chip–package interconnection. This is done by applying and patterning passivation layers that are between 2 and 6 µm thick. In advanced microprocessors, flip-chip technology is the leading technology for chip–package interconnection which involves fabrication of solder bumps within and on top of passivation layers to connect the underlying final metal layers. The total thickness of passivation/C4 structure above the chip metallization vary between 50 and 100 μm. In some other applications, provisions for peripheral wire bonding and TAB interconnections are provided on the chip.

1.4.1.1   Advanced chip interconnect metallization

Replacement of vacuum deposited aluminum by electroplated copper as the chip metallization brought about a paradigm shift in chip interconnect technology.[10,60] Leading chip manufacturers have now converted to electroplated Cu technology because of several advantages associated with copper wiring, and with the electroplating technology that has enabled the high volume manufacturing of Cu chips. The advantages of copper wiring include significantly low resistance, higher current carrying capability, and increased scalability. Furthermore, near bulk resistivity of Cu metallization can be obtained in sub-micron interconnects. Copper interconnects are fabricated by the dual-Damascene process, which is particularly amenable to a hierarchical scheme of high aspect ratio interconnects. Electroplating enables deposition of Cu in via holes and overlying trenches in a single step (Figure 1.5), thus eliminating a via/line interface and significantly reducing the processing cycle time. Compared to vacuum deposition processes, electroplated Cu provides improved superfilling capabilities, and abnormal grain growth phenomena. These properties contribute significantly to the improved reliability of electroplated Cu interconnects. Due to these reasons, and due to relatively less expensive tooling, electroplating is a cost-effective and efficient process for Cu interconnects. Several technological breakthroughs in electrochemical processing technologies took place that led to the implementation of Cu interconnect technology in the semiconductor industry. They include the development of a defect-free dual-Damascene electroplating process, the development and availability of high-volume manufacturing electroplating tools that are compatible with semiconductor manufacturing standards, and the development of chemical–mechanical polishing for Cu Damascene structure.[11] These aspects have been treated in greater detail in Chapters 2, 15, and 16.

1.4.1.2   Chip–package interconnection

Chip–package interconnection technologies currently used in the semiconductor industry include wire bonding, TAB, and flip-chip solder connection. Wire bonded electrical connections are created at the assembly stage by attaching a fine wire between each device I/O around the perimeter of the chip and its associated package pin. A typical TAB process uses bumped chips and planar tapes with inner lead bonding (ILB) and outer lead bonding (OLB) leads that are bonded to chip and substrate, respectively. Similar to wire bonding, TAB interconnection is peripheral but additionally involves bumping on the chip. Flip-chip interconnection, on the other hand, utilizes solder bumps deposited on wettable metal layers on the chip and

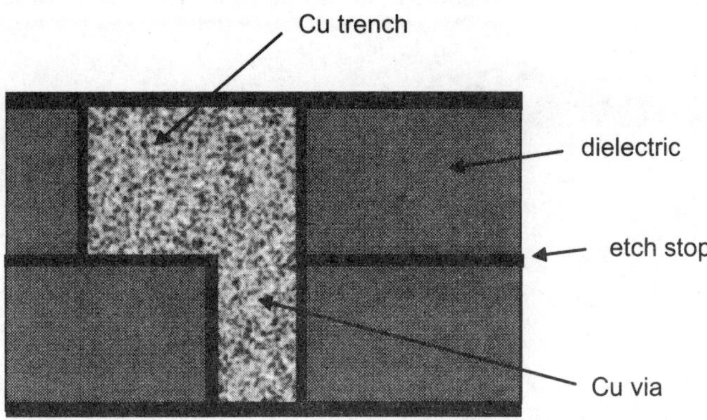

Figure 1.5 A schematic diagram of a copper chip interconnect fabricated by the dual-Damascene process. First a sandwich of two levels of dielectric and etch stop layers are patterned as holes for vias and trenches for lines. They are then filled by electroplating in a single step. Finally the excess metal is removed and wafer planarized by CMP.

a matching footprint of solder bumps on wettable metal layers on the substrate. Flip-chip interconnection is an area array configuration in which the entire surface of the chip can be covered with bumps for the highest possible I/O counts. The name flip-chip refers to the joining process that involves face-down soldering of chips to substrates.

Compared to wire bonding and TAB, the flip-chip technology offers some distinct advantages which include higher packaging density due to area array bonding (Table 1.2), uniform power and heat distribution, and shorter interconnect thus allowing fast signal response and low inductance. Availability of cost-effective electrochemically fabricated flip-chip technology is making it an increasingly popular chip–package interconnect technology (Figure 1.6). Flip-chip technology is expected to grow at the rate of 20% over the next five years.[61]

Table 1.2 Capabilities of different chip–package interconnection technologies.[57]

Technology	Description	Connection/chip
Wire bonding	Assembly step, peripheral	1,000
TAB	Assembly step, peripheral	1,200
C4 (flip-chip)	Wafer processing, area array	16,000

Flip-chip interconnect technology was first introduced by IBM in 1960s and the joining process was named controlled collapse chip connection (C4) technology.[62] IBM's original C4 technology was mainly meant for high-end applications involving ceramic packages and involved evaporation of both seed layers and high melting temperature PbSn (90–97% Pb) solders. However, with the increased demand of higher I/Os for consumer and mid-range products, which required a cost-effective C4 process with lower melting temperature eutectic solders (63Sn37Pb), limitations

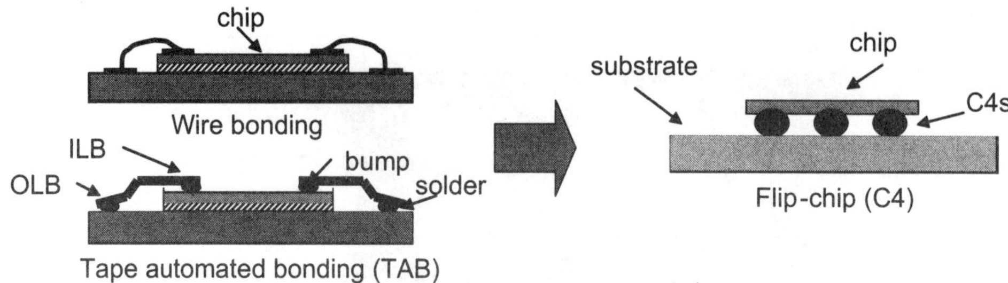

Figure 1.6 Chip–package interconnect technology trend; the industry is increasingly moving from wire bonding and TAB technology to flip-chip (C4) technology. The C4 technology is expected to grow at the rate of 20% over the next five years.[61]

of evaporated C4 process became apparent: (1) evaporation is an expensive process: the low-life metal mask, and waste disposal contribute significantly to the evaporated C4 cost; (2) evaporation is typically less than 5% efficient, with more than 95% of the evaporated material ending up on the evaporator wall and on the metal mask, which need to be cleaned and the Pb containing waste has to be disposed off by an expensive, regulated procedure; (3) evaporated C4s are not extendible to finer bump pitch and larger wafers (>200 mm); and (4) finally, and possibly most importantly, lower melting temperature Sn-rich solders cannot be evaporated due to a lower vapor pressure of Sn. On the other hand, electrochemical fabrication of C4s is an extremely selective and efficient process, which is extendible to a finer pitch, larger wafers, and a variety of solder compositions including some lead-free alloys.[9,63,64] These advantages, coupled with the advantages of area array interconnections, are making the plated C4 technology a widely preferable chip–package interconnection for a variety of products. Besides through-mask electroplating of solder alloys, the C4 fabrication involves careful etching of the underlying seed/ball-limiting metallurgy (BLM) layers. For some selected BLM layers such as phased CrCu, electroetching methods have been found to be the only means for their removal. This led to the development of manufacturing electroetching tools and processes.[54]

1.4.2 *First level packages*

First level packages are the building blocks of a microelectronic packaged system, which provide an interconnection between the PCB and the chip (Figure 1.7). As described earlier, these packages may carry a single chip module or a multichip module. These modules must have the desired number of wiring layers, provide thermal expansion compatibility with the chip, provide a thermal path for heat dissipation from the chip, and keep electrical noise and transmission delay to a minimum. Thin film multilevel wiring embedded in low dielectric constant thin film materials is important in the minimization of transmission delays. First level packages are categorized into two types: plastic packages and ceramic packages.

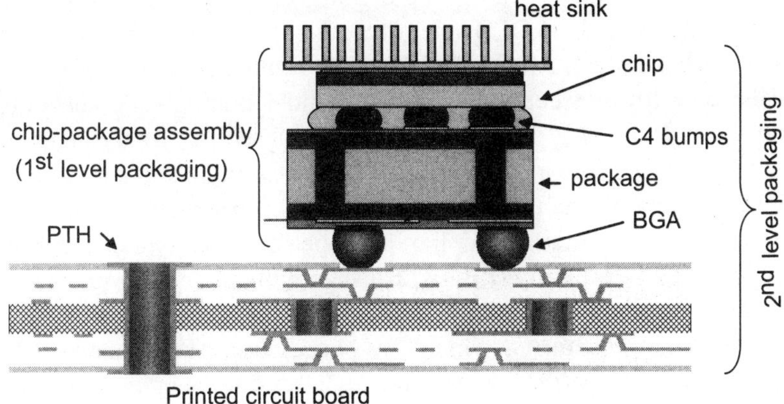

Figure 1.7 A schematic diagram showing an assembled chip–package unit (first level packaging), and further connection of the unit to a PCB (second level assembly). Also shown in the figure is a heat sink that is attached to the backside of the chip for heat dissipation. Electrochemical processing technologies have enabled miniaturization and sophistication of all of these components.

1.4.2.1 Plastic packages

Plastic or organic packages offer many advantages over ceramic packages in terms of size, weight, performance, cost, and availability. Tremendous improvements in hermeticity, and reliability of organic packages have been achieved since their early use in the 1970s. Organic packages are now used in a wide range of products spanning from consumer to low-end to high-end microelectronic assemblies and account for 97% of the first level package market.[65]

The earlier version of organic packages and substrate technologies were limited to injection-molded packages with metal lead frame substrates. Due to their low cost, injection-molded plastic package technology is still being used today for consumer and low-end microelectronics. The substrate technology is generally based on a single level metal frame, which is the foundation of the molded plastic package. The two most common lead frame metals employed by the industry are nickel–iron alloy (alloy 42, 42%Ni/58%Fe), and copper alloy.[65,66] In lead frame manufacturing, the metal sheet is formed into a metal interconnect by either stamping or chemical etch process. Stamping is a lower cost, high-volume process, whereas etching provides tighter design rules with higher routing density. The chip is bonded with a die attach adhesive onto the die paddle. Gold wire bonding is used to make the connections between the chip and the lead frame. The injection molding process is used to encapsulate the integrated circuit (IC) and bond wires. The outer leads are then solder screened or plated for board mount. While lead frame packages continue to meet the requirements of the low-density products, the limitations on feature size and routing density make them unsuitable for high-performance central processing units (CPUs) or application-specific ICs (ASICs).

TAB packages are another organic package technology for high lead count IC devices. Chemical etching is used in patterning a one-layer tape substrate. The substrate has exposed metal lead on both sides, the ILB side where the leads are bonded onto the chip and the OLB side where the I/O leads are bonded to the board. The ILB side is typically finished with electroless Ni–Au plating, whereas OLB leads are finished with tin or eutectic (63Sn37Pb) solder plating. The substrate is then processed through the ILB TAB bonding. The subsequent encapsulation process covers and protects the chip joints. The OLB leads of a single unit package are formed and mounted on board. The ultra-thin profile of TAB package is particularly suited for laptop market where minimization of package size is a key consideration.[66]

The plastic ball grid array (PBGA) package with an organic substrate technology has now become the technology of choice for IC packaging. It is based on the concepts of PCB manufacturing, which uses plated copper interconnect. Multilevels of metal interconnect in PBGA provide high routing density. The key design advantage of PBGA is that it provides the scalability to much higher I/O count due to the array structure of the solder ball (BGA) arrangement. Copper interconnects are fabricated by chemical etching of patterned copper laminates. Plastic pin grid array (PPGA) and plastic land grid array (PLGA) packages use multiple sheets of patterned layers that are stacked and laminated into a composite panel. Plated through-hole technology is used to form the interconnections. Solder mask is then screened to the surface. Organic land grid array (OLGA) and flip-chip pin grid array (FCPGA) are Intel's advanced packages, which use novel materials to minimize coefficient of thermal expansion (CTE) mismatch between silicon and organic substrate. Development of these denser but lighter packages was made possible through advances in electrochemical processing technologies including the ability to obtain micro-via plating coverage, fine wiring, and high aspect ratio through-hole plating.

1.4.2.2 *Ceramic packages*

Ceramic packages provide the highest wiring density of all packaging technologies. These high wiring densities are obtained by using cofired multilayer ceramic substrates. Cofiring of as many as 63 layers in full production and 100 layers in development have been reported.[57] Ceramic packages offer the highest reliability and are used in aerospace, defense, and leading-edge microprocessors. Indeed, hermiticity and exceptional dimensional stability of ceramic substrates make these packages superior to organic packages. Traditional ceramic packages used a typical combination of alumina ceramic with molybdenum or copper paste conductors. However, advanced packages now contain a stack of thin film layers on top of multiple ceramic layers.[67] Since high density of wiring is possible in thin film layers, the addition of thin film layers on ceramic packages reduces the total number of layers required in the package and improves the pulse transmission between chips. Electrochemical processing technologies played a significant role in the development of these packages. Some of these process steps include electroplating and etching to form vias and conductor wires, electroless plating of Co(P) as diffusion barrier layer, electroless deposition of Ni/Au on sintered molybdenum pads, and CMP for planar structures.

1.4.3 PCBs and cards

PCBs and cards provide the lowest cost packaging solution to the majority of the electronic requirements for global wiring and mechanical platform for electronic modules and components. Cards and boards use the same printed circuit technology except that the card is smaller in size, has fewer layers, and sometimes has coarser dimensions than boards. For the sake of simplicity, the term PCB will be used in the following text for printed cards as well as boards.

The insulating laminates used in PCBs are generally epoxy-glass based and some are phenolic laminates.[7,68] Depending on the function, and the product it is used for, the PCBs vary in thickness: 0.1 mm for pagers and calculators; 0.5 mm for notebooks, camcorders, and radios; and several millimeters for personal computers, workstations, and their accessories. The thickest multilayered PCBs are used in mainframe and office computers. The required interconnect length in PCBs increases with the complexity of functions it is required to perform. The PCBs of the 1950s contained a few centimeters of interconnection wiring per circuit board, which increased to kilometers in the 1990s.[7] This has been driven by the ever-increasing microprocessor performance and the integration level, and the concurrent need for increased I/O density. In response to these needs, the PCB industry followed an evolutionary path, which included significant advances in the development of electrochemical processing technologies that helped in the miniaturization of multilayered PCBs.

Both etching and plating technologies are used in the fabrication of PCBs. PCBs are made of a variety of structures. In their simplest form, PCBs are single and double sided. Copper plated insulting laminates are either patterned with photoresist and etched, or the panels are pattern plated by electrolytic or electroless copper. The predominant method for fabrication of in-plane lines and lands is the etching of copper foil bonded to the surface of an epoxy-glass cloth panel. The copper foil is produced by a continuous electrodeposition method, which enables close tolerances in width and purity. Foils as thin as 9 μm are commercially produced which are etched with precision to 25 μm lines.

The most common method of fabricating multilayered PCBs involves the following steps: (1) production of double-sided laminates or cores, (2) placement of partially cured prepeg between the cores to form a stack, and (3) lamination of them together into a single composite with two surface copper planes. The final drilling and plating operations connect the layers and cores vertically with plated through holes (PTH), and provide surface patterns for connections to the components. Figure 1.8 shows different types of plated through holes and their functions in a multilayered PCB. PTH technology enables interconnection of various layers through formation of vias and holes. Vias that interconnect internal layers only are called buried vias. Blind vias connect surface layers to internal layers. The sequential lamination approach is very common, which involves lamination of subassemblies with drilled and plated through holes interconnecting the subassemblies. For patterning, the electrodeposited photoresist process is becoming increasingly important in the

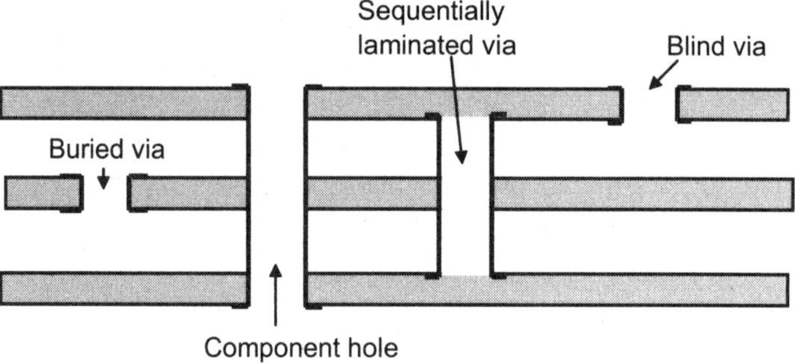

Figure 1.8   Different types of plated through holes (PTH) in a multilayered PCB. Buried
vias interconnect internal layers only.

advanced PCB industry because of its higher resolution and lower material cost than
dry-film resists.[7] The electrodeposited photoresist is applicable in both positive and
negative modes.

Evolution of PCB industry owes much to the development in electrochemical
processing technologies which enabled efficient fabrication of circuits on non-
metallic surfaces through the availability of high-volume plating and etching
technologies. PCB wiring requires high ductility copper, for which both electroless
and electrolytic baths have been developed.[69,70] Electroless plating provides good
throwing power; hence it is suitable for high aspect ratio multilayered boards with
densely packed inhomogeneous circuitry. The pyrophosphate electrolytic bath has
been used in the PCB industry for a long time due its good copper quality and good
leveling power.[71] However, much work in the development and application of acid
copper bath has led to its wide acceptance in the PCB industry due its ability to
provide high plating speed, finer lines, and higher aspect ratio holes.[72]

1.5   Future prospects

The impact of electrochemical processing technologies in different levels of microelectronic packaging has been briefly described in this chapter. These processing
technologies are expected to continue to enable further miniaturization of high-
performance chips, packages, and PCBs. While the fabrication of a 20 nm transistor
has been demonstrated, interconnecting these transistors will be one of the great
challenges that the electrochemical processing technologies will have to overcome
in few years. With the need for nanoscale metallization, and the need to integrate
them with ultra-low-k dielectric materials, electromigration issues are expected to be
the most serious reliability concern in chip level packaging. In addition, there is an
urgent need to develop barrier and cladding layers for passivation and adhesion
promotion. Electroless metal capping/barrier layers are expected to provide the
desired barrier properties so that lower effective interconnect capacitance can be
obtained with thinner (or even no) etch stop layers which have high dielectric
constant. Innovations in polishing methods will have to emerge in order to integrate

highly porous ultra-low-k dielectric materials. Development of the electropolishing based planarization approach must address some of the unique challenges associated with thin film metal removal that are different from the electroplating process.

Development of low melting temperature, Pb-free flip-chip technology for area array chip–package interconnection is another key area where electrochemical processing technologies are expected to have a tremendous impact. Identification of a compliant Pb-free solder that is electroplatable, development of an electroplating process, and development of a robust BLM material and its etching process are some of the challenges that the technology development activities are expected to face.

High aspect ratio PTH technology has had a tremendous impact in the fabrication of multilayered substrates, packages, and PCBs. Further miniaturization of these packages and boards is limited by the mechanical drilling technology, which has come to its limits. The PTH technology is expected to be replaced by photo via technology. New sequential processing techniques are expected to be used to fabricate interconnections similar to those used in semiconductor wafers. Currently, solder bumps are formed on these parts by screening solder pastes. However, increasing demands for finer bumps with finer pitch will be limited by the ability of screening metal masks. The screening technology is expected to be replaced by electroplated solders, which will easily fulfill the industry's fine dimensional demands.

References

1. Intel press release, Santa Clara, CA, June 11, 2001.
2. R.K. Springborn, Non-Traditional Machining Processes, ASME Publications, Philadelphia, PA (1967).
3. A.E. Cahill, *AES Proc.*, **44**, 130 (1957).
4. L.T. Romankiw, I. Croll, and M. Hatzakis, *IEEE Trans. Magn.*, **6**, 729 (1970).
5. T. Osaka, *Electrochim. Acta*, **42**, 3015 (1997).
6. L.T. Romankiw and D. Turner, eds., Electrodeposition Technology: Theory and Practice, PV 86–17, Electrochemical Society Proceedings, NJ (1987).
7. D.P. Seraphim, D.E. Barr, W.T. Chen, G.P. Schmitt, and R.R. Tummala, in *Microelectronic Packaging Handbook*, Part III, 2nd edition, R.R. Tummala, E.J. Rymaszewski, and A.G. Klopfenstein, eds., Chapman and Hall, NY (1997).
8. L.T. Romankiw, *Electrochim. Acta*, **42**, 2985 (1997).
9. M. Datta, R.V. Shenoy, C. Jahnes, P.C. Andricacos, J. Horkans, J.O. Dukovic, L.T. Romankiw, J. Roeder, H. Deligianni, H. Nye, B. Agarwala, H.M. Tong, and P.A. Totta, *J. Electrochem. Soc.*, **142**, 3779 (1995).
10. P.C. Andricacos, C. Uzoh, J.O. Dukovic, J. Horkans, and H. Deligianni, *IBM J. Res. Dev.*, **42**, 567 (1998).
11. M. Datta, *Electrochim. Acta*, **48**, 2975 (2003).
12. M. Schlesinger and M. Paunovic, *Modern Electroplating*, 4th edition, Wiley Interscience, NY (2000).
13. M. Datta and D. Landolt, *Electrochim. Acta*, **45**, 2535 (2000).
14. J. Newman, *Electrochemical Systems*, 2nd edition, Prentice Hall, Englewood Cliffs, NJ (1991).
15. J.O. Dukovic, *IBM J. Res. Dev.*, **37**, 125 (1993).
16. C. Madore, M. Matlosz, and D. Landolt, *J. Electrochem. Soc.*, **143**, 3927 (1996).
17. O. Kardos, *Plating,* **61**, 129, 229, 316 (1974).

18. S.S. Kruglikov, N.T. Kudriavtsev, G.F. Vorobiova, and A. Ya. Antonov, *Electrochim. Acta*, **10**, 253 (1965).
19. J. Dukovic and C.W. Tobias, *J. Electrochem. Soc.*, **137**, 3748 (1990).
20. D. Landolt, *Electrochim. Acta*, **39**, 1075 (1994).
21. N. Ibl, in *Comprehensive Treatise of Electrochemistry*, E. Yeager, J. O'M. Bockris, and B. Conway, eds., Vol. 6, 1, 133, 239, Plenum Press, NY (1982).
22. E. Matsson and J. O'M. Bockris, *Trans. Faraday Soc.*, **55**, 1586 (1959).
23. R. Winnad, *Electrochim. Acta*, **39**, 1091 (1994).
24. V.S. Donepudi, R. Venkatachalapathy, P.O. Ozemoyah, C.S. Johnson, and J. Prakash, *Electrochem. Solid-State Lett.*, **4**, C13 (2001).
25. D. Landolt, *J. Electrochem. Soc.*, **149** (3), S9 (2002).
26. T.P. Moffat, J.E. Bonewich, W.H. Huber, A. Stanishevsky, D.R. Kelly, G.R. Stafford, and D. Josell, *J. Electrochem. Soc.*, **147**, 4524 (2000).
27. C. Cabral, P.C. Andricacos, L.M. Cignac, and I.C. Noyan, *Adv. Metallization Conf. Proc.*, ULSI XIV, 81 (1998).
28. J. Horkans, I.C. Hsu Chang, and P.C. Andricacos, *IBM J. Res. Dev.*, **37**, 97 (1993).
29. Q. Lin, K. Sheppard, M. Datta, and L.T. Romankiw, to be published.
30. C.Y. Mak, *MRS Bull.*, 55, August (1994).
31. E.J. O'Sullivan, A.G. Schrott, M. Paunovic, C.J. Sambucetti, J.R. Marino, P.J. Bailey, S. Kaja, and K.W. Semkow, *IBM J. Res. Dev.*, **42** (5), 607 (1998).
32. Y. Okinaka and T. Osaka, in *Advances in Electrochemical Science and Engineering*, Vol. 3, H. Gerischer and C.W. Tobias, eds., VCH Verlagsgesellschaft mbH, Weinheim, Germany (1993), p. 55.
33. S. Lopatin, Y. Shacham-Diamand, V. Dubin, and P.K. Vasudev, in *Proceedings of the 14th Intl. VLSI Multilevel Interconnection Conference*, VMIC, p. 219, June 10–12 (1997).
34. C.A. Deckert, *Plat. Surf. Finish.*, 48, February (1995).
35. S. Krongelb, L.T. Romankiw, and J.A. Tornello, *IBM J. Res. Dev.*, **42** (5), 575 (1998).
36. J.W.M. Jacobs and J.M.G. Rikken, *J. Electrochem. Soc.*, **135**, 2822 (1988).
37. M. Datta and D. Harris, *Electrochim. Acta*, **42**, 3007 (1997).
38. H.K. Kuiken and R.P. Tijburg, *J. Electrochem. Soc.*, **130**, 1722 (1983).
39. L.M. Cook, *J. Non-Crystal. Solids*, **120**, 152 (1990).
40. S.R. Runnels and L.M. Eyman, *J. Electrochem. Soc.*, **141**, 1698 (1994).
41. C.-W. Liu, B.-T. Dai, W.-T. Tseng, and C.-F. Yeh, *J. Electrochem. Soc.*, **143**, 716 (1996).
42. J. Warnock, *J. Electrochem. Soc.*, **138**, 2398 (1991).
43. F. Preston, *J. Soc. Glass Technol.*, **11** (1927).
44. F.B. Kaufman, D.B. Thompson, R.E. Broadie, M.A. Jaso, W.L. Guthrie, D.J. Pearson, and M.B. Small, *J. Electrochem. Soc.*, **138**, 3460 (1991).
45. M. Datta and D. Landolt, *J. Electrochem. Soc.*, **122**, 1466 (1975).
46. M. Datta and D. Landlot, *Electrochim. Acta*, **25**, 1255 (1980).
47. M. Datta, *IBM J. Res. Dev.*, **37** (2), 207 (1993).
48. M. Datta, *IBM J. Res. Dev.*, **42** (5), 655 (1998).
49. D. Landolt, *Electrochim. Acta*, **32**, 1 (1987).
50. O. Piotrowski, C. Madore, and D. Landolt, *Electrochim. Acta,* **44**, 3389 (1999).
51. M. Matlosz, S. Sagino, and D. Landolt, *J. Electrochem. Soc.*, **141**, 410 (1994).
52. R. Vidal and A.C. West, *J. Electrochem. Soc.*, **142**, 2686 (1995).
53. R.V. Shenoy, M. Datta, and L.T. Romankiw, *J. Electrochem. Soc.,* **143**, 2306 (1996).
54. M. Datta and R. V. Shenoy, U.S. Patent, 5,543,032 (1996).
55. T. Ritzdorf and D. Fulton, Chapter 15, this volume.
56. J.V. Powers and L.T. Romankiw, U.S. Patent, 3,652,442 (1972).
57. E.J. Rymaszewski, R.R. Tummala, and T. Watari, in *Microelectronic Packaging Handbook*, Part I, 2nd edition, R.R. Tummala, E.J. Rymaszewski, and A.G. Klopfenstein, eds., Chapman and Hall, NY (1997).
58. J.H. Lau, ed., *Flip Chip Technologies*, McGraw Hill, NY (1996).
59. Intel press release, Hillsboro, OR, August 13, 2002.

60. D.C. Edelstein, *Tech. Dig. IEEE Intl. Electron. Devices Conf.*, 773 (1997); *IBM Res. Mag.*, No. 4, 16 (1997).
61. S. Prasad, *Advanced Packaging*, March 2001.
62. P.A. Totta and R.P. Sopher, *IBM J. Res. Dev.*, **5**, 226 (1969).
63. D. Clegg, R. Cole, J. Franka, D. Mitchell, and D. Wontor, *Advanced Packaging*, March 2001.
64. D.R. Frear, J.W. Jang, J.K. Lin, and C. Zhang, *JOM*, **53** (6), 28 (2001).
65. M.G. Pecht and L.T. Nguyen, in *Microelectronic Packaging Handbook*, Part II, 2nd edition, R.R. Tummala, E.J. Rymaszewski, and A.G. Klopfenstein, eds., Kluwer Academic Publishers, Boston (1999).
66. J.T. Breedis, *J. Metals*, A.I.M.E., 48, June (1986).
67. R. R. Tummala, P. Garrou, T. Gupta, N. Kuramoto, K. Niwa, Y. Shimda, and M. Terasawa, in *Microelectronic Packaging Handbook*, Part II, 2nd edition, R.R. Tummala, E.J. Rymaszewski, and A.G. Klopfenstein, eds., Kluwer Academic Publishers, Boston (1999).
68. R.F. Bonner, J.A. Asselta, and F.W. Haining, *IBM J. Res. Dev.,* **26** (3), 297 (1982).
69. G.C. Van Tiburg, *Plat. Surf. Finish.*, **71** (6), 78 (1984).
70. H. Houma and S. Mizushima, *Metal Finishing*, **82** (1), 47 (1984).
71. M. Janitz, C. Ogden, D. Tench, and R. Young, *Plat. Surf. Finish.*, **71** (1), 58 (1984).
72. D.A. Luke, *Circuit World*, **13** (10), 18 (1986).

Part II
Chip metallization

2 Electroplating process for Cu chip metallization

Valery M. Dubin, Harsono S. Simka, Sadasivan Shankar, Peter Moon, Thomas Marieb, and Madhav Datta

2.1 Introduction

Miniaturization of semiconductor device dimensions and search for new interconnect materials are mainly concerned with minimizing the interconnect RC delay while simultaneously increasing electromigration resistance. The objective is to reduce the resistivity and the length of the metal interconnect, and the permittivity of the insulator material. For many years aluminum has been the metal of choice for interconnects. Aluminum metallization layers are easily processed by physical vapor deposition (PVD), and reactive ion etching. These layers adhere well to SiO_2, and do not introduce deep electron energy levels into the silicon band gap. However, resistivity of aluminum is relatively high (2.66 $\mu\Omega$ cm) and because of its low melting point, its electromigration resistance is poor.

A remarkable change in interconnect technology took place in 1997 when the vacuum deposited Al was changed to electroplated Cu.[1] Based on its impressive performance, most of the leading chip manufacturers have now converted to electroplated Cu interconnect technology. Relative to comparable Al interconnect, Cu interconnect has several advantages which include significantly lower resistance (1.67 $\mu\Omega$ cm), higher current carrying capability, and increased scalability.[2] For similar dimensions, the electromigration lifetime of Cu interconnects is about 100 times longer than for Al lines. Cu metallization, therefore, supports much higher current density specifications and makes it extendible to finer dimensions and pitches. Furthermore, near bulk resistivity for Cu metallization can be obtained in sub-micron interconnects.

Varying needs of interconnect, from low capacitance to low RC to low resistance, require a hierarchical interconnect scheme consisting of lower interconnects at minimum possible pitch and thickness to minimize capacitance and maximize interconnect density.[3] The upper level interconnects are scaled uniformly, both vertically and horizontally, to maintain a constant capacitance per unit length while reducing resistance. The dual-Damascene process, described below, is particularly amenable to such a hierarchical scheme of high aspect ratio (i.e., height/width ratio of features, trenches, and vias) multi-level Cu interconnects.

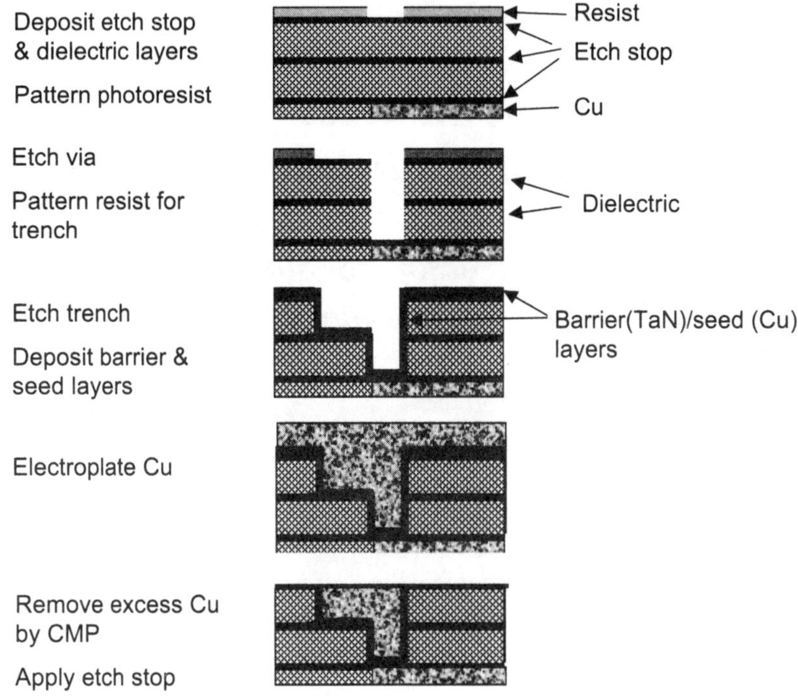

Figure 2.1   Process flow for fabrication of Cu interconnects by dual-Damascene method.

Compared to Al, Cu interconnects lack a self-limiting native oxide that is hard and stable. This aspect has both positive and negative connotations. On the one hand, this makes Cu interconnects more prone to corrosion and adhesion problems that must be controlled in the integration scheme. On the other hand, this property makes it relatively easy to obtain exceptionally low contact resistance for inter-level Cu vias.

Development of Damascene plating and Cu chemical–mechanical polishing (CMP) were among the key technological breakthroughs that led to the implementation of Cu interconnects. Damascene plating is another version of patterned plating in which the patterned dielectric remains intact and forms a functional part of the structure. A continuous seed layer covers the patterned dielectric. Plating occurs all over the surface thus creating challenges for void-free structure fabrication.[4] CMP is used for planarization and removal of excess "overburden" metal, and seed layers.

Cu interconnects are fabricated by the dual-Damascene process which is referred to a metallization patterning process by which two insulator (dielectric) levels are patterned, filled, and planarized to create a metal layer consisting of vias and lines.[4] A sandwich of two levels of insulator and etch stop layers are patterned as holes for vias and troughs for lines. They are then filled with a single metallization step. Finally, the excess material is removed and the wafer planarized by chemical mechanical polishing. While finer details of exact sequence of fabrication steps vary, the end result of forming a metal layer remains the same as shown in Figure 2.1. A key challenge for dual-Damascene patterning is to re-create a planar, non-reflective surface after the first patterning step to enable high resolution lithography during the second patterning step.[5] Another key challenge is to fill high aspect features with void-free copper.[4,6] This is achieved by using copper electroplating with superfilling (i.e., bottom-up fill) capability.

Electroplating enables deposition of Cu in via holes and overlying trenches in a single step thus eliminating a via/line interface and significantly reducing the process flow cycle time. Due to these reasons and due to relatively less expensive tooling, electroplating is a cost-effective and efficient process for forming Cu interconnects.

Compared to vacuum deposition processes, electroplated Cu provides improved superfilling (i.e., bottom-up fill) capabilities, and abnormal grain growth phenomena (i.e., room temperature grain growth). These properties contribute significantly to improved reliability of Cu interconnects. With the proper choice of additives and plating conditions, void-free, seam-free Damascene deposits are obtained which eliminate surface-like fast diffusion paths for Cu electromigration. Large grain size or bamboo-like structures of electroplated Cu have a positive impact on the grain boundary diffusion type of electromigration phenomena. The ratio of (111) grains to (100) grains for plated Cu is 4:1 vs. 59:2 for PVD Cu.[2] With the elastic modulus of Cu in (111) direction being 2.9 times higher than in (100) direction, these data indicate that plated Cu is relatively more compliant than vacuum deposited Cu films. Physical vapor deposition (PVD) and chemical vapor deposition (CVD) Cu films contain small grains with fewer twins, while the grain size and twin volume fraction of plated Cu films are significantly higher. Twin boundaries are known to be low diffusivity paths, thus making the electroplated films relatively less prone to electromigration. Furthermore, it has been shown that a decrease in line width decreases (111) texture and increases the twin volume fraction,[2] thus ensuring extendibility of plated Cu to finer dimensions.

Subsequent to deposition, plated Cu films show an increase in grain size as a function of time. The phenomenon, known as room temperature annealing, has been reported in the literature.[7] At room temperature, electroplated Cu films undergo a crystallization process in which grains grow in size from 30 to 100 nm to a few micrometers in 9 to 100 hours. This is accompanied by a decrease in sheet resistance (by 25%), a decrease in compressive stress to near zero values, and an increase in the Cu (111) texture. Elimination of electron scattering at grain boundaries and/or elimination of defects during growth are possible mechanisms of the observed decrease in resistance and compressive stress.[7] In order to effectively incorporate the advantages of the observed grain growth, an annealing step is introduced at an appropriate sequence in the integration scheme of interconnect fabrication.

2.2 Cu electroplating process development

One of the main objectives of electroplating process development for Cu interconnects is to obtain defect-free structures. Formation of voids is very common, particularly in narrow trenches, where deposition at the top of features can lead to formation of a void in the middle of the structure. Voids need to be eliminated to minimize electromigration issues and to ensure that the Cu vias and lines are free of trapped electrolytes. Another form of defect that is common in conformal plating is the formation of seams that can be as disastrous as the voids. In this section we report our research and development efforts to develop a robust high-yield manufacturable electroplating process for Cu interconnects.

2.2.1   Electroplating bath components

In general, the interconnect Cu plating bath is a commercially available chloride containing acid sulfate solution which contains a variety of tailored, proprietary additives. Different proprietary additives have been developed by different chip manufacturers, and electroplating tool vendors. For defect-free, super-filling deposition, several organic additives are added to the bath. Based on their function, these additives can be broadly categorized into two types: an anti-suppressor (ASUPP) and a suppressor (SUPP). The ASUPP additive is essentially an organic sulfur-containing compound (a mercapto species) while SUPP additive(s) in the bath may contain one or more components. The main SUPP component is generally a glycol, which acts to suppress the electrodeposition rate, especially in the presence of chloride ions (which also acts as a mild inhibitor). Some baths also contain a nitrogen-containing inhibitor that also functions as a leveler.

Through a proper choice of additives in the bath, copper electroplating can provide bottom-up fill or "super-fill" behavior resulting in the complete gap-fill of narrow Cu trenches and vias.[4,6] This is commonly assigned to the action of organic additives and chloride ions added in small amounts to the plating bath. Several mechanisms by which these additives lead to bottom-up fill have been proposed. An early model focused on the location-dependent growth rate derived from diffusion-limited consumption of an inhibiting/suppressing additive species.[4,8,9] Explanation of bottom-up fill phenomena, when fill is observed in an electrolyte containing an accelerator additive and a single high concentration suppressor, using this model is difficult.[10] As a result, a second model was proposed.[11–14] In this model, accelerated bottom-up deposition has been explained by a local accumulation of the accelerator species at the base of a feature, as the surface area within the feature decreases during deposition process. Recently, the bottom-up gap fill (superfill) of Cu electroplating with plating additives has been explained by accumulation of growth-accelerating additive species such as *bis*(sodiumsulfopropyl)disulfide (SPS) and its by-products such as mercaptopropane sulfonic acid (MPSA) at the bottom of the small features and slow diffusion of free MPSA out of these features.[15,16] The reduction of SPS to MPSA provides a possible catalytic pathway for copper deposition through the formation of cuprous thiolate.[17]

2.2.2   Bath characterization

To characterize the modes of action of plating additives, we employed traditional electrochemical experimental techniques, such as linear sweep voltammetry (LSV) and cyclic voltammetric stripping (CVS). In addition to these established methods, we have also developed a novel experimental technique to study electroplating chemistries and gap-fill mechanisms. This technique includes measurements of depolarization behavior as a function of time during plating on specially prepared patterned wafers.

Cyclic LSV characterizations were used to probe surface kinetics of the electroplating process, for baths with and without additives. Copper was electroplated on a rotating platinum disk electrode. Cathodic potential was cycled between -0.4 and -0.65 V. A mercury reference electrode and a copper counter electrode were used in these LSV measurements.

The CVS technique is useful to understand the role of each additive component in the plating process. A suppressor is an organic additive such as polyethylene glycol,[15] which suppresses or decreases the plating rate for a given potential. In contrast, the anti-suppressor is an organic component, which increases the plating rate.

Cathodic potential change with plating time during Cu electroplating on patterned wafers was studied using a PAR273 potentiostat, an Ag/AgCl reference electrode, and a pure copper counter electrode. The patterned wafers had been specially manufactured to contain large patterns of 0.2 µm wide, aspect ratio 2:1 trenches at a 0.4 µm pitch. Both high acid (i.e., H_2SO_4 concentration >150 g/l) and low acid plating bath (i.e., H_2SO_4 concentration <20 g/l) were investigated. Other inorganic components include copper sulfate (>10 g/l) and HCl (>10 ppm). Commercially available additives from Enthone Corporation and Shipley Corporation as well as the well-known SPS anti-suppressor and polyether-type suppressor were used in this study.

We performed cyclic LSV experiments of a typical ASUPP additive, SPS, as well as a polyether-type SUPP in high and low sulfuric acid concentration electrolytes. Depending on the additives present in the bath, the $I-V$ curve may or may not exhibit a hysteresis between the forward (cathodic direction) and reverse (anodic direction) potential sweep. No hystereses in LSV curves were observed when the only additive species present in the bath was the suppressor species (SUPP) (Figure 2.2). In contrast, a small hysteresis was observed when ASUPP species was present in the Cu plating bath (Figure 2.3). Current values in the cathodic sweep are lower than values in the subsequent anodic sweep. These hystereses depend on the additive concentrations, and are more pronounced when both SUPP and ASUPP species are present in the bath (Figure 2.2) and can be attributed to the action of an anti-suppressor, which can displace the suppressor on the surface and has complex surface reaction kinetics.[18] These interactions can result in "hump" formation over features, or the so-called "momentum-plating" or super-fill phenomena. Our results are consistent to recent LSV characterizations of the Cu/Cl/PEG/SPS[10] and Cu/Cl/PEG/MPSA[12,14] baths.

Results from CVS experiments, such as normalized measurements of the effect of a suppressor in Cu electrolyte, show that suppressor and leveler species inhibited Cu deposition while the anti-suppressor led to little impact on the observed stripping peak area Ar (Figure 2.4). In the case of fully suppressed deposition with a Cu electrolyte, the anti-suppressor has a more pronounced effect in accelerating Cu deposition and reducing the effects of the suppressor (Figure 2.5). This observation, in addition to LSV measurement results, supports the hypothesis that the anti-suppressor species in the plating bath can displace adsorbed suppressor species, and therefore enhance

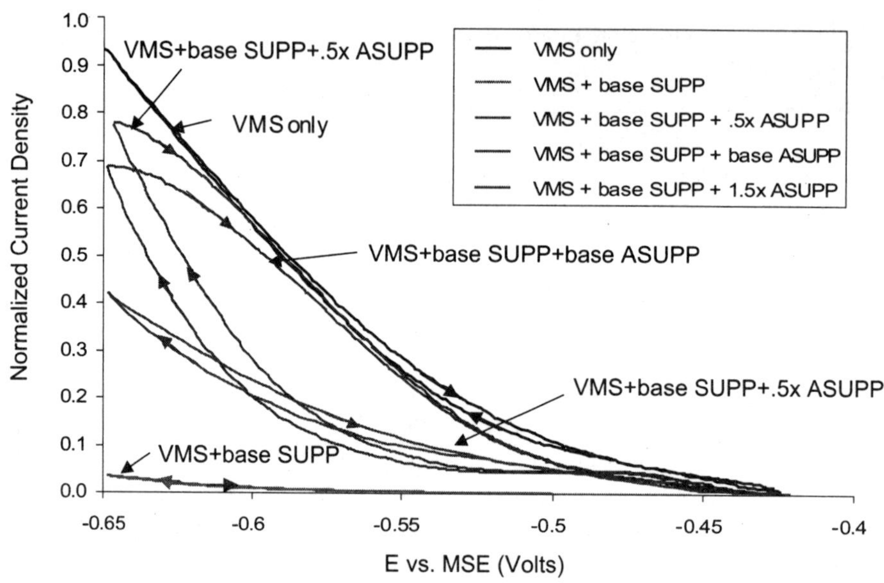

Figure 2.2   Cyclic voltammetry from LSV measurements for bath electrolyte (virgin make-up solution — VMS), bath electrolyte with suppressor (SUPP), and bath electrolyte with SUPP and antisuppressor (ASUPP). Scan rate is 10 mV/s.

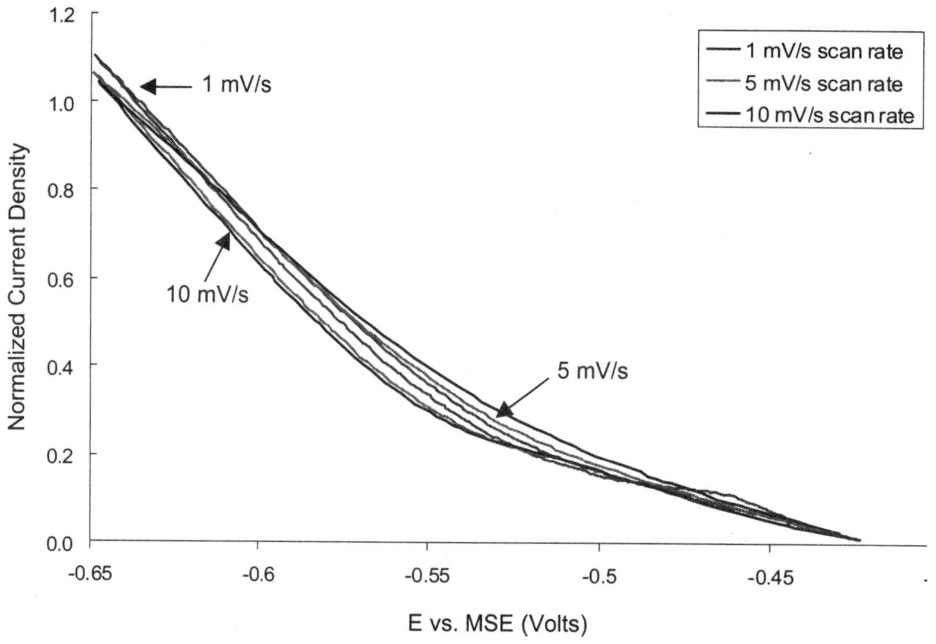

Figure 2.3   Cyclic voltammogram for ASUPP in Cu electrolyte bath.

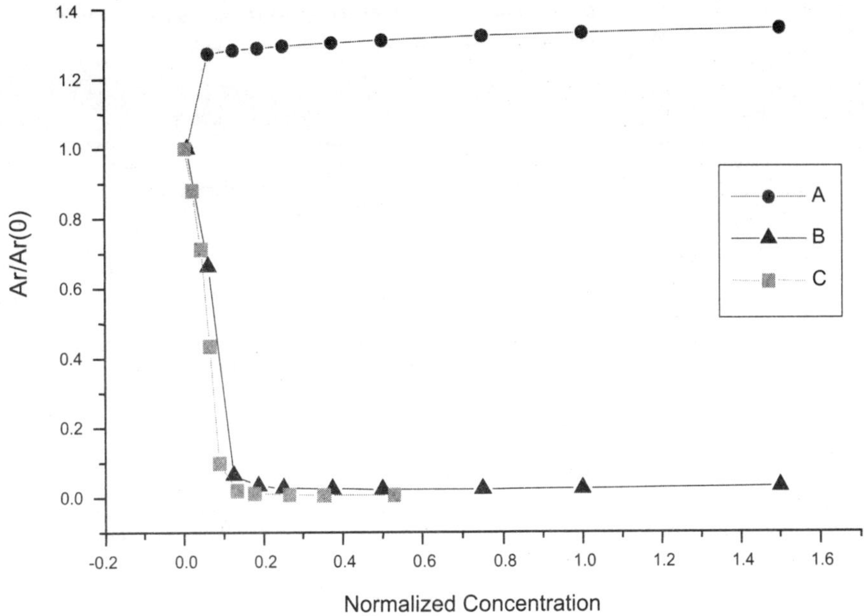

Figure 2.4   CVS response curves for ASUPP, SUPP, and leveler in Cu electrolyte. A = ASUPP, B = SUPP, C = leveler. Ar(0) is CVS stripping peak area in VMS.

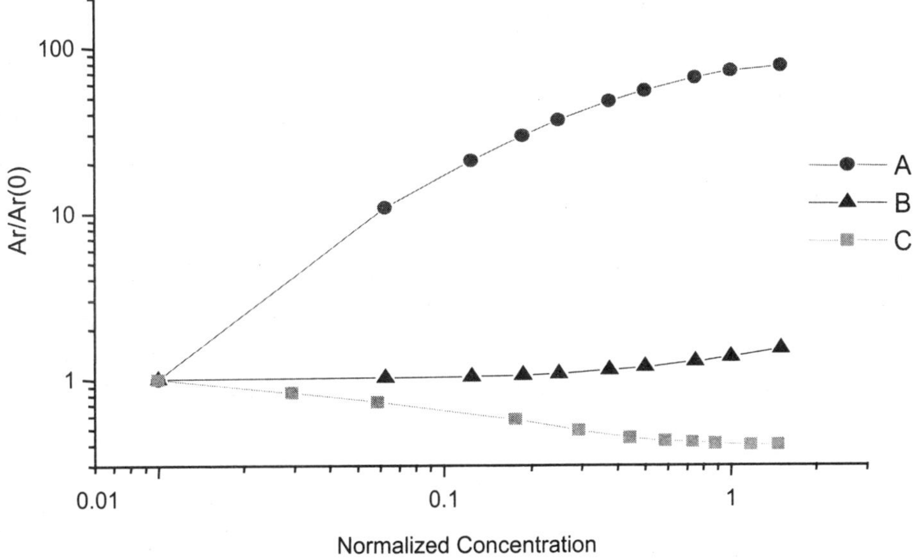

Figure 2.5   CVS response for ASUP, SUPP, and leveler in fully suppressed Cu electrolyte with polyethylene glycol. A = ASUPP, B = SUPP, C = leveler.

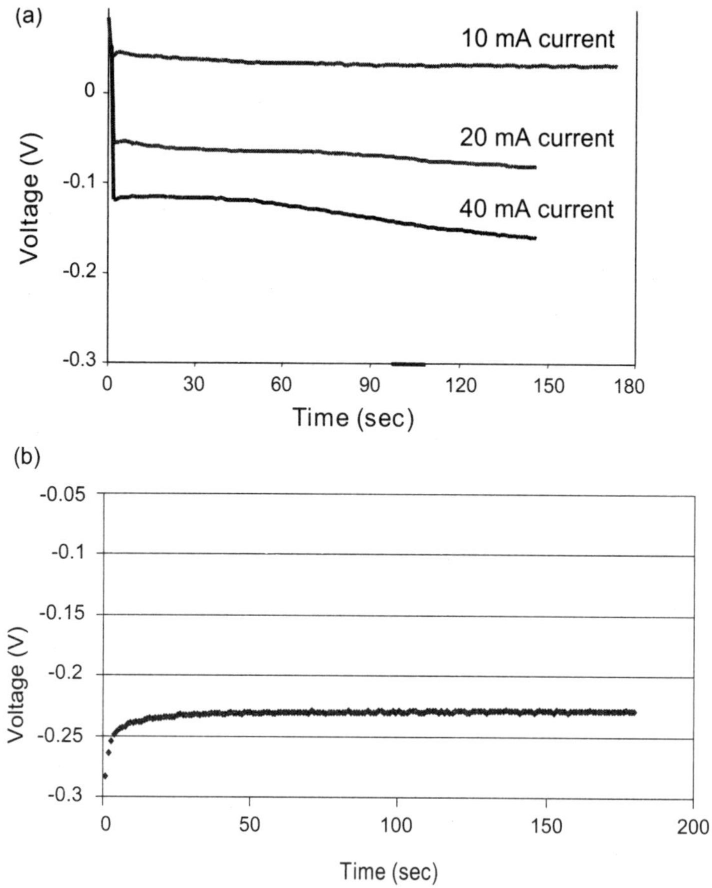

Figure 2.6 Cathodic potential response as a function of time during galvanostatic copper deposition in (a) high-acid, and (b) low-acid electrolytes containing additives (ASUPP, SUPP, and leveler) on blanket wafer surfaces at different current densities.

copper deposition rate. The suppressor did not provide additional suppression while the leveler species suppressed further electroplating. Figure 2.6 shows voltage–time curves for Cu electroplating on flat wafers, at a fixed applied current in high-acid (Figure 2.6a) and in low-acid (Figure 2.6b) concentration electrolyte with additives present. In both high- and low-acid concentration electrolytes, voltages were observed to start at more polarized values than those observed in the electrolytes without additives. This effect is attributed to the rapidly adsorbing suppressor and leveler additives. During the first 5 to 15 sec of deposition, the observed voltages become smaller, due either to morphology changes or increasing adsorption of the anti-suppressor additive species. Figures 2.7a and 2.7b show the voltage–time curves for Cu electroplating on patterned wafers at several fixed applied current values, with additives present in high- and low-acid electrolytes, respectively. As was observed for blanket wafers, the initial increase in polarization is larger than that during Cu electroplating without additives. This increase in cathodic potential corresponds to the suppression of Cu electroplating by additives. The cathodic potential decreases

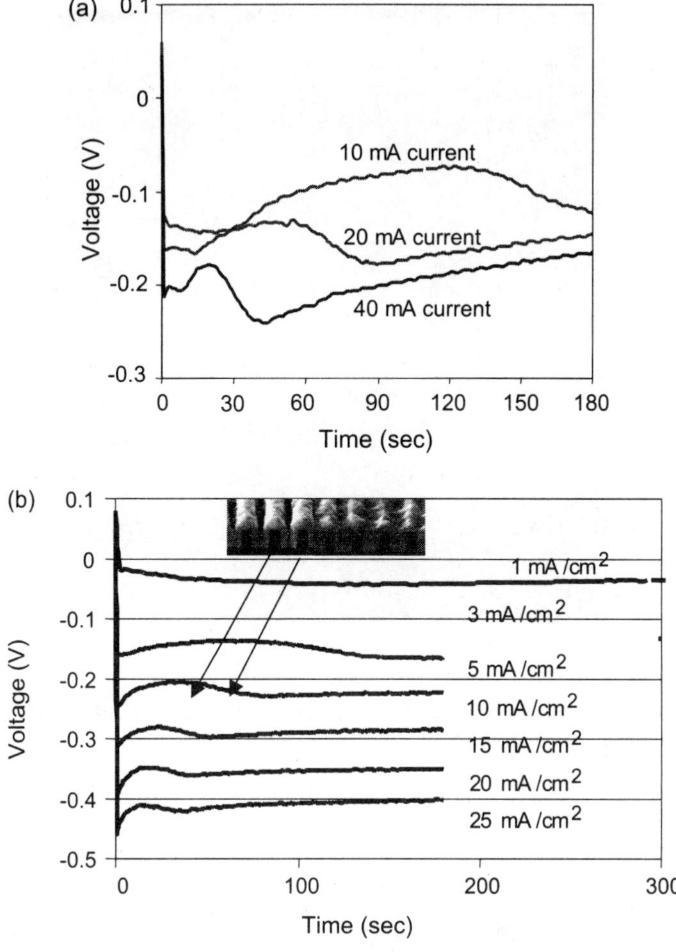

Figure 2.7 The change of polarization potential with plating time during galvanostatic Cu electroplating onto patterned wafers at different current densities in (a) high-acid and (b) low-acid electrolytes containing additives (ASUPP, SUPP, and leveler).

as plating continues, indicating a depolarization process. This depolarization is much larger and takes place over a longer period of time than that for plating on flat wafers. After this depolarization process, the cathodic potential peaks, increases in magnitude, and finally becomes constant.

2.3 Properties of electroplated Cu films

2.3.1 *Gap fill*

Figure 2.8 shows partial fill scanning electron microscope (SEM) pictures of sub-0.2 μm features. Bottom-up fill has been observed with the proper balance of plating additives. The EP bath chemistry and current wave form were optimized to

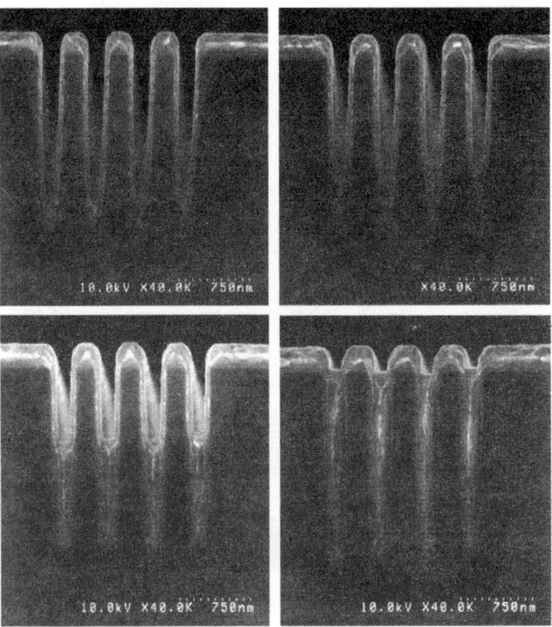

Figure 2.8   Demonstration of bottom-up fill for electroplated Cu in 0.2 μm trenches (AR
                 6:1). Notice: No copper growth on the field surface until the trenches are filled
                 with plated copper.

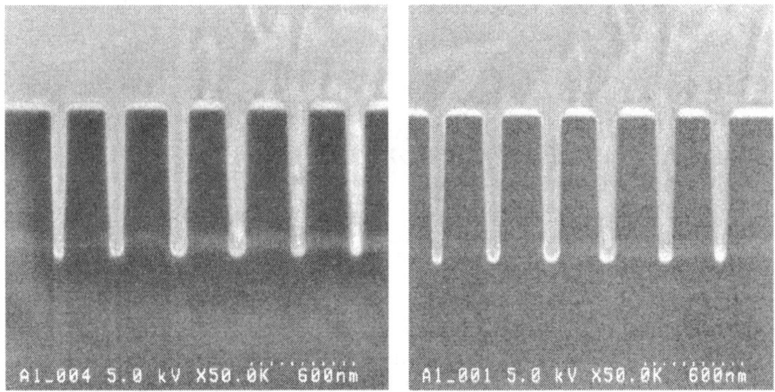

Figure 2.9   Demonstration of 0.07 to 0.1 μm EP Cu filling capability.

achieve complete fill of sub-0.1 μm trenches (Figure 2.9) with high aspect ratios (AR
> 10), thus demonstrating that the Cu electroplating process is extendable to 0.1 μm
technology.

2.3.2   Microstructure

We have developed a novel three-dimensional (3D) Winand diagram to describe the
electrodeposited film structure change as a function of current density, additive

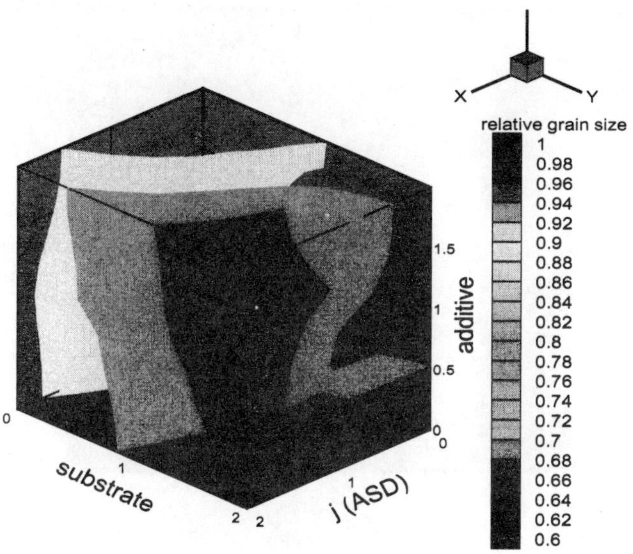

Figure 2.10 Winand diagram (3D): Grain size vs. plating current density, substrate texture, and additive inhibition strength.

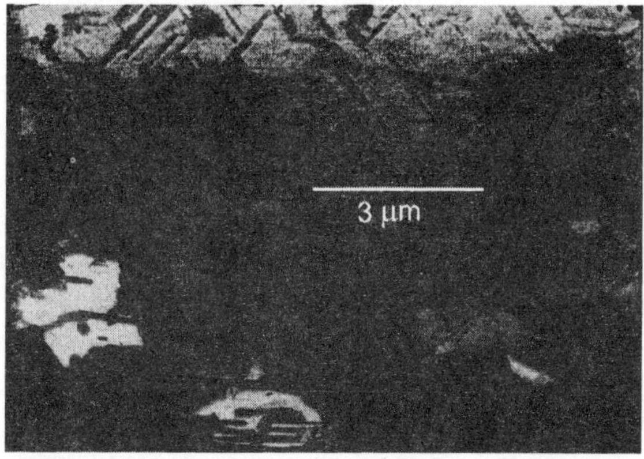

Figure 2.11 Ion channeling image of 1 µm thick electroplated Cu film.

inhibition strength, and substrate texture. It was observed that in the absence of additives, the grain size of the film was mainly controlled by the substrate microstructure; more textured substrate led to smaller grain sizes. High additive concentration, and highly textured substrate yielded the smallest grain size (Figure 2.10).

The self-annealed electroplated Cu grain structure contained the largest grains (the longest dimensions of any grain set including twins was >3 µm) with a large fraction of them being twinned grain boundaries (Figure 2.11). The electroplated films were strongly (111) textured with small but measurable (200), (220), and (311) components and with a random component of about 30%. The random component of the texture (i.e., other than $\langle 111 \rangle$ orientation) for electroplated Cu films was

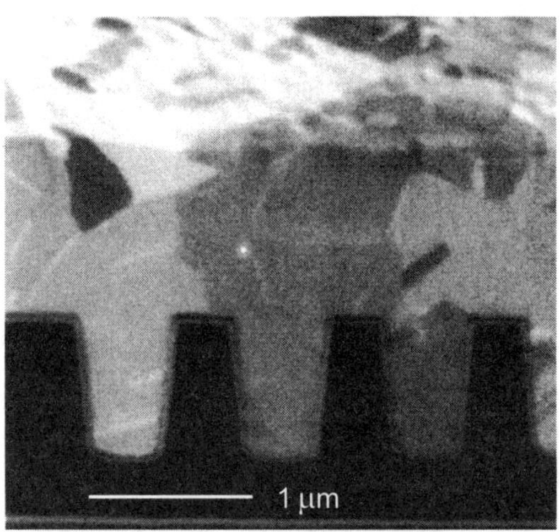

Figure 2.12   Ion-beam image of Cu-filled trenches.

measured to be about 3 to 4%. An angle for 95% of the $\langle 111 \rangle$ volume fraction (i.e., ω_{95}) was obtained in the range of 2 to 3 degrees. We also observe low tensile stress (in the range of 10^8 dyn/cm^2) in electroplated Cu films. Strong $\langle 111 \rangle$ texture, large grains, and low stress in plated Cu will improve the reliability related attributes of Cu metallization such as electromigration and stress voiding.[19]

We can also affect the grain structure and grain orientation by controlling the deposition conditions.[19] Copper grains occupying the entire trenches were found by ion-beam images for Cu-filled trenches under suitable deposition conditions (Figure 2.12). Grain sizes of electroplated copper increase further after low temperature annealing. The median grain size of electroplated copper was measured to be about 1 µm and the lognormal standard deviation is about 0.4 µm.

The elastic modulus was found to be about 150 GPa. Microhardness decreases from 1.5 to 1 GPa when the grain size increases.[19] If the grain size is large, a greater stress concentration is developed in the adjacent grain, and thus the applied stress needed to activate flow in this grain is relatively low, and vice versa. This is known as boundary strengthening and is described by the Hall–Petch relation.

Adhesion of plated Cu films to substrate was measured by pull test to be in the range of 200 kg/mm^2.[19] No failure was observed on plated Cu/sputtered Cu seed interface.

Microstructure variation as a function of plating current density was studied for Cu films plated without additives.[20] The (111) signal is decreased while the (200) signal is increased as the plating current density is increased. This degradation of the texture is probably due to the increased deposition rate under which the arriving ions do not have sufficient time to find equilibrium sites by surface diffusion, producing relatively more random film texture.

Figure 2.13 compares the grain size of films plated with and without additives. Additive inclusion seems to impede the grain growth in the plated films. Time-dependent microstructure variation was studied on the plated films with additives at

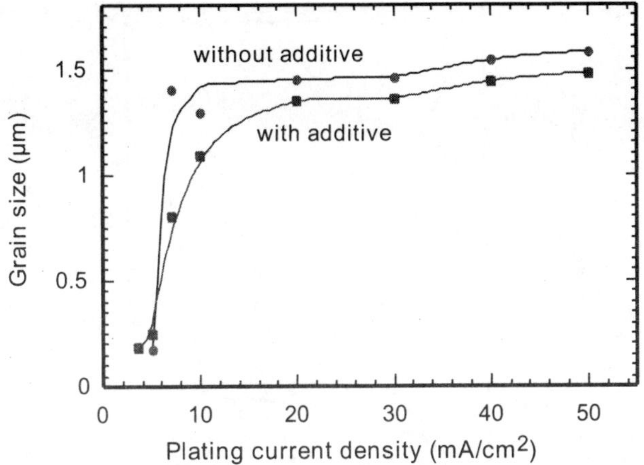

Figure 2.13 Dependence of grain size on plating current density for films plated with and without additives: 10 days aged films.

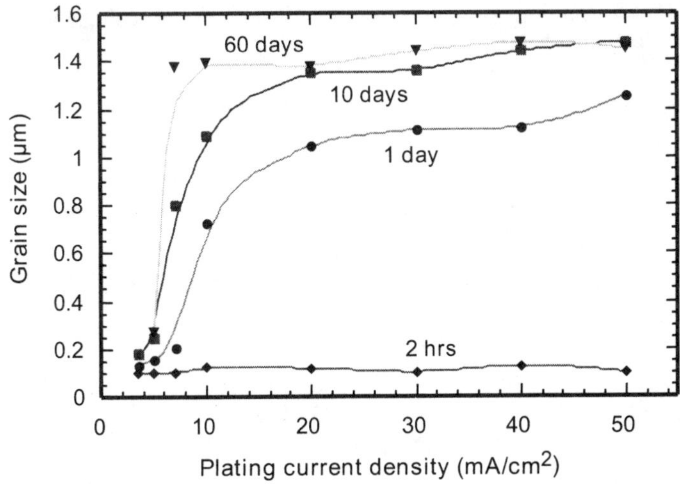

Figure 2.14 Grain size evolution for Cu films plated at different current densities from electrolytes containing additives.

different current density. The grain sizes are initially small in all films regardless of the plating current density. Films recrystallize faster as the plating current density is increased (Figure 2.14). This behavior is believed to be due to the higher energy state of films plated at higher current density, which may act as a driving force for faster recrystallization. These films have higher intrinsic stress introduced by higher defect density and more random texture as confirmed by Figures 2.15 and 2.16. It shows the decrease of (111) peaks and increase of (200) peaks with time, in all films plated with current densities ≤ 10 mA/cm^2. Only small changes of texture are observed in the films deposited at 5 mA/cm^2. The evolutions of (111) texture and Cu lattice constant as a function of plating current density are shown in Figure 2.17. The (111) texture

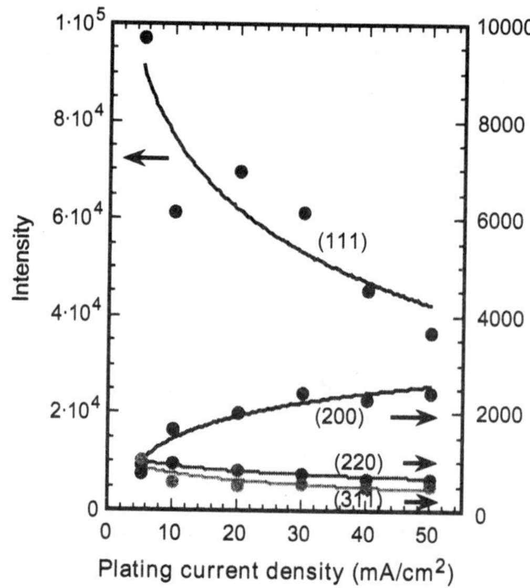

Figure 2.15 Texture variation with plating current density in films plated without additives: 30 days aged films.

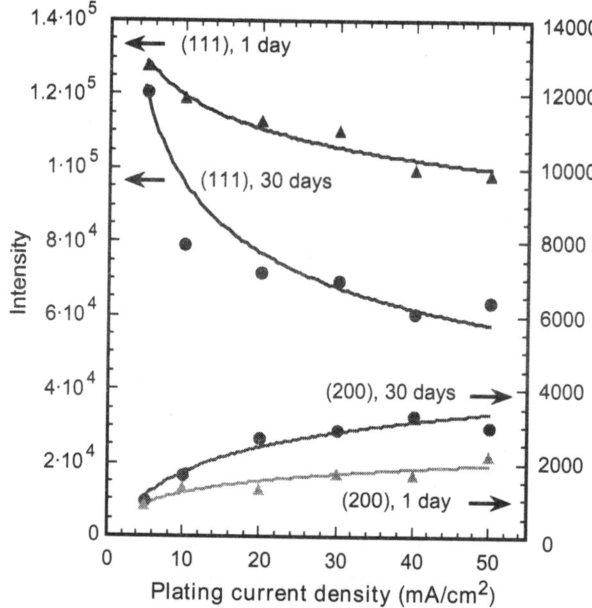

Figure 2.16 Texture variation with plating current density in films plated with additives.

in the films degrades faster with higher plating current density. This variation of texture indicates the transition of driving force from surface energy minimization to strain energy minimization by large tensile stress development in films due to large grain growth during self-annealing.

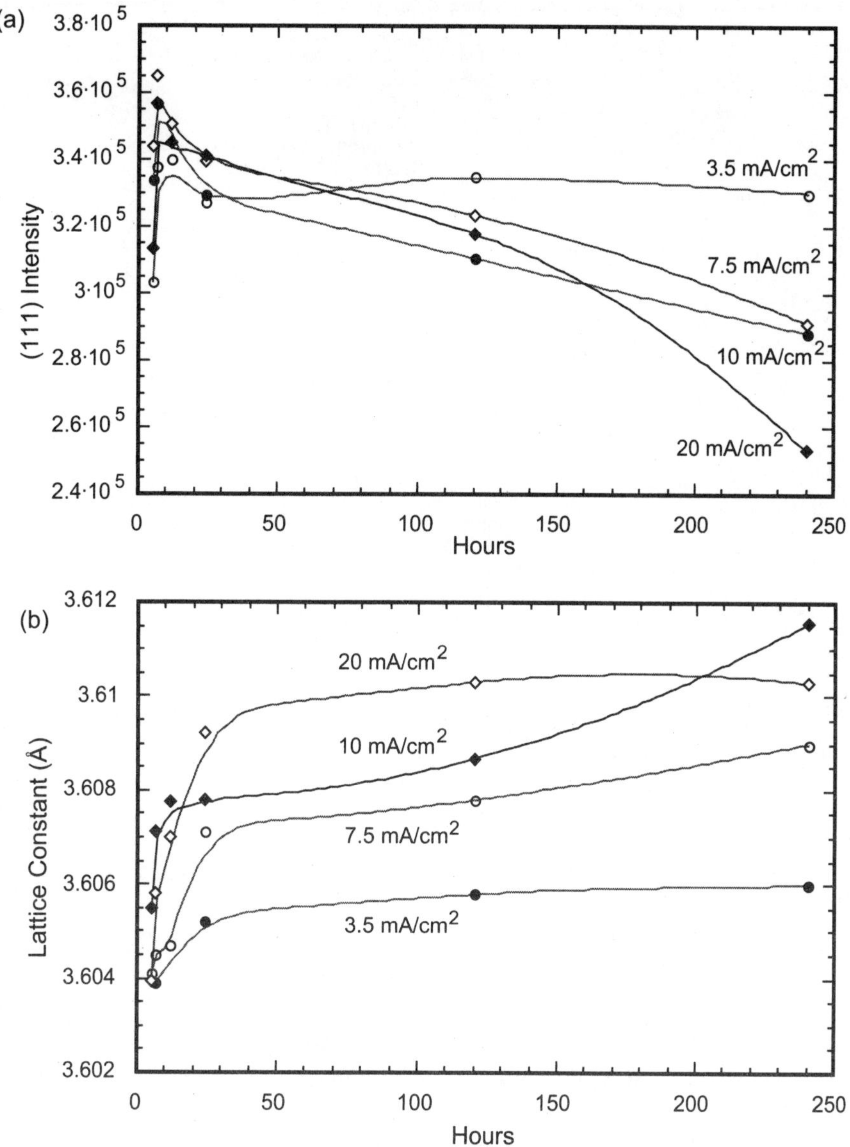

Figure 2.17 Evolution of (a) (111) intensity and (b) Cu lattice constant at room temperature for films plated with different current density and with additives.

2.3.3 Film composition

The impurities measured in our electroplated Cu films were Cl, S, C, and O. The concentrations of these impurities ranged from several to about 100 ppm, depending upon the plating chemistry and the current wave forms. The incorporation of these impurities is attributed to oxidation/reduction reactions of the plating additives.

Experimental Auger depth profiles of the Cu surface showed that the oxidation rate at room temperature first increased, and then reached a plateau after 6 to 8 hours.

                                                    Valery Dubin et al.

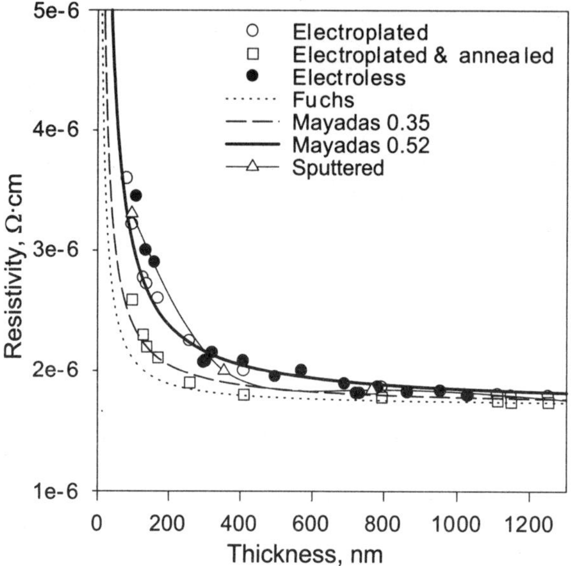

Figure 2.18 Resistivity of Cu thin films as a function of film thickness.

Up to 70% of the surface oxide was grown within the first 2 hours after Cu deposition. A $Cu/Cu_2O/CuO$ surface layer was formed due to copper oxidation.

2.3.4 *Electrical resistance*

Resistivity of electroplated Cu films was about 1.8 to 2.0 $\mu\Omega$ cm when the thickness of electroplated copper films exceeded 0.5 μm. The resistivity increases with decreasing copper thickness. The resistivity of electroplated Cu films can be further reduced (about 15%) by using a low temperature annealing cycle (Figure 2.18). Both Kelvin via and String (25,000 vias) via resistance were measured. Very low via resistance was demonstrated: 100% of vias (down to 0.35 μm size) have via resistance below 0.55 Ω.[19]

The resistivity of Damascene electroplated Cu lines was measured to be about 2.0 $\mu\Omega$ cm for the in-laid Cu line with the width down to 0.15 μm. The effective resistivity of in-laid Cu lines decreases (up to 15%) after low temperature annealing (<400°C).[18]

The electrical and thickness uniformity of electroplated Cu films was found to be in the range of 3%, 3σ for 300 mm wafers.

Comparison of the experimental data with different theoretical models describing the behavior of thin film resistivity as a function of film thickness is presented below. Two basic models were taken for the calculation: the Fuchs–Sondheimer model and the Mayadas–Shatzkes model.[21] The first model considers the surface and interface scattering of free electrons and the second model adds an additional term to account for grain boundary scattering.

The full Fuchs formula was used:

$$\rho = \rho_0 \left(1 - \frac{3}{2}(1-p)k^{-1} \int_0^{\pi/2} (1 - e^{-k/(\cos\theta)})\sin^3\theta\cos\theta\, d\theta \right)^{-1} \qquad (1)$$

where ρ_0 is the bulk value of the resistivity of material, $k = d/\lambda_0$ is the ratio of the film thickness d to the mean free path of electrons λ_0, and p is the scattering factor.

We assumed that the value of p was zero for both surfaces and interfaces. A theoretical study made by Zimman[22] shows that specularity parameter p approaches zero rapidly, meaning a nearly total inelastic scattering, when average height of the surface asperities is of the order of the wavelength of the electrons (about 0.5 nm).

The bulk resistivity value is 1.71 $\mu\Omega$ cm for the Cu. The value of the mean free path of electrons was obtained from the bulk resistivity and is equal to 39 nm for Cu.

The Mayadas formula is:

$$\rho = \frac{\rho_0}{3\left[\frac{1}{3} - \frac{1}{2}\alpha + \alpha^2 - \alpha^3 \ln(1 + 1/\alpha)\right]} \qquad (2)$$

where $\alpha = \dfrac{\lambda_0}{d}\dfrac{R}{1-R}$, and R is the reflection coefficient at the grain boundary.

The results of calculation and fitting for Cu are presented in Figure 2.18. It can be seen that experimental data for as-deposited electroplated and electroless Cu films correspond to the Mayadas formula with the $R = 0.52$. It shows that films have polycrystalline structure. This becomes clearer in comparison to the calculation made with the Fuchs formula. Fitting of the experimental data on the resistivity of annealed films is less accurate. The two regions can be distinguished. In the first region of thickness >250 nm, the behavior of the resistivity vs. thickness curve is described by the Fuchs's model. In the region of thin films <250 nm the experimental data correspond to the Mayadas model with $R = 0.35$. This fact requires discussion and may be explained by recrystallization, which runs in different ways for different films thickness. Thicker films are closer to the bulk copper, while thin films "feel" substrate and do not have enough ability to form single crystalline films.

More rapid increase of the resistivity of metal films at scaling is also determined by the surface and interface roughness. As indicated in Ref. 23, the Namba model taking into account a new parameter: roughness/mean free path ratio may help to describe the very thin films. A "barrierless" technology for metal interconnects is especially attractive for small line width structures where the barrier forms a substantial part of the line (Figure 2.19).

2.4　Electroplating modeling: An integrated simulation approach

Numerous publications on electroplating modeling are available in the literature, for example.[4,8–14] In general these analyses fall into two categories. The first is

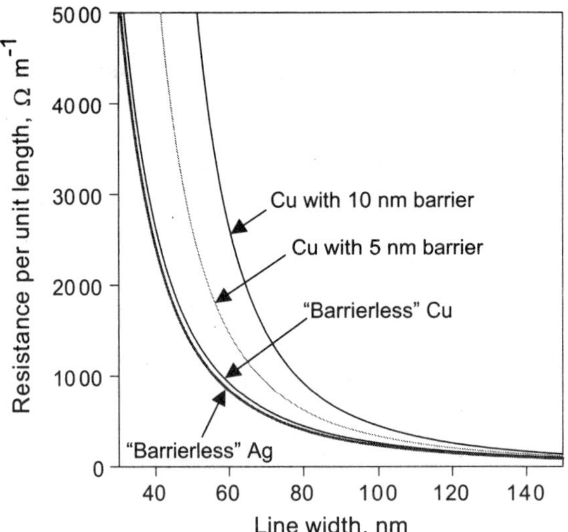

Figure 2.19 Effect of barrier layer on effective line resistivity.

simulations of macroscopic/wafer-scale phenomena, such as effects of reactor design and process conditions on plated film thickness uniformity. A more recent second category includes simulations of the feature-scale gap-fill process itself.[4,8–10,12–14,18] These shape evolution models attempt to relate the chemical actions of the additive to the various gap-fill behavior ("superfilling", conformal filling, and void formation). Interactions between these two length scales (wafer scale and feature scale) are mostly left unexplored in these studies.

Earlier electroplating models by Dukovic et al.[4,8] and West[9] assumed that the reactions of the additives are controlled by diffusion mass-transport while the metal deposition reaction is controlled by its kinetics. This means that by optimizing the reaction rates (by varying additive concentration and adjusting the current density of Cu deposition), it is possible to obtain a higher deposition rate at the bottom of the features. These models predict that growth rate variation along a feature is caused by diffusion and consumption of a suppressor additive, and do not include effects of accelerator additives. These simple models do not seem consistent with experimental observations, as illustrated in Refs. 19 and 12. Moffat et al.[12] evaluated the influence of three additives (chloride), polyethylene glycol (PEG), and 3-mercapto-1-propane-sulfonate (MPSA) on gap fill in deep trenches down to 90 nm in width. Comparisons of the gap fill obtained using binary additive combinations of Cl/PEG and Cl/MPSA against gap fill obtained using a combination of the three additives led them to conclude that a synergistic nature of the interaction between three species is required for void-free filling of small features.[12] A more detailed additive model is therefore required. Chemical interactions of these additives during electroplating, and their roles in controlling the metal deposition rate and gap-fill behavior, however, are not understood.

One particular phenomenon related to additive interactions that previous models were not able to explain is the so-called "momentum plating." This occurs when the

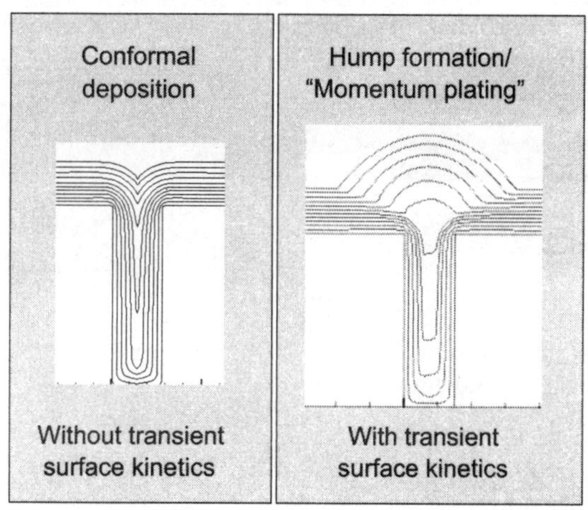

Figure 2.20 Representation of conformal deposition (left) and "momentum plating" phenomenon (right). Modeling shows that transient surface kinetics need to be considered
to reproduce the momentum plating phenomenon.

superfilling behavior in a feature leads to a protrusion in the plated film profile,
compared to the field areas where there are no features present (Figure 2.20). A model,
which includes both a suppressor and an accelerator additive, had been proposed,[10]
but does not reproduce the momentum plating phenomenon under any conditions.
Newer models[12–14,18] assume that superfilling is due to local accumulation of
accelerator at the bottom of the feature, due to reduction in the surface area during
deposition. While able to reproduce momentum plating, these models do not consider
interactions between suppressor and accelerator, and rely on somewhat arbitrary
assumptions. For example, the model by West et al.[18] assumes that surface site
density varies with deposition, while others[12,14] assume that variation of the
accelerator surface concentration depends on local surface curvature, as well as on
an empirical form for current density variation with accelerator concentration. These
assumptions make comparisons with electrochemical characterization data difficult
due to lack of precise correlation; hence they limit their applicability to general
electroplating additive systems.

Recently we have developed an integrated model to investigate interactions
between the additive species, their effects on deposition rates and uniformity, and
gap-fill behavior.[24] The methodology used to develop this integrated model is
summarized in Figure 2.21. Simulations include the effects and phenomena at both
wafer scale and feature scale. This approach leads to a consistent integrated model
for investigations and optimizations of reactor configurations, bath chemistry, and
process conditions to achieve desired thickness uniformity and good gap-fill
behavior. A multi-species tertiary current distribution model, based on a finite
element method (FEM) has been developed to investigate and optimize the effects of
the plating cup configuration, fluid flow profiles, and mass transfer effects. This
model was used to investigate fluid flow patterns in a rotating-wafer electroplating
cup, for different solution flow rates and wafer rotation rates. Flow recirculation cells

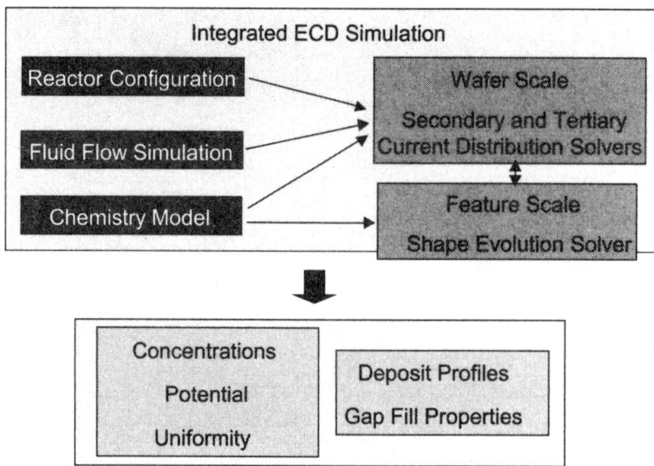

Figure 2.21 Integrated electroplating modeling methodology, including wafer-scale and feature-scale solvers.

that affect thickness uniformity were identified. Simulations have shown that the design of the plating cup strongly affected the fluid flow and species mass-transport in regions close to the wafer, leading to across-wafer thickness variation, as was illustrated in Ref. 15. Strategies for improving thickness uniformity for a given bath and plating cup design can be identified using this model without requiring extensive experiments to characterize the system.

We have also developed a new multi-species model for gap-fill and momentum plating based on mass conservation of additives in the electrolyte and surface. The model, as summarized in Figure 2.22, includes the important physical phenomena leading to momentum plating, particularly additive diffusion in the electrolyte, and transient potential-dependent surface reactions. The goal of the model is to provide an accurate and flexible method to relate electrochemical data to gap-fill properties for a given electroplating bath. It also provides insights into ways to reduce local thickness variations caused by momentum plating, which can be detrimental for the subsequent CMP processes.

The model includes a detailed chemical species for the actions of the additives, with interactions between adsorbed suppressor and accelerator species consistent with observations from electrochemical characterizations of various additive systems. Kinetic parameters for the model were extracted by comparing model predictions with LSV data for different additive concentrations. The chemical mechanism was incorporated into the feature-scale shape evolution model, which uses a surface-based technique such as a boundary element method (BEM) to solve species diffusion equations in the electrolyte domain above the features to be plated:

$$\nabla^2 C_i = 0 \tag{3}$$

subject to boundary conditions involving surface reactions:

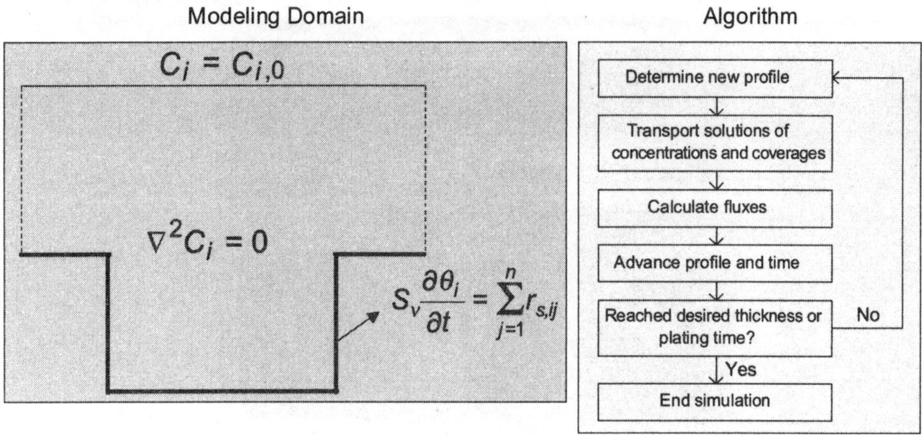

Figure 2.22 A summary of detailed gap fill that include effects of multiple additive species and transient surface kinetics. The model is being used to determine gap-fill behavior including momentum plating in Cu features.

$$S_v \frac{\partial \theta_i}{\partial t} = \sum_{j=1}^{n} r_{s,ij} \tag{4}$$

where C_i = concentration of species i in the electrolyte, S_v = copper surface site density; q_i = surface coverage of species i, $r_{s,ij}$ = rate of change of species i in surface reaction j, and n = total number of reactions. The profile evolution is calculated using a Eulerian–Lagrangian based approach,[25] in which the moving surface is represented by a set of line segments moving over a fixed grid. The position of the moving front is tracked by updating the locations of the individual segments according to the local current density values obtained from the BEM solver. The model does not require assumptions of how additive concentrations change with surface shape or its curvature; instead it is based on the well-defined adsorption-based surface kinetic formalism, and accounts for time-dependent changes in additive concentrations due to mass-transport and surface reactions.

Applications of the model to the two component additive systems: Cu/Cl/PEG/MPSA[3,12] and Cu/Cl/PEG-like suppressor/SPS-like suppressor[11] will be discussed. An example of the applications of the feature-scale shape-evolution model in relating additive chemical mechanism and gap-fill behavior is described in Ref. 24. An additive model that represents a multi-component Cu/Cl/PEG-like/SPS-like bath[18] has been developed and used in shape evolution simulations to predict gap fill for different size trenches (Figure 2.23). Figure 2.23a shows SEM pictures of three different size features (from left to right: 1.7, 0.9, and 0.3 µm) at an early stage of plating (0.08 µm nominal thickness). Figure 2.23b shows SEM pictures of the same features for 0.24 µm nominal deposit thickness. Deposit profiles predicted by the model from 0 to 0.24 µm nominal thickness were plotted in Figure 2.23c. It is noted here that these simulations used the actual seed layer profiles as a starting point, as

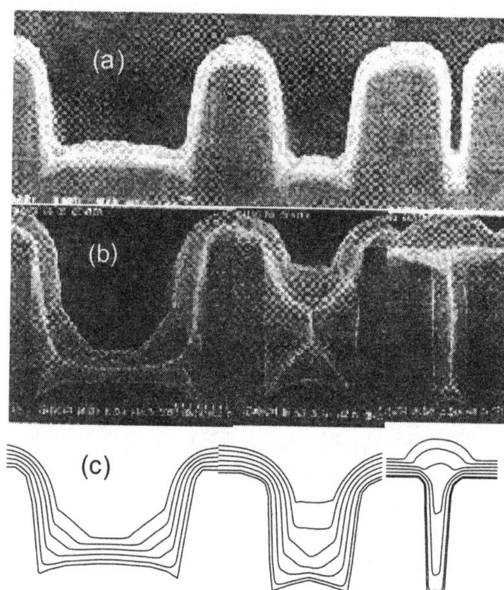

Figure 2.23 Comparison of model predictions and SEM data for the Cu/Cl/PE-like/SPS-like electrolyte.[15] (a) SEM profile at 0.08 μm thickness, (b) SEM profile at 0.24 μm thickness, (c) model predictions for profiles between zero and 0.24 μm thickness. SEM pictures are taken from West et al.[18] Trench widths are left to right: 1.7, 0.9, and 0.3 μm.

extracted from SEM micrographs. This allowed realistic seed layer thickness variation along the feature to influence the shape evolution during the electroplating process, and enabled accurate investigations of gap-fill behavior. The use of an idealized (i.e., conformal) seed layer profile in a gap-fill model often leads to inaccurate gap-fill behavior, and does not allow investigations of the plated profile sensitivity on the seed layer coverage.

Excellent agreement was achieved between the SEM profiles and the shape evolution model results. Both model (the third intermediate profiles in Figure 2.23c) and data show that for the 0.9 and 1.7 μm trenches, initial deposition at the bottom corners is faster relative to the rest of the surface, leading to corner profiles at about a 45 degree angle from the vertical direction. The model also predicts bottom-up fill behavior for the narrower trench (0.3 μm), followed by momentum plating of the deposit (bump formation) before the wider trenches are filled, consistent with data. This illustrates a problem associated with momentum plating where the final Cu profile can include widely varying thickness, depending on the density and size of features filled. The model described in this work has been used to aid optimizations of the process conditions and additive chemistry to address this issue.

The integrated model is also useful to investigate effects of reactor-scale phenomena on feature-scale plating performance. In general, as the underlying seed layer thickness decreases in next generation copper interconnect technologies, and as wafer diameter increases from 200 to 300 mm, less uniform plating on the reactor scale is to be expected, due to the finite electrical resistance of the seed layer. For non-patterned wafers, these effects lead to thickness variation across wafer, where thickness at the wafer edge is typically higher than that at the wafer center. For patterned wafers, these effects lead to undesired variations in gap-fill behavior across the wafer. Interactions between these two length scales are often difficult to quantify, complicating optimization efforts. The linking between the reactor-scale and feature-scale effects was investigated using the integrated model described above. The model was used to investigate current density variations across patterned wafers and gap-fill behavior for features at different wafer locations. This led to understanding of the process window for the combination of process conditions, reactor hardware, and bath chemistry. In addition, it contributed to plating recipe optimizations for patterned wafers.

2.5   Reaction mechanisms: Role of additives

Copper oxidation on the anode can be described by a two-step reaction. The first step is a fast oxidation of Cu to Cu^+. The rate-limiting step is the oxidation of Cu^+ to Cu^{2+}. Accumulation of Cu^+ and CuCl precipitation on the anode is likely to occur in the presence of Cl^-, resulting in a so-called "black" film formation on the surface of the anode, which contained CuCl, O, S, and P. We observed that the additives had significantly lower consumption rates when the anode was protected by this black anode film. The copper ion reduction on the cathode, in the absence of additives, can also be described by a two-step reaction:

$$Cu^{2+} + e^- \rightarrow Cu^+ \qquad E_0 = 0.153 \text{ V} \tag{5}$$

$$Cu^+ + e^- \rightarrow Cu \qquad E_0 = 0.52 \text{ V} \tag{6}$$

In this case, the rate-limiting step is the reduction of Cu^{2+} to Cu^+. Cu^+ did not accumulate and CuCl precipitation was not as likely in the presence of Cl^-, due to its low solubility.

The suppressor additive, which is typically a member of the polyalkylene glycol family, is a wetting agent. It can be adsorbed on the copper surface and form ordered polymeric structures in the presence of chloride ions, depending on the applied potential. The suppressor reduces the copper plating deposition rate, as can be seen from CVS, LSV, and depolarization studies, most likely by competing with and blocking Cu ions from adsorbing on the surface.

The accelerator (anti-suppressor) enhances deposition rate of copper, most likely by catalyzing adsorption of copper ions on the surface through a surface complex intermediate. It can reduce the effects of the suppressor by adsorbing on the copper surface and providing active growth sites for copper, or by reducing blockage of growth sites by the adsorbed suppressor. The accelerator can decompose to other strongly accelerating species. Kinetics of adsorption and desorption of the anti-suppressor species are typically slow compared to surface reactions of other additives; this transient effect must be accounted for to achieve an accurate model.[18] Due to its slow kinetics, accelerator species coverage can increase during the deposition process in the feature, where there is a progressive decrease in surface area, resulting in possible higher surface coverage of accelerators or their by-products on the surface of the feature compared to the flat surface. This result in a bottom-up fill behavior where deposition rates close to the bottom of the feature are enhanced relative to those of the flat surface.

A leveler can also be added to the plating bath to provide a "leveling" effect (i.e., produce deposits relatively thicker in small recess and relatively thinner on the peaks). The levelers exhibit mass transfer sensitive adsorption, and accumulate preferentially near the most negatively charged or high mass transfer sites of the cathode.

Depolarization effects observed during trench fill are easily interpreted as resulting from accumulation of accelerator and its by-product within the trenches. The surface concentration of the accelerator becomes larger at the bottom of features and decreases the plating potential needed to provide required plating current. The decrease in polarization while filling patterned wafers is observed even though the actual surface area being plated is decreasing by over a factor of two as the features fill. Without the effect of the accumulating accelerator this would be expected to *increase* polarization by at least 50 mV. When the trenches are filled, the leveler, which is a suppressor, increases the polarization by replacing the accelerator on the surface where it has a higher mass transfer availability. The voltage–time curves are also current dependent. Depolarization only occurs with plating currents in the intermediate range of 5 to 25 mA/cm^2 where bottom-up fill is observed. Depolarization begins and is completed more quickly as current density is increased. Likewise, when plating at fixed voltage the depolarization begins and is completed more quickly at higher applied voltages. Rapid depolarization follows simply from the rapid accumulation of the accelerator, which takes place with more rapid initial filling at higher currents. In leveler-containing systems the rapid depolarization is followed by more rapid polarization as levelers adsorb on filled feature surfaces. A leveler provides a leveling effect which reduces the within die non-uniformity (WIDNU).

Based on the depolarization, LSV, and CVS studies, we can conclude that the following steps occur during the trench fill: (1) nucleation of electroplated copper on PVD films at a high overpotential due to polarization caused by suppressor adsorption; (2) conformal growth of copper during initial deposition and at low current densities — this step is usually used to repair discontinuous PVD Cu seed layer; (3) bottom-up fill which is distinguished by cathodic depolarization from otherwise suppressed values; bottom-up fill occurs because of ASUPP and its by-product accumulation due to rapid decrease in the surface, especially at the bottom corners of narrow features and SUPP replacement by accelerating species; (4) planarization of filled trenches due to replacement of the accelerator on the surface with leveler and suppressor. Since these adsorption/desorption reactions occur on the time scale comparable with gap fill, "hump" formation can be found over small features, especially if the leveler is not present in the plating additive system.

Additives also play an important role in regulating the film properties. The ASUPP, often called a brightener, reduces the grain size by adsorbing on the growing grain surface and promoting nucleation of new grains, which are smaller than the thermodynamic equilibrium grain size. At room temperature these films experience grain growth, with a decrease in random texture and an increase in surface roughness (in the range of 10 nm). We found that the incorporation of S into the plated copper was proportional to the concentration of ASUPP in plating bath, while incorporation of carbon in deposited copper increased with increasing total oxidizable carbon (TOC) level in the bath. The impurity level also depended on the concentration of sulfuric acid in the plating bath.

2.6   Manufacturability issues

Implementation of Cu plating process in high-volume chip manufacturing is extremely challenging. The process has to meet all the requirements of a high yielding, reliable microprocessor manufacturing. These requirements include: wafer to wafer uniformity of plating with desired electrical and electromigration properties, excellent adherence of Cu lines/vias to the dielectric, and desired reflectivity. In order to meet these criteria, a strict control of bath components is essential. In addition the integration scheme must include attention to proper selection of a barrier/adhesion/ seed layer between dielectric and the plated Cu, and a reliable chemical–mechanical polishing (CMP) process. Some of these aspects are discussed in the following sections.

2.6.1   On-line bath metrology

A closed-loop control system is needed to maintain the bath component concentrations within process specifications. The additive concentrations are measured using electrochemical and analytical techniques and are adjusted by dosing solutions containing additives to the bath as needed. An on-line bath metrology has been developed to maintain concentrations of additives in the plating bath within the specified control limits to ensure process stability with respect to gap-fill performance, film properties, and impurities content. The analysis capability of the on-line bath metrology was developed with $p/t < 0.3$ for inorganic and organic components in the Cu plating bath (the p/t ratio is defined as the 6σ of the measurement precision p divided by the process range t).

Measurements of the inorganic bath components (Cu^{2+}, H_2SO_4, and Cl^-) are achieved using the potentiometric titration method. Organic additive components can be measured by CVS and high-performance liquid chromatography (HPLC) methods.

Figure 2.24 shows a typical voltammogram of an organic measurement using CVS. The potential is cycled between two extreme potentials so that metal deposition and stripping takes place and the area of the metal stripping peak is measured as a function of the additive concentration. Increasing the concentration of organic

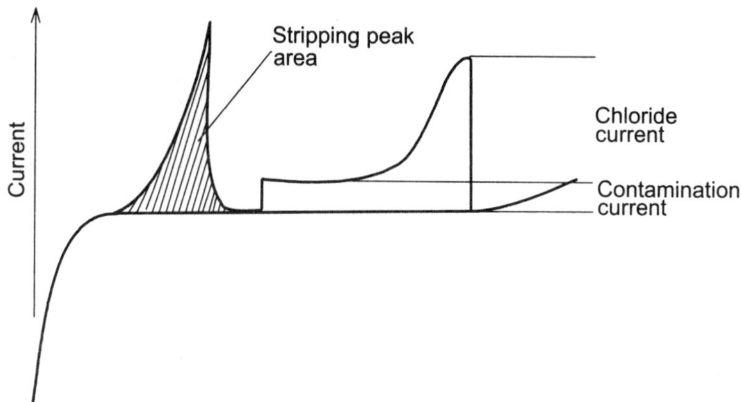

Figure 2.24 A typical CVS voltammogram for measurement of organic components in Cu electroplating bath.

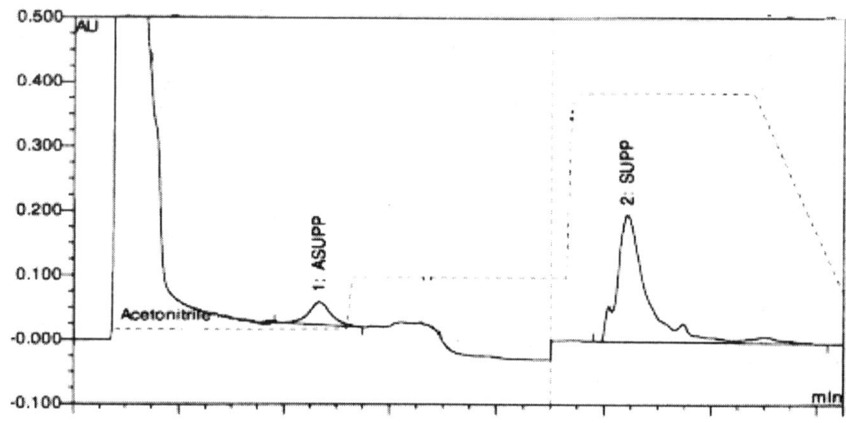

Figure 2.25 Chromatogram of organic additives in copper electroplating bath measured by HPLC.

suppressors causes the stripping peak area to decrease whereas addition of an anti-suppressor has the opposite effect. By correlating the additive concentration to the peak area, it is possible to construct a response curve for the additive concentration.

An alternative method of measuring additive concentrations is HPLC. The anti-suppressor and suppressor can be measured using a reverse phase column coupled with a UV detector and effluent made of water and acetonitrile. Figure 2.25 shows a typical chromatogram for a plating bath with organic additives. Peak 1 corresponds to the anti-suppressor and peak 2, to the suppressor. Additional peaks may appear as organic by-products formed during plating.

By using the above techniques, it is possible to measure the concentration of all bath components. In a bleed-and-feed system, concentrations of the inorganic components remain stable, but concentrations of the organic additives require frequent adjustments due to their degradation. Additive concentration changes occur through the following mechanisms: (1) baseline degradation, where the additives decompose with time; (2) per pump degradation, which occurs during solution recirculation; and (3) consumption during electrochemical processing. An algorithm has been developed to maintain the additive concentration as a function of processing conditions, but daily bath analysis is required in order to confirm and adjust the bath composition as required. Figure 2.26 shows typical trend charts for the inorganic components of Cu plating bath. Figure 2.27 shows the trend charts of organic components in the plating bath. Analytical methods have been qualified to meet $p/t < 0.3$ and accuracy within 5%. The concentrations of inorganic and organics ingredients in Cu plating bath have been maintained within specification limits with $Cpk \geq 1.5$. Cpk is a common measure of process capability (a capable process has $Cpk \geq 1.3$).

2.6.2 *Process integration*

The key process integration challenges for copper electroplating process are related to minimizing yield-limiting defects, establishing a void-free deposition while maintaining a reasonably planar surface, and preventing copper contamination of transistors. These three issues, and the copper reliability issues to be discussed in the next section of this chapter, must be overcome in order to achieve a useful copper electroplating process for copper chip metallization.

Defects in electroplated copper can be caused by several different sources including contaminants on the seed surface, copper corrosion, and particles.[26,27] The seed surface contamination defect mode arises due to the extreme sensitivity of the electroplating process to surface chemistry. These defects can either inhibit deposition and form pits or voids or enhance deposition and form mounds in the copper. One of the most common of these defect modes, called "swirls,"[26] is caused by seed surface contaminants that inhibit wetting of the electroplating solution leading to microscopic air bubbles that adhere to the wafer surface and eventually form voids or pits in the copper. These defects can be suppressed by maintaining a clean seed surface, adding surfactant to the electroplating solution and careful design of the electroplating tool. Copper corrosion-induced defects are also a major concern during copper electroplating but can be suppressed by proper choice of wave form during plating and proper design of the post-plating rinse hardware.

Forming void-free copper deposition in complex dual-Damascene patterns requires bottom-up fill as described previously. However, bottom-up fill can form a non-planar surface due to the pattern-dependent deposition rate that is intrinsic to bottom-up fill. It is desirable to make the final copper surface after copper electroplating as planar as possible to minimize the challenge to the subsequent chemical–mechanical planarization step. These conflicting requirements for gap fill and planarity can be satisfied through careful optimization of the electroplating additive concentrations and the design rules that govern the dimensions of the dual-Damascene patterns.[26]

Trace amounts of copper will damage CMOS transistors and must be rigorously avoided during wafer processing until the transistors are protected by appropriate

Valery Dubin et al.

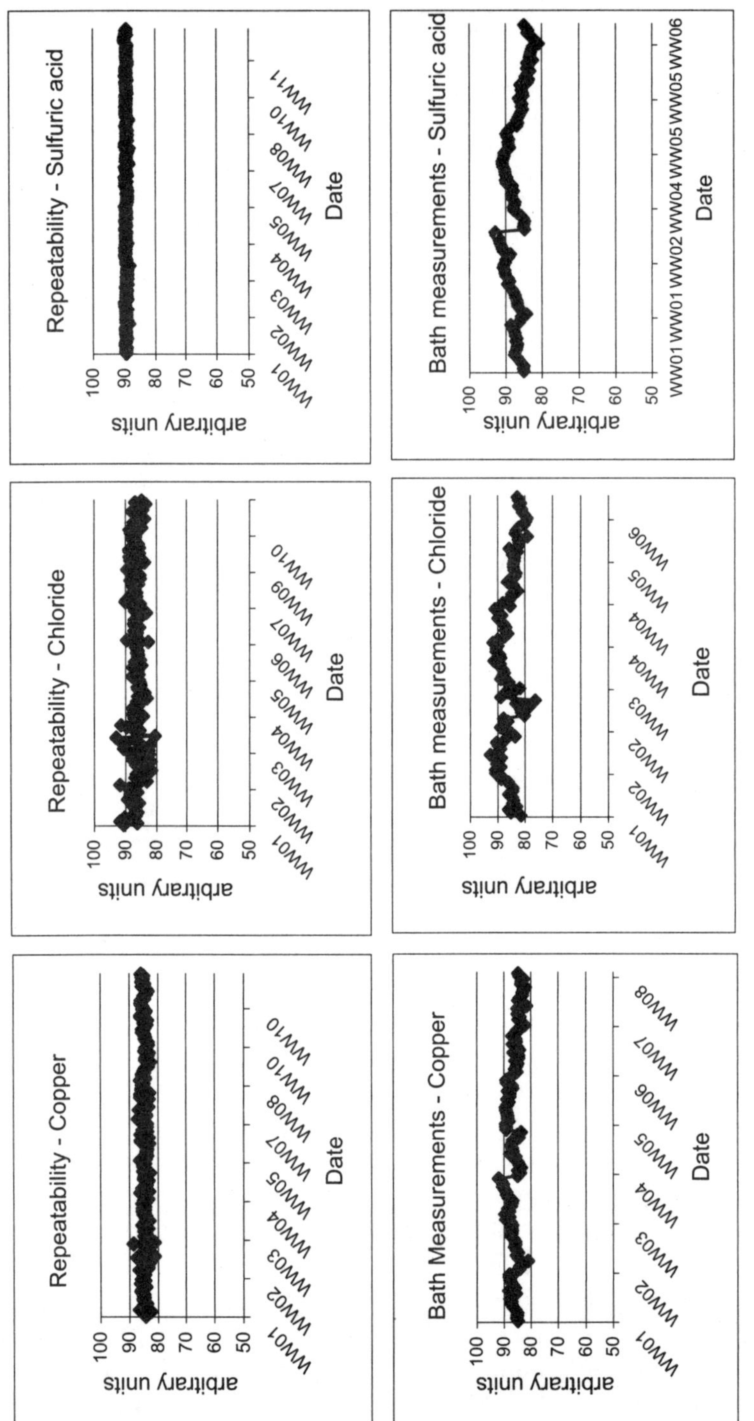

Figure 2.26 Trend charts of inorganic components in the plating bath measured by on-line bath analyzer with $p/t < 0.3$ and the following *Cpk* values: Cu: 3.5, Cl: 3.6, acid: 3.3. Where *Cpk* is min {[(Mean − LowSpecLimit)/3StdDev], [(UpperSpecLimit − Mean)/3StdDev]}.

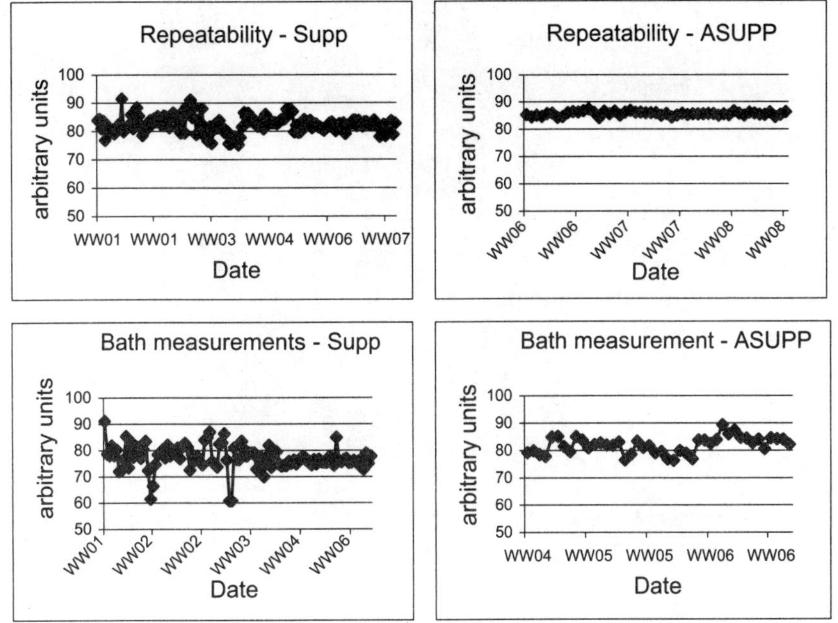

Figure 2.27 Trend charts of organic components in the plating bath measured by on-line bath analyzer with $p/t < 0.3$ and the following *CpK* values: ASUPP: 1.8, SUPP: 1.5.

diffusion barriers. Therefore, copper contamination on the backs and edges of wafers after copper electroplating can pose a serious manufacturing problem by cross-contaminating other process tools, which, in turn, can contaminate wafers that have exposed transistors. The key approaches to solving these problems involve cleaning the wafer surfaces after copper electroplating, maintaining clean surfaces at subsequent steps, and, in some cases, isolating process tools to avoid cross-contamination.[28]

The above developments have enabled the development and manufacturing of advanced interconnects. Figure 2.28 shows a cross section of 90 nm generation, with seven layer Cu interconnects embedded in low-k carbon doped oxide (CDO) dielectric developed by Intel Corporation.[29] The defect-free fabrication of complex interconnect structures shown in Figure 2.28 demonstrates the exceptional strength of electroplating technology. The combination of copper interconnects and low-k dielectric meets the performance and manufacturing goals of Intel's next generation products. Intel chips using this interconnect technology were introduced to the market in January 2004.

2.7 Reliability issues

By implementing Cu into high-volume manufacturing the bulk of reliability work was on determining the electromigration performance of the integrated material and on making sure that the barrier layers prevented Cu diffusion into the dielectric

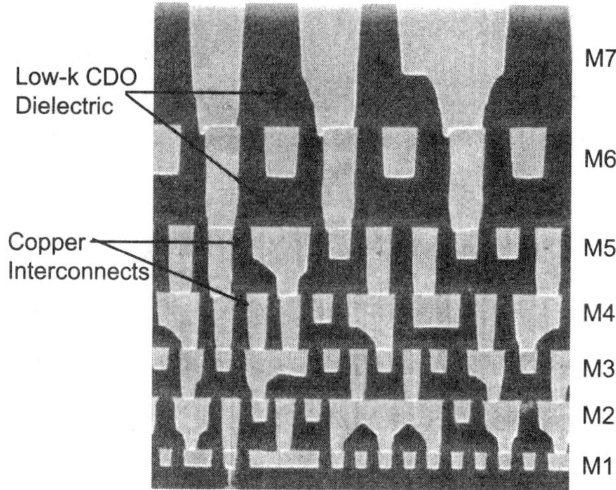

Figure 2.28 A cross section of 90-nm generation Cu interconnect developed by Intel Corporation.[22] Reproduced with permission from Intel Corporation.

layers. The focus of this section will be on the impact of the change to electroplated Cu on electromigration.

As metal line dimensions decrease, the need for current carrying capability of the line remains roughly constant. This forces about a factor of 2 increase in current density with each process generation. It was determined early on that this would not be possible with traditional Al-based interconnect systems. Literature reports based on single level Cu wires reported increases in electromigration lifetime on the order of 100× which translates to a 10× increase in current density carrying capability.[30] Early optimism about Cu electromigration faded as the material integrated in a manufacturable process indicated that these types of gains would not materialize. A comparison of Cu electromigration performance as compared to the performance of Al with Cu doping and shunts is shown in Figure 2.29. Gains in lifetime were not the 100× projected by single level work, but approximately 30×.

One of the reasons for this apparent disconnect between early development and practice is that the Al interconnect was a highly developed system. Companies used Cu doped Al with additional electrical shunt layers such as Ti and TiN. The Cu doping increased fail lifetimes, and the shunt layers provided an alternative current path around voids to also increase lifetime at the expense of effective line resistance. To ensure that resistance reductions achieved with the switch to Cu were maintained, no electrical shunting layer was developed for the Cu system. The barriers used to prevent Cu out-diffusion from lines were not sufficiently conductive to be an electrical shunt.

The lack of shunt made the system fail when only small voids were formed. Figure 2.30 shows the various possible failure locations in integrated Cu systems. Three basic modes are seen: a small void underneath a via (Mode 1), a larger void above a via (Mode 2), and a small void at the bottom of the via (Mode 3). As can be seen by the graph in Figure 2.30, the Mode 3 failures were at early times in the failure

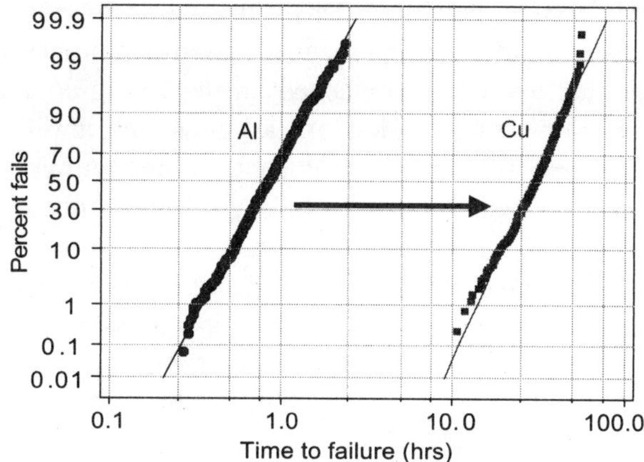

Figure 2.29　Comparison of normalized fail times for Al-based interconnect with Cu doping and shunts to first generation Cu. All structures tested are metal lines terminated by vias.

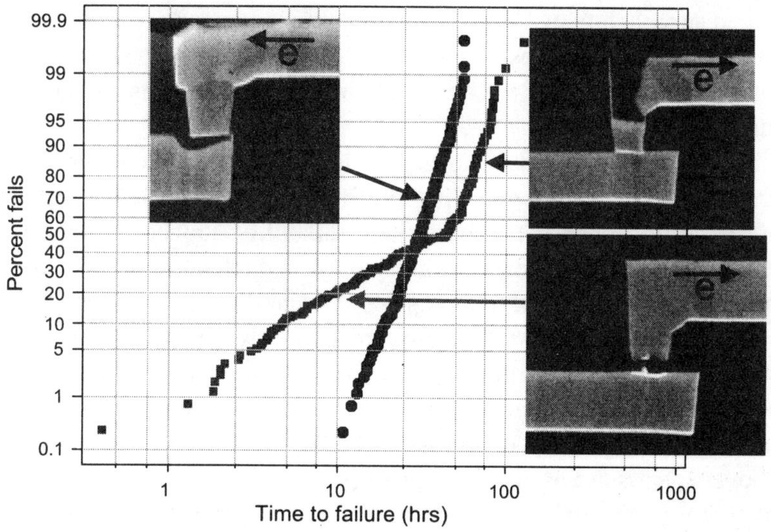

Figure 2.30　Relative failure times for different void locations and stress directions at equivalent stress conditions for Cu metallizations.

distribution and varied greatly. These types of failure modes have been seen in the literature from many companies.[31] The amount of Mode 3 type failures can be modulated by electroplating conditions at the initiation of via filling.

All of the modes of failure create smaller voids than what were seen in Al-based technologies. This means that only small amounts of material need to diffuse to cause fail. So while Cu has inherently a slower rate of diffusion as compared to Al, less material needs to diffuse to cause failure.

The other reason for shorter failure times was that the dominant diffusion path for Cu electromigration in the dual-Damascene architecture was the top surface (or the CMP surface) of the Cu metal line. This is in contrast to grain boundaries being the dominant diffusion path in Al interconnects. Various processing effects at this interface in Cu lines had a dramatic impact on Cu electromigration lifetime. With the dominant diffusion path being the top of the metal line, the effects of different electroplating conditions had very little impact on electromigration performance. Plating conditions that caused radically different grain structures did not change electromigration. The key area where plating has an impact is by growing in initial voids. If the fill traps voids, or if it does not completely fill the high aspect ratio bottoms of vias, these areas serve as nucleation points for electromigration voids and lower the lifetime of the lines. As the integration technology for Cu becomes more mature and better control of the top surface of the Cu line is demonstrated, the effects of the grain structure will probably come back into play.

2.8   Concluding remarks

A review of ongoing efforts on the development electroplating process for Cu interconnects has been presented in this chapter. Use of electroanalytical techniques, and an integrated approach to mathematical simulation have generated an understanding of the specific role of additives in the bath. These results have been successfully employed in the optimization of the electroplating bath and process parameters, which meets current industry goals.

With an increasing trend towards finer lines and finer pitch, and with the need to integrate Cu with low-k and ultra-low-k dielectric materials, electromigration issues are expected to be the most serious reliability concern in chip making. In addition, copper has poor adhesion to most low-k and ULK dielectric materials. Consequently, there is a need to develop barrier and cladding layers for passsivation and adhesion promotion. The barrier layers are currently deposited by PVD process, which runs into problems of providing continuous coverage of sidewalls in small features with a high aspect ratio. Efforts are also needed to develop conformal electroless processes for Cu seed layer deposition to replace the PVD process, or to repair the seed layers deposited by the PVD process.

Use of low-k dielectric materials in advanced interconnects necessitates the use of low pressure during chemical mechanical polishing (CMP), making the CMP process an extremely low throughput process. In the case of low-K dielectric materials, even "soft" CMP leads to deformation and delamination of Cu/ULK dielectric interface. Therefore, there is an urgent need for novel approaches to planarization in which there is no physical contact or minimal contact pressure to the wafer.

Acknowledgments

The authors strongly acknowledge Y. Shacham-Diamond and V. Bogush from Tel-Aviv University for resistivity simulation of thin metal films, Roger Brewer from Portland Technology Development for bath metrology data as well as our colleagues in Logic Technology Development, Intel Corporation for their help and support in this work.

References

1. D.C. Edelstein, Tech. Digest, IEEE Intl. Electron Devices Conferences, 773 (1997); *IBM Res. Mag.*, **4**, 16 (1997).
2. R. Rosenberg, D.C. Edelstein, C.-K. Hu, and K.P. Rodbell, *Annu. Rev. Mater. Sci.*, **30**, 229 (2000).
3. D.C. Edelstein, G.A. Sai-Halasz, and M.J. Mii, *IBM J. Res. Dev.*, **39**, 383 (1995).
4. P.C. Andricacos, C. Uzoh, J.O. Dukovic, J. Horkans, and H. Deligianni, *IBM J. Res. Dev.*, **42**, 567 (1998).
5. M. Hussein, S. Sivakumar, R. Brain, B. Beattie, P. Nguyen, and M. Fradkin, "A Novel Approach to Dual Damascene Patterning," International Interconnect Technology Conference 2002, Proceedings of the IEEE 2002 International, pp. 18–20 (2002).
6. V.M. Dubin, C.H. Ting, and R. Cheung, International VLSI Multilevel Interconnect Conference, VMIC Catalog No. 97 IMSIC-107, Santa Clara, CA, p. 69 (1997).
7. C. Cabral, P.C. Andricacos, L.M. Cignac, and I.C. Noyan, Adv. Metallization Conf. Proc., ULSI XIV, 81 (1998).
8. J.O. Dukovic, *IBM J. Res. Dev.*, **37**, 125 (1993).
9. A.C. West, *J. Electrochem. Soc.*, **147**, 227 (2000).
10. Y. Cao, P. Taephaisitphongse, R. Chalupa, and A. West, *J. Electrochem. Soc.*, **148**, C466 (2001).
11. J. Reid and S. Mayer in "Proc. of the Adv. Metallization Conf.," edited by M. Gross et al., MRS, Warrendale, PA (2000), p. 53.
12. T. Moffat, J. Bonevich, W. Huber, A. Stanishevsky, D. Kelly, G. Stafford, and D. Josell, *J. Electrochem. Soc.*, **147**, 4524 (2000).
13. J. Reid, S. Mayer, E. Broadbent, E. Klawuhn, and K. Ashtiani, *Solid State Technol.*, **43**, 86 (July 2000).
14. D. Josell, D. Wheeler, W. H. Huber, and T.P. Moffat, *Phys. Rev. Lett.*, 87 (2001).
15. V.M. Dubin, C.D. Thomas, N. Baxter, C. Block, V. Chikarmane, P. McGregor, D. Jentz, K. Hong, S. Hearne, C. Zhi, D. Zierath, B. Miner, M. Kuhn, A. Budrevich, H. Simka, and S. Shankar, International Interconnect Technology Conference, IEEE Catalog Number 01EX461, San Francisco, CA, p. 271 (2001).
16. R.A. Binstead, J. Wu, R. Mikola, J.M. Calvert, Abstract ECS Meeting, Philadelphia, PA (2002).
17. E. Farndon, F. C. Walsh, and S.A. Campbell, *J. Appl. Electrochem.*, **25**, 572 (1995).
18. A.C. West, S. Mayer, and J. Reid, *Electrochem. Solid State Lett.*, **4**, C50 (2001).
19. V.M. Dubin and R. Cheung, "Cu electroplating and Cu electroless plating" in *Cu Interconnect Technology*, Realize Corp., ed., Japan (1998).
20. H. Lee, S.D. Lopatin, D.Y. Kim, V.M. Dubin, and S.S. Wong, "Effect of Plating Current Density and Solution Additive on the Mictrostructure and Recrystallization rate of Electroplated Cu Films," in Proc. 16th International Advanced Metallization Conf., Orlando, FL (1999); "Advanced Metallization and Interconnect Systems for ULSI Applications in 1999" ULSI XV MRS Issue (1999).
21. K. Fuchs, *Proc. Cambridge Philos. Soc.*, **34**, 100 (1938); A.F. Mayadas, and M. Shatzkes, *Phys. Rev. B*, **1**, no. 4 (1970).
22. J. M. Ziman, *Electrons and Phonons*, Oxford, London (1962).
23. H.-D. Liu, Y.-P. Zhao, G. Ramanath, S.P. Murarka, and G.-C. Wang, "Thickness dependent electrical resistivity of ultrathin (<40 nm) Cu films," Thin Solid Films **384**, 151 (2001).
24. V.M. Dubin, R. Brewer, H. Simka, and S. Shankar, Characterization and control of Cu electroplating chemistries for on-chip metallization, Future Fab International, Issue 13, p. 244–251 (2002).
25. W. Shyy, H.S. Udaykumar, M.M. Rao, and R.W. Smith, in *Computational Fluid Dynamics with Moving Boundaries*, Taylor & Francis, London, pp. 195–200 (1995).
26. P. Singer, "Progress in copper: A look ahead," *Semiconductor Int.*, **25**, no. 4, 46 (2002).

27. S. Varadarajan, D. Kalakkad, and T. Cacouris, "Understanding and reducing copper defects," *Semiconductor Int.*, **25**, no. 5, 125 (2002).
28. M. Itoh et al., "The cleaning at a back surface and edge of a wafer for introducing Cu metallization process," *IEEE Trans. Semicond. Manuf.*, **13**, 300 (2000).
29. M.T. Bohr, "Intel's 90 nm technology: Moore's law and more," *Intel Developer Forum* (Fall 2002).
30. J.R. Lloyd and J.J. Clement, Electromigration in copper conductors, *Thin Solid Films*, **262**, 135 (1995).
31. J. Gill et al., 2002 Reliability Physics Symposium Proceedings, pp. 298–304 (2002).

3 Electroless barrier and seed layers for on-chip metallization

Valery M. Dubin, Sergey Lopatin, Amit Kohn, Nick Petrov, Moshe Eizenberg, and Yosi Shacham-Diamand

3.1 Introduction

Many studies were carried out to develop the practice of chemical autocatalytic electroless plating for numerous applications such as printed circuit boards, magnetic storage media, and metallization of plastics. Brenner and Riddell[1] first reported autocatalytic chemical deposition of cobalt and nickel films from aqueous solutions of the metal chlorides or sulfates on a catalytic surface. Many other works followed and demonstrated the application of this method for the deposition of both pure metals, like Ni, Co, Cu, and Ag, and their multicomponent alloys.[2,3]

The electroless plating technique is especially appealing because of low processing temperature, intrinsic process selectivity, self-aligning, initiation of plating on small isles of catalytic material, and conformal deposition in high aspect ratio features. Moreover, the possibility of electroless Cu deposition for ultra-large-scale integration (ULSI) metallization was demonstrated elsewhere.[4–8]

Several problems related to electroless plating technique need to be resolved to make this process manufacturable. They are mainly related to solution instability resulting in spontaneous solution decomposition and plate out. The design of deposition equipment is especially important to prevent plating bath decomposition. The deposition equipment, which allows spray or single use bath, is of special interest. Electroless plating is also a slow deposition process in comparison to electroplating with deposition rate of <50 nm/min to achieve reasonable quality of the film (i.e., low surface roughness, or low inclusion of H_2). However, it is not a concern for thin film deposition (<50 nm), and thus high throughput equipment can be built. Electroless plated films also contain H_2, and therefore care should be taken to reduce H_2 incorporation into the films and/or hydrogen removal by annealing.

Renewed interest in the electroless deposition process in the last decade is due to its promising applications in ULSI and microelectro-mechanical systems (MEMS) fabrication technologies.[4–8] The electroless deposition technique may be implemented for the formation of both conductive lines as well as diffusion and/or cladding layers.

3.1.1 *Electrolessly deposited barriers and cladding layers for Cu interconnects*

Cu electroplating is the method of choice to fill dual-Damascene structures because of low cost, low processing temperature, high quality of deposits, and ability to provide void-free filling of high aspect ratio features due to the bottom-up mechanism.[9,10,19] However, copper easily corrodes, diffuses rapidly into silicon, drifts into SiO_2 and other dielectrics, and forms weak chemical bonds with low-k materials and polymers.[11] In addition, copper has poor adhesion to most dielectrics, and there is a lack of a feasible dry etching method for copper. Consequently, copper needs both barrier and cladding layers for passivation and adhesion promotion. A SiN or SiC layer is typically used as the etch stop and passivation layer for copper. SiN/Cu or SiC/Cu interface is a major electromigration voiding site. Selective metal capping with electroless metals was proposed as one of the ultimate solutions to this problem.[12–18,68] Electroless metal capping/barrier layers should also provide barrier properties against copper and oxygen diffusion so that lower effective interconnect capacitance can be obtained with thinner or even no SiN (SiC) layers which have high dielectric constant.[68]

Electrolessly deposited metals such as thin Ni–P[12], Co–P, and Ni–Co–P[13] films were investigated as the barrier/cladding materials for copper metallization. The former electroless barriers for Cu were effective up to a temperature of 400°C. It was proposed that the introduction of a third alloying element, especially refractory metal (W, Mo, Re) ion, may improve barrier efficiency.[14,15] NiB, NiReP, and NiWP show ability to serve as a barrier and cladding layers without significant diffusion of copper into the barrier and no Cu line resistance increases up to 400–500°C.[16] Electroless CoWP thin films showed an ability to function as effective barriers/cladding layers for Cu metallization at temperatures up to 550°C.[13–18] Electroless barrier layers can be formed on physical vapor deposition (PVD) or chemical vapor deposition (CVD) deposited Co (or Ni) seed layers as well as directly on dielectric layers (such as SiO_2). A self-assembled monolayer of silane-based compounds followed by treatment in a Pd-containing activator were utilized as the adhesion promoter/catalyzed layer.[16]

We will discuss the results of electroless Co deposition for cladding/barrier applications with hypophosphite or dimethylaminoborane as reducing agents. Pourbaix's diagrams for various aqueous systems were calculated to estimate the thermodynamic capability of the electroless Co-alloy plating. An electrochemical study of the partial reactions of reducing agent oxidation and metal ions reduction using electroless plated Co–P and Co–W–P electrodes was conducted. The rate-limiting step for electroless Co deposition was determined to be the oxidation of the reducing agent. Tungsten was incorporated in the film up to 12 at. % when tungsten phosphoric acid was used as a tungsten source in the plating bath.

Electroless Co film properties have been studied by AFM, SEM, TEM, XRD, EDS, and RBS. It has been demonstrated that 10 nm thick electroless CoWP films are continuous and can be conformally deposited in 30–40 nm features. The 30 nm thick electroless CoWP films have low surface roughness (<5 nm) and high conductivity (<30 $\mu\Omega$ cm). No interdiffusion was observed by RBS, AES, and MOS capacitor measurements in CoWP/Cu structures during annealing at 500°C. The improved diffusion barrier properties relative to PVD cobalt may be explained by one of the following microstructural phenomena: (1) the so-called solute effect, phosphorus and tungsten impurities create a "stuffed" boundary, blocking the high rate of copper diffusion through the grain boundaries; (2) the ε-cobalt grains are embedded in an amorphous matrix, thus eliminating the short-circuit paths of grain boundaries in regular polycrystalline materials.

3.1.2   *Electrolessly deposited seed layers for on-chip Cu interconnects*

Copper electroplating requires a seed layer. Seed layer is typically deposited by PVD technique which is known to have poor step coverage and overhang issues. Electroless deposition can provide conformal Cu films to be used as a seed layer or to repair a discontinuous PVD seed layer. It has been demonstrated in previous works [8,20–22] that the electroless deposition technique has low cost, inherent selectivity, and excellent conformality of deposition as well as ability to deposit superior electroless Cu films on very thin seed layers (<10 nm).

The electroless Cu deposition process can be better understood by using an interdisciplinary approach and concepts developed in thermodynamics, chemistry, and electrochemistry.

The technology of electroless Cu deposition can be improved by using novel reliable seeding processes, optimization of deposition solutions and operating conditions, using advanced electroless deposition tools, integration, and defect-data collection for electroless Cu-based metallization processes.

We will review in this work electroless Cu deposition thermodynamics, mechanism, electrochemistry, kinetics, and mass-transport. Electroless copper deposition from formaldehyde- or glyoxylic acid-based solutions with ethylenediaminetetraacetic acid as a complexing agent was investigated for seed layer and PVD seed repair applications. Electroless Cu deposition is a thermodynamically favorable and kinetically inhibited process, with two electrochemical reactions including anodic oxidation of a reducing agent and cathodic reduction of metal ions occurring simultaneously, with a multi-step catalytic redox mechanism, an Arrhenius type of rate equation, and mass-transport limited reaction in narrow and deep features such as sub-half micron trenches and vias of high aspect ratios (>3).

We will discuss experimental procedure and will describe a single-wafer electroless Cu deposition system which has been designed at Cornell University.[23] A single-wafer electroless Cu deposition system design is reviewed. An electroless Cu deposition solution and operationg conditions have been optimized to obtain electroless Cu films at high deposition rate (~75–120 nm/min), with low resistivity ($\rho < 2$ μΩ cm), low surface roughness ($R_a \sim 10$–15 nm for ~1.5 μm thick deposits) and good within wafer uniformity (STD DEV <3%). We will also review a spray electroless deposition system described in more detail elsewhere.[24]

Experimental results on the study of electroless Cu film properties and integration issues of the electroless Cu process into ULSI interconnects will be discussed. Two different seeding processes including sputtered Cu seed layer with Al protection layer and contact displacement Cu deposited seed will be described. The dry seeding method on sputtered Cu/Al bi-layers was developed to provide protection of Cu catalytic properties from passivation by using an Al sacrificial layer and to obtain uniform initiation and blanket growth of electroless Cu by *in situ* Al dissolution in the plating bath. Electroless Cu films blanket deposited on sputtered Cu/Al seed layer were conformal with 100% step coverage. The wet seeding method was developed with Cu contact displacement deposition on a TiN diffusion barrier to provide electroless copper plating.

3.1.3   *Electroless metal deposition mechanism*

Because of the limited diffusivity of ionic species, conformal deposition of a barrier and seed layers is a critical step in fabrication of deep submicrometer interconnects for high aspect ratio features. From this point of view, the investigation of basic mechanisms and processes involved into electroless barrier formation is of primary interest.

It is widely accepted that electroless plating proceeds along the electrochemical mechanism as a simultaneous reaction of cathodic metal deposition and anodic oxidation of a reducing agent.[25,26] All these postulations were made based on the Wagner–Traud[27] mixed potential theory of corrosion processes.

Four principal reaction mechanisms have been proposed for the reduction of binary metal-phosphorus, for example, NiP, alloys with hypophosphite: atomic hydrogen mechanism, hydride ion mechanism, metal hydroxide mechanism, and electrochemical mechanism. The first, second, and third are based on chemical reactions, which lead to island-type growth and agglomerated structures in the initial stage of film deposition. Only electrochemical reaction on a conductive surface (i.e., surface between catalytic sites) leads to a uniform growth in the initial stage of film deposition. The theory of electroless deposition is based on the mechanism of the processes occurring at the electrode in a system with two different simultaneous electrochemical reactions: reduction of metal ions M^{n+} (cathodic reaction) and oxidation of reducing agent Red (anodic reaction).[26] The partial electrode reactions in the system are:

1. Reduction of metal ions

$$M^{n+} + ne^- \xrightarrow{\text{Catalytic surface}} M \tag{1}$$

2. Oxidation of reducing agent

$$\text{Red} \xrightarrow{\text{Catalytic surface}} Ox + ne^- \tag{2}$$

In order to establish basic mechanisms of electroless plating (i.e., metal reduction and reducing agent oxidation), calculate instantaneous deposition rate, investigate the role of electroless plating solution components, and confirm mixed potential theory, many studies of independent partial reactions were carried out.[26,28–32] Good agreement between the measurement of instantaneous deposition rate and that calculated from partial reaction data in the case of Cu,[26,30] Ni–P,[28,30] Au[29,30], and Co–P[30–32] was obtained. An introduction of a third element in the electroless deposition system may change significantly the behavior of both anodic and cathodic reactions and the complete process.

3.2 Electroless deposition of barriers and cladding layers

Electroless CoWP films were first obtained by Pearlstein and Weightman.[33] They added tungstate ions into a Brenner's chloride bath and studied basic chemistry of the process. Gorbunova et al. investigated material properties and process characteristics of CoWP reduction from similar chloride electrolyte.[34] It was suggested that CoWP reduction follow the electrochemical mechanism. The incorporation of the W into the films was explained in terms of so-called induced co-deposition of refractory metals with iron group metals that was proposed elsewhere.[35,36] Finally, an electrochemical study of the chloride-based CoWP electrolyte using a rotating disk electrode (RDE) was performed.[37] The electrochemical character of the reaction was revealed. Nevertheless, at present there is a lack of data on the basic processes that occur during the reduction of CoWP.

3.2.1 Electrochemical mechanism of electroless Co deposition

Mechanism of electroless ternary CoWP alloy deposition can be described by a set of electrochemical oxidation–reduction reactions. It is a more complicated system compared to the binary alloy deposition due to different catalytic activities of the co-deposited metals at the surface.

Anodic oxidation of hypophosphite in both acidic and basic solutions occurs according to the following reactions:

$$H_2PO_2^- + H_2O = H_2PO_3^- + 2H^+ + 2e^- \tag{3}$$

$$H_2PO_2^- + OH^- = H_2PO_3^- + H + e^- \tag{4}$$

In basic ethylenediamine (EDA) solution a hydrogen-free reaction may occur as follows:

$$H_2PO_2^- + 3OH^- = HPO_3^{2-} + 2H_2O + 2e^- \tag{5}$$

If borohydride or dimethylaminoborane are used as the reducing agents, the following anodic oxidation reaction can occur:

$$(CH_3)_2NH\,BH_3 + 3H_2O = (CH_3)_2NH_2^+ + H_3BO_3 + H^+ + 2H_2 \tag{6}$$

$$NaBH_4 + 4OH^- = B(OH)_4^- + 2H_2 \tag{7}$$

Borohydride and dimethylaminoborane electroless baths are unstable. However, these baths allow direct deposition of cobalt and nickel on copper without activation steps. Further discussion will be limited to the hypophosphite-based electroless Co bath.

Anodic oxidation reactions lead to the formation of electrons, which are consumed in the cathodic part of the electrochemical reaction (metal ion reduction). Cathodic reduction reactions of Co and P co-deposition in acidic solution can be described as:

$$Co^{2+} + 2H_2PO_2^- + 2H_2O \rightarrow Co + 2H_2PO_3^- + H_2 + 2H^+ \tag{8}$$

$$4H_2PO_2^- + 2H^+ \rightarrow 2P + 2H_2PO_3^- + H_2 + 2H_2O \tag{9}$$

$$H_2PO_2^- + H_2O \rightarrow H_2PO_3^- + H_2 \tag{10}$$

Cathodic reactions of Co and P co-deposition in basic solution can be described as:

$$Co^{2+} + 2H_2PO_2^- + 4OH^- \rightarrow Co + 2HPO_3^{2-} + H_2 + 2H_2O \tag{11}$$

or

$$Co^{2+} + H_2PO_2^- + 3OH^- \rightarrow Co + HPO_3^{2-} + 2H_2O \text{ (EDA solution)} \tag{12}$$

$$4H_2PO_2^- \rightarrow 2P + 2HPO_3^{2-} + H_2 + 2H_2O \tag{13}$$

$$H_2PO_2^- + OH^- \rightarrow HPO_3^{2-} + H_2 \tag{14}$$

Tungsten is co-deposited with CoP in a small amount, up to 12% of composition. Tungsten itself is not catalytically active for the reduction reaction and W cannot be electrodeposited in the pure state from aqueous solutions.

Two conditions must be met to deposit ternary alloy:

1. At least one of the metals can be deposited independently (in our case Co).

2. Deposition potentials must be close together or must be capable of being brought close together by using complexing agents. An electrode (wafer surface on which alloy is deposited) can have only one potential at a time, so that for three reactions to take place simultaneously at the electrode, they must

take place at the same potential. The Nernst potentials for the three different cathodic processes (E_{Co}, E_P, and E_W for Co, P, and W reduction, respectively) can be approximately equal to each other in the simultaneous reaction:

$$E^{\circ}_{Co} + (RT/nF)\ln(a_{Co}{}^{n+}/a_{Co}) + P_{Co} \approx E^{\circ}_{P} + (RT/nF)\ln(a_{P}{}^{n+}/a_{P}) + P_{P} \approx$$
$$E^{\circ}_{W} + (RT/nF)\ln(a_{W}{}^{n+}/a_{W}) + P_{W} \qquad (15)$$

Where E°_{Co}, E°_{P}, E°_{W} are the constant characteristics of the metal electrode for Co, P, and W, respectively; R is the gas constant; T is the absolute temperature; n is the valence change; $a_{Co}{}^{n+}$, $a_{P}{}^{n+}$, $a_{W}{}^{n+}$ are the activities of the Co, P, and W ions, respectively; a_{Co}, a_{P}, a_{W} are the activities of the Co, P, and W metals, respectively; P_{Co}, P_{P}, and P_{W} are the polarizations for Co, P, and W, respectively.

In the simplest case, for Co, P, and W co-deposition in CoWP alloy film, polarizations and constant characteristics of the metal electrode for Co, P, and W must already be about equal to each other and minor adjustments in the activities and concentrations of the Co, P, and W will allow E_{Co} to equal E_P and E_W.

The practical approach to equalize the Nernst potentials shifting them to more negative values is to regulate metal activities by complexing. A complexing agent forms complexes with metals of varying stability. The electroless CoWP deposition reaction can follow the electrochemical theory in a three-component system as a combination of the oxidation of hypophosphite and the reduction of Co, W, and P complex ions.[37,38]

The electrochemical theory of alloy depositions has not yet advanced far enough to lead to exact equilibrium deposition conditions of the electroless CoWP deposition process, and they are determined by an empirical approach. Typical controlled variables of the electroless process are temperature, hydrodynamic conditions, pH, and type and concentrations of reactants.[39]

3.2.2 *Thermodynamic analysis of electroless Co plating*

Metals that are thermodynamically stable in water are candidates for autocatalytic reduction (i.e., electroless metal deposition) in aqueous media. Such metals are Fe, Cd, Tl, Co, Ni, Sn, Pb, Bi, Re, In, Cu, Hg, Pd, Pt, Ag, and Au. However, reactions that are possible according to thermodynamic considerations are retarded due to the reaction's kinetics and the overall deposition rate is very low. Moreover, the product of autocatalytic metal reduction does not necessarily form a continuous metal layer. It may be a highly porous deposit or a metal powder.

Electroless deposition has many stages, and there are many parameters that govern the overall reaction kinetics. Among these parameters are the following:

1. Transport phenomena of the reactive species in the liquid

2. Transport phenomena of the reactive species at the solid/liquid interface

3. Transport phenomena of adsorbed species on the interface

4. Nucleation kinetics

5. The kinetics of the electrochemical and chemical reactions that take place at the liquid and solid/liquid interface

The functionality of a reducing agent depends on the pH and temperature of the solution. Under particular conditions the redox potential of the reducing agent can significantly deviate from its standard value.[31,32,40] In the case of electrochemical oxidation of hypophosphite the oxidation reaction can be written as in Reaction 3.

The equilibrium potential of Reaction 3, E_{eq}, is given by

$$E_{eq} = -0.504 + \frac{2.3RT}{F}\text{pH} + \frac{2.3RT}{F}\log\left[\frac{H_2PO_3^-}{H_2PO_2^-}\right] \tag{16}$$

where R is the gas constant, T is the temperature, and F is Faraday constant.

The difference between the redox potential of the reducing agent and that of the metal, ΔE, is related to the reaction rate. Typically, when ΔE is too large, one can get fast formation of highly dispersed reduction products. To inhibit the reaction, electroless plating baths contain ligands that form strong complexes with metal ions. The complexing of metal ions causes ΔE to decrease due to the shift of the redox potential of the metal ion–metal pair to a more negative region. This can be demonstrated using the Nernst equation

$$ML_m^{n+} + n\text{e} \rightarrow \text{M} + m\text{L} \tag{17}$$

$$E_{eq} = E^0_{ML_m/M} + \frac{2.3RT}{nF}\ln\beta_m + \frac{2.3RT}{nF}\ln[M^{n+}] \tag{18}$$

where $E^0_{ML_m/M}$ is the standard redox potential of Reaction 17, n is the number of electrons involved, and β_m is stability constant of the complex.

Ligands also prevent formation of insoluble metal hydroxides in strongly alkaline media.

Pourbaix's diagrams were used to predict the thermodynamic capability of the autocatalytic Co (W, P) deposition. Such diagrams were constructed for the following systems at 90°C:

1. Co, cobaltous ions, Co-citrate complex in water

2. W-water

3. Soluble phosphorus compounds–water

The diagrams were constructed assuming that the water product at 90°C is equal to $\sqrt{K_w} = a_{H^+} = 6.17\times10^{-7}$ and all redox potentials at 90°C were recalculated with respect to their temperature coefficients.[41,42]

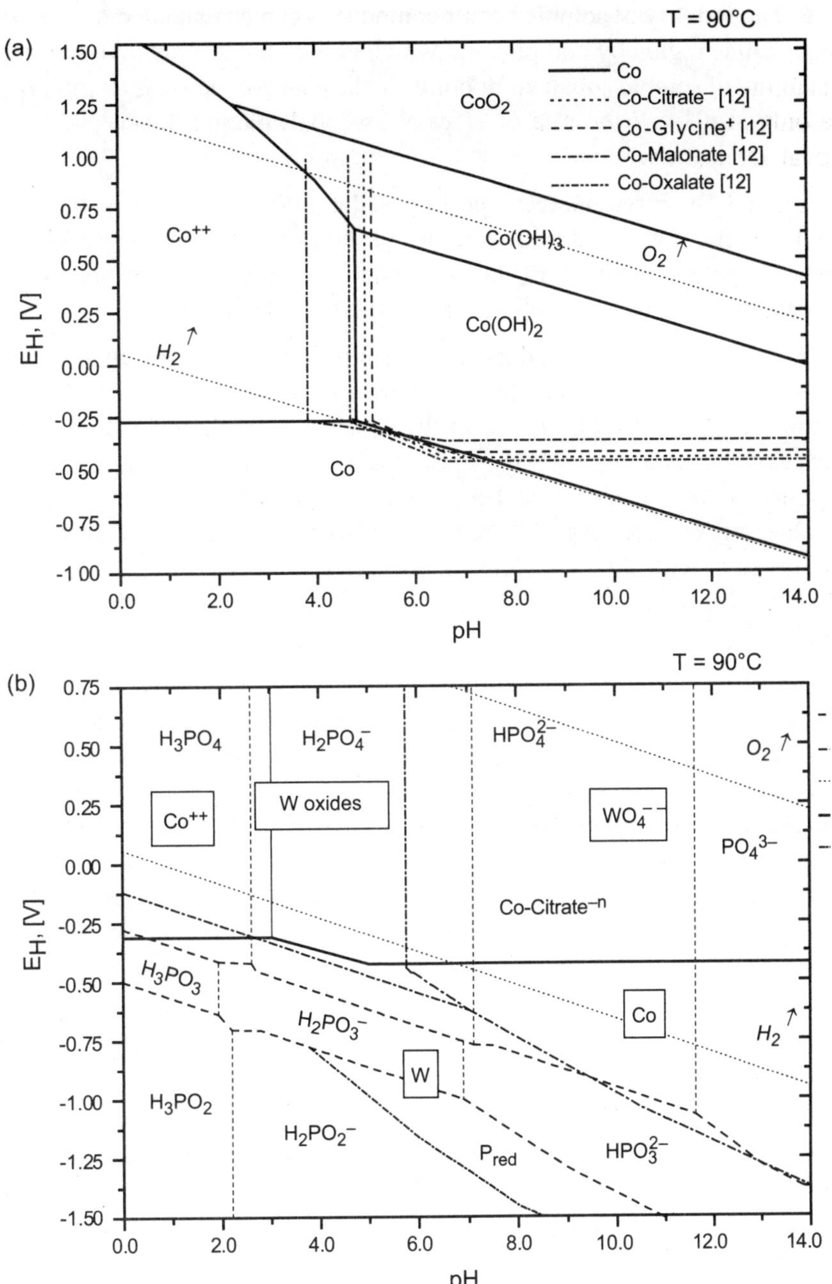

Figure 3.1   Pourbaix's diagrams of the systems (a, top) — Co–water, complexed Co–water and (b, bottom) — Co–Co-citrate, phosphorus, tungsten–water at 90°C.

The thermodynamic stability conditions of cobalt and its derivatives that can exist in the presence of complexant-free water or aqueous solutions are presented in Figure 3.1a (solid lines). This diagram shows that Co forms insoluble hydroxides at pH ~ 5. The stability of soluble cobalt compounds can be extended into alkaline and strong alkaline region by complexing, which is shown in Figure 3.1a (dashed lines). The stability of soluble cobalt compounds in the alkaline region is an important issue since only at pH > 7 the rate of electroless cobalt plating is acceptable for most practical applications.[1–3,34]

Figure 3.1b represents the superimposed equilibrium diagrams of the systems (1) and (2) onto the equilibrium diagram of the system (3) at the working concentrations, temperature, and pH (Table 3.1). Such a diagram has been constructed in the aforementioned manner to study the autocatalytic reduction of Co(W, P) alloys.

Thermodynamically, hypophosphite is a powerful reducing agent at almost all pH, since the domain of its stability lies below that of water. Hypophosphites are stable in the aqueous solution and in the absence of oxidizing agents. Only in the presence of catalysts, such as Cu, Ni, Pd, Co, is the reducing capability of the hypophosphite revealed. Under these conditions hypophosphites may reduce the salts of the noble metals (Au, Ag, Pt, Pd), and also arsenious, cupric, cobaltous, and nickelous salts with the deposition of the corresponding metal, being themselves oxidized to phosphates.[40,43] At elevated temperatures hypophosphites can oxidize to form phosphorus.[43]

Tungsten equilibrium potential–pH diagram at 90°C is presented in Figure 3.1b (dash dotted line). The information is calculated assuming the absence of complexing substances and substances forming insoluble salts with tungsten. From this figure, tungsten is seen to be a base metal, as its domain of stability lies completely below that of water. In the presence of alkaline solutions, it has a slight tendency to decompose water with the evolution of hydrogen, dissolving as tungstic ions. In the presence of noncomplexing acid solutions, it tends to become covered with dioxide WO_2 or a higher oxide.[40] According to Figure 3.1b, it should be possible to deposit tungsten from alkaline solutions of tungstates. Nevertheless, in practice, pure tungsten cannot be deposited from aqueous solutions. However, it is co-deposited with iron group metals (Ni, Co, Fe) by the so-called induced codeposition mechanism.[35,36]

The reduction of cobalt–tungsten–phosphorus alloy is thermodynamically feasible when redox potentials of Co, W, and oxidation products of $H_2PO_2^-$ are located in the domain of their stability. While increasing the pH, the oxidation potential of the reducing agent is shifted to the more negative values increasing its reducing power. At pH > 5, cobalt redox potentials are located in the stability domain of its hydroxides (Figure 3.1a, solid lines). The region of the cobalt reduction is significantly extended when the soluble cobalt complexes are present in solution (Figure 3.1a, dashed lines). At the same time, at high pH the ΔE between the partial reactions of metal ion reduction and $H_2PO_2^-$ oxidation is increased. The electroless deposition of Co(W, P) in the present investigation was carried out at pH 9. Figure 3.1b shows that reduction of Co(W, P) alloy at this pH is thermodynamically favorable since the redox potentials of all three reacting components lies in the domain of their stability.

Table 3.1   Compositions of electroless Co (W, P) plating bath.

Component	Concentration, M	Co-P	Co–W–P(1)	Co–W–P(2)
$CoSO_4 \cdot 7H_2O$	0.0082–0.33	×	×	×
$Na_3C_5H_5O_7 \cdot H_2O$	0.492	×	×	×
H_3BO_3	0.502	×	×	×
$NaH_2PO_2 \cdot 2H_2O$	0.02–0.82	×	×	×
$Na_2WO_4 \cdot 2H_2O$	0.0082–0.33		×	
$H_3[P(W_3O_{10})_4]$	0.005–0.02			×
KOH		To adjust pH		
Temperature 90±5°C				

3.2.3   Electroless Co deposition solutions

Electroless CoWP or CoWB alloys can be deposited from aqueous solutions using hypophosphite and borane dimethylamine as reducing agents, respectively. The deposition rate of the electroless CoWB alloy with borane dimethylamine is higher than that of CoWP with hypophosphite, but solution stability with borane dimethylamine as reducing agent is very low.

The electroless Co bath contains Co salt (such as cobalt chloride or cobalt sulfate), the source of W ions (ammonium tungstenate or tungsten phosphoric acid), complexing agents (e.g., tartrate, citrate, ethylenediaminetetraacetic acid (EDTA), malic acids, ethylenediamine, triethanolamine, or ammonia), buffer agents (such as NH_4Cl or boric acid), reducing agents (including hypophosphite, dimethylamine borane (DMAB) and borohydride) as well as pH adjuster (NH_4OH, potassium hydroxide, or tetramethyl ammonium hydroxide) with additions of ductility promoter and surfactant.[39]

Examples of CoWP bath compositions are presented in Table 3.1.

CoWP films are deposited from the electroless solution at temperatures in the range of 55 to 95°C and pH 8.5 to 10.5. The deposition was initiated on a Co seed for a blanket deposition and on copper surfaces by an activation step in an aqueous solution of palladium chloride, hydrochloric acid, and hydrofluoric acid for selective deposition.[39]

3.2.4   Electroless Co experimental techniques

Polarization experiments were carried out on CoP or CoWP working electrodes. The working electrodes were prepared by electroless plating of copper foil. A platinum wire was used as an auxiliary electrode. The potential of the working electrode was measured vs. Ag/AgCl electrode via the Luggin capillary. *I–E* curves were obtained from linear sweep voltametric measurements. EG&G Princeton Applied Research model 273 potentiostat/galvanostat was used for voltametric measurements.

In order to simulate partial cathodic or anodic reactions, the investigation of the electrolytes in the absence of either the reducer or the cobalt ion was carried out.

Polarization measurements were made in electroless plating baths with various pH, temperature values, and various concentrations of main electrolyte components. The pH of the solution was measured by electronic pH-meter Orion model 720A.

To perform the gravimetrical measurements, electroless Co alloys plating of Cu foil was conducted at 90°C during 1 h. Electroless deposition of cobalt alloys was initiated galvanically by using Al wire at the beginning of the plating. The weight of the samples was determined by electronic microbalance Presica 62A. Mixed potentials of electroless deposition were measured vs. an Ag/AgCl electrode in the subsequent electrolyte. These measurements were conducted over 30 min to ensure the steady state of the solution and the overall reaction. The film composition was analyzed by X-ray photoemission spectroscopy (XPS) in a Physical Electronics PHI model 590A system. The standard compositions of the baths and the experimental conditions are summarized in Table 3.1.

Quantitative energy dispersive spectroscopy (EDS) using a link ISIS system (Oxford Instruments) was used to evaluate the cobalt, phosphorus, and tungsten atomic concentration in the layers. In order to analyze the thin films, low-energy (5 keV) electron excitation of the layer was used which ensures that the X-ray photons originate from the $Co(W, P)$ film only. The EDS analysis correction procedure used was ZAF. The measurements were found to be in agreement when compared with Rutherford backscattering spectrometry, Auger electron spectroscopy (AES — Physical Electronics PHI model 590A) and additional EDS measurements ($\pi-\rho-z$ correction and an EDS system attached to a TEM). The crystalline structure and texture of the thin films were determined by a powder X-ray diffractometer (XRD — Philips PW-1820 diffractometer) in the Bragg–Brentano geometry. The microstructure of the films were examined by cross-section samples in a transmission electron microscope (TEM — JEOL 2000FX, JEOL 3010 UHR) using bright field (BF), dark field (DF), and selected area diffraction (SAD). The surface morphology of the layers was examined in a high resolution scanning electron microscope (HRSEM, LEO system, model Gemini 982 FEG). The resistivity of the $Co(W, P)$ layers was measured by four-point probe.

Electroless CoWP deposition on a Cu surface at a relatively low temperature of 65°C was achieved by an activation of the Cu surface in a Pd-contained solution. An activation step is needed to initiate electroless CoWP deposition at low temperature and start the autocatalytic reaction. $PdCl_2$ in an HF solution was used as the activation solution. The substrate is first placed in the activation solution for a short period of time, during which Pd nucleates on the surface. The sample is then placed in the CoWP plating solution, where CoWP is deposited on the Pd sites and Cu surface between sites. Once there is a CoWP layer formed the autocatalytic reaction takes over.

Four steps were involved in the uniform CoWP deposition on Cu:

1. Wet Pd activation of Cu surface prior to CoWP deposition formed Pd catalytic sites on a Cu surface improving uniformity of CoWP nucleation and decreasing the temperature of following electroless CoWP deposition.

2. Rinsing in warm deionized (DI) water increased the temperature of the substrate before treatment in the electroless CoWP solution by an intermediate step.

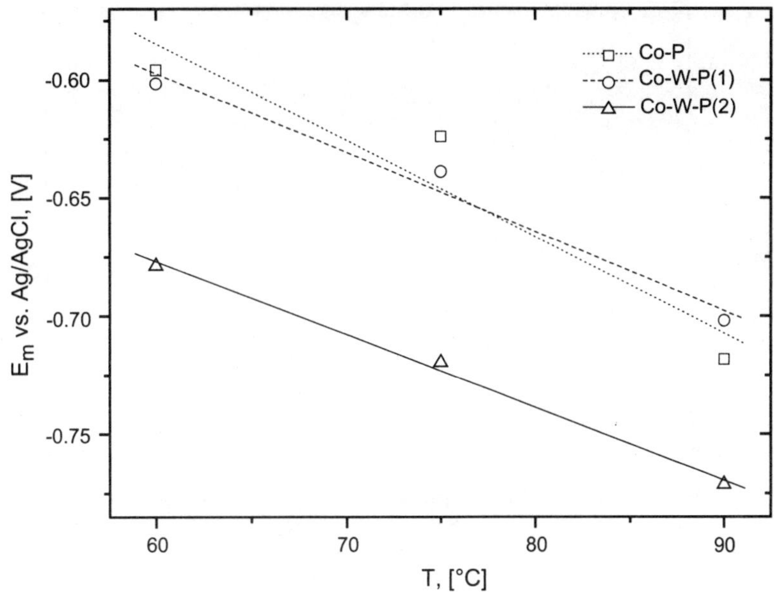

Figure 3.2   Dependence of the mixed potential on the temperature for electroless Co plating.

3. Slow low-temperature CoWP plating formed a thin well-adhered CoWP film on the Pd-activated Cu surface.

4. Rinsing in warm DI water decreased the temperature of the substrate by an intermediate step.

Two thin Cu seed layers were studied for this process. These Cu films were deposited by electroless deposition and PVD and these films had different grain sizes: large for electroless film and small for PVD. *Ex-situ* surface morphology of the Cu seed layers, Pd-activated Cu layers, and as-deposited CoWP films were observed by atomic force microscopy (AFM).

3.2.5   *Polarization studies of electroless Co deposition*

The data presented in this section were obtained from polarization measurements of the electroless Co plating electrolytes.[44] The measurements were carried out in the absence of either Co^{2+} (partial anodic curve) or $H_2PO_2^-$ (partial cathodic curve). The intersection of these two polarization curves represents, in terms of the mixed potential theory,[27] mixed potential (ordinate) and deposition rate (abscissa).[26]

3.2.5.1   *Temperature effect*

The function of the mixed potential vs. temperature is presented in Figure 3.2. It can be seen that it shifts cathodically with temperature growth. The CoWP(2) mixed potential is more negative than those of CoWP(1) and Co–P. For the last two solutions, the mixed potential values are almost the same. Behavior of the mixed

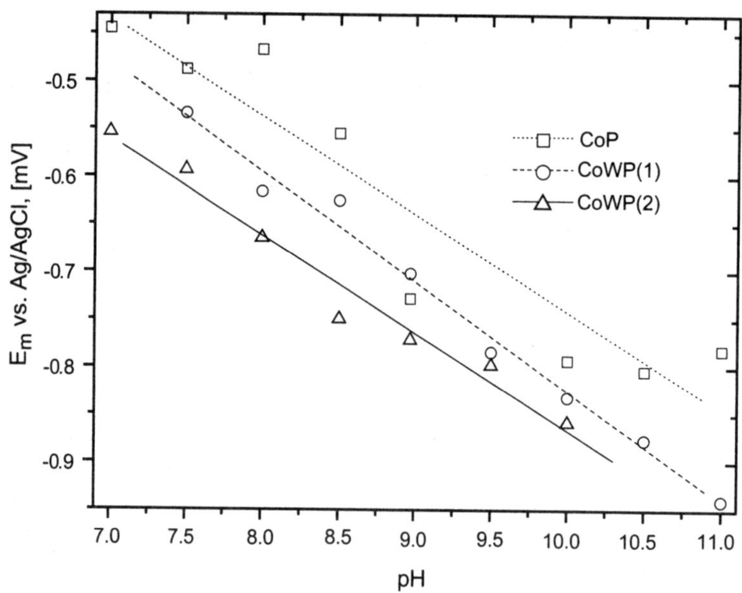

Figure 3.3   Dependence of the mixed potential on the pH for electroless Co plating.

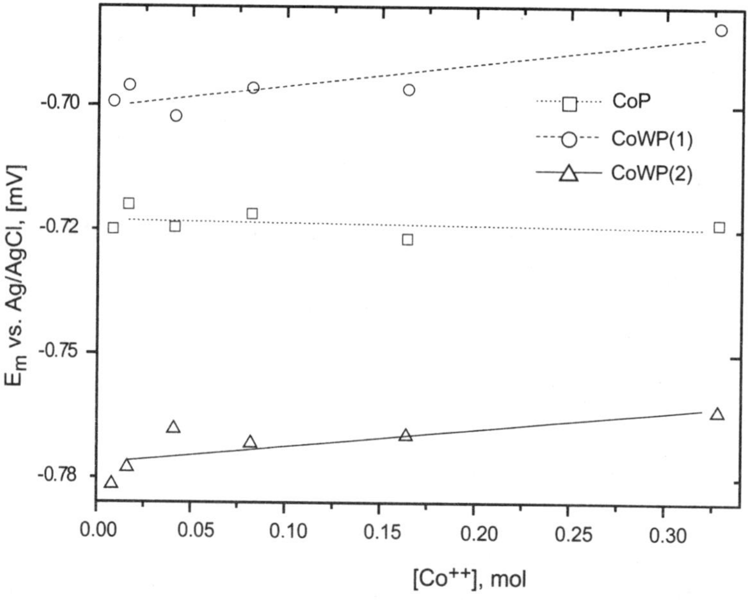

Figure 3.4   Dependence of the mixed potential on the $[Co^{2+}]$ for electroless Co plating.

potential against the temperature is similar to that observed in Ref. 34. At low temperatures the process is slow and may not start below 50°C. Working potential reaches the value of the mixed potential only at the temperature range of 90 to 95°C. The quality of deposits was better at higher deposition temperatures than at lower temperatures.

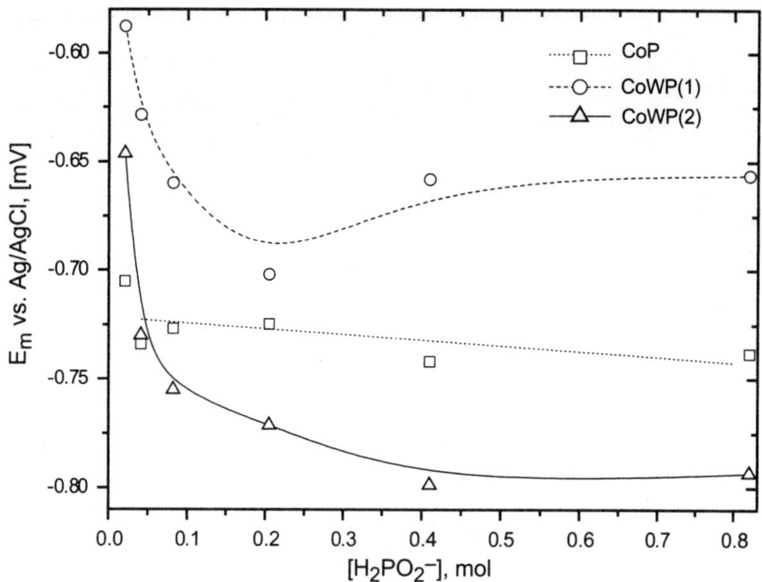

Figure 3.5 Dependence of the mixed potential on the $[H_2PO_2^-]$ for electroless Co plating.

3.2.5.2 pH effect

Figure 3.3 shows the dependence of the mixed potential vs. the pH of the solution. Experimental data shows that it shifts cathodically with increasing pH for all three solutions. The mixed potential values of Co alloys were changed in a noble direction in the following order: CoWP(2), CoWP(1), CoP. The pH has the most significant influence on the mixed potential of the process. The mixed potential changes by approximately 400 mV. A similar trend was observed in work.[45] The working value of the mixed potential (MP) is achieved for all systems in the vicinity of pH 8.5 to 9.5. At pH < 7, the rate of deposition is infinitesimally low. At pH > 10, the system becomes unstable, and spontaneous decomposition of hypophosphite may begin along with the homogeneous cobalt reduction in the volume of the electrolyte.

3.2.5.3 Concentration effect

The effect of $[Co^{2+}]$ on the behavior of the investigated systems is presented in Figure 3.4. The most negative mixed potential was measured for the CoWP(2) as compared to the other two electrolytes. However, the mixed potential of the Co–P is more negative than that of CoWP(1). The mixed potential for CoWP systems is slightly shifted anodically with $[Co^{2+}]$, while that of Co–P has approximately the same magnitude. The concentration of Co^{2+} affects mainly the mixed potential of CoWP alloys while for CoP its deviation is not pronounced. Mixed potential shifts to the more noble direction with cobaltous ion concentration. This behavior may indicate the mechanism of the Co reduction and co-deposition of the W.

Figure 3.5 represents the effect of $[H_2PO_2^-]$ on the mixed potential of electroless Co alloy plating systems. It can be seen that the mixed potentials for CoWP and CoP behave differently. In the case of CoP the linear function of mixed potential vs.

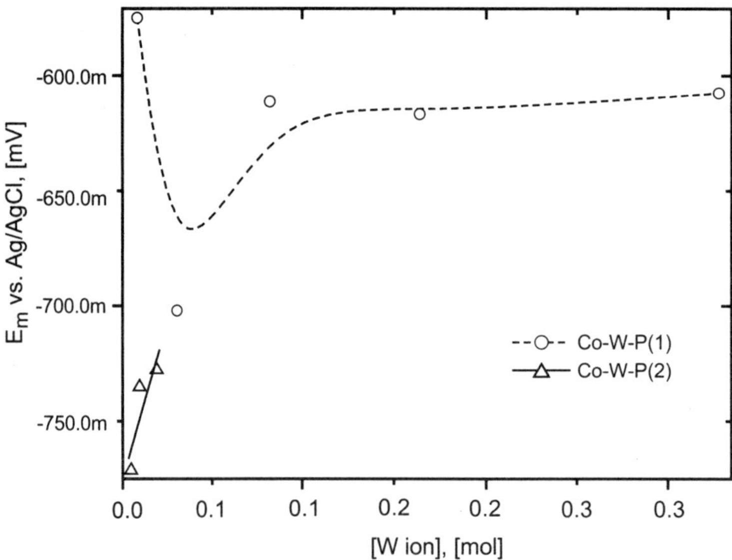

Figure 3.6   Dependence of the mixed potential on the [W ion] for electroless Co plating.

$[H_2PO_2^-]$ with slight slope is observed. The value of the mixed potential has a minor cathodic shift with hypophosphite concentration. For the CoWP the values of mixed potential are less negative at $[H_2PO_2^-] < 0.169$ mol. When $H_2PO_2^-$ concentration exceeds 0.169 mol, the mixed potentials shift sharply cathodically. At higher hypophosphite concentrations, mixed potential reaches saturation and changes slightly.

It can be concluded that the concentration of both cobalt ions and the reducing agent influence primarily the behavior of the CoWP systems (Figures 3.4 and 3.5). However, even though the effect of hypophosphite is similar for all baths, it is relatively more pronounced than that of Co^{2+}. This fact can indicate that the rate-determining step for the chemical Co-alloys reduction is the oxidation of hypophosphite. This conclusion is in a good agreement with the results of a number of investigations on electroless Co-alloys deposition.[31,34,37,45,46]

Figure 3.6 presents the effect of tungsten ion precursors on the electrochemical behavior of electroless Co–W–P plating systems. If Na_2WO_4 is a tungsten source, the mixed potential shifts cathodically with concentration change from 0.0082 to 0.03 mol (working value). Further $[WO_4^{2-}]$ increase causes anodic shift of mixed potential value that remains approximately the same. When $H_3[P(W_3O_{10})_4]$ is used, the mixed potential is shifted linearly in anodic direction. The effect of the W-ion source, which is observed in this study, is similar to that studied in Refs. 33 and 34 in the case of Na_2WO_4 (see Table 3.1). When tungsten phosphoric acid is utilized, the mixed potential is more negative than that of Co–W–P(1). This effect can be explained from the chemical properties of $H_3[P(W_3O_{10})_4]$. In an alkaline medium it dissociates releasing $12\,WO_4^{2-}$ groups as

$$H_3[P(W_3O_{10})_4] + 27\,OH^- \rightarrow PO_4^{3-} + 12\,WO_4^{2-} + 15\,H_2O \qquad (19)$$

Table 3.2   Average deposition rates of electroless Co alloys and mixed potentials.

Cobalt alloy	Average deposition rate $(mg\ h^{-1}\ cm^{-2})$	Average mixed potential $(V\ vs.\ Ag/AgCl)$
Co–P	1.12	−0.75
Co–W–P(1)	1.34	−0.702
Co–W–P(2)	1.00	−0.78

Large numbers of tungstate ions cause, on the one hand, a cathodic shift of MP; on the other hand, they give higher amounts of tungsten in the deposit.[47]

3.2.5.4   Deposition rate

The average rates of electroless deposition of Co alloys determined gravimetrically (by weighting before and after deposition), at 90°C, from the solutions referred to in Table 3.1, after 1 h plating, are shown in Table 3.2. These values were obtained if the time of deposition is counted from the instant of immersion of the Cu foil into the solution. The average values of the mixed potential during Co alloys deposition measured for 30 min are presented in Table 3.2. For comparison, the mixed potentials and deposition rates for different Co alloys were also determined by means of the mixed-potential theory from the Evans diagrams. Partial cathodic and anodic I–E curves at 90°C generated in the previous section were used for this purpose. The results of the electrochemical measurements are shown in Figure 3.7. The

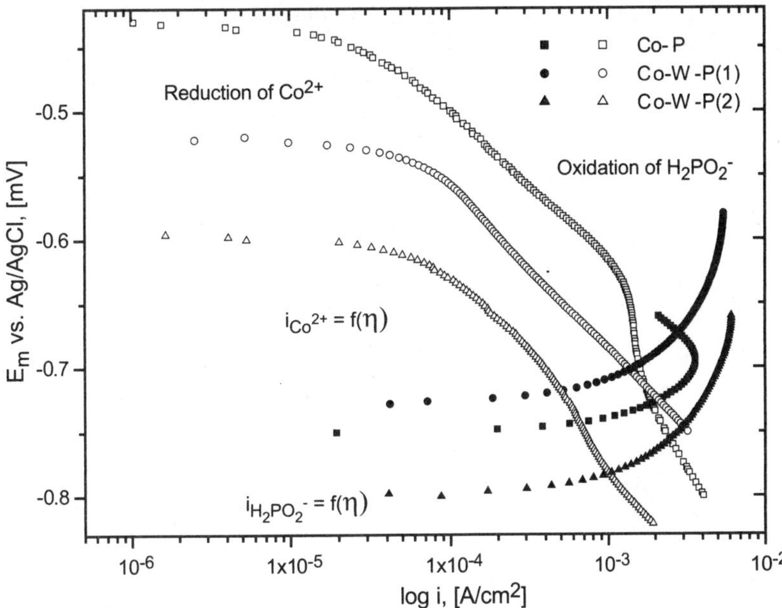

Figure 3.7   Current-potential curves for reduction of Co^{2+} ions and for oxidation of $H_2PO_2^-$ combined into the Evans diagram for the electroless deposition system of CoP, and CoWP.

coordinates of intersection represent the deposition rate in A·cm^{-2} and the mixed potential in volts vs. Ag/AgCl electrode are presented in Table 3.3. The rate of deposition, w, was calculated from the deposition current, i_{dep}, using Faraday's law

$$w[\text{mg·h}^{-1}\,\text{cm}^{-2}] = i_{dep}[\text{mA cm}^{-2}] \times 1.099 \qquad (20)$$

Table 3.3 Electrochemically measured deposition rate and mixed potential of electroless Co alloys.

Cobalt alloy	Mixed potential (V vs. Ag/AgCl)	Deposition current i_{dep} (mA cm^{-2})	Calculated deposition rate, w (mg·h^{-1} cm^{-2})
Co–P	−0.728	1.9	2.09
Co–W–P(1)	−0.702	1.374	1.51
Co–W–P(2)	−0.782	1.05	1.15

Those values are also presented in Table 3.3.

3.2.6 *Electroless Co film properties*

3.2.6.1 *Composition of electroless CoW films*
The composition of the electroless deposited Co-alloy films was determined by using XPS. The results of this measurement are shown in Table 3.4. The films consist mostly of Co in the range of 85 to 95 at. %. In the case of CoP alloy, phosphorus content is about 5 at. %, and when tungsten is added, the phosphorus concentration lowers to approximately less than 4 at. %.

Table 3.4 Composition of electroless Co alloys.

Component	Content, at. %		
	Co–P	Co–W–P(1)	Co–W–P(2)
Co	94–95	92–95	85
W	—	2.5–4.5	12–13
P	5–6	2–4	2–3

Tungsten concentration in the solid depends on the source of tungsten ions. When $H_3[P(W_3O_{10})_4]$ is used as the tungsten source, its concentration in the film exceeds 12 at. %. When sodium tungstate is applied as the source for W, its concentration is about 2 to 4 at. %. The analysis of Co–W–P films indicated that tungsten in the film is present in the form of W and WO_2. The corresponding XPS energy spectra are demonstrated in Figure 3.8.

Reaction of the tungsten co-deposition can be described in terms of so-called induced co-deposition that was proposed in Refs. 35 and 36. According to the proposed model, refractory metal (e.g., molybdenum or tungsten) in the anionic form of

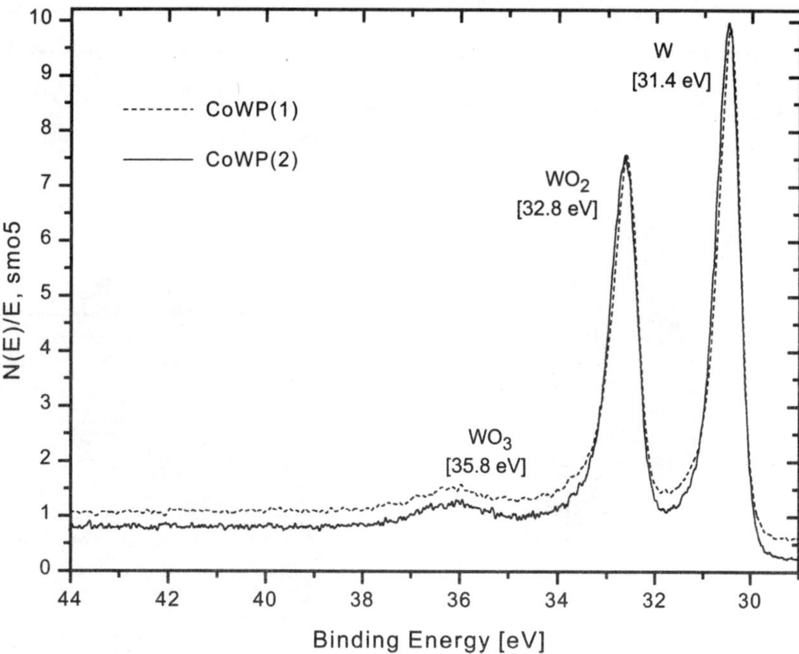

Figure 3.8 XPS spectra of electroless CoWP films.

MoO_4^{2-} or WO_4^{2-} in an electrochemical deposition bath could form an intermediate complex with the main metal and complexant. This complex is adsorbed on the electrode surface, and the iron group metal acts as a catalyst for the reduction of the refractory metal. This co-deposition process may go in parallel with the reduction reaction of the iron group metal in the electrochemical deposition. Gorbunova[34] suggested that the intermediate tungsten–citrate complex at the substrate surface dissociates with the formation of WO_2^{2+} and this ion, in turn, is reduced to W. Taking into account all the above considerations, the overall tungsten co-deposition reaction, according to Ref. 34, can be presented as

$$WO_2^{2+} + 6H_2PO_2^- + 4H_2O \rightarrow W + 6H_2PO_3^- + 3H_2 + 2H^+ \tag{21}$$

The composition of the films was determined by XPS technique and the corresponding spectra are demonstrated in Figure 3.8. This figure indicates that the films contain tungsten in a form of pure tungsten metal as well as WO2. This result supports the mechanism of tungsten co-deposition during electroless CoWP plating that was proposed in Ref. 34. The appearance of tungsten dioxide in the solid can be explained by discharge of WO_2^{2+} ion at the electrode surface with the electrons that were released in Reactions 3 to 5. From the thermodynamic point of view (Figure 3.9), the electroless plating of Co (W, P) alloys can be described as the deposition of the main alloy component, Co. W deposition may be represented as "underpotential" deposition. Moreover, the mixed potential and open-circuit potential (OCP) of the

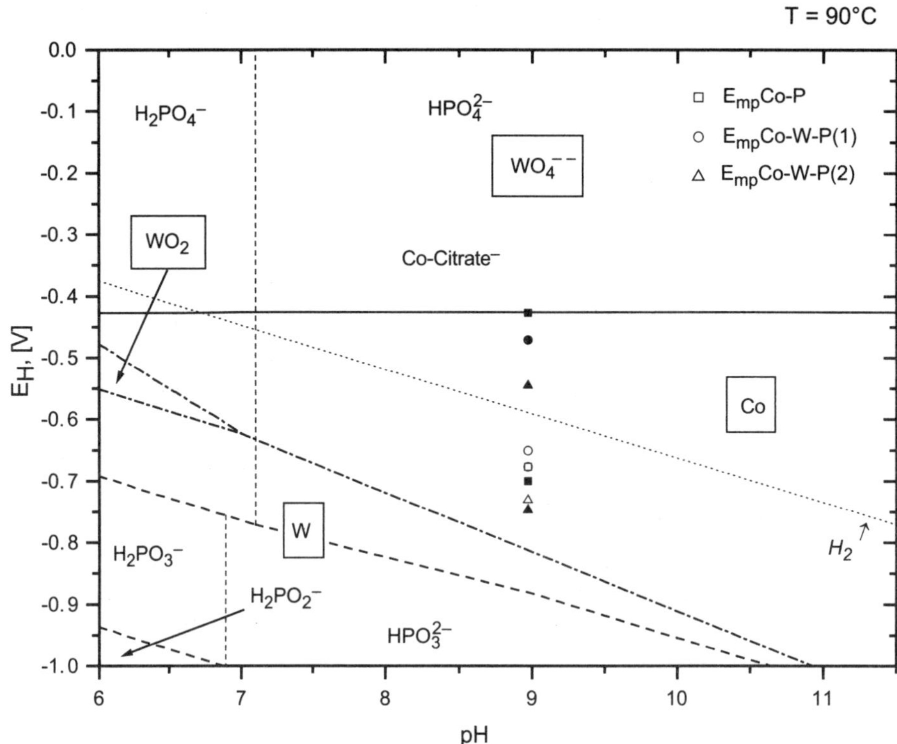

Figure 3.9 Pourbaix's diagram of the Co, P, W – water system representing mixed and open circuit potentials for the autocatalytic chemical reduction of CoP, CoWP alloys. Up filled figures are OCPs of Co^{2+} reduction and low-filled figures are OCPs of $[H_2PO_2^{-}]$ oxidation in the electroless plating electrolyte. Open figures are the mixed potentials.

hypophosphite oxidation for the CoWP(2) plating are closer to the potential of WO_4^{2-}/W than those of the CoWP(1). This may explain the higher tungsten content in the deposit that was prepared from CoWP(2) bath. For all alloys, the mixed potential is close to the OCP of $H_2PO_2^{-}$ oxidation. This phenomenon indicates that the oxidation process of the reducing agent is the rate-determining step in the autocatalytic chemical reduction of Co alloys.

3.2.6.2 *Surface morphology and microstructure*

AFM image of the morphology of electroless Cu film before CoWP deposition (Figure 3.10a) showed that the surface of Cu has low roughness and Cu grains are large (about 750 nm). AFM image of the electroless CoWP film surface after deposition on electroless Cu (Figure 3.10b) showed that the CoWP film has small grain sizes (about 10 nm).

Figures 3.11a to c show three-dimensional (3D) AFM images of the morphology of as-received PVD Cu film, surface after Pd activation of the PVD Cu for 3 sec in an acidic solution and CoWP thin film morphology after an electroless deposition on the activated Cu surface, respectively. Electroless CoWP film has high-quality

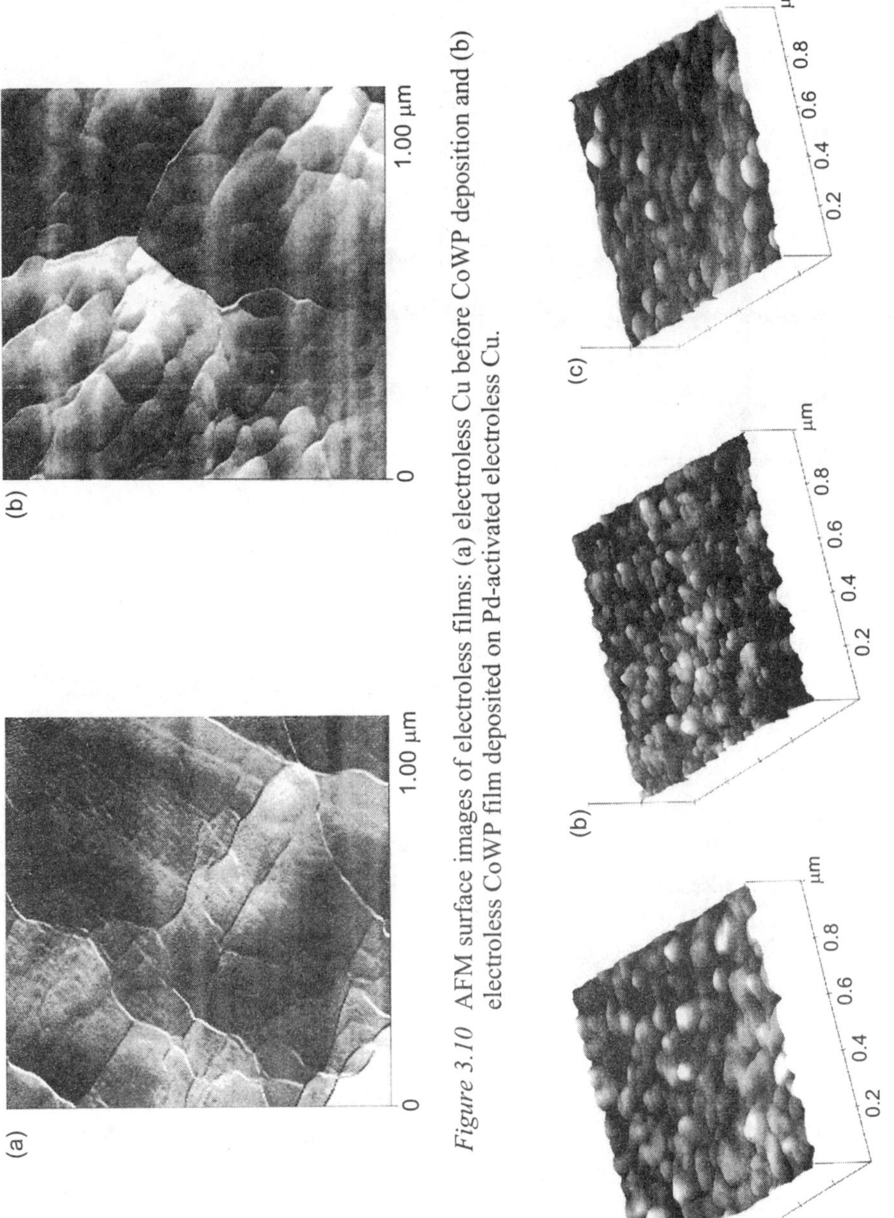

Figure 3.10 AFM surface images of electroless films: (a) electroless Cu before CoWP deposition and (b) electroless CoWP film deposited on Pd-activated electroless Cu.

Figure 3.11 3D AFM images of surface morphology of (a) as-received PVD Cu, (b) Pd-activated Cu, (c) electroless CoWP on Pd-activated Cu.

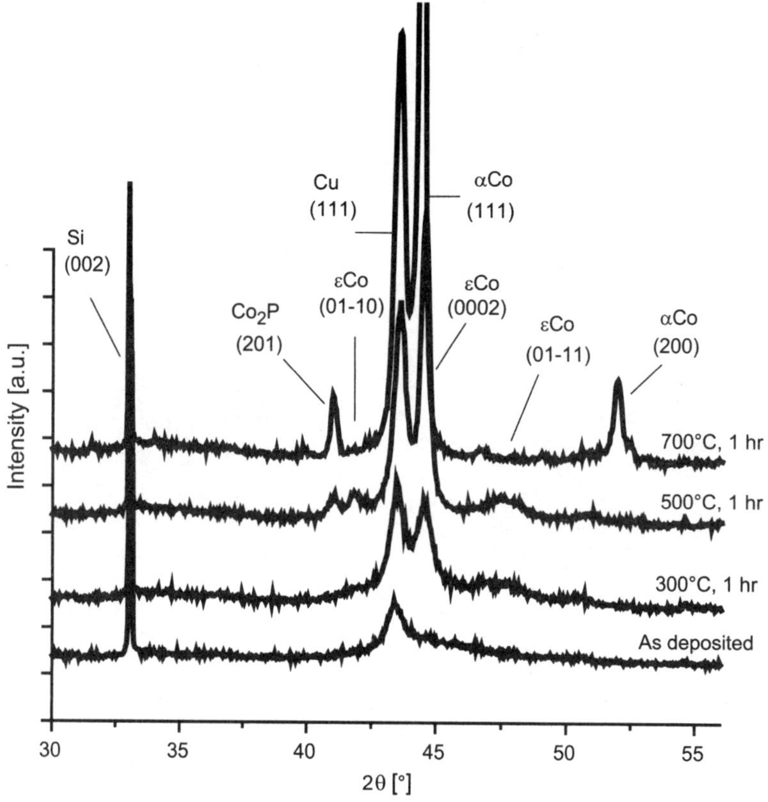

Figure 3.12 XRD patterns of electroless deposited CoWP layers on a copper catalytic seed
layer, as-deposited and after thermal anneal for 1 h at 300, 500, and 700°C.

morphology of the surface and a small grain size. AFM sectional analysis of CoWP
film grain size showed that grain size is between 5 and 10 nm.

Electroless CoWP deposition forms uniform films with high-quality morphology and small grain sizes on both electroless and PVD Cu seed layers. The electroless deposited and PVD Cu films had different grain sizes: 750 and 150 nm,
respectively. The grain size of as-deposited CoWP film was around 7 nm on both
seed layers. It is probably because the grain size of as-deposited CoWP film does not
depend on the grain size of seed Cu layers and is mostly dependent on the nucleation
size of Pd sites and distance between the Pd sites on the Cu surface.

The crystalline component of the as-deposited layers was the h.c.p. phase of cobalt (ε-Co) regardless of the seed layer utilized. The grain size is on a nanometric scale
based on the full width half maximum (FWHM) of the XRD peaks (several degrees).
After annealing, grain growth was observed according to the increased intensities of
the XRD peaks of the ε-Co phase and the decreasing FWHM (Figure 3.12). The layers
exhibit a high degree of preferred orientation of the <0001> direction or c-axis
perpendicular to the substrate based on the relative intensity of the XRD peak origi-
nating from the {0002} plane. The degree of preferred orientation on the copper seed
layer is stronger than on the cobalt seed layer. Transition to the f.c.c. (α-Co) phase

occurred in these layers above 500°C. A relatively small fraction of the Co_2P phase compared to the total phase content is detected after thermal treatment at 500°C.

3.2.6.3 Resistivity of electroless Co

The resistivity of electroless CoWP deposits depends on the solution temperature.[39] When the deposition temperature decreases from 85 to 70°C, the resistivity of electroless CoWP deposits decreases from about 30 to 20 $\mu\Omega$ cm. Annealing at a temperature of 400°C for 30 min in vacuum decreases the bulk resistivity from 27 to 20 $\mu\Omega$ cm for a deposition temperature of 75°C. Annealing at 300°C for 30 min in vacuum reproduced a similar change in bulk resistivity.

The high resistivity, 30 to 40 $\mu\Omega$ cm [17,39,67] of the as-deposited Co(W, P) films compared to 6 $\mu\Omega$ cm of bulk cobalt is attributed to the impurities, small grain size, and presumed high defect density of the film. The mean free path of electrons in cobalt at room temperature is approximately 10 nm.[65] As the grain size of as-deposited Co(W, P) is in this range, the high resistivity may be due, in part, to electron scattering at the grain boundaries.

3.2.6.4 Barrier properties of electroless Co

It has been shown in works [17,67] that Co(W,P) layers, 100 nm thick, are effective capping layers, blocking copper diffusion up to 500°C, as measured by AES. According to Ref. 66, 30 nm thick Co(W,P) layers block copper diffusion up to 500°C as determined by capacitance-voltage and transient capacitance measurements of MOS capacitors. It is shown in Ref. 68 that electroless CoWP is also an effective barrier for oxygen. Thus, electroless cobalt can be used as a barrier as well as a cladding/passivation layer to directly deposit and pattern low-*k* film on Cu lines plated with electroless CoWP (with no high-*k* SiN or SiC passivation and etch stop layers).[12–18,68]

In comparison, evaporated cobalt barriers fail as diffusion barriers at less than 400°C. The difference in the diffusion barrier quality of PVD cobalt and electroless Co(W, P) can be explained by the structure and chemical nature of the grain boundaries as well as the degree of crystallinity and grain size.[17,67] The electroless films have a nanocrystalline structure. The h.c.p. cobalt grains may be embedded in an amorphous matrix which eliminates the short-circuit paths of the grain boundaries and improves the barrier properties. The diffusivity of copper atoms and ions is mainly along the boundaries. It is experimentally known that impurities can significantly improve the diffusion barrier properties of polycrystalline thin films. The term "stuffed barrier" has been used to refer to the diffusion blocking effect of impurities in the grain boundary.[61] Hono and Laughlin[62] have demonstrated this effect in electroless Co(P) using an EDS analysis in TEM. When the electron probe (diameter of less than 20 nm) was located at the triple point of grain boundaries, a phosphorus peak was detected. However, spectra from the center of individual grains did not show a phosphorus peak.

In summary, the proposed microstructure of as-deposited electroless Co(W, P) is grains of crystalline ε-Co, sized ~10 nm while phosphorus and tungsten reside in the grain boundaries. The oxygen content measured in the films is less than 1 at. %, which excludes the possibility that phosphorus or tungsten is mainly in the form of oxides. The phosphorus and tungsten may be up to a few monolayers thick enveloping the cobalt grains, creating stuffed grain boundaries.

3.2.6.5 *The influence of thermal treatment on the microstructure*

At room temperature, the h.c.p. phase of cobalt is energetically favorable compared to the f.c.c. phase. The h.c.p. to f.c.c. transition occurs in bulk cobalt at 422°C.[63] However, the free energy difference between the h.c.p. and f.c.c. phases is small, −9.2 J/mol. The phase transformation is sluggish, influenced by impurities and stress. In PVD cobalt thin films, as-deposited films are usually comprised of α-Co phases as a result of stress.[64] The PVD cobalt used as seed layers for electroless deposition showed this phenomenon, according to XRD measurements. In electroless deposited Co(W, P), the h.c.p. phase remains stable up to 500°C. The dominant microstructural evolution during these thermal treatments is grain growth. A complete transition from the h.c.p. to the f.c.c. phase is observed after a thermal treatment at 700°C. This result is significantly different from PVD cobalt thin films in which the transition starts at 380°C.[64] When a Co(W, P) sample was subjected to an additional 10 h anneal at 500°C, the XRD pattern and resistivity of the film were similar to those obtained after the initial 1 h anneal and only trace amounts of the α-Co phase were identified. This indicates that the phase transformation delay is probably due to the relatively large amounts of phosphorus and tungsten impurities and not stress. The delay in the phase transition stabilizes the crystal structure up to relatively high temperatures and is therefore beneficial for thin films applied as capping layers or diffusion barriers.

3.2.7 *Selective electroless Co deposition on in-laid Cu lines*

Thin CoWP was selectively deposited for capping the Cu surface.[15,39,49] Good selectivity was reported for electroless Co deposited onto Cu lines with Pd-activation. An advantage of electroless CoWP deposition, its selectivity, was achieved by the activation of the patterned metal lines vs. the non-activated silicon dioxide surface. The amount of activation is controlled by the Pd and HF concentrations, and the activation time. Increasing the HF concentration and activation time led to removal of the Cu lines from the trenches, since the silicon dioxide was etched along the sidewalls.

HF concentrations in the range of 20 to 100 ml/l with 3 sec of activation resulted in non-uniform Co deposits. Chemical–mechanical polishing (CMP) treated Cu is usually covered with a layer of copper oxides or copper-organic component complexes that are removed in the activation solution by HF. Also the CMP process modifies the silicon dioxide surface that is etched in diluted HF. Therefore, optimizing the HF concentration is critical to CoWP uniformity and selectivity.

Selective and uniform growth of electroless CoWP on Cu lines was observed after activation for 2 to 3 sec with 150 to 200 ml/l HF concentrations. Optimized HF concentration and activation time are about 200 ml/l (pH = 3) and 2 sec, respectively.

Optimized Pd activation resulted in selective CoWP deposition on Cu line surfaces at 65°C. Three critical steps are involved in this process:

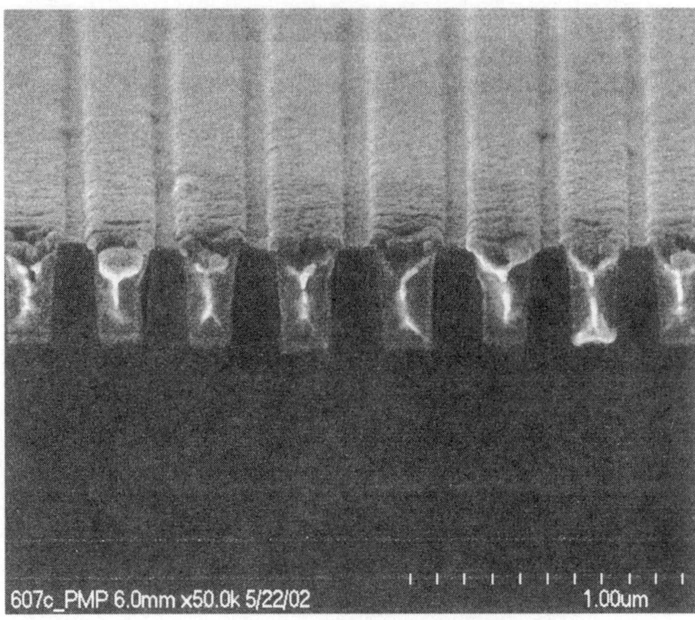

Figure 3.13 SEM cross section of selective Co films on 0.2 μm wide Damascene Cu lines.

1. Removal of the copper oxide (or Cu-BTA) on Cu lines and etching of the modified layer on silicon dioxide surface among Cu lines with HF

2. Surface activation with a high density of Pd nucleation sites on Cu line surfaces

3. Uniform initiation of the autocatalytic electroless CoWP deposition only on the Cu lines

An example of selective electroless Co cladding layer on Damascene 0.2 μm Cu lines is presented in Figure 3.13. No Co deposition between Cu lines was observed.

It was reported in work[68] that line-to-line leakage on comb structures (300 μm × 600 pieces = total 18 cm long 0.32 μm wide copper lines with spacing of 0.21 μm at 44 chips on 8 in. wafers) has exactly the same distributions before and after CoWP deposition to prove perfect selectivity. It was found that removal of metal ions out of the dielectric surface to a level below detection limit (10^{10} atoms/cm^2) prior to electroless plating is needed to achieve the perfect selectivity. After electroless plating, the metal particles generated in the plating bath and deposited on the dielectric surface can be removed in the same manner as removal of CMP slurries in post-CMP cleaning, because the metal particles are not embedded in the dielectric layer.

Figure 3.14 shows that the electroless deposited CoWP protection layer is continuous across the top of the Cu lines and extends slightly down the higher sidewalls. Figure 3.14 demonstrates a unique capability of the electroless Co deposition to form conformal film inside very narrow features. A conformal 10 nm thick CoWP layer was formed on the sidewalls of 30 to 40 nm wide seam of aspect ratio about 5:1.

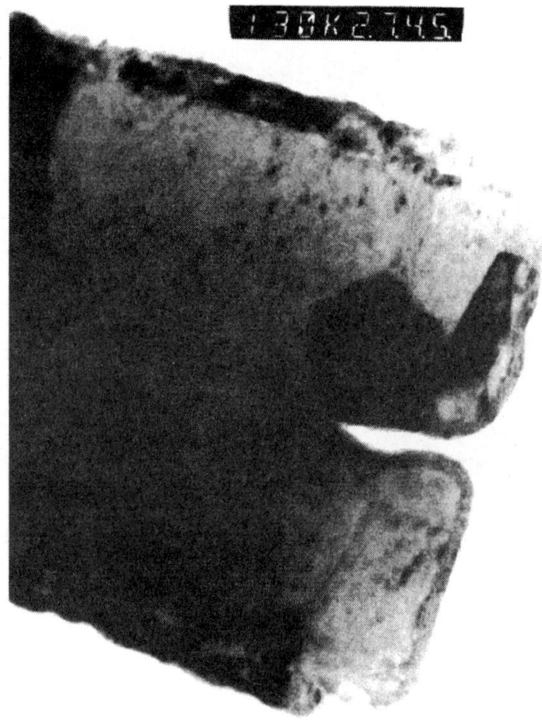

Figure 3.14 TEM cross-section of the thin conformal CoWP film inside of 30 to 40 nm wide seam on the top of 0.4 μm wide in-laid Cu line.

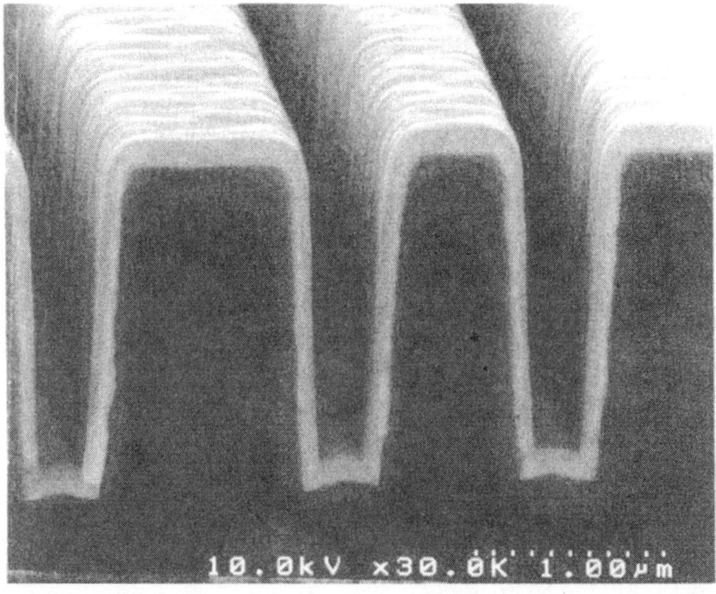

Figure 3.15 SEM cross section of conformal CoWP films in 0.35 μm trenches with aspect ratio of 4.5:1.

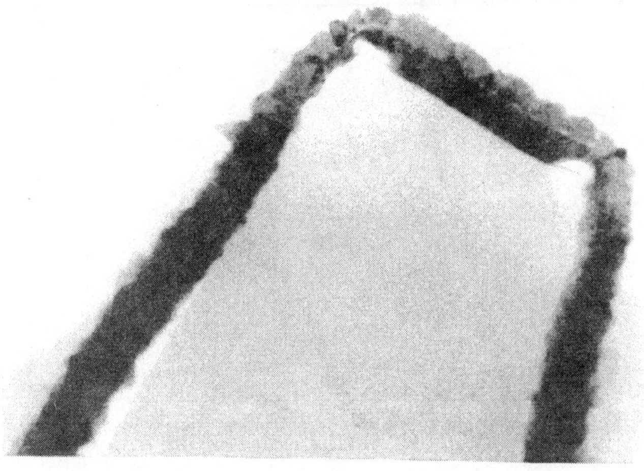

Figure 3.16   TEM cross section of the thin electroless CoWP film on sputtered Co at the top of the trench.

3.2.8   *Conformal electroless CoWP barrier deposition into the trenches*

The CoWP barrier was deposited on a Co seed layer on a test chip with a periodic structure of via contacts and trenches.[39,49] The SEM cross section of electroless CoWP barrier on sputtered Co seed layer in high aspect ratio trenches is presented in Figure 3.15.

A Co seed layer with thickness measured on the field of 30 nm was sputtered using a collimator. The CoWP alloy was deposited from heated aqueous solution with addition of ductility promoter and surfactant RE 610. A profile of seed layer covered with thin electroless CoWP film in sub-half-micron trenches was studied by SEM and TEM cross sections.

TEM cross-sectional profile of CoWP/Co films is presented in Figure 3.16. The 335 to 350 nm wide trenches with an aspect ratio of 5:1 were covered with a thin (thickness ~50 nm) conformal electroless CoWP film. A high magnification image of the same pattern showed the high quality of the coating on the sidewalls. A very conformal electroless CoWP deposition in such narrow trenches was observed. Similar results were observed for trenches greater than 350 nm width with an aspect ratio less than 4:1.

Figures 3.15 and 3.16 demonstrate conformal films of the electroless CoWP after deposition on Co. The thickness and surface roughness of the electroless CoWP at the top edges, lower sidewalls, and bottom of the trenches play an important role in the following Cu deposition onto CoWP barrier. Very uniform and smooth thin barrier CoWP films are electroless deposited. This leads to a void/seam-free Cu filling of sub-half-micron trenches with very high aspect ratios and vertical sidewalls.

The AES depth profiles of the CoWP/Cu/CoWP/SiO$_2$ structure formed by electroless CoWP barrier deposition was compared with the AES profile of this structure annealed at 500°C for 60 min. The comparison of the two spectra showed

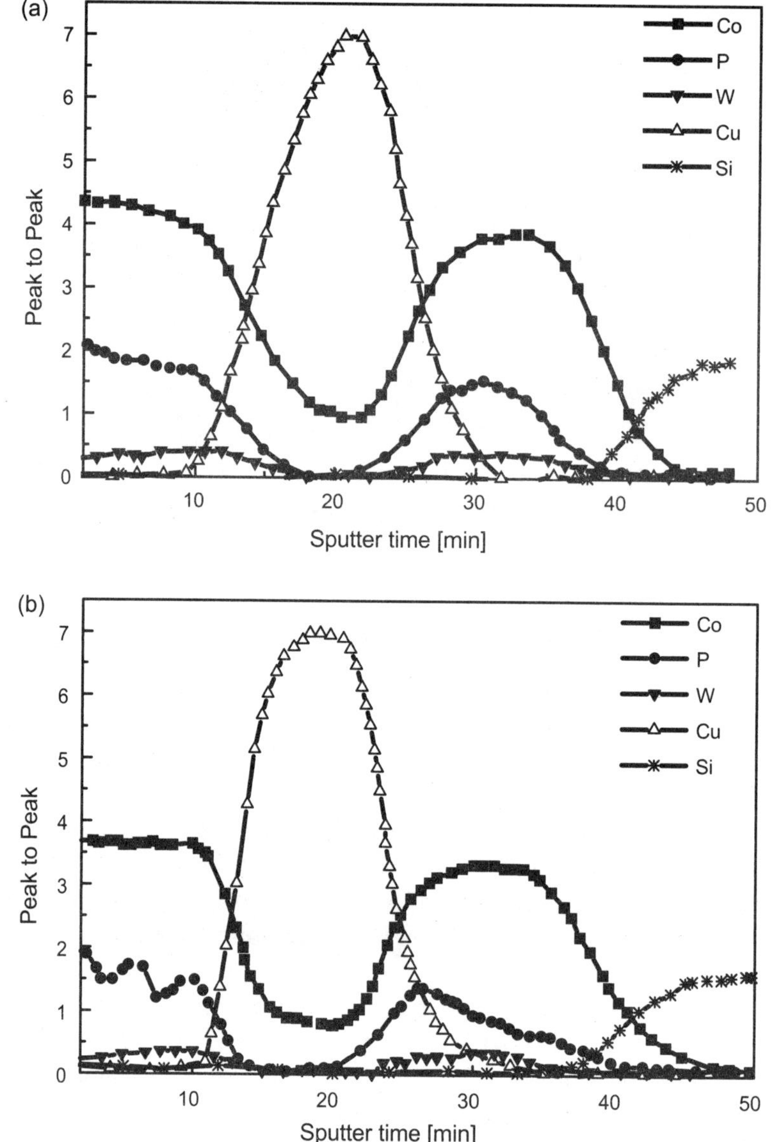

Figure 3.17 AES depth profiles of CoWP/Cu/CoWP/SiO$_2$/Si samples as deposited (a) and
          after annealing at 500°C for 1h (b).

that annealing did not result in significant redistribution of P and Co in the CoWP/Co
interfaces (Figure 3.17).

The process flow for electroless CoWP barrier and Cu deposition on Si
substrate was described in Ref. 49: *in-situ* ion-beam cleaning of the Si surface and
sputtering of a thin Co seed layer; electroless CoWP deposition, rinsing in DI water
and drying in N$_2$ flow; annealing at 300°C for 30 min in vacuum (10^{-7} torr); electroless Cu deposition,[7,8] rinsing in DI water and drying in N$_2$ flow; annealing at 400°C
for 60 min in vacuum 10^{-7} torr. The RBS spectrum of the Cu/CoWP/Co/Si structure

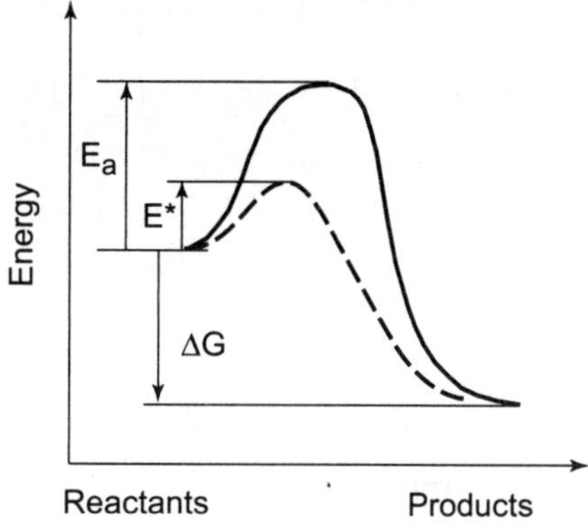

Figure 3.18 Schematic Gibbs free energy diagram. Notice: the activation energy (E_a) of the exothermic reaction ($\Delta G < 0$) decreases on the catalytic surface (E^*).

was formed by electroless CoWP barrier/electroless Cu deposition and annealed at 400°C for 60 min in vacuum 10^{-7} torr, and a simulation of this spectrum showed no interdiffusion in deposited films. The thickness of the Co, CoWP, and Cu films were approximately 20, 200, and 200 Å, respectively.

3.3 Electroless deposition of copper seed layer

3.3.1 Electroless Cu thermodynamics

Electroless Cu deposition is a thermodynamically favorable process since the sum ($E°$) of the standard redox potentials of the oxidation and reduction reactions is positive (i.e., the reaction free energy is negative, $\Delta G° = -nFE°$).[26,50–51]

While electroless Cu deposition is a thermodynamically favorable process, spontaneous solution decomposition does not occur since the electroless Cu plating process is kinetically inhibited. The activation energy for this process was estimated to be ~0.63 eV.[50] The Gibbs free energy diagram illustrates the potential barrier (E_a) which exists between the reactants and the products and prevents spontaneous solution decomposition (Figure 3.18). The potential barrier can be reduced by the formation of reactive intermediate species on the catalytic surface having a lower activation energy (E^*).

3.3.2 Electroless Cu deposition mechanism

The overall reaction of electroless Cu deposition in a formaldehyde-based plating solution with EDTA as a complexing agent is[51]

$$[CuEDTA]^{2-} + 2HCHO + 4OH^- \rightarrow Cu^0 + 2HCOO^- + 2H_2O + H_2 + EDTA^{4-} \quad (22)$$

Electroless Cu deposition reaction in a glyoxylic-based bath can be written as

$$[CuEDTA]^{2-} + 2HCOCO_2H + 4OH^-$$

$$\rightarrow Cu^0 + 2COOCO_2H^- + 2H_2O + H_2 + EDTA^{4-} \qquad (23)$$

The electroless Cu deposition mechanism was discussed in detail elsewhere[52] and includes the following steps:

a. Formation of methylene glycol anions

b. Dehydrogenation reaction on a catalytic surface (Cu, Au, Pt, Pd, Rh, Ag)

c. Hydrogen evolution reaction

d. Copper complex (with EDTA) reduction reaction

Hydrogen is one of the products of electroless Cu deposition reactions. Care must be taken to prevent hydrogen inclusion in deposits. This is accomplished by adding surfactants to the electroless Cu solution.

3.3.3 *Electroless Cu electrochemistry*

An electroless Cu plating system is basically a shorted galvanic cell with time-varying microscopic local anodes and cathodes on a catalytic surface. The catalytic layer is needed to initiate electroless Cu deposition. The role of the catalytic layer is to catalyze the dehydrogenation reaction and the hydrogen evolution reaction as well as serve as conductive material to transport electrons from local microanodes (anodic reaction of reducing agent oxidation) to local microcathodes (cathodic reaction of metal ion reduction). Thus the seed layer has to be non-oxidized and non-contaminated in order to serve as a catalytic and conductive layer.

Current-potential curves for a system with two different simultaneous electrochemical reactions are presented in Figure 3.19 as Tafel plots (this is the case for charge-transfer limited redox reactions). The equilibrium electrode potential for the oxidation (anodic) half reaction ($E_{RED}{}^{eq}$) has to be more negative than the equilibrium electrode potential for the reduction (cathodic) half-reaction ($E_{Me}{}^{eq}$). The electrode potential for the oxidation half-reaction becomes more negative when the pH of the solution is increased.[50] In an electroless copper deposition process with formaldehyde as a reducing agent, only basic solutions with pH > 11 can be used.

The difference between redox potentials of the reducing agent and metal should not be very large in order to obtain thin metal films without spontaneous metal deposition in the solution. The stability of the electroless metal deposition solution can be increased by decreasing the difference between $E_{Me}{}^{eq}$ and $E_{RED}{}^{eq}$, for example, by using stronger (with larger stability constant) complexing agents (a more negative $E_{Me}{}^{eq}$), or decreasing the pH (a less negative $E_{RED}{}^{eq}$).[50] EDTA which has a high stability constant for complex formation with Cu ions ($pK \sim 18.7$) and negative redox potential ($E_{Me}{}^{eq} \sim -0.216$ V) for the reduction of $[CuEDTA]^{2-}$ is commonly used as a complexing agent to provide stable electroless Cu solutions with high deposition rate.

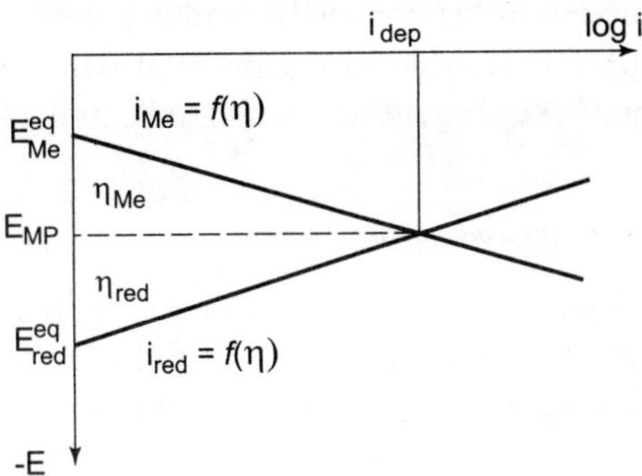

Figure 3.19 Current-potential curves (Tafel plots) for the system with two simultaneous anodic (oxidation) and cathodic (reduction) reactions.

3.3.4 *Electroless Cu deposition kinetics*

The overall rate law for the electroless Cu deposition reaction presented by Equation 22 can be written as

$$R = K[C_{Cu^{2+}}]^a[C_{HCHO}]^b[C_{OH^-}]^c[C_{EDTA^{4-}}]^d \qquad (24)$$

where K is the observed rate constant at a given temperature and steps a, b, c, and d are the reaction orders for the reactants.

The plating rate of electroless Cu can be affected by concentration of copper ion $[C_{Cu^{2+}}]$, formaldehyde $[C_{HCHO}]$, hydroxide ion $[C_{OH^-}]$, and complexing agent $[C_{EDTA^{4-}}]$ as well as temperature (T).

The fact that the reaction rate increases when the temperature is raised implies that the rate constant (K) of Reaction 22 has increased. The increase in K was found in Ref. 53 to obey an Arrhenius equation. From the slope of the Arrhenius plot, an activation energy of 60.9 kJ mol^{-1} (0.63 eV) was estimated.[53]

The values of the reaction orders of the reactants depend on the experimental conditions. The authors of Refs. 2 and 54 reported that when the concentration of Cu^{2+} $[C_{Cu^{2+}}]$ is between 0.015 and 0.1 mol l^{-1}, $0.05 \leq C_{HCHO} \leq 0.5$, $0.07 \leq C_{OH^-} \leq 0.3$ mol l^{-1}, $0.03 \leq C_{EDTA^{4-}} \leq 0.4$ mol l^{-1}, and the temperature is 50°C, the values of the reaction orders (steps a to d) are 0.78, 0.13, 0.02, and 0.02, respectively. However, for other ranges of chemical conditions, for example, $0.003 \leq C_{Cu^{2+}} \leq 0.03$, $0.03 \leq C_{HCHO} \leq 0.4$, $0.5 \leq C_{OH^-} \leq 2.0$ mol l^{-1}, and at 21°C solution temperature, the values of the reaction order (steps a to c) are 0.37, 0.0, and 0.37, respectively.

From the reaction stoichiometry, hydroxide ion can affect the plating rate kinetically; the plating rate is higher at higher pH. However, hydroxide ion can affect the plating rate thermodynamically. Different complexing agents have different complexing ability with the copper ion, and the complexing ability is sensitive to the pH of the bath. Usually, the higher the pH, the stronger the complexing ability. Thermodynamically, the reduction of the copper(II) complex ion is more difficult at higher pH.[2,12] An optimum pH for electroless Cu deposition in formaldehyde-based solution with EDTA as the complexing agent was found to be in the range of 12.5 to 12.6.

3.3.5 *Electroless Cu mass-transport*

The electroless Cu deposition reaction is electron-transfer limited if deposition occurs on a flat surface,[26] for example, on the field of the wafer. The deposition current on the field of a wafer $[J_{deposition\,(field)}]$ can be expressed as

$$J_{deposition\,(field)} = J_{anodic} = J_{cathodic} \qquad (25)$$

where J_{anodic} and $J_{cathodic}$ are the anodic and cathodic current densities for the simultaneous reactions of anodic oxidation of reducing agent and cathodic reduction of metal ions, respectively.

The mass-transport of reactants in small features such as deep sub-half-micron vias and trenches becomes the limiting step in electroless Cu deposition. The thickness of the diffusion layer at the solution/copper interface is larger in trenches/vias than that on a flat surface (i.e., on the field of the wafer) because of diffusion layer overlapping on the sidewalls of the trenches. Therefore the thickness of the diffusion layer increases up to the trenches/vias depth, and the electroless Cu deposition in deep ($\sim$1 μm or deeper) sub-half-micron trenches/vias becomes controlled by diffusion of reactants through the thick diffusion layer. In this case the deposition current density in trenches can be expressed as:[56]

$$J_{deposition\,(trench)} = J_{diffusion} = zFD(C_o - C_s)/\delta_{dif} \qquad (26)$$

where $J_{diffusion}$ is the diffusion current density; z is the charge of ions; D is the diffusion coefficient of ions; C_o is the concentration of ions in the bulk of deposition solution; C_s is the concentration of ions at the outer Helmholtz plane at the electrode–solution interface; δ_{dif} is the thickness of diffusion layer.

Since the electron-transfer process is expected to be faster than diffusion of ions in the solution, we can expect that $J_{diffusion} < J_{cathodic}$ and $J_{deposition\,(trench)} < J_{deposition\,(field)}$.

The higher deposition rate at the top of trenches/vias than that at the bottom can lead to poor conformality of deposition.

The degree of conformality can be increased by reducing the deposition rate (i.e., by making $J_{diffusion}$ closer to $J_{cathodic}$). However, this will result in a decrease in throughput and will not completely eliminate seams. $J_{diffusion}$ can also be increased by increasing the concentration of reactants (i.e., increasing C_o, Equation 26), but this results in a decrease of solution stability.

The other way to increase the degree of conformality is to increase $J_{diffusion}$ by increasing the diffusion coefficient of reactants taking part in the plating reactions. The diffusion coefficient (D°_i) of species in the solution can be expressed as[56]

$$(D^\circ_i) = kT/6\pi\nu r_i \qquad (27)$$

where T is the solution temperature, v is viscosity, and r_i is the radius of ion in the solution.

The diffusivity increases with increasing the solution temperature and decreasing the viscosity.

3.3.6 Electroless Cu deposition solutions

Electroless Cu deposition was initiated by dry and wet activation processes. A sputtered Cu seed layer with an Al protection layer was used in a dry activation process for electroless Cu deposition. The Cu seed layer was deposited onto the base metal by contact displacement in the wet activation process for electroless Cu plating.

Contact displacement deposition of Cu seed on TiN base metal was performed at room temperature for 1 to 600 sec in an activation solution containing Cu^{2+} cations (0.001 to 1 mol/l) and F^- anions (0.001 to 5 mol/l) as the main components.

The electroless Cu plating bath contained cupric sulfate (0.016 to 0.08 mol/l), EDTA (0.04 to 0.2 mol/l), and formaldehyde (0.13 to 1 mol/l). Stabilizer and surfactant were also added to the plating bath to increase the bath stability and to decrease the surface tension of the solution. The electroless Cu plating bath was operated at 30 to 80°C and pH in the range of 11 to 13 (adjusted by tetramethyl-ammonium/potassium hydroxide).

3.3.7 Electroless Cu experimental techniques

Electroless Cu deposition has been performed with a single wafer Cu deposition tool with capability for an 8 in wafer as well as smaller wafer diameters and wafer fragments. This tool allows both immersion and spray deposition techniques.[8] The schematic diagram of the system is presented in Figure 3.20. It mainly consists of

1. Process chamber/spin processor
2. Holding tank with pH/mV/temperature meter, dissolved-O_2 electrode, and plating monitor probe with sensor crystal (to measure the electroless Cu deposition rate by quartz-crystal microbalance)
3. Liquid to liquid heat exchanger with immersion circulator, heater exchange coils, and open bath
4. Magnetic drive pumps and filters
5. Air-operated valves, fittings, and piping
6. Sequential process controller

Electroless plating can be also accomplished with a spray processor where atomized droplets or a continuous stream of an electroless plating solution is sprayed on the surface of the substrate.[24] The electroless plating solution can be prepared by mixing reducing agents and a metal stock solution prior to the spraying. A suitable plating solution spray rate can be as low as 100 ml/min. Thus the stability of the

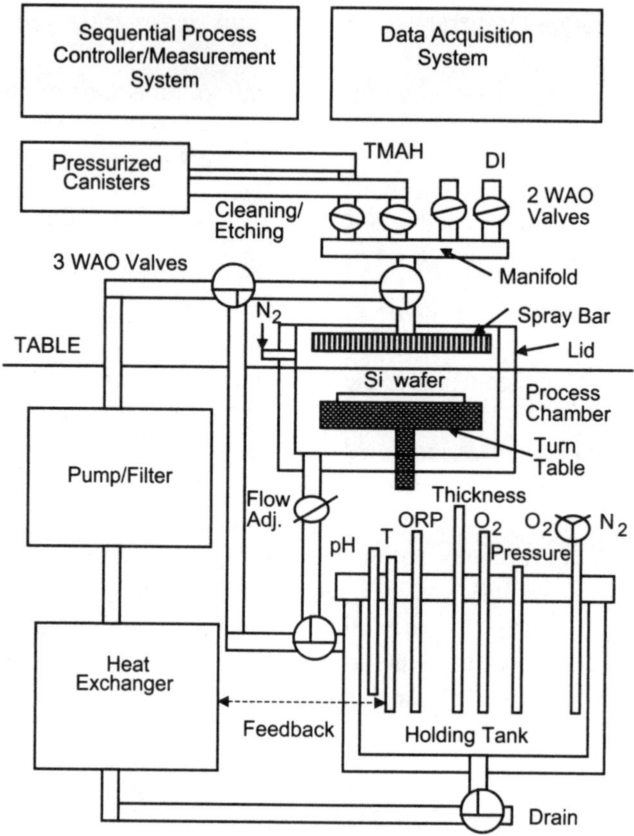

Figure 3.20 Schematic diagram of electroless Cu deposition system.

electroless plating bath is no longer a concern since a single use bath can be used. The spray process can be used to form metal films as thin as 10 nm, and these films have low resistivity approaching bulk values, low surface roughness (few nanometers), and excellent electrical and thickness uniformity.[24]

Blanket electroless Cu deposition was performed for via and trench filling applications on a 10 to 55 nm thick Cu seed layer deposited by sputtering (dry activation process) or contact displacement (wet activation process). An Al sacrificial layer (10 to 35 nm thick) was sputtered without breaking the vacuum after Cu seed sputtering. Collimated Ti (25 nm)/ TiN(40 nm), uncollimated Ti (30 nm and 60 nm thick), and uncollimated Ta (40 nm thick) were used as diffusion barrier/adhesion promoter layers. The nominal trench (via) depth in a SiO_2 layer ranged from 1 to 2 μm. The designed widths of trenches were 0.3/0.5/0.8/1.0/1.5/2/5/10/15 μm. The nominal via sizes were 0.35 and 0.5 μm.

Selective electroless Cu deposition was performed into 0.35 and 0.55 μm via holes in ~1.1 μm SiO_2 interlevel dielectric on patterned ~0.6 μm thick AlCu/TiN activated by contact displacement Cu deposition.

Table 3.5 Electroless Cu deposition rate and resistivity vs deposition solution parameters.

Increasing parameter	Deposition rate change	Resistivity change
$CuSO_4 \cdot 5H_2O$ concentration	↑	↔
HCOH concentration	↑	↑
EDTA concentration	↔	↔
pH (12.2–12.8)	Maximum at pH 12.5–12.6	↔
pH (11.8–12.2)	↑	↓
pH (>12.8)	Unstable solution	Unstable solution
Temperature	↑	↓

↔ No strong effect; ↑ increases; ↓ decreases.

The step coverage (degree of conformality) of the electroless Cu deposition process was investigated by cleaved and focus ion beam (FIB) SEM cross sections. The thickness of electroless Cu films was determined by SEM and surface profilometry. The electrical resistivity of the Cu films was measured by a four-point probe. The surface roughness and electrical uniformity were studied by Alpha-step® 200 and VersaProbe® VP10, respectively.

Contaminations were measured by SIMS in electroless Cu films. The thickness measurements of blanket Cu films were performed by XRF.

3.3.8 *Electroless Cu film properties*

Electroless Cu film properties were studied as a function of bath composition and deposition conditions (pH, temperature). The results of this study are summarized in Table 3.5.

Deposition rate and resistivity of electroless Cu deposits increased slightly with increasing concentration of cupric sulfate and formaldehyde (Table 3.5). Similar results have been obtained with glyoxylic acid as a reducing agent. The resistivity is practically independent of pH in the range of 12.3 to 13, while the deposition rate has a maximum at pH 12.5 to 12.6. When the pH is lower than 12, the deposition rate increases and the resistivity decreases with increasing pH. The solution stability decreases when pH increases above 13. EDTA concentration does not significantly influence the deposition rate and resistivity (Table 3.5).

We observe that temperature has the strongest effect on deposition rate and resistivity. Deposition rate increases and resistivity decreases with increasing the deposition solution temperature (Figure 3.21). The high deposition rate (up to 120 nm/min) of low resistivity ($\rho \leq 2\ \mu\Omega$ cm) electroless Cu films can only be achieved at a deposition temperature in the range of 75 to 80°C.

A pH value larger than 12 is required to achieve stable deposition. The pH value decreases as the temperature increases. The oxidation–reduction potential (ORP) is also a function of temperature. The dependence of the pH and the ORP as a function of the solution temperature is shown in Figure 3.22. There is a large drop in the ORP at about 64 to 65°C, below which the Cu deposition rate decreases significantly.

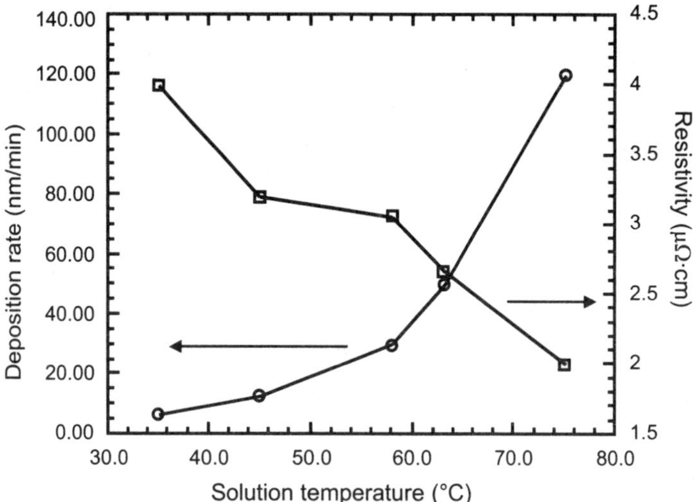

Figure 3.21 Deposition rate and resistivity of electroless Cu vs. solution temperature in plating bath containing 7.5 g/l CuSO$_4$·5H$_2$O, 10 g/l HCHO, 14 g/l EDTA, 0.04 g/l RE 610 and at pH ~ 12.5.

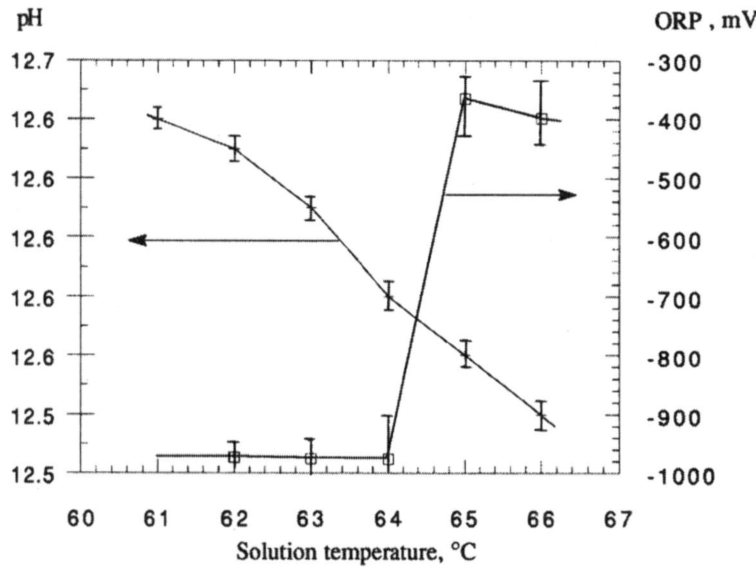

Figure 3.22 Oxidation–reduction potential and pH vs. solution temperature.

A small amount of surfactant (such as RE 610, polyethyleneglycol, NCW-601A, Triton® X-100) and stabilizer (e.g., Neocuproine, 2,2′-dipyridyl, CN⁻, Rhodamine) can be added to the copper plating solution in order to control surface tension and to retard hydrogen inclusion in the deposits as well as to increase solution stability. Surfactants also have a favorable effect on the surface roughness of electroless Cu films. For example, electroless Cu films were deposited with low surface

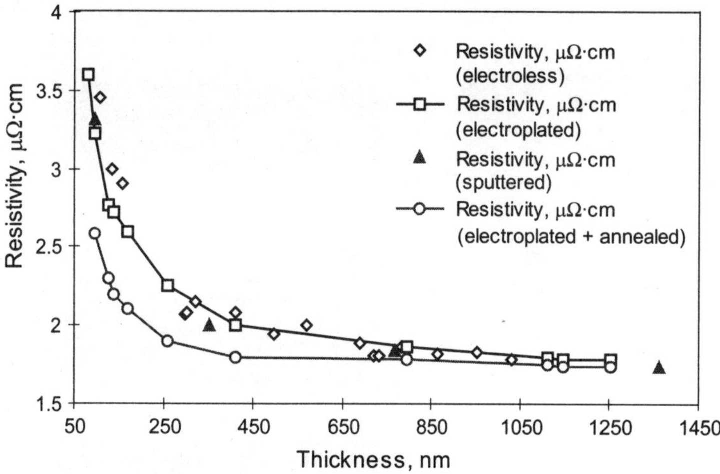

Figure 3.23 Resistivity of Cu films vs. copper thickness.

roughness ($R_a \sim 10$ to 15 nm for 1.5 μm thick films) when about 0.001 to 0.01 wt. % surfactant was added to the plating bath.

We found that deposition from the solution containing 5 to 10 g/l $CuSO_4$ $5H_2O$, 5 to 10 g/l HCHO, 14 to 26 g/l EDTA, 0.01 to 0.1 g/l RE 610 at 75 to 80°C solution temperature and pH in the range of 12.5 to 12.6 provided high deposition rate and good quality deposits. The plating rate in this solution is affected only by the concentration of copper sulfate and the concentration of formaldehyde. The values of the reaction orders a and b (Equation 23) are ~ 0.9 and ~ 0.3, respectively. For example, the deposition rate in the solution containing 7.5 g/l $CuSO_4 \cdot 5H_2O$ and 10 g/l HCHO was measured to be about 120 nm/min, while in the solution containing 5 g/l $CuSO_4 \cdot 5H_2O$ and 5 g/l HCHO the deposition rate was about 75 nm/min.

The major contaminant was found to be potassium (up to 90 ppm) and hydrogen (up to 80 ppm). Using TMAH instead of KOH (to adjust pH >11.8 for plating bath) results in the reduction of alkaline metal contaminations of Cu films below the SIMS detection limit. Carbon was also incorporated into electroless Cu film with a level up to about 10 ppm. Other contaminants such as oxygen, S, Cl, and Al were found to be in as-deposited electroless Cu films in the range of several parts per million.

Electrical uniformity (sheet resistance uniformity) of blanket deposited electroless Cu films was measured on 6 inch wafers to be in the range of 1 to 3%. Thickness uniformity was about in the same range as electrical uniformity. Figure 3.23 shows resistivity of as-deposited blanket electroless Cu film depends on film thickness. Resistivity below 1.8 μΩ cm was obtained when the thickness of electroless Cu films exceed 1 μm. The resistivity increases when thickness of electroless Cu films decreases. The same behavior was found for sputtered Cu films (Figure 3.23).

Average surface roughness (was measured by using profilometry) also increases with increasing film thickness (up to 20 nm for 2 μm thick Cu films).

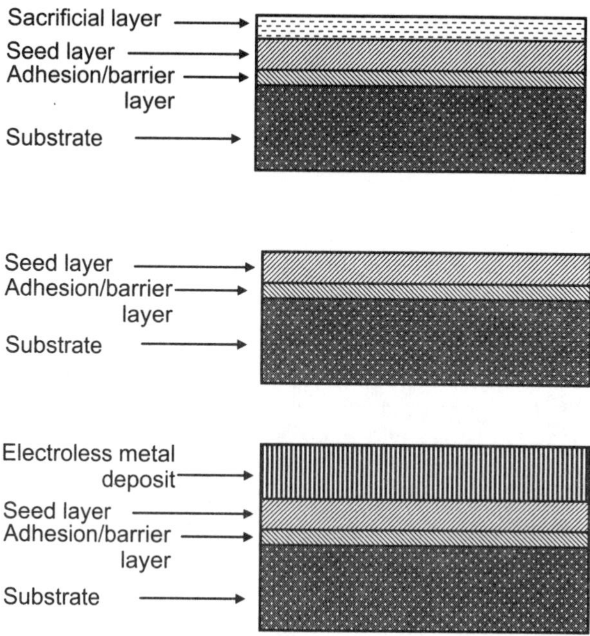

Figure 3.24 Dry activation process of electroless Cu deposition with sputtered Cu seed and Al protection layer. Notice: (1) Al is deposited on Cu seed to protect its catalytic properties, (2) the Al layer is dissolved in the basic electroless Cu bath, (3) Cu is electrolessly deposited onto non-oxidized and non-contaminated catalytic seed layer.

3.3.9 *Electroless Cu deposition on sputtered Cu catalytic layer with Al protection layer*

The seeding method with an Al protection layer has been developed.[8] This seeding method includes

1. The deposition of a Cu seed layer by sputtering or evaporation in vacuum
2. The deposition of a sacrificial thin Al layer without breaking the vacuum
3. Etching the Al layer in the electroless Cu plating bath followed by electroless Cu deposition (Figure 3.24)

Since the Al sacrificial layer is etched away in the same electroless Cu plating bath as for electroless Cu deposition without wafer transfer, the catalytic Cu surface is not exposed to air and is not oxidized before Cu deposition starts and acts as an excellent catalytic material for deposition. Without the Al sacrificial layer the copper seed layer quickly oxidizes when exposed to the atmosphere and does not form a good catalytic surface. Additionally, aluminum has a high affinity for oxygen and tends to combine with any oxygen which may be adsorbed onto the seed layer. Good adhesion of the Cu deposits to the seed layer is obtained because surface contamination of the seed layer during processing is prevented by the Al sacrificial layer.

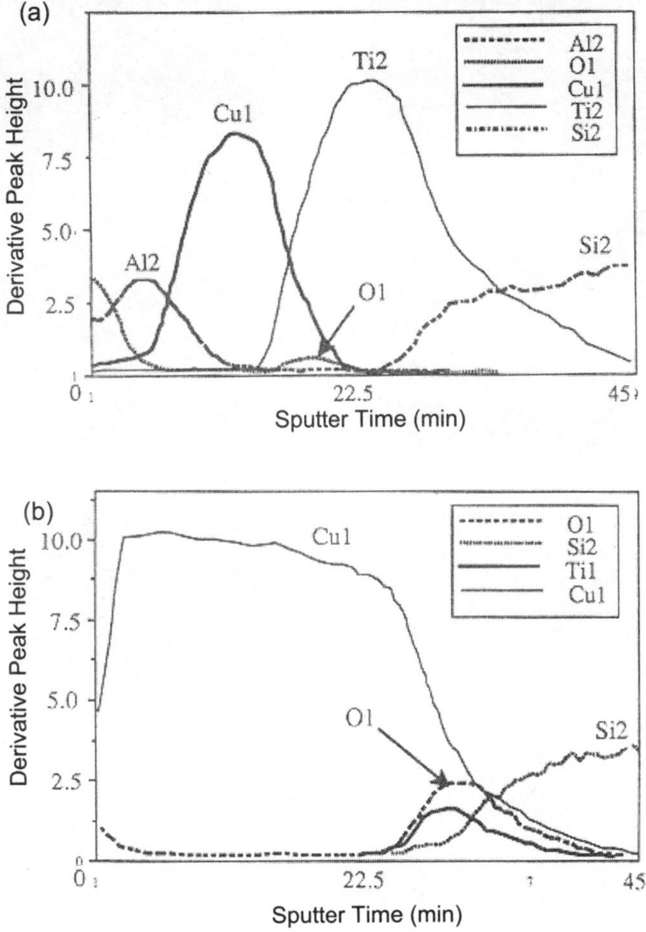

Figure 3.25　AES spectra of Al/Cu/Ti multilayer film on Si substrate (a) and electroless Cu film deposited onto protected Cu seed layer with Ti (b). Notice: no surface oxide on Cu seed layer, surface oxide on Al sacrificial layer, and some surface oxide on Ti adhesion layer (a); complete dissolution of Al sacrificial layer in basic electroless Cu bath (b).

The sputtered Cu/Al seed layer before and after electroless Cu deposition was studied by AES (Figure 3.25). We observe that (1) no oxide is formed on a sputtered Cu surface when it is protected by an Al layer (Figure 3.25a); (2) the Al protection layer is completely dissolved in the electroless Cu plating bath (Figure 3.25b). An oxide-free sputtered Cu surface provides uniform initiation of electroless Cu deposition. The incubation period for deposition on the sputtered Cu seed layer with Al protection layer was found to be ~1 sec in the plating solution at 75°C.

It was also found that the electroless seed plating depends on adhesion of the sputtered seed layer to the diffusion barrier, step coverage of the sputtered seed layer, and electroless Cu deposition rate. Poor adhesion of the sputtered seed layer to the diffusion barrier results in electroless Cu film delamination from the sidewalls of trenches/vias and on the field of the wafer, void/seam formation, and unfilled

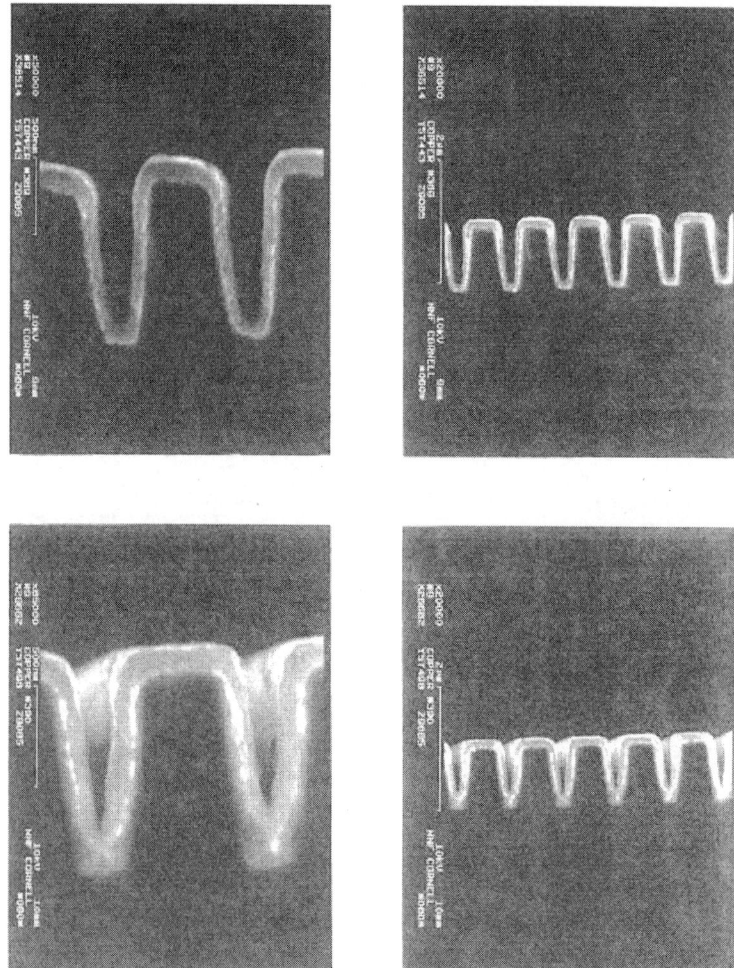

Figure 3.26 Demonstration of conformal electroless Cu deposition with about 100% step coverage. Cleaved SEM cross sections of ~0.3 to 0.5 μm trenches of AR (~3) plated with ~0.14 μm electroless Cu on sputtered Ti(30 nm)/Cu(30 nm)/Al (25 nm).

trenches/vias. Poor step coverage of the sputtered Cu seed (~10 nm thick Cu films on the field of the wafer) results in void formation in the Cu-filled trenches/vias, especially at the bottom and lower sidewalls of the trenches.

Annealing of the seed/barrier layer system at 300°C in vacuum ($<10^{-7}$ torr) for 15 min improves adhesion of the seed layer. Increasing the thickness of the sputtered Cu seed up to 30 nm on the field of the wafer leads to sufficient Cu seed layer deposition on the lower sidewalls and at the bottom of trenches/vias to initiate electroless Cu deposition in sub-half-micron vias of high aspect ratios (>4).

Step coverage of the electroless Cu films was found to be about 100% which results from uniform initiation and conformal growth in sub-half-micron trenches (Figure 3.26) as well as in features of larger sizes (Figure 3.27).

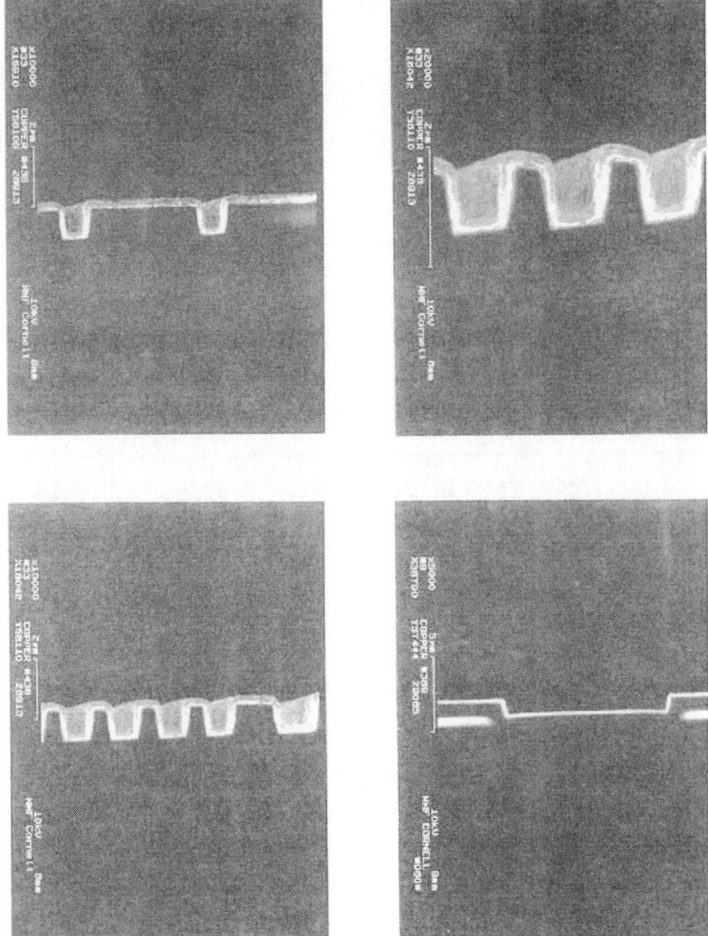

Figure 3.27 Cleaved SEM cross sections of wide (>1 μm) trenches plated with ~0.75 μm electroless Cu seeded by collimated sputtering Ti(25 nm)/TiN(40 nm) adhesion/barrier layer + conventional sputtering Cu(35 nm)/Al(20 nm) seed/protection layer.

We did not observe any difference in electroless Cu step coverage as a result of different materials being used for a diffusion barrier/adhesion promoter (i.e., Ti/TiN, Ti, or Ta) when sputtered Cu/Al films were used as a catalytic layer.

We found that electroless Cu line resistivity for 0.45 μm wide trenches of aspect ratio 3:1 was about 2.8 μΩ cm for as-deposited copper. Thin electroless Cu films of about 0.22 μm thickness need to be deposited to fill 0.45 μm wide trenches. Due to small grain sizes (<0.2 μm) the resistivity of electroless Cu lines of 0.4 μm width increases up to 2.8 μΩ cm (Figure 3.28). This result is in good correlation with the measurements of the resistivity for as-deposited 0.2 μm thick blanket electroless Cu films (Figure 3.23). The resistivity of copper lines/films increases with decreasing line width/film thickness due to decreasing grain size as well as increasing surface scattering contribution to the line resistance.

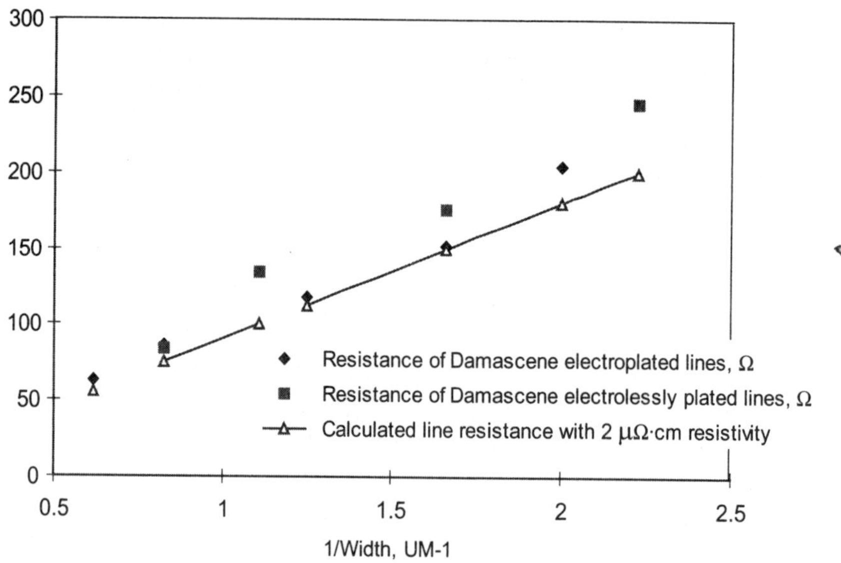

Figure 3.28 The resistivity of Damascene as-deposited electroless Cu lines vs. 1/width.

3.3.10   Electroless Cu deposition on contact displacement Cu seed

Blanket electroless Cu can be plated on a Cu catalytic layer deposited by contact displacement. Integration issues related to electroless Cu deposition on a contact displacement Cu catalytic layer for seed layer formation include poor adhesion of the contact displacement Cu catalytic layer to diffusion barrier, base metal consumption, and plating rate dependence on the via sizes. However, blanket electroless Cu deposition into trenches on a contact displacement Cu catalytic layer results in electroless Cu film delamination from the sidewalls of the trenches. Contact displacement Cu deposition time on a TiN(40 nm) diffusion barrier was found to be a crucial parameter which influences electroless Cu via filling capability. For example, voids were found in Cu-filled vias if contact displacement Cu deposition time was increased by 1 min in a solution containing 200 g/l $CuSO_4 \cdot 5H_2O$, 20 ml/l HF (48%), and 0.1 g/l surfactant. This results from etching (consumption) the TiN layer.

3.3.11   Electroless Cu deposition for PVD Cu seed repair

The seed layer deposited by PVD is discontinuous in high aspect ratios features thereby leaving areas of barrier exposed which leads to voids in these structures during the electroplating fill step. Electroless plating has been used to repair discontinuous PVD seed layer prior to the electroplating fill step.[57–60] Complete gap fill of sub 0.1 μm features of AR 10:1 with electroless Cu seed repair and electroplated Cu fill was demonstrated. Comparison of electrical results show similar parametric performance between wafers processed through the standard process steps with PVD seed layer and copper electroplating fill vs. electroless PVD Cu seed repair followed by Cu electroplating fill.

3.4 Conclusions

The electroless metal deposition process for liner/barrier and cladding layers as well as for Cu seed and PVD seed repair applications potentially provides many advantages, among which are the selectivity to dielectrics, possibility of an amorphous barrier alloy deposition, alloy properties enhancement by the addition of a third component, high conformity, low resistivity, and low cost.

Electrochemistry and thermodynamics of electroless Co were reviewed. The mixed potential theory was verified for the autocatalytic chemical reduction reaction of CoWP alloys with hypophosphite. It was carried out by comparison of the average plating rate and the mixed potential determined during the deposition process with those that were calculated from the electrochemical measurements. The mechanism of tungsten-induced co-deposition is confirmed on the basis of the electrochemical and XPS measurements. Proceeding from Pourbaix's diagrams, the thermodynamic feasibility of the CoWP electroless deposition is illustrated.

The electroless CoWP barrier offers low thin-film resistivity (20 to 25 $\mu\Omega$ cm), good step coverage in the high aspect ratio trenches–vias and selective cladding layer formation on the top of in-laid 0.2 μm Cu lines. Good extendibility of the electroless ternary alloy deposition in very narrow features with thin films of extremely small, nano-scale dimensions was observed. A conformal 10 nm thick CoWP layer was formed on the sidewalls of 30 to 40 nm wide seam of an aspect ratio about 5:1 on the top of 0.4 μm wide in-laid Cu line.

Electroless reaction of ternary CoWP alloy deposition leads to a uniform growth in the initial stage of the film deposition. AFM images of the electroless CoWP film surface and sectional analysis showed small grain sizes of CoWP films. Thermal stability of electroless CoWP/Cu films investigated by RBS, AES, SIMS, and MOS capacitors was demonstrated up to 500°C.

The proposed microstructure is of ε-Co grains, ~10 nm in diameter in which the grain boundaries are enriched by elemental phosphorus (8 to 10 at. %) and tungsten (2 to 3 at. %). The improved diffusion barrier properties relative to PVD cobalt may be explained by one of the following mechanisms:

1. Phosphorus and tungsten impurities creating a "stuffed" boundary, blocking the high rate of copper diffusion through the grain boundaries
2. ε-Cobalt grains embedded in an amorphous matrix, thus eliminating the short-circuit paths of grain boundaries in regular polycrystalline materials

The phosphorus and tungsten impurities stabilize the ε-cobalt phase, postponing the phase transition $\varepsilon \rightarrow \alpha$-Co to above 500°C, a significantly higher temperature compared to the transition temperature of 422°C in bulk cobalt.

Excellent selectivity, diffusion properties (preventing copper and oxygen diffusion), and good adhesion of CoWP thin films allow electroless cobalt to be used as the barrier as well as the cladding/passivation layer to directly deposit and pattern low-*k* film on Damascene Cu lines plated with electroless CoWP (with no high-*k* SiN or SiC passivation and etch stop layers) in order to improve electromigration resistances and reduce effective dielectric constant of Damascene copper metallization.

Electroless Cu deposition processes are controlled by charge-transfer reactions (i.e., kinetics) on a flat surface, while the diffusion of reactants (i.e., mass-transport) in narrow and deep features such as sub-half-micron trenches/vias of high aspect ratios (>3) is the rate-limited step of electroless Cu deposition.

An electroless Cu seed deposition process was developed including blanket electroless Cu deposition on a sputtered Cu catalytic layer with an Al protection layer and a blanket electroless Cu deposition on a contact displacement Cu catalytic layer with TiN diffusion barrier.

Electroless Cu films deposited at up to 120 nm/min plating rate in solutions with optimized plating chemical environment exhibit low resistivity ($\rho < 2$ $\mu\Omega$ cm for as-deposited Cu films), low surface roughness ($R_a \sim 10$ to 15 nm for 1.5 μm thick films) and good electrical uniformity (STD DEV < 3%).

Excellent step coverage (up to $\sim 100\%$) of electroless Cu films was demonstrated for blanket electroless Cu plating in sub-half-micron trenches (down to 0.3 μm width) of high aspect ratios for electroless Cu deposition on sputtered Cu films with an Al protection layer.

Excellent uniformity, conformality, and low-surface roughness of electroless Cu films allow these films to be used as the seed layer as well as to repair discontinuous PVD Cu seed in order to provide void-free sub-0.1 μm gap fill.

Acknowledgments

The authors are grateful to Professor E. Gileadi, Dr. A. Inberg, Dr. E. Sverdlova, and Dr. L. Zhu from Tel-Aviv University, and Dr. L. Burshtein of the Wolfson Applied Materials Research Center, Tel-Aviv University, for their technical support and comprehensive discussions.

References

1. A. Brenner and G. Riddell, *Proc. Am. Electroplaters Soc.*, **34**, 156 (1947).
2. V.V. Sviridov, T.N. Vorob'eva, T.V. Gaevskaya, and L.I. Stepanova, *Chemical Metal Deposition from Aqueous Solutions*, V.V. Sviridov, Editor, University Publishers, Minsk, in Russian (1987).
3. Electroless Plating—Fundamentals and Applications, G.O. Mallory and J.B. Haydu, Editors, American Electroplaters and Surface Finishers Society, Orlando, FL (1990).
4. C.H. Ting and M. Paunovic, *J. Electrochem. Soc.*, **136**, 456 (1989).
5. V. Dubin, *J. Electrochem. Soc.*, **139**, 633 (1992).
6. J.S.H. Cho, H.-K. Kang, S.S. Wong, and Y. Shacham-Diamand, *MRS Bull.*, **18**, 31 (1993).
7. Y. Shacham-Diamand and V. Dubin, *J. Microelectron Eng.*, **33**, 47 (1997).
8. V.M. Dubin, Y. Shacham-Diamand, B. Zhao, and P.K. Vasudev, *J. Electrochem. Soc.*, **144**, 898 (1997).
9. D. Edelstein et al., International Electronic Device Meeting Technical Digest, IEEE, Washington, D.C., p. 773 (1997).
10. V.M. Dubin et al., International VLSI Multilevel Interconnect Conference, VMIC Catalog No. 97 IMSIC-107, Santa Clara, CA, p. 69 (1997).
11. S.-Q. Wang, *MRS Bull.*, **19**, 30 (1994).

12. M. Paunovic, P.J. Bailey, R.G. Schad, and D.A. Smith, *J. Electrochem. Soc.*, **141**, 1843 (1994).
13. E.J. O'Sullivan, A.G. Schrott, M. Paunovic, C.J. Sambucetti, J.R. Marino, P.J. Bailey, S. Kaja, and K.W. Semkow, *IBM J. Res. Dev.*, **42**, 607 (1998).
14. V. Dubin, Y. Shacham-Diamand, B. Zhao, P.K. Vasudev, and C.H. Ting. U.S. Patent 5695810. Dec. 9, 1997.
15. S. Lopatin, Y. Shacham-Diamand, V. Dubin, and P.K. Vasudev, in Proceedings of the 14th International VLSI Multilevel Interconnection Conference, VMIC, p. 219, June 10–12 (1997).
16. T. Osaka and T. Yokoshima, 201st ECS Meeting, Philadelphia, PA, May 12–17, Abstract 509 (2002).
17. A. Kohn, M. Eizenberg, Y. Shacham-Diamand, and Y. Sverdlov, *Mater. Sci. Eng., A*, **302**, 18 (2001).
18. Y. Shacham-Diamand, B. Israel, and Y. Sverdlov, *J. Microelectron Eng.*, **55**, 313 (2001).
19. V.M. Dubin et al., IEEE 2001 International Interconnect Technology Conference, San Francisco, June 4–6, IEEE Catalog Number: 01EX461, p. 271 (2001).
20. P.L. Pai and C.H. Ting, *IEEE Electron. Device Lett.*, **10**, 423 (1989).
21. C.H. Ting and M. Paunovic, *J. Electrochem. Soc.*, **136**, 457 (1989); C.H. Ting, M. Paunovic, P.L. Pai, and G. Chiu, *J. Electrochem. Soc.*, **136**, 462 (1989).
22. Y. Shacham-Diamand, *J. Micromech. Microeng.*, **1**, 66 (1991).
23. Y. Shacham-Diamand, V. Dubin, C.H. Ting, B. Zhao, P.K. Vasudev, U.S. Patent 5830805, Nov. 3, 1998.
24. Y. Shacham-Diamand, V. Nguyen, and V. Dubin, U.S. Patent 6065424, May 23, 2000.
25. M. Saito, *J. Met. Finish. Soc. Jpn.*, **16**, 447 (1965); M. Saito, *J. Met. Finish. Soc. Jpn.*, **17**, 14 (1966).
26. M. Paunovic, *Plating*, **55**, 1161 (1968).
27. C. Wagner and W. Traud, *Z. Elektrochem.*, **44**, 391 (1938).
28. M. Paunovic, *Plat. Surf. Finish*, **70**, 62 (1983).
29. Y. Okinaka, *J. Electrochem. Soc.*, **120**, 739 (1973).
30. I. Ohno, O. Wakabayashi, and S. Haruyama, *J. Electrochem. Soc.*, **132**, 739 (1973).
31. I. Ohno and S. Haruyama, *Surf. Technol.*, **13**, 1 (1981).
32. I. Ohno and S. Haruyama, *J. Met. Finish. Soc. Jpn.*, **20**, 979 (1981).
33. F. Pearlstein and R. F. Weightman, *Plating*, **54**, 714 (1967).
34. *Physico-chemical Origins of the Chemical Reduction of Cobalt*, K.M. Gorbunova, Editor, Science Publishing, Moscow, in Russian (1974).
35. A. Brenner, *Electrodeposition of Alloys*, Vol. 1–2, Academic Press, New York (1963).
36. E.J. Podlaha and D. Landolt, *J. Electrochem. Soc.*, **144**, 1672 (1997).
37. Y.S. Kim, S. Lopatin, and Y. Shacham-Diamand, in Proceedings of the IEEE. Cornell Conference on Advanced Concepts in High Speed Semiconductor Devices and Circuits, p. 192, Aug. 4–6 (1997).
38. S. Kim, S. Lopatin, and Y. Shacham-Diamand, in MRS Proc. ULSI XIII, p. 445 (1997).
39. S. Lopatin, Y. Shacham-Diamand, V. Dubin, P.K. Vasudev, J. Pellerin, and B. Zhao, in MRS Proc. Electrochemical Synthesis and Modification of Materials, Vol. 451, p. 463 (1996).
40. M. Pourbaix, *Atlas of Electrochemical Equilibria*, Pergamon Press, London (1966).
41. A.J. de Bethune and N.A. Swendeman Loud, Table of Standard Aqueous Potentials and Temperature Coefficients at 25°C, from *Encyclopedia of Electrochemistry*, C.A. Hampel, Editor, Reinhold Publishing, New York (1964).
42. A.J. Bard, R. Paresons, and J. Jordan, Standard Potentials in Aqueous Solutions, IUPAC (1985).
43. V.I. Yudevich, L.B. Sokolov, and B.I. Ionin, *Russ. Chem. Rev.*, **49**, 46 (1980).
44. N. Petrov, Y. Sverdlov, and Y. Shacham-Diamand, *J. Electrochem. Soc.*, **149**, C187 (2002).
45. L. Cadorna, P. Cavallotti, and G. Salvago, *Electrochim. Metallorum.*, **1**, 177 (1966).

46. B. Kaznachey, M. Shuvalova, and G. Sadakov, in *Trudy VNIITR*, Proceedings of the Russian Scientific-Research Institute of Television and Broadcasting, **3**, 187 (1966).
47. Y. Shacham-Diamand, Y. Sverdlov, and N. Petrov, *J. Electrochem. Soc.*, **148**, C162 (2001).
48. Y. Shacham-Diamand et al., in Electrochemical Technology Applications in Electronics III, L.T. Romankiw, T. Osaka, Y. Yamazaki, and C. Madore, Editors, PV 99–34, p. 102, The Electrochemical Society Proceedings Series, Pennington, NJ (2000).
49. S. Lopatin, Y. Schacham-Diamand, V. Dubin, Y.S. Kim, and P.K. Vasudev, in Multilevel Interconnect Technology II, M. Graef and D.N. Patel, Editors, Vol. 3508, p. 65, International Society for Optical Engineering Series, Santa Clara, CA (1998).
50. Y. Shacham-Diamand, V.M. Dubin, and M. Angyal, *Thin Solid Films*, **262**, 93 (1995).
51. R.M. Lukes, *Plating*, **51**, 1066 (1964).
52. Y. Shacham-Diamand, *J. Micromech. Microeng.*, **1**, 66 (1991).
53. J.J. Pesek and O.R. Melroy, *J. Phys. Chem.*, **89**, 4338 (1985).
54. A. Molenaar, M.F.E. Holdrinet, and L.K.H. van Beek, *Plating*, **61**, 238 (1974).
55. A. Hung and K.-M. Chen, *J. Electrochem. Soc.*, **136**, 72 (1989).
56. V.L. Kubasov and S.A. Zarezkij, *Fundamentals of Electrochemistry*, Khimiya Publishers, Moscow, Russia (1976).
57. V.M. Dubin and C.H. Ting, U.S. Patent 5913147, June 15 (1999).
58. L.L. Chen, U.S. Patent 6197181, March 6 (2001).
59. S. Gandikota et al., IEEE 2001 International Interconnect Technology Conference, San Francisco, June 4–6, IEEE Catalog Number: 01EX461, p. 30 (2001).
60. T. Andryushchenko and J. Reid, IEEE 2001 International Interconnect Technology Conference, San Francisco, June 4–6, IEEE Catalog Number: 01EX461, p. 33 (2001).
61. M.-A. Nicolet, *Thin Solid Films*, **52**, 415–443 (1978).
62. K. Hono, D.E. Laughlin, *J. Magn. Magn. Mat.*, **80**, L137–L141 (1989).
63. K. Ishida and T. Nishizawa, *Bull. Alloy Phase Diagrams*, **11**, 555–560 (1990).
64. B. Cabral, Jr., K. Barmak, J. Gupta, L.A. Clevenger, B. Arcot, D.A. Smith, and J.M.E. Harper, *J. Vac. Sci. Technol. A*, **11**, 1435–1440 (1993).
65. K.I. Chopra, *Thin Film Phenomena*, McGraw-Hill Book Company, NY, p. 369 (1969).
66. I. Barak, M.Sc. Thesis, Tel-Aviv University, Tel-Aviv, Israel (1999).
67. A. Kohn, M. Eizenberg, Y. Shacham-Diamand, B. Israel, and Y. Sverdlov, *J. Microelectron Eng.*, **55**, 297 (2001).
68. Y. Segawa et al., Adv. Metallization Conf (AMC-2001), Oct. 9–11, Montreal, Canada (2001).

4 Alternative materials for ULSI and MEMS metallization

Yosi Shacham-Diamand, Nathan Croitoru, Alexandra Inberg, Yelena Sverdlov, Valery Dubin, and Vadim Bogush

4.1 Introduction

The rapid development of integrated circuits is the enabling technology for the modern miniaturized electronics systems revolution. This is achieved by scaling of the critical dimensions while increasing the chips size.[1] In the year 2004 it is expected that transistors with 70 nm long gates will be available for random logic circuits. The corresponding interconnects scaling raises technological and material problems. Interconnects scaling for conventional Al metallization technology is a problem because of the narrow-line resistivity increase and electromigration problems. One way to avoid rapid resistivity increase while downscaling the lateral dimension of interconnects is keeping their vertical dimensions, for example, increasing the aspect ratio (height/width) of the features of the metal lines. However, such an approach increases cross-talk, delay, and noise as the distance between the lines decreases[2] and their vertical dimension does not decrease accordingly. This is the reason why the investigation of new materials and methods for interconnects with higher electromigration resistance and lower resistivity for sub-100-nm lines is of great importance.

Multilevel interconnects technology is the key for signals routing in ultra-large-scale integration (ULSI) circuits. Typical interconnect schemes include adhesion layer, barrier layer, conducting layer, capping layer, and possibly anti-reflective coating. Currently, aluminum and copper are used for on-chip interconnects. Al metallization uses mostly physical vapor deposition (PVD) for the conductor and PVD or chemical vapor deposition (CVD) for the barriers. In Cu technology electroforming (i.e., electro- or electroless plating) seems to be the technology of choice for the conductors; and PVD, CVD, or electroless deposition for the barrier layers.

Sub-quarter-micron metallization is achieved by using embedded Cu technology (i.e., Damascene technology). There are some challenges that have to be met in sub-100 nm Damascene technology. Several physical phenomena, which appeared in this case, will lead to a resistivity (ρ) value higher than that of the bulk material.

One of the main reasons for the narrow line resistivity increase is the interface effect when the line lateral dimension (d) or vertical dimension (h) become smaller than the electron mean free path in the metal. In this case there is scattering of the free electrons on the walls of the films. Theoretical studies of the ρ dependence on dimensions (d or h) were made by Fuchs[3] and Sondheimer.[4] Two simplified analytical solutions of this relation for the case of entirely diffuse and specular scattering of the electrons on the walls of the film were obtained from the general relation. For entirely inelastic scattering and d or $h < \lambda$ (λ = mean free path of electrons), the relation is:[9]

$$\rho_0/\rho = 3k/4[\ln(1/k) + 0.4228] \tag{1}$$

where $k = d/\lambda$ or $h/\lambda < 1$.

In the case of both $k_1 = d/\lambda < 1$ and $k_2 = h/\lambda < 1$ the simplified equation is:[9]

$$\rho_0/\rho = 3/4(k_1 + k_2)\ln(2.414) + 0.414\,(k_2{}^2 + k_1{}^2)/4 \tag{2}$$

As seen in the case of a thin wire with both dimensions of cross section (d and h) smaller than λ, there appears a non-linear dependence on k_1 and k_2, different from relation 1.

In addition to the problem of electrons scattering on the walls, in the case of very thin films, the structural and morphological properties have to be considered. For example, the calculations of Fuchs were made for a perfect two-dimensional film but in the process of film formation, at the first monolayers clusters may appear, which will produce deviation from a perfect two-dimensional film. Also grains of different dimensions separated by variable distances may be formed as a function of method and parameters of deposition. Theoretical studies of polycrystalline grain structures, which are not taken into account in the Fuchs calculation, were added in several later publications.[5–7]

Deviations from the perfect two-dimensional film will have an essential influence on the ohmic resistance making it non-ohmic due to percolation, tunneling, or thermoionic emission.[8] The conductivity as a function of substrate coverage p and of p_c (the critical value of p when percolation appears), is expressed by the equation:[9]

$$\sigma = a(p - p_c)^s \tag{3}$$

where a is a fitting constant and s is the critical exponent (e.g., values of about 1.2 to 1.3 were obtained for s experimentally).

The technology of very thin film deposition usually introduces defects, which create films with resistivity higher than that of the bulk. Recently, intensive studies of electroless deposition for application in microelectronics and micro-electro-mechanical systems (MEMS) technology were performed.[10–19] Electroless deposition is relatively simple and offers high selectivity that allows self-aligned deposition and hence saves lithography steps. This method is also very versatile and can produce multi-component films. In addition, many metals and alloys can be produced with good quality thin films on high aspect ratio vias and trenches by electroless (autocatalytic) deposition. It should be noted that alloys can be deposited by sputtering, but it does not yield good step coverage and trench and via filling for high aspect ratio in the sub-100-nm technology.

Analyzing the deposition parameter (e.g., rate, type of solution, annealing) effects on the formation of two-dimensional films is useful to optimize the process in order to achieve the lowest possible resistivity. Taking into account that by scaling down the thin film metal thickness below 100 nm the film microstructure changes drastically, structural and morphological studies have to be made together with electrical characterization of the films. For example, measurements of resistivity dependence on temperature together with atomic force microscopy (AFM) studies may give information about the deviations from ohmic behavior of the film and may be used as a criterion for quality of coverage. Changing rate of deposition or introducing additives in solution may improve the coverage and reduce the deviation from a two-dimensional film.

In this work we look at using electroless (chemical) methods to deposit Cu and Ag alloys for ULSI and MEMS applications. Analysis of recent results indicate that electroless deposited Cu and alloys or Ag–W films are good candidates for ULSI interconnect technology. Cu and Ag alloys are under study for ULSI metallization. It is proposed that lightly doped metals will improve their reliability with minimal compromising of the conductivity. In the next sections we discuss the deposition methods and electrical and material properties of thin film Cu and Ag alloys deposited by the electroless method. We present Ag–W alloy and various Cu alloys. Finally we discuss the future of sub-100 nm technology in the light of recent scientific and technological advances.

4.2   Ag and Ag–W metallization

Silver has the highest bulk conductivity[20] and is being re-considered as a candidate for sub-100 nm metallization. Using pure Ag is a problem since it corrodes or tarnishes in air and in many solutions, does not adhere well to SiO_2 and most low-K dielectrics, and diffuses rapidly in silicon.[21] Those problems can be overcome by alloying the silver with tungsten. Ag–W can be deposited by electroless deposition which yields low-conductivity metal layers that do not corrode.

Several studies were performed to reduce the oxidation and to increase the reliability of Ag thin films deposited by sputtering,[22] evaporation,[23] and electroless[11–12] methods for application in ULSI interconnect. Reliable Ag films formed by electroless deposition were obtained by introducing tungsten into the silver film to improve the film's morphology, increase its corrosion stability, and decrease the thin film resistivity. The Ag–W film deposition is compatible with most packaging and ULSI interconnect applications. The Ag–W films have shown improved surface coverage, high resistance to corrosion in air, and lower electromigration than that for Al. The developed deposition parameters of Ag–W allowed good adhesion of Ag–W to Si and SiO_2[21] to be obtained. The electrical, optical, and mechanical properties of the Ag–W films are promising for interconnect applications and as barrier, anti-reflective, or protective layers.[11–12,21]

In this chapter we will describe the methods of Ag–W thin films, electroless (autocatalytic) deposition, and thin film material and electrical characteristics. The films are compatible with common very large scale integration (VLSI) and MEMS processing techniques. Finally, we refer to the properties of very thin films that can be integrated with advanced sub-100 nm interconnect technology.

The resistivity of relatively thick electroless Ag layers, $d > 300$ nm, is $\rho = 1.6 \times 10^{-6}\ \Omega$ cm which is close to that of bulk silver. This value drastically increases for film thicknesses below 300 nm and depends on the substrate nature, activation method, and process parameters. To find a way to lower the resistivity of very thin layers is one of the goals in advanced thin film technology. The morphology, structure, and electrical properties of the electroless Ag and Ag–W layers as a function of the deposition parameters and activation methods are presented here. The study enabled us to optimize layer resistivity and adhesion and to develop advanced deposition procedure.

We assumed that the large increase in thin-film resistivity was due to micro-defects that are caused by the proximity to the substrate and other scattering centers that appear at the first stages of deposition. The nature of the scattering should be studied and techniques to eliminate or reduce the scattering effect in ultra-thin films should be devised.

4.2.1 *Electroless Ag alloy*

The thin films were deposited by electroless method on silicon wafers with or without thermal silicon dioxide (SiO_2) layers using wet palladium[11,22–25] or sputtered metal seed activation.[26–27] Substrates were cleaned and etched prior to the deposition as described in Ref. 11. The wafers were rinsed between and after the step in deionized (DI) water.

The thin films were deposited from ammonium acetate[11,12] and benzoate[28–30] Ag complex based solutions. Hydrazine hydrate was used as a reducing agent in both baths. Na_2WO_4 was added to the silver electroless plating bath as a source of tungsten to prepare Ag–W films. Minute quantities of additives were introduced for improving the brightness and smoothness of the deposit.

The baths were designed to improve the deposition selectivity and achieve the best deposition on activated surfaces. The silver reduction was in the volume (homogeneous), on the activated and non-activated surfaces (e.g., cell walls). However, the deposition on the non-activated surfaces did not adhere to the substrate and was completely rinsed away in DI water. Unlike the ammonium-acetate solution, benzoic acid provides a much more stable deposition process when the homogeneous nucleation appears only after a relatively long time (more than 20 to 30 min).

Lower concentration of the complexing agent did not complex all the silver ions and this affected the Ag–W deposition rate and increased the film resistivity. A high content of the complexing agent in the solution causes the film conductivity to decrease, probably due to partial adsorption of the organic molecule or its parts, on the surface. An excess of the reducing agent in the electrolyte results in faster silver reduction in the volume of solution, grain size increase, and impairment of the films resistivity.

The concentration of the initial solution components, silver ions, complexing agent, and reducing agent was varied in order to optimize thin (<100 nm) Ag–W film resistivity. Ag–W films deposited from such solutions have a good adhesion both to the dry and wet activated SiO_2 and show quite good surface coverage.

The influence of a small amount of additives such as potassium nitrate, RE-610, and Triton X-100^{TM} on the film resistivity was observed. The temperature dependence of the electroless process kinetics and Ag–W film electrical properties was studied in the 18 to 40°C range using a temperature-controlled cell.

Deficiency of hydrazine hydrate, as well as the absence of Na_2WO_4, renders the metal deposition process on a SiO_2 surface impossible. Evidently, Na_2WO_4 is acting as a catalyst for silver reduction from the benzoate complex. This is probably the main reason why pure Ag film was not deposited on SiO_2 from this solution.

4.2.2 Ag–W films properties and characterization

4.2.2.1 Composition

The properties of electroless deposited layers were analyzed and characterized. The tungsten concentration in the film depends on the composition of the deposition bath. The relative amount of tungsten in the solid increases as the tungstate ion concentration in the solution increases. The concentration of tungsten in the solid obtained from ammonium acetate solution reached a maximum when the molar ratio $[WO_4^{2-}]/[Ag^+]$ was about unity. The data in Figure 4.1 are presented for solution with a 0.03 M silver ion concentration. We can see that as the molar concentration of the tungstate ions in the electrolyte reach that of the silver, the tungsten relative content in the film reaches about 3.1 at. % and becomes almost independent of the amount of tungstate ion in the bath. Oxygen was also present in the Ag–W film. The oxygen atomic concentration was approximately three times larger than that of the tungsten as shown in Figure 4.1.

For the benzoate solution three different areas can be selected on Figure 4.2: positive slope, saturation and further increase of the tungsten content. The dependence of the oxygen content as a function of the Na_2WO_4 amount in the electrolyte behaves differently from that for tungsten content. In the first region, below 0.01 M Na_2WO_4, both the tungsten and oxygen concentrations increased. In the second region, Na_2WO_4 concentration >0.01 to ~0.03 M, the tungsten quantity in deposit is saturated, but the oxygen content decreased. In the third region, >0.03 M Na_2WO_4, the amount of both W and O increased. We assume that there are differences in the mechanisms of the film deposition at low- and high-tungstate concentration. These differences are governed by catalytic effect of the sodium tungstate. For lower molar concentration of tungsten ions in the solution, porous Ag–W layers are formed with slow deposition rate. This results in sorption of oxygen into the deposit from the solution. Elevating the sodium tungstate concentration from 0.01 to 0.03 M increases

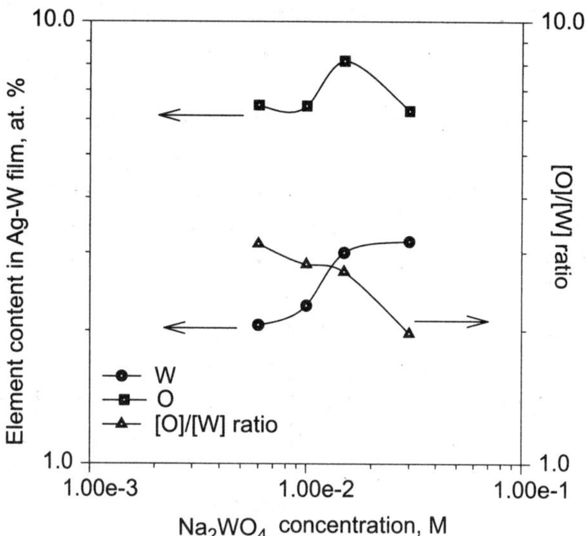

Figure 4.1 Tungsten and oxygen atomic concentrations and [O]/[W] ratio and in electroless Ag–W deposit as a function of the sodium tungstate molar concentration in the ammonium acetate solution (0.03 M $AgNO_3$).

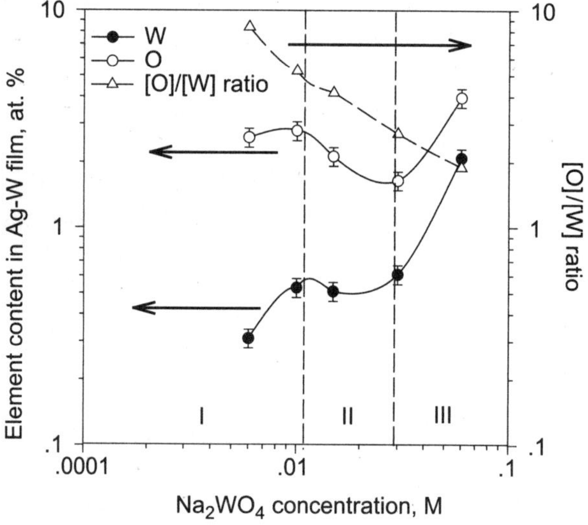

Figure 4.2 Tungsten, oxygen atomic content, and [O]/[W] ratio in electroless Ag–W film as a function of the sodium tungstate molar concentration in a benzoate solution (0.03 M $AgNO_3$).

the film deposition rate (Figure 4.3), and reduces the oxygen introduction into the silver. In this case, only oxygen bounded to deposited tungsten is included in the solid. The tungsten content in the film reaches 0.61 at. % and becomes almost independent of the tungstate ion concentration in the solution. The concentration of W and O in the film increases simultaneously as the [O]/[W] atom ratio decreases

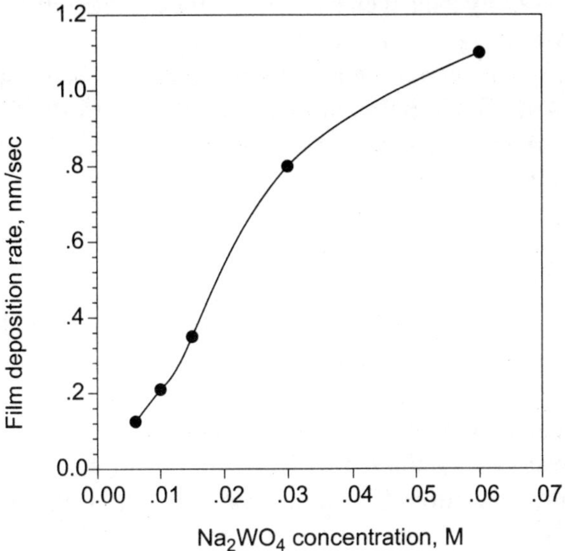

Figure 4.3   The film deposition rate vs. sodium tungstate concentration in a benzoate solution.

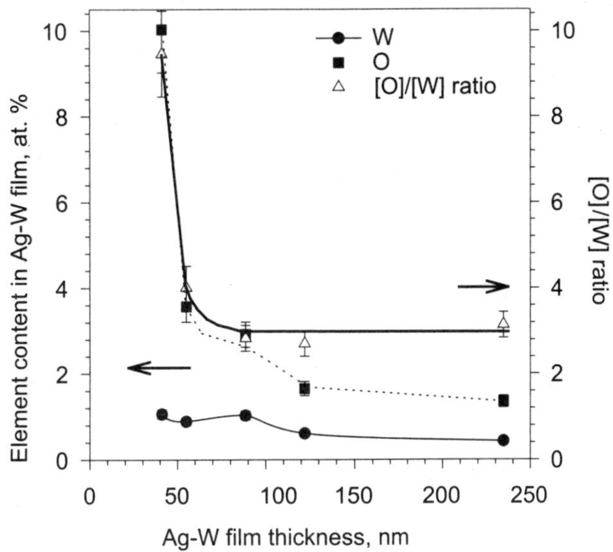

Figure 4.4   Tungsten and oxygen atomic concentrations and [O]/[W] ratio in electroless Ag–W deposit (Regime II) as a function of film thickness (benzoate solution).

continuously down to ratio of 2 when the Na_2WO_4 content increases above 0.03 M. In this concentration, apparently the tungsten is introduced into the deposit in WO_x form, where $x \geq 2$. Taking into account that at tungsten concentration higher than 0.03 M the deposition reaction became mostly homogeneous after a very short time, behavior of the film composition as a function of Na_2WO_4 content in the solution may explain the film resistivity data, which show that the conductivity of the Ag–W films decreases with lower oxygen concentration.

The tungsten and oxygen content in the deposits were also a function of the film thickness (Figure 4.4). The [O]/[W] ratio increases dramatically for thin layers. This may be due to adsorbed oxygen. It has a significant effect on the thin film resistivity and should be eliminated in the future.

4.2.2.2   Deposition rate

The diluted ammonium acetate solutions were found to be suitable to produce thin metal layers ($d < 300$ nm). The initial deposition rate was about 2.3 nm/sec for solutions with $AgNO_3$ concentration in the range of 0.008–0.05 M. However, the film thickness reached a maximum of 70 and 100 nm for the solutions with 0.008 and 0.01 M $AgNO_3$, respectively. For higher concentration of $AgNO_3$ saturation was not observed for 3 min deposition. The saturation was due to the presence of homogeneous deposition in the volume that competed with the heterogeneous, auto-catalytic electroless deposition on the substrate.

The deposition rate was found to be a function of the tungstate ion concentration. Adding 0.03 M of sodium tungstate to the solution reduced the deposition rate from 2.1 to 1.7 nm/sec.

As distinct from ammonium acetate solution, introduction of Na_2WO_4 to the benzoate deposition bath increased the Ag–W deposition rate (Figure 4.3). As a matter of fact, Ag heterogeneous deposition was not obtained without adding the tungstate ion to the solution. We assume that Na_2WO_4 catalyzes the Ag reduction reaction either directly or indirectly by one of its products after dissolution in water at high pH.

The film deposition rates from benzoate solution were significantly lower than that from the ammonium acetate one. The deposition rate was a strong function of the electrolyte pH and the temperature (Table 4.1).

Table 4.1   The deposition rate as a function of the pH and temperature.

pH	Deposition rate (nm/sec)	
	$T = 18°C$	$T = 40°C$
8.5	0.42	–
9.17	0.29	0.98
9.75	0.167	–

4.2.2.3   Morphology and texture

The incorporation of W into the deposit decreased the Ag–W film roughness slightly. The roughness varied between 7.5 and 8 nm and it was slightly lower than that of pure silver films (Figure 4.5). Increasing the tungsten concentration results in the improvement of the film uniformity, as observed by optical and electron

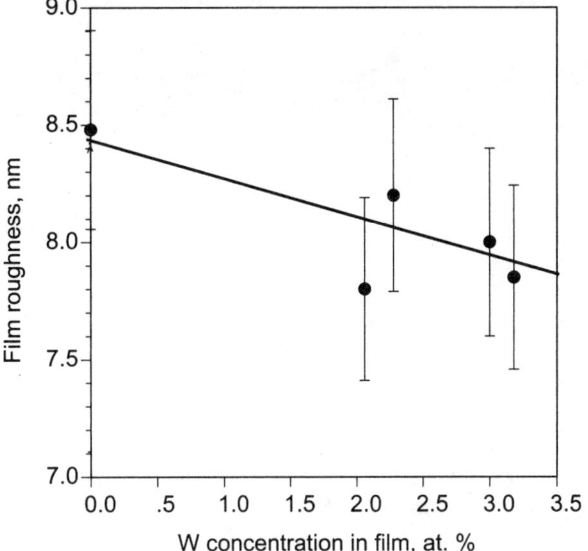

Figure 4.5   Thin film roughness as a function of the W concentration in the film deposited from the ammonia acetate solution.

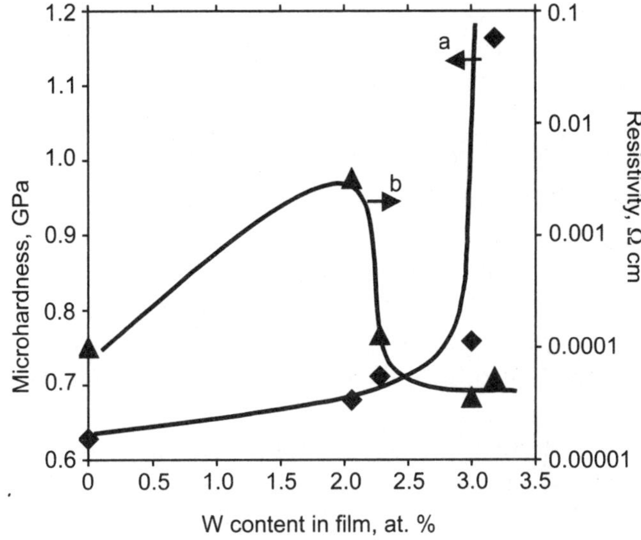

Figure 4.6   Ag–W thin film resistivity and microhardness as a function of W content in the film.

microscopy. The island structure after a short deposition time appears as smaller and denser islands for Ag–W while the islands of the Ag were larger and sparser. Ag–W films showed smaller dimension grains and better surface coverage than Ag films. The initial growth pattern of the islands in both samples is mostly three-dimensional crystallites (i.e., there is both lateral and vertical growth).

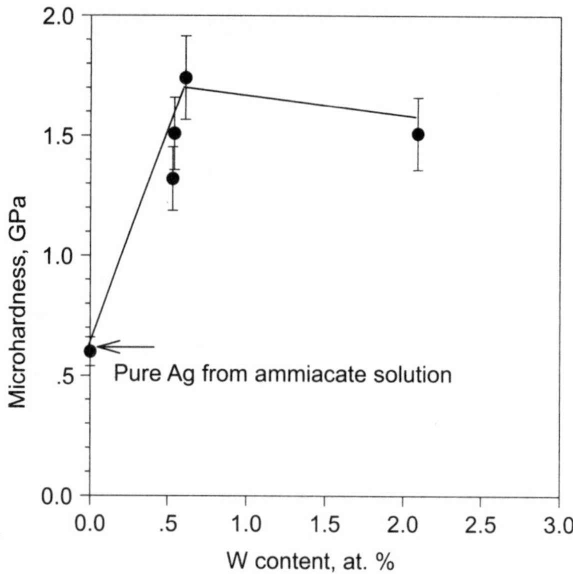

Figure 4.7   Microhardness of Ag–W thin film deposited from a benzoate solution.

Surface morphology of thin metallic films may affect their electrical, mechanical, and optical properties. The Ag–W morphology was found to be a function of the activation process, the deposition chemistry, and the deposition parameters.[24,26–27]

4.2.2.4   Mechanical properties

The microhardness of 90 nm thick Ag–W films from ammonium acetate solution with different tungsten content was measured. We found that generally the microhardness increases as the tungsten concentration rises (Figure 4.6). The variation was rather moderate up to 3 at. %. For higher concentrations the microhardness rises almost by a factor of two relative to pure silver. The graph includes also the resistivity of the same film. We will refer to the thickness and concentration dependences of the Ag–W film resistivity in the next section.

The mechanical properties of Ag–W from benzoate solution were tested for various film thicknesses and compositions. The thickness of the analyzed films was in the range from 90 to 120 nm. The layer microhardness increases when the W content in the deposit increases up to 1 at. % and achieves saturation for higher tungsten concentration (Figure 4.7). Such microhardness dependence on the tungsten introduction in the film may be due to including the harder component into the silver and improving the film morphology.

4.2.2.5   Resistivity

There are several main factors that influence Ag–W film resistivity: composition and the thickness of the film, roughness, coverage, grain sizes, and electrical uniformity.

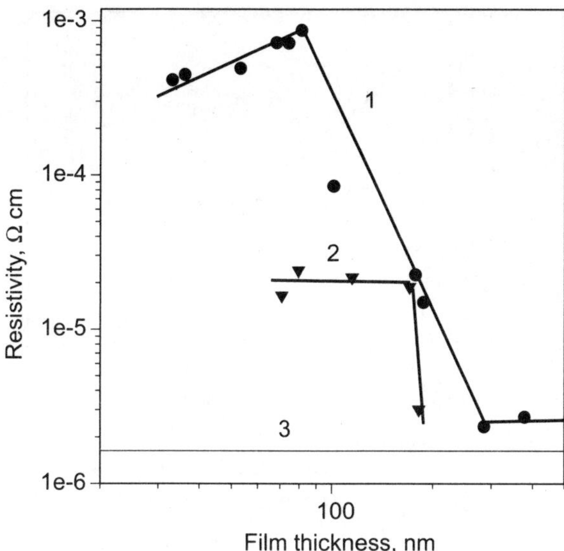

Figure 4.8   Silver film resistivity (ρ) as a function of film thickness (d) for pure Ag (1), Ag–W (1.5% W) films (2) and bulk Ag (3).

The effective resistivity (ρ) of Ag and Ag–W layers deposited from ammonium acetate solution was measured on Pd-activated Si wafers. The thin-film resistivity was measured as a function of thickness. Ag–W thin film resistivity showed a strong dependence on the $AgNO_3$ and the Na_2WO_4 concentrations, especially for below 100 nm thickness films. Ag and Ag–W films were deposited using solution with 0.01 M of $AgNO_3$ for Ag films, and with 0.05 M of $AgNO_3$ and 0.012 M of Na_2WO_4 for the Ag–W films. Figure 4.8 presents the resistivity, for both Ag and Ag–W, as a function of the layer thickness. The bulk resistivity of pure silver is presented for reference.

The resistivity (ρ) of thick Ag films ($d > 300$ nm) was in the range of 2 mΩ cm, as compared to ~1.6 mΩ cm for pure bulk silver. The resistivity of thinner Ag films increases abruptly from ~2×10^{-6} Ω cm at $d = 300$ nm up to ~8×10^{-4} Ω cm for $d = 80$ nm. The resistivity remains practically constant ($4 \times 10^{-4} \le \rho \le 8 \times 10^{-4}$ Ω cm) for thinner layers down to 40 nm.

As expected, the electrical resistivity was a strong function of the W concentration (Figure 4.6). The resistivity of Ag–W layer with 2 at. % of tungsten is higher than that of pure 90 nm thick electroless silver. Further increase of the tungsten amount up to 3 at. % reduces the electrical resistivity of the Ag–W.

The temperature coefficient of the resistivity (TCR, $\alpha = 1/R \times dR/dT$) for these films was also determined as a function of the film composition (Figure 4.9). The TCR (α) of thin electroless Ag was slightly higher than that of the bulk silver and decreases as the W content increases from $\alpha = 6.11 \times 10^{-3}$ K^{-1} for pure silver to $\alpha = 1.63 \times 10^{-3}$ K^{-1} for electroless Ag–W film with 3.2 at. % of tungsten.

For a benzoate bath the thin film resistivity was characterized as a function of all the solution parameters: temperature, pH, and composition.

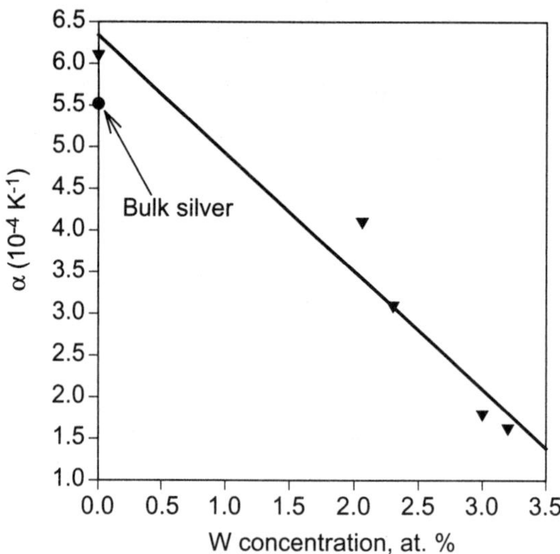

Figure 4.9 The thermal coefficient of resistivity (TCR) $\alpha = \dfrac{1}{R}\dfrac{\mathrm{d}R}{\mathrm{d}T}$ as a function of tungsten concentration in the solid.

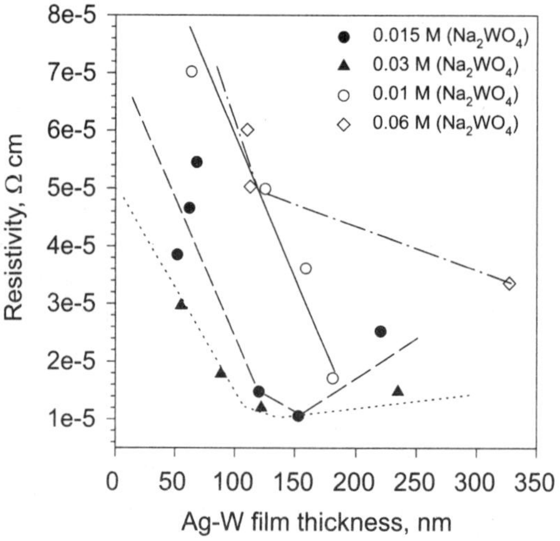

Figure 4.10 Thin film resistivity as a function of thickness for various sodium tungstate concentrations in the benzoate solution (0.03 M AgNO₃).

The film resistivity was high for low sodium tungstate concentration (less than 0.01 M) (Figure 4.10). Increasing the tungsten content in the solution up to 0.03 M decreases the resistivity, while the resistivity starts to rise again from concentrations above 0.03 M. Films deposited from solution with tungstate less than 0.01 M were discontinuous, whereas films obtained from electrolyte with 0.015 to 0.03 M of Na_2WO_4 had good coverage, smoothness, and good uniformity, which are the reasons for their higher conductivity. Higher concentration of sodium tungstate in the

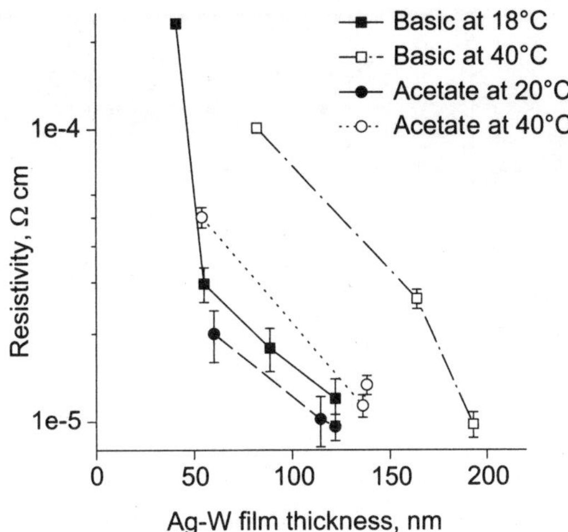

Figure 4.11 Ag–W resistivity as a function of thin film thickness for various deposition temperatures.

solution may cause a resistivity increase due to the higher W introduction in the solid as well as high film roughness. Note that higher tungstate ion concentration causes the deposition rate to increase and in some cases trigger homogeneous deposition in the volume that may introduce impurity inclusions into the deposit.

Another important parameter is the temperature that affects the nucleation rate, grain growth, and other parameters that are related to the resistivity. Ag–W layers were produced at room (18°C) and higher temperatures (up to 40°C). Further increase of the temperature results in unstable deposition following the rapid Ag reduction in the volume of solution. It also drastically increases the Ag grain sizes and decreases the layer density. All these factors decrease the film conductivity (Figure 4.11). The solution pH also had a significant effect (Figure 4.12). Generally, higher pH is required to achieve the Ag–W lower resistivity.

Additives are used to improve the solution stability and the thin film properties, such as adhesion, optical reflection and roughness, grain size, and coverage. It was shown that common surfactants and brighteners, like saccharin and sodium dodecyl-sulfate, did not improve the electrical resistivity of Ag–W electroless deposits.[30] Other additives, such as ammonia acetate, potassium nitrate, RE-610, and Triton X-100[TM] have a strong effect on the Ag–W resistivity and deposition rate at room temperature (Figure 4.13). Ammonium acetate and RE-610 improve Ag–W film morphology and lower resistivity. Re-610 serves also as a wetting agent that allows deposition onto a sub-micron feature in dual-Damascene structure when it is necessary to fill the sub-0.1 μm trenches with high aspect ratio.[30]

4.2.2.6 *Reflection of Ag–W films*

The reflection of the Ag–W films was studied in the infrared and visible ranges. Ag–W deposits were notable for higher reflection than similar Ag films. For example

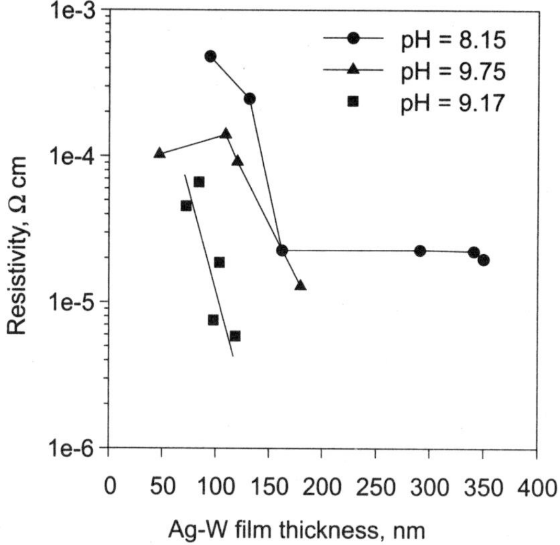

Figure 4.12 Ag–W resistivity as a function of film thickness for various benzoate solution pH values.

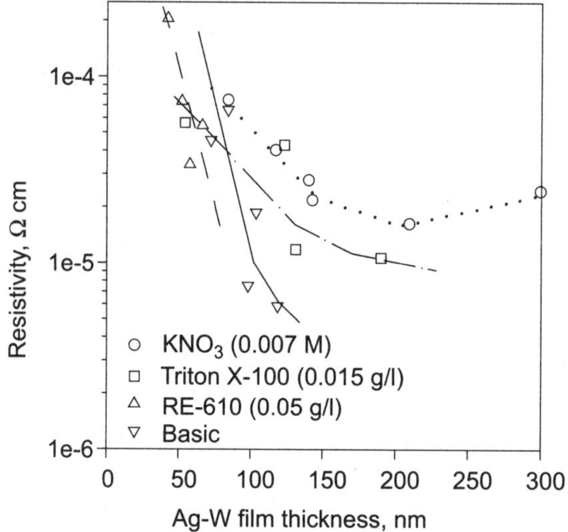

Figure 4.13 Ag–W resistivity as a function of the thickness for various additives in benzoate solution.

(Figure 4.14), we present the reflection of such a thin film deposited on Si in the 2 to 12 μm spectral range. The data were measured using Fourier transform infrared (FT-IR) spectrometry that was calibrated with a standard gold mirror. The Ag–W reflection increases when the tungsten content in the film rises up to a value of 0.61 at. %. Further increase of the W concentration causes the reflection to reduce. This result is correlated with structural and resistivity measurements. The high reflection of Ag–W films is probably due to the low film roughness.

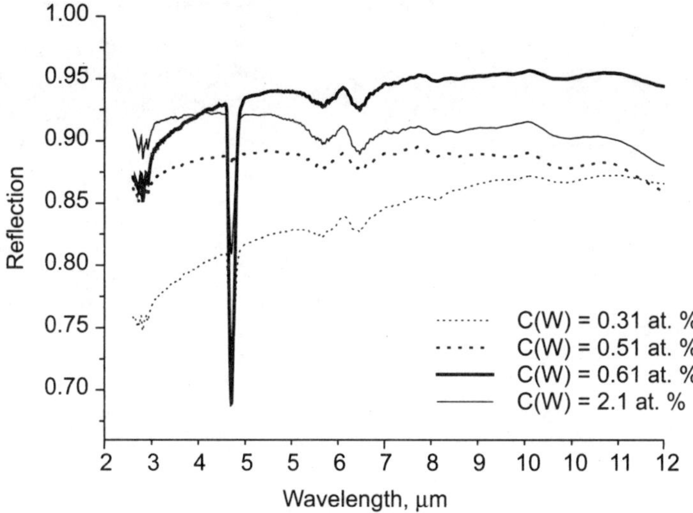

Figure 4.14 Infrared reflection of Ag–W films as a function of W concentration in the film.

4.2.2.7 MOS capacitors

MOS capacitors with electroless (ELS) and ion beam sputtered (IBS) Ag–W were formed in order to study the Ag–W integrity in silicon interconnect schemes. The main goal was to define the Ag diffusion and drift in silicon dioxide. Both capacitances vs. voltage $(C–V)$ and capacitance vs. time $(C–t)$ measurements were performed. The flat band voltage of ELS capacitors was 0.5 V before and after 250°C annealing while for IBS capacitors it was $\cong 1$ V as is and 0.25 V after the same annealing for 30 min.

Transient capacitance analysis indicates that there is no Ag effect on the minority carriers lifetime after 250°C annealing for 30 min. AES analysis of thermally stressed (TS) capacitors showed that the penetration of silver onto silicon dioxide was about a factor of 10 lower than that of Cu.

4.2.2.8 Annealing and corrosion stability

Vacuum annealing at temperatures up to 470°C was applied to study the annealing effect on the electrical resistivity and thermal stability of deposited Ag–W films. As seen in Figure 4.15, Ag–W resistivity decreases due to annealing at 350°C more than twice and achieves the value of 6 to 7 $\mu\Omega$ cm for films thinner than 60 nm. Thicker layers (~120 nm) have demonstrated a resistivity of 4 $\mu\Omega$ cm that is only twice higher than that of bulk silver at room temperature. We assume that the agglomeration of the grains and metal recrystallization are because the annealing decreases the Ag–W resistivity. Further heating up to the 470°C caused cracks of the film to appear and resulted in silver diffusion into the substrate. Thus, it was shown that vacuum annealing significantly improves Ag–W thin film resistivity. The Ag–W deposits have demonstrated corrosion stability in air up to 250°C.[24] The available value of the

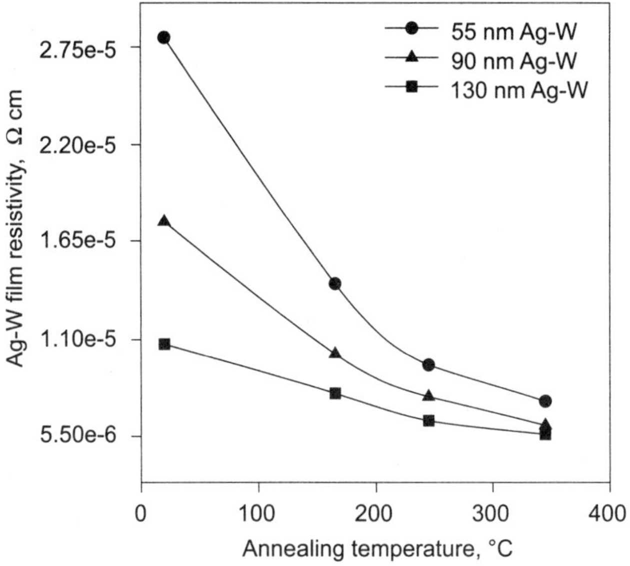

Figure 4.15 Annealing temperature effect on the Ag–W film resistivity.

Ag–W thin film (less than 100 nm) electrical resistivity as well as high corrosion stability make electroless a very promising technology for microelectronics application.

4.2.2.9   *Electroless Ag–W deposition — discussion*

The actual model of the simultaneous electroless co-deposition of silver and tungsten from aqueous solutions is not fully understood. The large difference in the reduction potentials of those two metal ions indicates that the exact chemical path that describes the deposition may take into consideration the existence of single metal complexes as well as complexes that include both metals.[24,28] X-Ray photoemission spectroscopy (XPS) and X-ray fluorescence spectroscopy (XRF) analyses of the films deposited from the silver solution, containing tungstate ions, indicate that the tungsten is included into the silver.

The chemical mechanism of tungsten co-deposition with silver is not yet resolved in this investigation. However, it may be assumed that the tungsten is included as a solid in the form of WO_3 or as a combination of tungsten oxides with various compositions. This assumption is based on previous XPS quantitative results which have shown that the oxygen concentration in the Ag–W film is three times that of the tungsten.[11] At the same time secondary ion mass spectrometry (SIMS) and transmission electron microscope (TEM) analysis have not shown any silver oxides. Current studies are focused on the thin film properties.

The introduction of the tungsten in the silver deposit significantly affects the solid film properties and enables thinner and smother layers to be deposited than these of the pure Ag with lower electrical resistivity. Optical and scanning surface morphology analyses have shown that adding tungsten to the film reduces the terrace-like growth of pure silver and smooths the surface at both nanometer and micrometer scale.[25] These structural particularities have an influence also on the electrical resistivity. Small quantities of W added to the silver (up to 2 at. %) lead to the disordering of film structure, and, as a result, the electrical resistivity increases in comparison to the silver bulk. However, the homogeneous tungsten distribution in film results in the structure ordering and decreases its resistivity. The electrical resistivity drastically reduces and stabilizes with the following increase of the W content (W $\geq$ 2.3 at. %). The microhardness value was also affected by the presence of W. It increased up to 1.2 GPa for the W content of ~2 at. %.

These data correlate with the results of the optical structural investigations. The W atoms or clusters are distributed quite homogeneously in the structure, particularly at the grain boundaries, which provides an increase of microhardness for the Ag–W composition.

Scanning probe analysis indicated that Ag–W films have much denser nucleation site concentrations and smaller roughness than those of pure silver. Such changes in structure affect the scattering process that determines electron transport in the solid. It may cause a decrease in the electrical resistivity by one order of magnitude when compared to pure silver film.

The resistivity temperature coefficient dependence on the solid-film composition may support this observation. The determined value of α/α_{bulk} for the Ag–W layer with 3.2 at. % of tungsten agrees well with the theoretical calculation for the case of a "thick" film where the thickness is larger than the free mean path using a fully inelastic scattering model.[24]

The higher reflection of the deposited Ag–W films in comparison with the pure silver sample may also be the result of the smaller roughness and better homogeneity of the surface.

Adding tungsten to the silver film improves the corrosion resistivity in air for at least up to 200°C. Such structural stability of the layers will allow their use as a corrosion protective or high reflecting stable coatings.

4.3　Cu alloys metallization

In the last decade Cu has been intensively studied as a candidate for improved interconnects. Recently the use of copper as the interconnect material has been demonstrated for VLSI using either electro-plating[31] or electroless plating.[10]

The basic advantages of Cu metallization in comparison to Al are the lower resistivity and higher electromigration resistance.[31–37] The existing electroplating and electroless Cu technology have shown essential advantages in filling of the trenches that is very important for dual-Damascene technology.[31] These methods, due to the high selectivity, allow self-aligned deposition and the possibility to produce multi-component systems.

One of the well-known Cu disadvantages, the very high diffusion in silicon, reduces the reliability of the circuits containing these devices. Supplementary barriers and cap films, used to avoid diffusion of Cu into silicon, made the metallization technology more complex and expensive. Recently, studies of Cu with Al and

                                    Yosi Shacham-Diamand et al.

Cr alloys thin film deposited by sputtering[36] and electroless deposited Cu with Sn, Zn, Ni, and Co alloys were performed.[38] The exact solution for Cu or Cu alloy applications in ULSI has not yet been obtained at this initial stage. The sputtered Cu alloy films did not show a sensible reduction of Cu surface diffusion and had problems with films roughness and stress. The electroless deposition of Cu with additional metals has shown increased resistivity and, as a positive element, improved adhesion to the substrate in comparison to the pure Cu.

Along with these applications electroless Cu was proposed for seed[39] and seed repair[40] application for dual-Damascene schemes in sub-0.1 μm geometries. Most of the recent works refer to pure Cu. However, electroplating of Cu alloys offers Cu-like films with electrical properties similar to pure Cu but with improved adhesion, corrosion resistance, and reliability. For instance, sputtered Cu–Cr, Cu–Al,[36] and PVD Cu (Sn, In, Zr)[41] were evaluated for Cu diffusion, adhesion, and electromigration reliability and showed improved characteristics as compared to unalloyed Cu. Binary alloys also affect the crystalline structure and the thin film texture, but this dependence has not been yet characterized.

In this chapter we present the results of the investigation of electroless Cu alloys with Ni, Co, Zn, and Sn deposition for ULSI applications. Alloyed Cu films are suggested to improve the reliability of Cu metallization. First, it is assumed that they can improve electromigration resistance and stress voiding resistance as compared to pure Cu. Second, copper alloys are expected to have better corrosion resistance and may reduce the need for a capping layer. Third, it is thought that some alloying elements enhance the copper adhesion to dielectrics and barriers and reduce Cu line delaminating.

4.3.1   Electroless Cu-alloy deposition

Alloying Cu with other elements is rendered possible by the electroless deposition technique. It is known from literature that Cu alloys with Ni,[42,38] Cd,[43–44] Au,[45] Pb, Bi, Sb, Zn,[42,46,38] or Sn,[38,44] can be obtained using electroless deposition. Resistivity of electroless deposited films was very high.[42] It should be stressed that the resistivity of electroless deposited pure Cu layers far exceeds the resistivity of bulk Cu ($1.7 \times 10^{-6} \, \Omega$ cm). Electrodeposition of the alloys is the result of simultaneous electro-reduction of two or more components from solution.[47–50] Processes of electroless co-deposition of metals obey the same thermodynamic rules as those for electro-deposition.[42,51,52] For electroless deposition, the kinetics of electron diffusion to the metal surface are governed by the rate of anodic oxidation of the reducing agent, which, in turn, depends on the catalytic properties of the surface. Thus, there are two ways to produce electroless alloys which are defined by the metal catalytic properties in a given bath with a specific reducing agent.[40] We define the basic metal, which forms most of the conducting metal matrix (i.e., Ag or Cu and the doping metal). The basic metal deposition process is autocatalytic. The deposition of the doping metal can be either autocatalytic or not. The doping metals can be reduced where the electrochemical conditions allow it or deposited by other chemical processes that may occur either on the deposited metal surface or at the solution.

1. Both metals are catalytic.

   In this case the deposition of both metals is catalyzed at the solid–liquid interface. The alloy composition depends on the partial cathodic reaction of each metal. The concentration of the doped metal varies in a very wide range. Electroless deposition of Cu–Ni alloy films with dimethylamineborane (DMAB) as a reducing agent,[53] and Cu–Au[45] with formaldehyde fall into this group. All these metals — Cu, Ni, and Cu, Au — are good catalysts for the oxidation of DMAB and formaldehyde, respectively.[45]

2. One metal is catalytic, and the other metal is non-catalytic.

   In this case the deposition of the doping (or alloying) metal does not occur due to an autocatalytic reduction at the liquid–solid interface. The metals that can be included under this classification are reduced by other ways. They can be deposited on the surface by electroplating as long as the electrochemical potential of the surface is sufficiently negatively charged during the reduction of basic metal. Alloys of Cu and Ni with catalytic inert metals (Cu–Cd, Cu–Pb, Ni–Re–P, and Ni–Re–B) are examples.[53]

In some cases the doping metal retards the reduction of the basic metal. In those cases the doping metal concentration in the alloy should be limited to prevent it from completely inhibiting the electroless deposition process. For example, Sb inhibits the deposition in electroless Cu-deposition with formaldehyde as a reducing agent. Sn and Zn act also as strong inhibitors for Cu deposition. Including Zn or Sn in the deposited films is best explained by the electrochemical reduction on the Cu surface. Most of the electrolessly deposited Cu alloys with formaldehyde or glyoxylic acid as the reducing agents belong to this classification as non-autocatalytic doping (e.g., Cu Zn, Sn, Co, Ni, Cd, Pb, Sb).

Generally, the higher the content of non-catalytic metal in the film the lower is the deposition rate. Inactive metal may precipitate as a discrete phase and block the catalytic surface of the active metal. In this case a considerable decrease of the deposition rate is observed. Such films typically contain low concentrations of the inactive metal. When both metals form the homogeneous phase, the formed film may contain a higher amount of the inactive metal. In some cases the depositing alloy may be more catalytically active than either individual metal. This effect increases the deposition rate of the alloy.

During electroless deposition of Cu–Me alloys with formaldehyde or glyoxylic acid as the reducing agents, Cu ions are reduced autocatalytically. The added inactive metal does not catalyze the oxidation of the reducing agent. Consequently, it cannot be reduced separately; it can only be co-deposited with catalytic metal. The alloying metal might be reduced according to the mechanism as is the main metal (i.e., with the electrons released from the oxidation of the reducing agent). The other possible mechanism is "jump-started" reaction due to the contact with the catalytic metal that has a higher redox potential.

The content of the diluted metal alloys may vary as a function of the metal concentration in the solution, concentration of the reducing agent, pH of the solution, and deposition temperature. Typically, an increase of the inactive metal content in the deposited film reduces the activity of the deposit surface and, in turn, inhibits the electroless deposition process.

Deposition rate and composition of the solid are determined by relative fluxes of various components to the reaction interface. The net flux of the component depends on:

1. Potential difference between the solution and the depositing metal
2. Transport of the species in the solution and on the metal–liquid interface
3. The reaction kinetics at the solid–liquid interface

This approach is correct when the various components are deposited via a pure electrochemical process on the cathode. Typically this takes place when the metals are in an ionic state in the solution and can be deposited independently. This is not always the case, especially for the co-deposition of the main metal with a refractory metal, and phosphorus or boron. In this case the co-deposition process is a mixture of an electrochemical process on the cathode and a chemical process in the solution. For example, the deposition of noble metal (Ni, Co, Cu, etc.) with refractory metal (W, Mo, Re, etc.) is assumed to have two parallel processes:[10]

1. Reduction of the noble metal complex with the complexing agent
2. Reduction of the noble metal complex, the alloying element, and the complexing agent

A cathodic current is balanced by an external current or by the anodic current due to the oxidation of the reducing agent.

4.3.2   Electroless Cu alloys

The seed layers used for electroless deposition were either a thin catalytic metal film (Co or Cu) on oxide or Pd activated Si wafer. The samples were prepared as follows:

1. *Thin metal films.* The sputtered multi-layers films were made of a 10-nm Al sacrificial layer on a 10-nm Cu or Co seed layer on a 10-nm Ti adhesion layer. The layers were deposited by ion-beam sputtering using three targets on the same chamber so the samples were in a high vacuum ($<10^{-7}$ torr) during the whole structure deposition. The multi-layer seed deposited on a 100 nm thermal oxide on Si substrate. Just prior to the Cu–Me deposition the wafers were etched in strong alkaline media in order to remove Al sacrificial layer.
2. *Si wafers* (*p*-type, 10 Ω cm resistivity). The pre-deposition procedure included cleaning and surface activation in Pd activation solution.[38]

The thin films were analyzed by XPS, Auger electron spectroscopy (AES) methods in a Physical Electronics PHI model 590A tool, and SIMS technique. The sheet resistance and resistivity of the deposited films were measured by In-Line-Four Point Probe and the thickness was determined by a Tencor Alpha-step 500 profilometer. The topography and average height profiles were obtained by Digital Instruments AFM (Auto Probe CP, Park Scientific Instrument). All the reported measurements were carried out at room temperature.

4.3.2.1　Electroless deposition of Cu with doping of Co and Ni for ULSI applications

Cu alloy layers (Cu–Ni, Cu–Co) 20 to 140 nm thick were deposited from several electroless baths. Their compositions and operation conditions are listed in Table 4.2.

Table 4.2　Cu-Ni (Co) electroless plating baths compositions and operation conditions.

Solution A (A1 – for Ni, A2 – for Co)		*Solution B*	
Chemicals	*Concentration* $mol \cdot dm^{-3}$ *(g/l)*	*Chemicals*	*Concentration* $mol \cdot dm^{-3}$ *(g/l)*
$CuSO_4 \cdot 5H_2O$	0.028 (7.0)	$CuSO_4 \cdot 5H_2O$	0.023 (6.0)
$NiCl_2 \cdot 6H_2O$	0.0084 (2.0)	$NiCl_2 \cdot 6H_2O$	0.014 (3.5)
(or $CoSO_4 \cdot 7H_2O$)		K–Na tartrate	(30)
K–Na tartrate	0.0797 (22.5)	NH_4OH	0.357 (50 ml)
Na_2CO_3	0.0189 (2.0)	NaOH	0.225 (9.0)
NaOH	0.1125 (4.5)	$NaBH_4$	0.026 (1.0)
CH_2O	0.321 (26 ml)	RE-610	0.004
pH	12.1–12.5	pH	12.6 – 12.9
Temperature	room	Temperature	40°C

　　Solutions A1 and A2: Electroless Cu–Ni and Cu–Co alloys. Copper sulfate was used as the source of Cu ions. K–Na tartrate was used as the complexing agent. Nickel chloride and cobalt sulfate were used as the Ni and Co ions supply. The reducing agent was formaldehyde. The solution pH was in the range of 12.1 to 12.5 at room temperature. Carbonate ions usually increase the stability of the tartrate solutions and the deposition rate.

　　Solution B: Electroless Cu–Ni alloy. Copper ions are complexed with tartrate. Nickel chloride was used as the source of Ni ions and sodium borohydride was exploited as the reducing agent.

　　Cu alloys were deposited on different seed layers as shown in Table 4.3. It was found that the doping level depends on seed layer type.

4.3.2.2　Electroless deposition of Cu–Zn and Cu–Sn alloys

The Cu–Sn and Cu–Zn layers were deposited from alkali electroless baths. Their composition and operating conditions are listed in Tables 4.4 and 4.5.

　　Electroless Cu–Zn alloy was obtained from a novel nitrate $Cu(NO_3)_2$ solution with Rochelle salt as a complexing agent and $ZnCl_2$ as a source of Zn ions. The reducing agent was formaldehyde and the solution pH was kept in the range 12.3 to

Table 4.3   Concentration of doped metal in deposits obtained from electrolytes
             A1 and A2 as a function of seed layers.

Cu alloy	Seed	Doping metal content in the film, at. %
Cu–Ni	Co	2.96
	Si	3.59
Cu–Co	Co	3.1
	Si	3.36
	Cu	2.43

12.5 at room temperature. In our work concentration of Zn ions in the electrolyte was
changed in the range from 0 to 0.0017 M. Concentration of Cu ions was constant at
0.062 M. Molar ratio [Zn ions]/[Cu ions] in solution of 0.027 is a maximum. Further
increase of Zn concentration in solution tends to termination of electroless
deposition.

Table 4.4   The composition of the Cu–Zn deposition solution.

Chemicals	Concentration g/l (M)
$Cu(NO_3)_2$	15 (0.0621)
K–Na tartrate	30 (0.106)
$ZnCl_2$	0.5 (0.0017)
CH_2O	200 ml/l (2.47)
Na_2CO_3	10 (0.094)
NaOH	20 (0.5)
pH	12–12.5
Temperature	Room

Table 4.5   Bath composition and operating conditions for electroless deposition
             of Cu–Sn films (solutions Sn-1 and Sn-2).

Chemicals	Concentration g/l,(M)	
	Sn-1	Sn-2
$CuSO_4 \cdot 5H_2O$	6.75 (0.027)	6.75 (0.027)
$SnCl_2 \cdot 2H_2O$	0.9 (0.004)	0–0.9 (0.004)
EDTA	11.7 (0.04)	0.1
NH_4OH	–	(4.0)
CH_2O	32 ml/l (0.4)	–
$NaBH_4$	–	(0.027)
NaOH	to pH = 13	to pH = 13
Temperature	40–45°C	55–65°C

Electroless Cu–Sn alloy was deposited from copper sulfate solutions (solutions
Sn1 and Sn2) that contain ethylenediaminetetraacetic acid (EDTA) as a complexing

agent and tin chloride as a source of Sn ions. The reducing agents were either form-aldehyde or sodium borohydride ($NaBH_4$). The solution pH was ~13 and temperature was 40 to 45°C for an electrolyte with formaldehyde as a reducing agent (electrolyte Sn1) and 58 to 63°C for an electrolyte with borohydride as a reducing agent (electrolyte Sn2). Sn concentration in the bath was changed in the range from 0 to 0.004 M, whereas Cu concentration in work solution remained constant (0.027 M). Sn concentration of 0.004 M (0.9 g/l of $SnCl_2$) is a limiting value for this electrolyte.

Cu–Zn and Cu–Sn alloys with different content of doped metal were obtained by electroless method. The Sn and Zn doped Cu layers had a grayish tinge.

It should be recalled that Zn and Sn compounds hydrolyze in aqueous solutions. This process can be described as follows:

$$SnCl_2 + H_2O \rightarrow Sn(OH)Cl + HCl \tag{4}$$

$$SnCl_2 + 6H_2O \rightarrow SnCl_2 \cdot 3Sn(OH)_2 + 6HCl \tag{5}$$

$$SnCl_2 + 2H_2O \rightarrow Sn(OH)_2 + 2HCl \tag{6}$$

To avoid this process $ZnCl_2$ and $SnCl_2$ should be dissolved in a small (but known) volume of hydrochloric acid before adding to the electrolyte. An excess of hydrochloric acid shifts the equilibrium in the previous reactions to the left with further formation of double salt:

$$SnCl_2 + 2HCl \rightarrow H_2SnCl_2 \tag{7}$$

Growth rate. The deposition rate depends on the electrolyte, the substrate material, and the concentration of the added elements (Co, Ni, Zn, Sn) in solution. The kinetics of Cu–Co and Cu–Ni electroless deposition are presented in Figure 4.16 for electrolytes A1 and A2. In our experiments the concentration of Co or Ni ions was 0.0084 M (2 g/l). Over the course of the first 12 min, the kinetics of the process for Co and Ni were different. The thickness of Cu–Co films was about 40 nm on Si substrate and on Co seed after 5 min of deposition. Thickness of Cu–Ni layers deposited for 5 min on Pd-activated Si and Co seed was about 70 and 100 nm, respectively. Thickness of Cu–Co and Cu–Ni films after 20 min of deposition was approximately the same for every seed. The reason for this effect may be due to the sparse distribution of the concentration of the reducing agent and establishing the saturation.

Figures 4.17 and 4.18 demonstrate the kinetics of Cu–Zn and Cu–Sn deposition processes. The kinetics of Cu–Sn and Cu–Zn electroless deposition are different. Thickness of Cu–Zn films was about 150 to 200 nm after 5 min of deposition (Figure 4.17). The deposition rate of Cu–Sn deposition process was much higher after some time. It can be seen from Figure 4.18 that the thickness of Cu–Sn layers was 350 to 400 nm on Cu and Co seed, but the thickness was much lower on silicon substrate

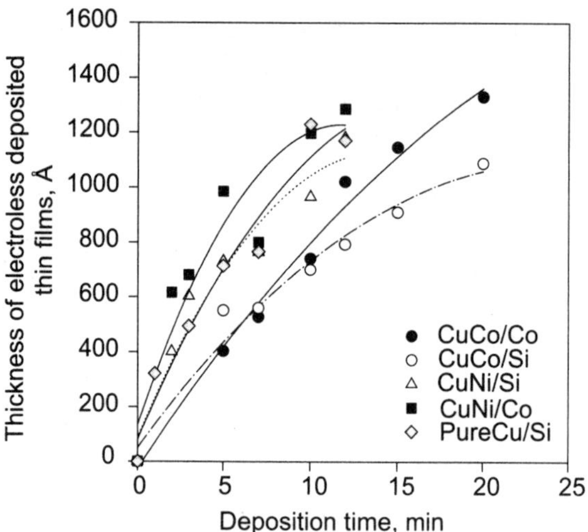

Figure 4.16 Thickness of electroless deposited Cu–Co and Cu–Ni layers as a function of deposition time.

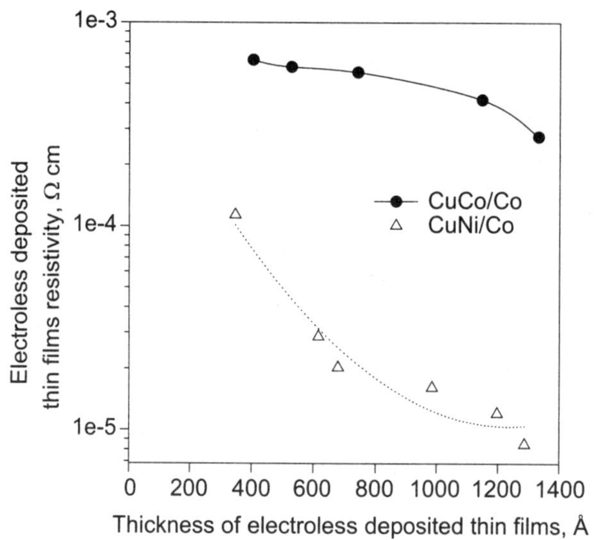

Figure 4.17 Resistivity of electroless deposited Cu–Co and Cu–Ni layers as a function of the film thickness.

(about 100 nm). Thickness of Cu–Zn layers during 5 to 6 min of deposition process was very similar for different substrates (Si, Cu, and Co seeds). Thickness of chemical deposited Cu–Zn films on Co seed and Si surface was very different after 12 min of deposition and reached 400 and 600 nm, correspondingly.

Passivation of the Cu. The best Cu films are found from tartrate electrolytes. In fact, passivation of Cu takes place in tartrate electrolytes. The passivation is assisted by the following factors:

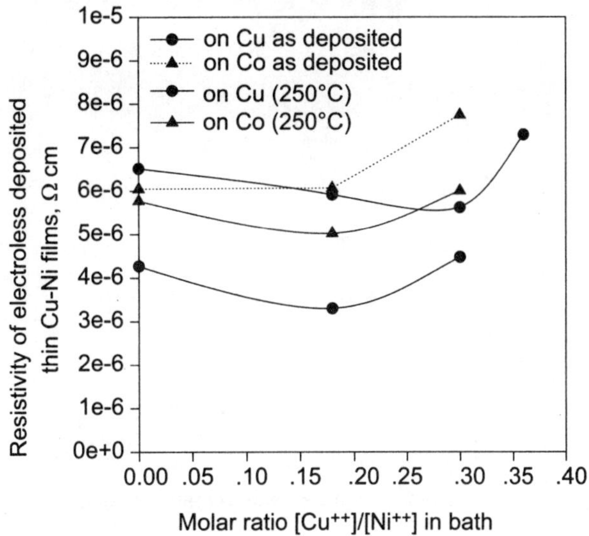

Figure 4.18 Resistivity of thin Cu–Ni films as a function of molar ratio $[Ni^{++}]/[Cu^{++}]$ in bath.

- low pH values (<12.3)
- exposure of Cu surface to the oxygen from the air
- increased temperature of the solution

Due to those effects the process of Cu–Me deposition from tartrate solutions might be performed at pH values 12.3 and higher (no more than 13) at room temperature and without stirring. In the solutions with EDTA as a complexing agent, passivation of the Cu surface was not observed. The addition of the foreign ions, for instance, Ni or Co, reinforces the Cu passivation in tartrate electrolytes. Passive Cu films can be reduced to pure Cu by formaldehyde in strong alkaline solutions, for example, 1.4 M NaOH and 0.3 M formaldehyde. It was found that chemically and electrochemically reduced or passivated layers consist of Cu_2O. In a matter of minutes the thickness of the passive layer may reach a magnitude of 0.01 to 0.03 μm.[42] Cu_2O is produced as a result of Cu^{2+} reduction from the solution rather than by the oxidation of the surface.

Comment. It is pertinent to note that the decrease of pH value to 11 in Cu electrolyte with Ni or Co ions involves a gel formation. It is assumed that the polymerization process takes place with participation of formaldehyde and Ni or Co compounds.

4.3.3 Material properties and characterization

Resistivity. The resistivity and average resistance of the obtained films was measured by In-Line Four Point Probe. In Table 4.6 we summarize the data on the effect of composition on resistivity. The resistivity of deposited films depends on the film thickness. Figures 4.19 and 4.20 demonstrate average resistance and resistivity of electroless obtained Cu–Sn films as a function of film thickness. We can see here

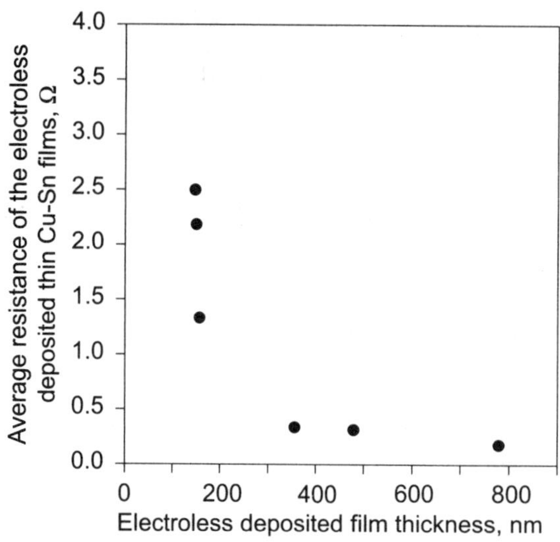

Figure 4.19 Average resistance of electroless deposited Cu–Sn films as a function of film thickness. Sn concentration of electrolyte is 0.9 g/l.

that the resistivity increases drastically with decreasing film thickness below 200 nm. Resistivity of Cu–Sn layers more than 200 nm thick remains unchanged, about 1.5×10^{-5} Ω cm. The resistivity of Cu–Zn films depends inversely on thickness as opposed to Cu–Sn layers. Figure 4.21 shows the effect of Cu–Zn film thickness on

Table 4.6 The resistivity and resistivity increase per 1 at. % alloy in units of [$\mu\Omega$ cm at. $\%^{-1}$] of electroless films as a function of the ratio [Cu^{2+}]/[Me^{n+}] in the solution.

Solution	Film type	Thickness (nm)	Seed	ρ, $\mu\Omega$ cm	$\Delta\rho$ at. $\%^{-1}$, $\mu\Omega$ at. $\%^{-1}$
Cu–Zn	Pure Cu	230	Co	2.62	
Cu–Zn	Cu–Zn (2.95%)	400	Co	7.5	1.65
Cu–Sn 1	Pure Cu	160	Co	5.79	
Cu–Sn 1	Cu–Sn (1.5%)	300	Co	9.3	1.38
A	Pure Cu	160	Co	5.7	
A	Pure Cu	200	Cu	6.5	
A1	Cu–Ni (2.96%)	150	Co	18	3.16
A2	Cu–Co (3.1%)	130	Co	32	4.5

its resistivity. The resistivity of layers less than 250 nm thick linearly increases with the decrease of Cu–Zn film thickness. The resistivity of layers 250 nm thick and up is constant at about 7×10^{-6} Ω cm. The concentration of doped metal (Sn or Zn) depends strongly on the film resistivity. Figure 4.22 demonstrates the value of resistivity as a function of concentration molar ratio [Cu ions]/[Sn ions] in baths. In

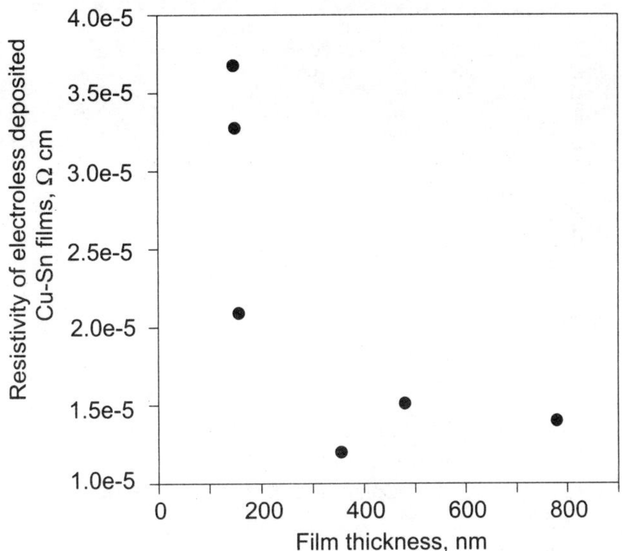

Figure 4.20 Resistivity of electroless deposited Cu–Sn films as a function of thickness. Sn concentration in bath – 0.9 g/l.

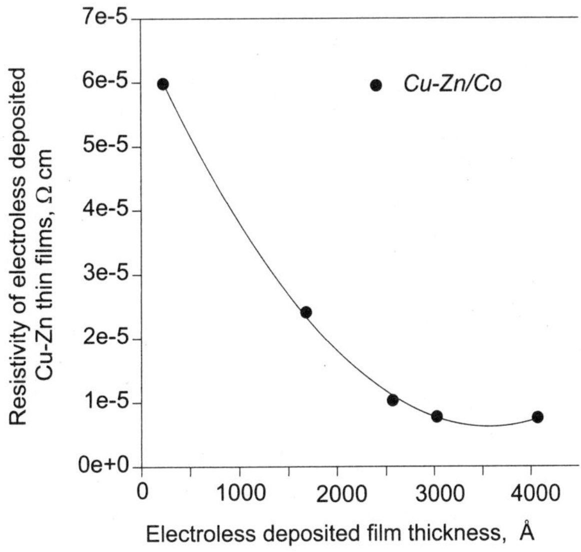

Figure 4.21 Resistivity of electroless deposited Cu–Zn thin films as a function of film thickness.

the case being considered, increasing the molar ratio in the electrolyte tends to increase the Cu–Sn layer resistivity. This effect we can see for both Cu and Co seeds.

The concentration of doped metal (Ni or Co) strongly affects the film resistivity. Figures 4.23, 4.24, and 4.25 demonstrate the value of Cu–Me film sheet resistance (Rs) and resistivity as a function of concentration molar ratio [Cu ions]/[Me ions] in

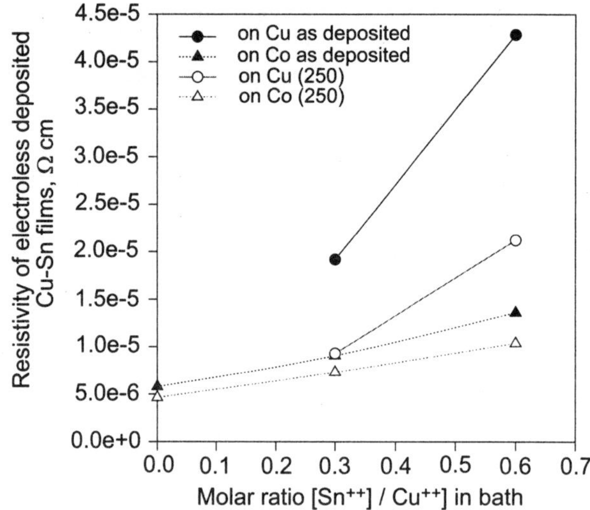

Figure 4.22 Resistivity of electroless deposited Cu–Sn films as a function of molar ratio $[Sn^{++}]/[Cu^{++}]$ in bath.

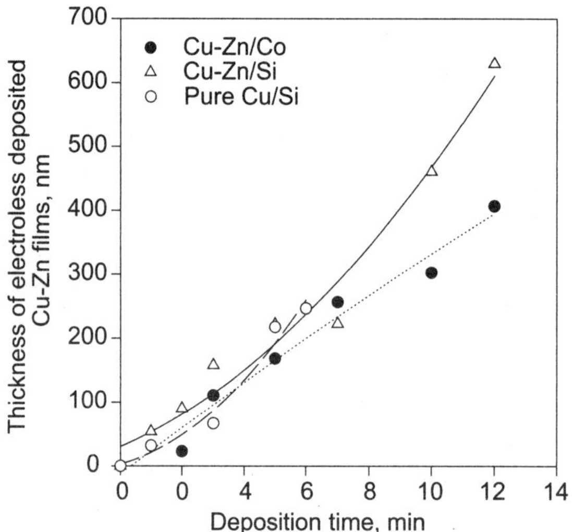

Figure 4.23 Thickness of electroless deposited Cu–Zn thin films as a function of deposition time (Zn ion concentration in electrolyte is 0.0017 mol/l).

the baths. It is seen that Cu–Me film resistivity increases with molar ratio of the metal ions in the electrolyte. For the Cu–Ni films this effect is negligible. For the Cu–Co layers the tendency of the ρ increase is revealed but its character is not clear (Figure 4.24). Figure 4.26 shows the average resistivity of deposited films as a function of thickness. For both doping metals (Ni and Co) Rs value decreases with increase of the film thickness. One can see that sheet resistance of Cu–Ni films is significantly lower than that of Cu–Co films.

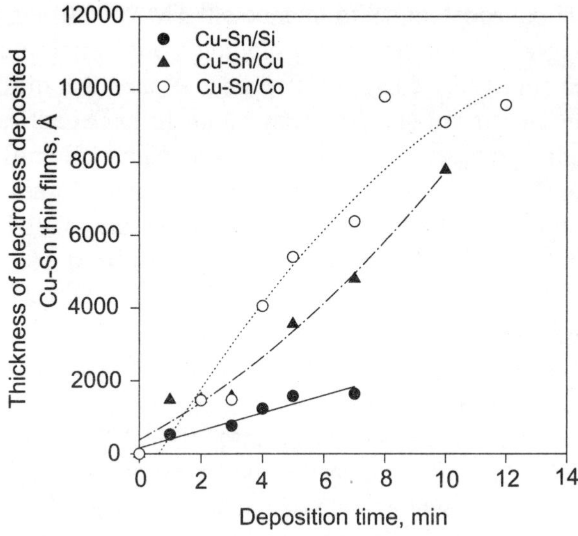

Figure 4.24 Thickness of electroless deposited Cu–Sn thin films as a function of deposition time.

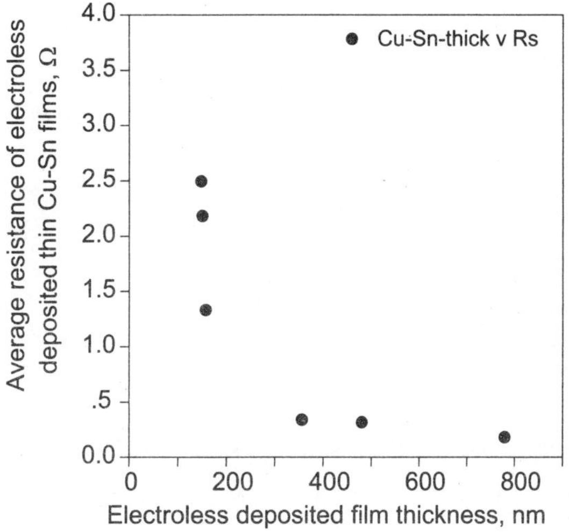

Figure 4.25 Average resistance of electroless deposited Cu–Sn films as a function of film thickness, nanometer Sn concentration in electrolyte.

The annealing of the deposited films also affects the resistivity. Annealing at 250°C during 45 min in vacuum decreases the resistivity of the films; at the same time they are not oxidized. Figures 4.22 and 4.23 demonstrate resistivity of Cu–Co, Cu–Ni, and Cu–Sn films as a function of temperature.

Composition. The composition of the deposited Cu–Me films was determined by using XPS, AES, and SEM techniques. The results of the experiments obtained by each method were similar. The concentration of doped metal (Zn or Sn) in the

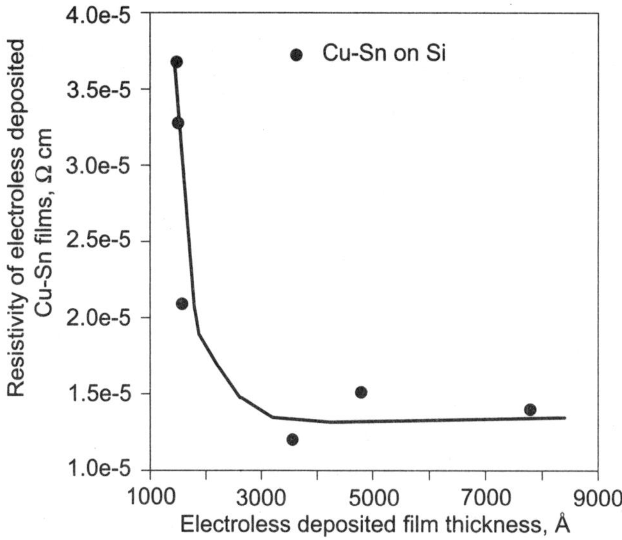

Figure 4.26 Resistivity of electroless deposited Cu–Sn films as a function of thickness. Sn concentration in bath – 0.9 g/l.

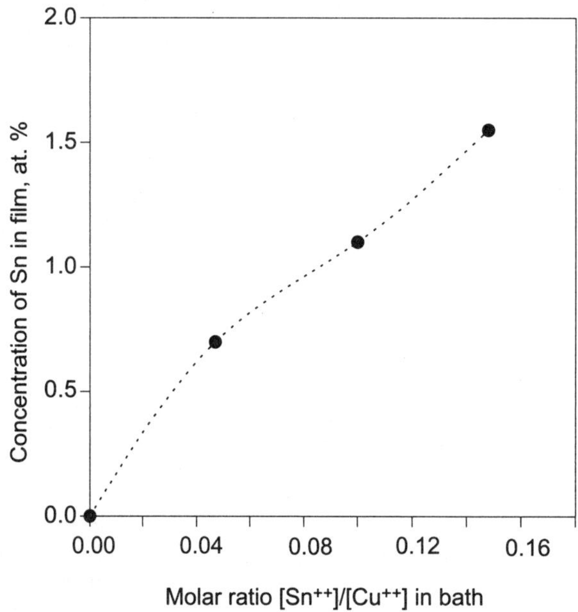

Figure 4.27 Concentration of Sn in electroless deposited films as a function of molar ratio $[Sn^{++}]/[Cu^{++}]$.

deposit depends on the concentration of its salt in the electrolyte. When the concentration of added metals salt in the bath increases, the content of doped metal in electroless deposited layers also grows. For example, when the $ZnCl_2$ concentration in the bath is 0.2 g/l (0.0007 M), the content of Zn in deposited layer is 0.85 at. %. When the concentration of the zinc chloride in solution is 0.5 g/l (0.0017 M), Zn concentration in film is 2.5 at. %. This effect may be seen in the case of Sn. Figure 4.27 demonstrates the effect of molar ratio [Sn ions]/[Cu ions] in solution on Sn concentration in film. Sn concentration in the layer increases in the range from 0 to 1.6 at. % with increase of molar ratio from 0 to 0.15.

In contrast, Ni(Co) content in the solids was fixed for every underlayer. The films were deposited from the electrolytes with 0.008 M concentration of doped metal ions (Ni or Co). The concentration of doping metals (Ni or Co) was deducted for different seed layers (Si, Cu, and Co). This effect is presented in Table 4.3.

It is known that adhesion of electroless deposited pure Cu layers to smooth surfaces of Si and SiO$_2$ is very poor. Doping of metal (Ni, Co, Zn, Sn) improves adhesion of layers to smooth surfaces, such as to silicon. So, the addition of 2 at. % of nickel to the precipitate increases the adhesion 1.5 times.[42] Adhesion of electroless deposited Cu–Ni, Cu–Co, Cu–Zn, and Cu–Sn layers to silicon surface is very good.

Corrosion resistance. Corrosion resistance of electroless deposited Cu alloys is considerably beyond the corrosion resistance of pure Cu films. Pure Cu layers oxidize after annealing at 150°C during 45 min in air. Cu alloys did not oxidize at the indicated conditions.

Surface imaging. The texture of electroless deposited Cu–Me layers was observed by AFM and SEM methods. It is seen that the surface morphology of the solid depends on the substrate material, kind of used electrolyte, type of the deposited films, and concentration of doped metal.

We present surface AFM scans of the layers Cu–Sn, Cu–Zn, deposited on the same Co seed. Figures 4.28a and 4.28c show that surface scans of Cu–Sn and Cu–Zn, films deposited on Co seed are different. Figures 4.28a to 4.28d demonstrate the surface morphology of Cu–Sn, Cu–Zn films deposited on different underlayers: Pd activated Si and Co seed. In this case we can see that the structure of electroless deposited Cu–Sn films depends on the underlayer. SEM characterization of the films is given in Figure 4.29.

AFM surface scans of Cu–Ni and Cu–Co formed on the same Si surface indicate that the surface morphology of these layers is identical at work concentration of doped metal in electrolyte — 2 g/l (0.0084 M). This demonstrates the effect of the electrolyte nature on the texture of solids when the concentration of the doped element is very low. The surface morphology of Cu–Ni, Cu–Co films deposited on Pd-activated Si from electrolytes with different concentration of doped metal demonstrates the effect of Me content on the deposited structure. The increase of Me concentration in solution and in deposit tends to coarsening. However, the effect of the underlayer on the structure of electroless deposited Cu–Me films is the decisive factor. As an illustration we present SEM images of Cu–Co and Cu–Ni films deposited on different underlayers: Co seed, Cu seed, and Pd-activated Si (Figures 4.30a to 4.30e).

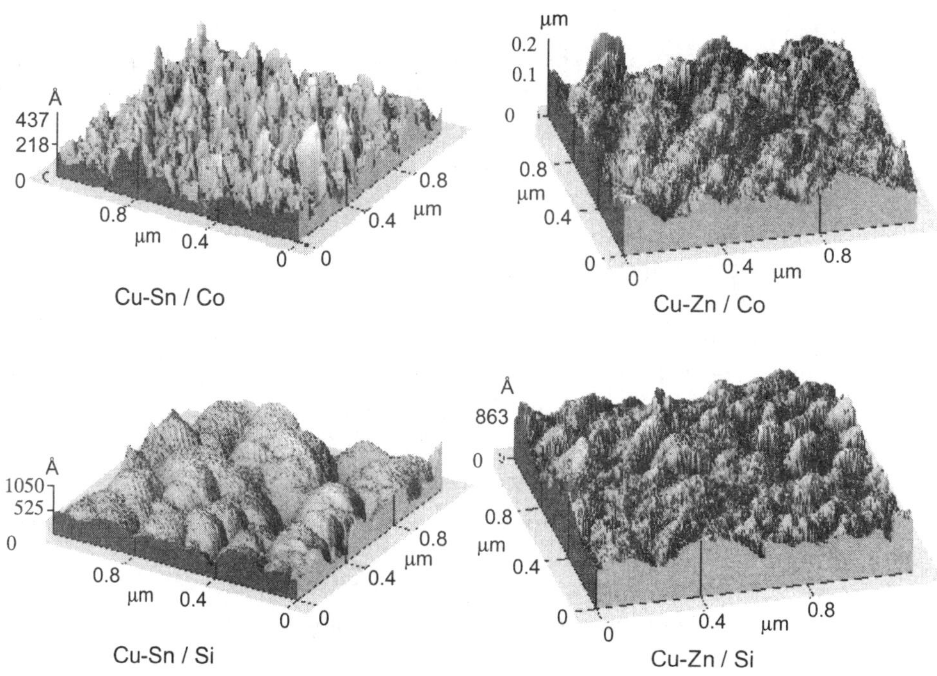

Figure 4.28 AFM surface views of Cu–Zn and Cu–Sn layers electroless deposited on Co
seed and Si substrate.

4.4 Summary and conclusions

Thin electroless Cu and Ag alloy films with sufficiently low resistivity and suitable
for ULSI interconnect applications were obtained. It was demonstrated that the
electroless deposition technology of Cu and Ag may be a candidate for ULSI metallization. However, all other thin film deposition methods (i.e., electroplating, sputtering, evaporation, and chemical vapor deposition) may also be considered for
conductor alloy deposition. Electroforming seems to be the most suitable due to the
low temperature deposition and the availability of a variety of soluble doping ions.
Electroforming deposition tools exist in the industry and research laboratories and
can be easily modified for alloy deposition.

In summary, there are few alternative technologies to the conventional metallization method. The alternative technologies are either extensions of existing technologies, such as Ag and Cu alloys, or new approaches. It seems that electroforming,
either by the electro- or electroless technologies, will serve a lead role in the research
and development of many of the new alternative technologies.

4.5 Future alternative directions

The intensive studies in the field of ULSI interconnect technology will probably
succeed in a couple of years to resolve the current metallization problems, but the
development of the new nanotechnologies has already started due to further

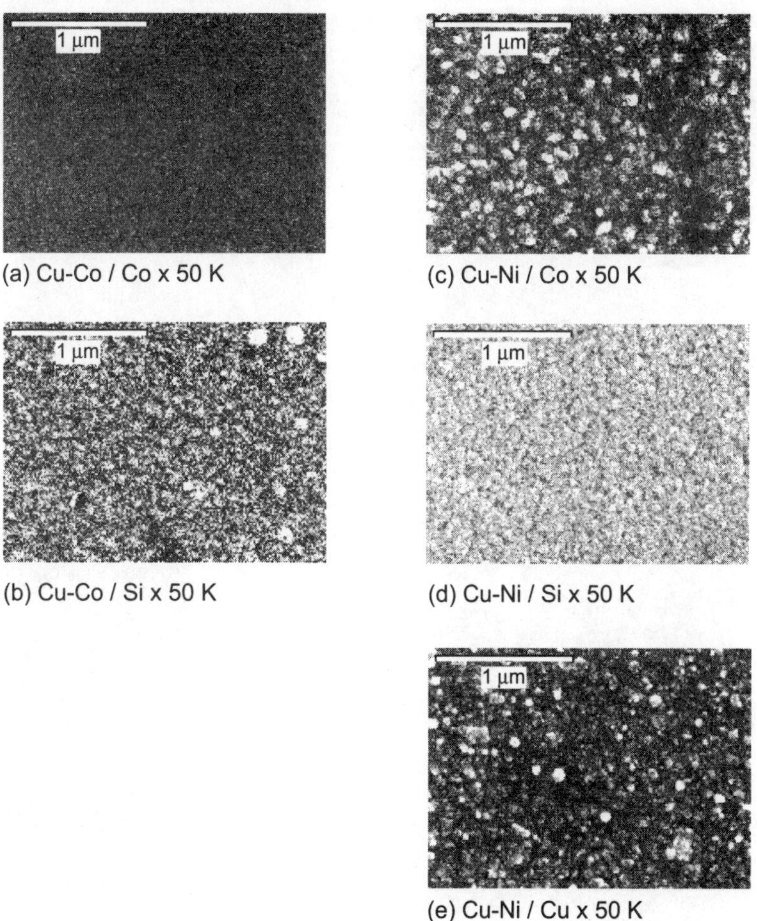

Figure 4.29 SEM surface images of Cu–Ni and Cu–Co films deposited on Co seed and Si substrate at resolution of 50 K.

increased demands to the density of electronic devices and passive elements in the circuits. Several results in possible future interconnects technologies are presented here.

The studies in the field of mesoscopic devices have advanced rapid and new single molecule transistors that are already in the last stages of development. Electronic circuits made of these devices will demand nanoscale interconnects. In this direction, it has to be taken into account that metal nanothick ($d \sim 0.5$ to 10 nm) films of high uniformity have already been made.[54] The typical oscillatory characteristic of resistivity on thickness for the quantum effect was observed for such films. The first theoretical study of quantum size effects in metal films was made by Sandomirski,[55] who has obtained the oscillatory behavior in the condition of ideal uniformity of surface. A complete analysis of the quantum transport was published by Trivedi and Ashcroft,[56] who also considered the effects of scattering which produce level broadening. To observe the quantum effect, it is necessary that the spacing between the energy levels is larger than the broadening produced by scattering (e.g.,

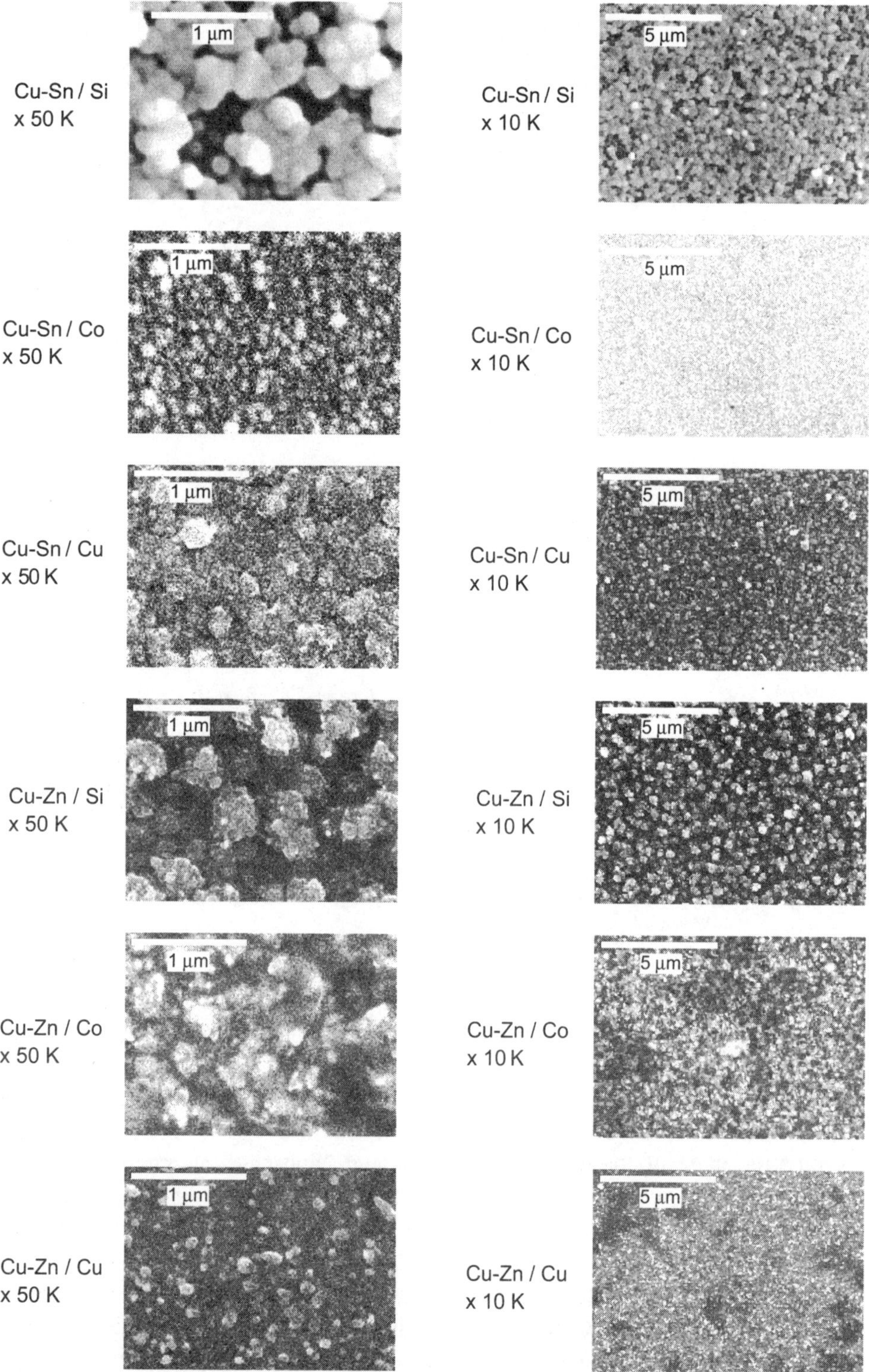

Figure 4.30 SEM images of the Cu–Zn and Cu–Sn thin films electroless deposited on different underlayers.

electron–electron or electron–impurity). For this reason measurement of conductivity has to be made at very low temperatures. The thin films, which showed the quantum size effect, could be considered as a two-dimensional electron gas (2DEG) under bias voltage.[57] In this case the measured conductance $G = I/V$, where I is the current intensity, is

$$G = n2e^2/h = nG_0 \tag{8}$$

where $G_0 = (12.9 \ \mathrm{K\Omega})^{-1}$ is the quantum conductance, n is the number of allowed propagation modes, and h is the Planck constant.

Recently, several ideas were investigated to solve practically interconnect applications of future nanocircuits by developing metal nanowires using chemical methods of deposition. Because of the very recent beginning of these studies, only partial results have been published. We shall show here three examples of Ag nanowires, which were obtained.

1. Single crystal and polycrystalline Ag nanowires array (AgNWA) that were made by template, using an alumina membrane.[58] The AgNWA were obtained by electrodeposition of Ag into the bores of the template. The diameter of wire was $d = 20$ to 50 nm, which may correspond with the scaling need for future nanodevices and circuits but there was no indication of stability (corrosion) and the value of resistivity was not mentioned in Ref. 58. The indicated behavior of the temperature coefficient of resistivity seems not to be reliable.

2. The second paper[59] describes Ag nanowires of $d = 30$ to 50 nm obtained by reducing silver nitrate with ethylene glycol. XRD analysis has shown single crystal fcc structure. Electrical measurements have shown a resistivity $\rho \sim 10^{-5} \ \Omega$ cm (which is only one order higher than the bulk Ag).

3. In the paper,[60] Ag wire arrays with $d = 0.4$ nm of a single wire were studied. A method to avoid corrosion in air of Ag for such wires was found. The high stability and uniformity of these wires were due to confining the Ag in self-asembled organic nanotube arrays, which act as a shield protecting Ag against oxidation. Calculation of the electronic structure has shown that the Ag nanowire is metallic due to three crosses of the s band with Fermi level, which suggest the existence of three conducting channels. No electrical properties of these nanowires are indicated, considering the very small diameter ($d < 1$ nm), their high structural quality, and detection of quantum properties for these Ag nanowires.

Another interesting material with possible future applications in the interconnect of nanocircuits is the single wall carbon nanotube. Theoretical calculations of the energy band structure have shown that carbon nanotubes can behave as metals or semiconductors depending on their diameter.[61] From current–voltage measurements at temperatures $5 \ \mathrm{K} \le T \le 300 \ \mathrm{K}$ it was shown that metallic behavior appears for bundles of nanotubes and the corresponding resistivity was $\rho = 1.6 \times 10^{-2} \ \Omega$ cm at

5 K and $\rho = 6.5 \times 10^{-3}$ Ω cm at 300 K. The variation of ρ with temperature in the interval 0.1 K $\leq T \leq 1$ K is not linear and a proper explanation was not found.

Conductance quantization (CQ) in metallic wires (1 nm long and ~ 0.2 nm wide) at room temperature was observed.[62] A real CQ effect requires the existence of ballistic transport, which was not demonstrated in several papers where CQ was claimed. In paper,[63] CQ and ballistic transport were demonstrated simultaneously. Measurements of conductance, using a mobile contact in Hg, have given the periodic values $\sigma = nG_0$ (staircase shape), which shows that the transport time is connected with the quantum values of σ.

The single wall carbon tubes have shown impressive potential applications for novel nanodevices with the developing fabrication of the field effect and the single electron transistor.[64–66] This demands the acceleration of studies of nano-interconnect technology for circuits made with these devices, which will be developed soon. Multi-walled carbon nanotubes (MWNT) have shown the ability for transportation of very high and stable current densities ($j \sim 10^7$ A/square), that make them possible good conductors.[67]

References

1. L. Baldi, B. Franzini, D. Pandini, and R. Zafalon, *Microelectronic Eng.* **55** (2001) 11–18.
2. R.H. Dennard, F.H. Gaensslen, H.Yu. Rideout, E. Bassous, and A.R. Le Blanc, *IEEE J. Solid State Circuits*, **SC-9** (1974) 2561.
3. K. Fuchs, *Proc. Cambridge Phil. Soc.*, **34** (1938) 100.
4. F.H. Sondheimer, *Phys. Rev.*, **80** (1950) 401.
5. M.S.P. Lucas, *J. Appl. Phys.*, **36** (1965) 1632.
6. S.B. Soffer, *J. Appl. Phys.*, **38** (1967) 1710.
7. A.F. Mayadas and M. Shatzkes, *Phys. Rev. B*, **1** (1970) 1382.
8. E.M. Baskin and M.V. Entin, *Microelectronic Eng.*, **50** (2000) 335.
9. N.G. Einspruch, S.S. Cohen, and G.Sh. Gildenblat, VLSI Microelectronics Microstructure Science, Vol.15, *VLSI Metallization*, Academic Press, Orlando, 1987, p. 24.
10. Y. Shacham-Diamand, V. Dubin, and M. Angyal, *Thin Solid Films*, **262** (1995) 93.
11. Y. Shacham-Diamand, A. Inberg, Y. Sverdlov, and N. Croitoru, *J. Electrochem. Soc.*, **147** (2000) 169.
12. A. Inberg, E. Ginsburg, Y. Shacham-Diamand, N. Croitoru, and A. Seidman, *J. Microelectronic Eng.* **65** (1–2) (2002) 197.
13. M.J. DeSilva, V. Dubin, and Y. Shacham-Diamand, Advanced Metallization for Copper Deposition for ULSI Conference, MRS Publications, Portland, Oregon, Oct. (1995).
14. F.P. Perlstein, R.F. Weightman, and R. Wick, *Metal Finishing*, **61** (1963) 77.
15. Lopatin, Y. Shacham-Diamand, V. Dubin, and P.K. Vasudev, Proceedings of the 191st Meeting of Electrochemical Society. Montreal, Symposium F1, Canada (1997).
16. S.P. Murarka, Metallization: theory and practice for VLSI and ULSI, p. 250, Butterworth-Heinemann, Boston (1993).
17. Y. Shacham-Diamand and V.M. Dubin, *Microelectronic Eng.* **33** (1997) 47.
18. Y. Shacham-Diamand and S. Lopatin, *Electrochimica Acta*, **44** (1999) 3639.
19. J.P. Chu, C.J. Liu, C.H. Lin, T.N. Lin, and S.F. Wang, *Material Chemistry and Physics*, **72** (2001) 286.
20. R.C. Weast and M.J. Astle (Eds.), *Handbook of chemistry and physics*, 61st Edition, CRC Press, Boca Raton, 1980–1981, p. E-86.
21. T.L. Alford, D. Adams, T. Laursen, and B.M. Ullrich, *Appl. Phys. Lett.*, **68** (23) (1996) 3251.

22. Y. Wang and T.L. Alford, *Appl. Phys. Lett.*, **74**, N 1 (1999) 52.
23. Y. Wang, T.L. Alford, and J.W. Mayer, *J. Appl. Phys.*, **86** (1999) 5407.
24. A. Inberg, Y. Shacham-Diamand, E. Rabinovich, G. Golan, and N. Croitoru, *Thin Solid Films*, **389** (2001) 213.
25. A. Inberg, L. Zhu, G. Hirschberg, A. Gladkikh, N. Croitoru, Y. Shacham-Diamand, and E. Gileady, *J. Electrochem. Soc.*, **148** (2001) C784.
26. Y. Shacham-Diamand, Y. Sverdlov, N. Petrov, Li Zho, N. Croitoru, A. Inberg, E. Gileadi, A. Kohn, and M. Eizenberg, in: L.T. Romankiw, T. Osaka, Y. Yamazaki, and C. Madore (Eds.), Electrochemical Technology Applications in Electronics III, Honolulu, Hawaii, October 20–22, 1999, The Third International Symposium Proceedings 99–34 (1999) 102.
27. Y. Shacham-Diamand, A. Inberg, Y. Sverdlov, and N. Croitiru, in: L.T. Romankiw, T. Osaka, Y. Yamazaki, and C. Madore (Eds.), Electrochemical Technology Applications in Electronics III, Honolulu, Hawaii, October 20–22, 1999, The Third International Symposium Proceedings 99–34 (1999) 172.
28. A. Inberg, Y. Shacham-Diamand, E. Rabinovich, G. Golan, and N. Croitoru, *J. of Electronic Mater.*, **30**, (4) (2001) 355.
29. A. Inberg, V. Bogush, N. Croitoru, V. Dubin, and Y. Shacham-Diamand, The Electrochemical Society Meeting Abstracts 2002-1 (2002) 493.
30. A. Inberg, V. Bogush, N. Croitoru, V. Dubin and Y. Shacham-Diamand, *J. Electrochem. Soc.*, **150** (5) (2003) C285.
31 P.C. Andricacos, C. Uzoh, J.O. Dukovic, J. Horcans, and H. Deliganni, *IBM J. Res. Develop.*, **42** (1998) 567.
32. J.M.E. Harper and K.P. Rodbell, *J. Vac. Sci.Technol.*, **B15** (1997) 763.
33. M. E. Gross and C. Lingk, Proc. Adv. Metallization Conf., Colorado Springs, CA (1998) p. 55.
34. M. Bruggermann, A. Masten, and P. Wissmann, *Thin Solid Films*, **406** (2002) 294.
35. M. Hauder, J. Gstottner, W. Hansch, and D. Schmitt-Landsiedel, *Appl. Phys. Lett.*, **78**, (6) (2001) 838.
36. M. Hauder, W. Hansch, J. Gstottner, and D. Schmitt-Landsiedel, *Microelectronic Eng.*, **60** (2002) 51.
37. J. Baumann, C, Kaufmann, T. Gessner, H. Koenigsmann, A. Bartzsch, and P. Gilman, Proc. Advanced Metallization Conf., MRS, Montreal, Canada (2001), p. 221.
38. Y. Shacham-Diamand and Y. Sverdlov, Proc. Advanced Metallization Conf. MRS, Montreal, Canada (2001), p. 67.
39. S. Gandikota, C. McGuirk, D. Padhi, S. Parikh, J. Chen, A. Malik, and G. Dixit, Proceedings of the 2001 IITC, pp. 30–32, June (2001).
40. T. Andryuschenko and J. Reid, Proceedings of the 2001 IITC, pp. 33–35, June (2001).
41. C.P. Wang, S. Lopatin, A. Marathe, M. Buynoski, R. Huang, and D. Erb, Proceedings of the 2001 IITC, pp. 86–88, June (2001).
42. M. Shalkauskas and A. Vashkjalis, *Chemical metallization of plastics*, Chemistry Publishing, Leningrad (1985) (in Russian).
43. F. Pearlstein, *Plating Surf., Fin.,* **70** (Oct. 1983) 43.
44. N.N. Balashova et al., *Electrochemistry*, **9** (1973) 23.
45. A. Molenaar, *J. Electrochem. Soc.*, **129**, 9 (1982) 1917.
46. A.Y. Vashkjalis, Proceedings Resp. Conf. Electrochemists Lit. SSR, Vilnius, Oct. 1973, Vol. 2, pp. 159 – 163, Vilnius (1973) (in Russian).
47. G. Mallory, *Trans. Inst. Metal Finish.*, **52** (1974) 156.
48. K. Aoki, O. Takano, and S. Ishibashi, *J. Metal Finish. Soc. Jpn*, **30**, 3, (1979) 126–131.
49. O. Takano, *J. Metal Finish. Soc. Jpn*, **34**, 6 (1983) 316.
50. M. Lelental, *J. Electrochem. Soc.*, **22**, 4 (1975) 486.
51. J. O'M. Bokris and A.K.H. Reddy, *Modern electrochemistry*, Plenum Press, New York (1970).
52. E. Gileadi, *Electrode kinetics*, VCH Publishers, New York (1993).

53. V.V. Sviridov, M.I. Ivanovskaia, and L.I. Stepanova, *J. Sci. Appl. Photogr. Cinematogr.* **28**, 6 (1983) 406 (in Russian).
54. G. Fischer and H. Hoffmann, *Solid State Commun.*, **35** (1980) 793.
55. V.B. Sandomirski, *Sov. Phys.-JETP*, **25** (1967) 101.
56. N. Trivedi and N.W. Ashcroft, 38 (1988) 12298.
57. M.Garcia-Martin, M. del Valle, J.J. Saenz, J.L. Costa-Kramer, and P.A. Serena, *Phys. Rev.*, **B62** (2000) 11139.
58. J. Zhang, X. Wang, X. Peng, and L. Zhang, *Appl. Phys. A*, **75** (2002) 485.
59. Yugang Sun and Younan Xia, *Advanced Materials*, **14** (2002) 833.
60. Byung Hee hong, Sung Chul Bae, Chi-Wan Lee, Sukmin Jeong, and Kwang Kim, *Science*, **294** (2001) 348.
61. J.P. Issi, L. Langer, J. Heremans, and C.H. Olk, *Carbon*, **33** (1995) 941.
62. B.J. van Wees et al., *Phys Rev. Lett.*, **60** (1988) 848.
63. S. Frank, P. Poncharal, Z.L. Wang, and W.A. Heer, *Science*, **280** (1998) 1744.
64. T. Hertel, R. Fasel, and G. Moos, *Appl. Phys. A*, **75** (2002) 449.
65. S.J. Tans, A.R.M. Verschueren, and C. Dekkr, *Nature*, **393** (1998) 49.
66. R. Martel, T. Schmidt, H.R. Sea, and T. Hertel, *Ph. Avouris. Appl. A*, **73** (1998) 2447.
67. S. Frank, P. Poncharal, Z.L. Wang, and W.A. de Heer, *Science*, **280** (1998) 1744.

Part III

Chip–package interconnect

5 Tape carrier and development trend

Osamu Yoshioka and Akira Chinda

5.1 Trends in semiconductor packages

Technical innovations in semiconductor devices are eye-catching, while advances are remarkable in the integration, functionality, and speed of such devices due to progress in micro-processing technology. Package technology is becoming an important issue, with the driving force in network development to high functionality and to maximize the porous electronics function of semiconductors. Demand is also high for high-density packaging for smaller, higher functionality electronics. To meet these requirements, various variations of packages are being developed.

The purpose of packages is to support integrated circuit (IC) chips that pursue even more advanced functions and allow them to fulfill their purposes fully. With the entry of small, thin equipment into the market, packaging technology is becoming more important.

Figure 5.1 indicates the trends in the development of semiconductor packages. Currently in wide use are plastic packages. Plastic packages are easy to mass-produce and have low cost, and they have risen greatly in thermal resistance due to the enhancement of mold resins and passivation technology. Plastic packages have therefore come to account for no less than 90% of semiconductor packages. In the market of plastic packages, dual in-line packages (DIPs) and zigzag in-line packages (ZIPs), consisting of boards with lead pins inserted into them, used to be part of the mainstream in the beginning. With the entry of small, thin packages and increase of their packaging densities, small on-line packages (SOPs), thin SOPs (TSOPs), quad flat packages (QFPs), and other packages have been developed consecutively. Since the latter half of the 1990s, ball grid arrays (BGAs) and other area terminal types have been playing a central role in small, large-capacity packages. BGAs are packages where solder balls as external terminals are neatly arranged on the back of the package.

BGAs have been reduced to sizes comparable to chips, that is, to chip size/scale packages (CSPs) as illustrated in Figure 5.2. In response to demand for highly functional equipment, multi-chip modules (MCMs) have been developed, consisting of multiple chips on a single package, and also stacked packages, consisting of thin packages laminated one on top of another. As an ultimate design, multi-stacked packages have been developed, consisting of boards, each of which houses many kinds of components.

To allow such semiconductor electronics to use their functions to the maximum, package technology is becoming an important element. Technical demand is

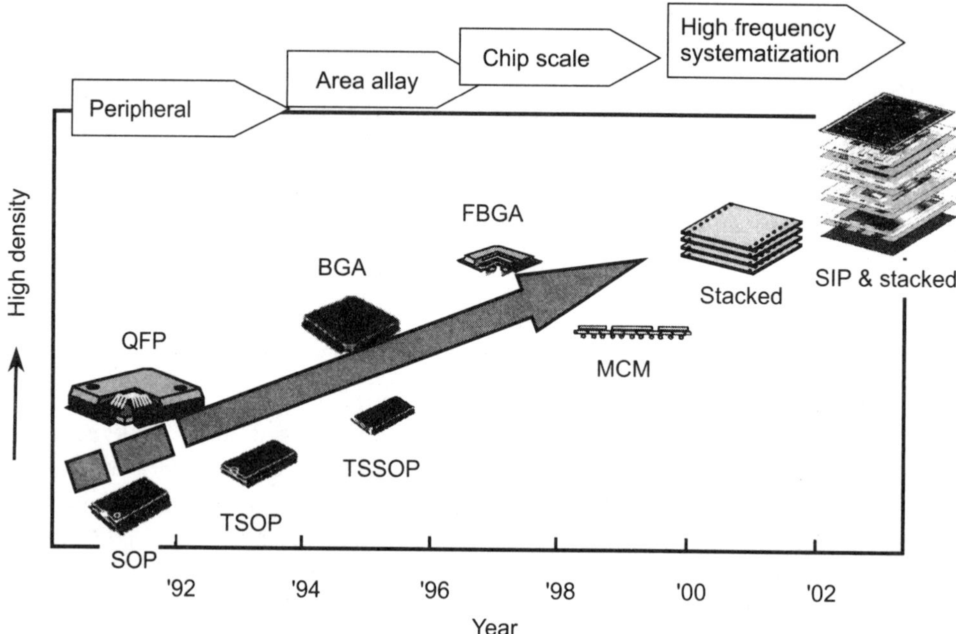

Figure 5.1   Technology trend of semiconductor package.

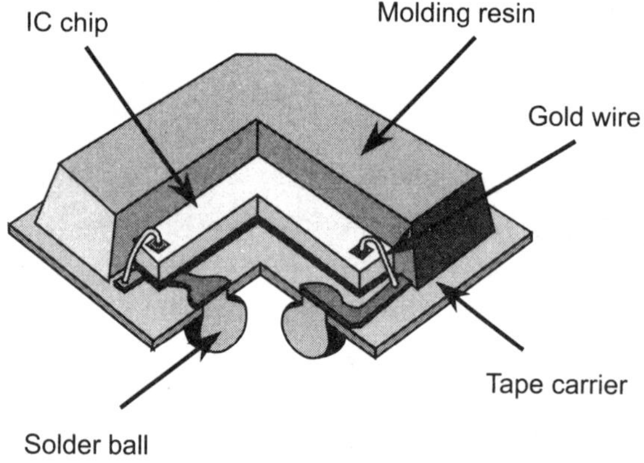

Figure 5.2   Structure of CSP/BGA.

becoming even higher for boards housing such semiconductors. For that reason, the trends in joint development are growing higher and higher among manufacturers of lead frames, IC devices, materials, and other equipment.

5.2   Tape carriers

These days, the trend is toward the application of tape carriers as LSI-mounting boards. Tape carriers have been growing in production with the development of

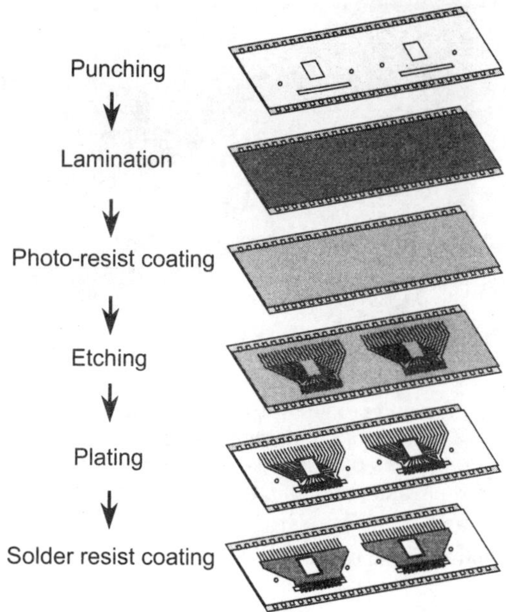

Figure 5.3   Process flow of tape carrier.

liquid crystal drivers (LCDs). Their manufacturing process is illustrated in Figure 5.3. A typical tape carrier has a three-layer structure, where a copper foil and a piece of tape are attached together with an adhesive. As tape material, a polyimide base is most commonly used, because of its heat resistance, chemical resistance, and mechanical strength. First, coat a piece of tape made of a polyimide base with a thermosetting adhesive; then make punches in it: sprocket holes, device holes, and outer lead holes. Next, attach copper foils together with an adhesive while heating them (laminating them). Apply a photoresist on top of it, and expose and develop it, thus forming a chemical-resistant resist film in a desired pattern. Spray on it an etchant (such as ferric chloride or copper chloride solution) to decompose and remove unnecessary copper, thus removing the resist film and making a microscopic copper foil pattern. Whenever necessary, apply an insulating resist film to cover the portions where no plating is needed. Then apply a plating material required for assembling the package. The tape carrier is now complete.

The tape carrier is continuously produced in reels throughout the process. It therefore has such excellent characteristics as high productivity, possibility of forming a fine wiring pattern, and uniformly finishing of patterns and plating (Figure 5.4).

Copper foils used as tape carriers fall into two categories: electrolytic and rolled foils. Electrolytic foils are made by depositing copper on a stainless steel or titanium rotary drum, which is the cathode, and by peeling it off and using it as a continuous foil. The surface in contact with the electrode is smooth and glossy, and it is therefore called the S (shiny) surface. That surface is normally a circuit surface. The surface which is in contact with the electrolyte and where copper is electrodeposited is called the M (matt) surface. It has a considerably rough shape. After that, deposit fine

Figure 5.4   Characteristic of tape carrier.

particles of copper and copper oxide onto the plating surface; then develop a chemical film on it in order to make it resistant to chemicals and heat. Electrolytic copper foils are rough on the surface, therefore being affine with adhesives, which produces high lead peel strength. However, crystals are grown in the direction of thickness, thus being slightly vulnerable to repeated stresses.

On the other hand, rolled copper foils are made by hot or cold rolling; the copper strip is then in thin foils. They are then given fine particles of copper composite and a treatment of corrosion inhibitors. Rolled copper foils are more flexile and easier to anneal than electrolytic copper foils.

Today, most tape carriers use electrolytic copper foils. That is because electrolytic copper foils can be easily made into thin films, are stronger, and have no material directionality, which is an issue in etching. In rolled copper foils, copper crystals develop in the direction of rolling (lengthwise). Etching may therefore result in a difference in the amount of etching between the rolling and the lateral direction. Another difference to note is that leads in the rolling direction are low in bending stresses.

The tape carriers are made by laminating copper foils and polyimide films, with thermal-setting adhesion. The anchoring effect and adhesion strength are higher in general when the adhesion surface (rear surface) is rough. However, when the back of the copper foil is excessively rough, copper residues are easier to develop on the lower side (adhesion side) of the fine pattern after etching, resulting sometimes in a decline in lead linearity. For that reason, copper foils for fine pattern etching are made of low-profile foils with inhibited matt-side roughness. However, the lower the roughness of the matt surface is, the less the expected anchoring effect. A decline may therefore be noted in adhesiveness. To increase adhesiveness, therefore, copper foil manufacturers are developing copper foils for fine patterns with various chemical film treatments on the adhesion surface.

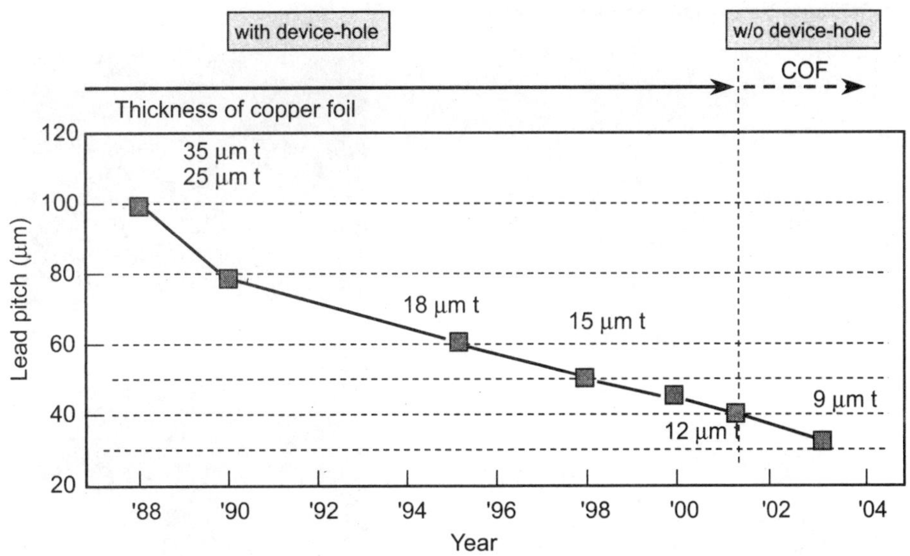

Figure 5.5   Transition of fine patterning of tape carrier.

Demand is growing even further for the fine patterning of tape carriers. Figure 5.5 indicates the trends toward a fine copper wiring pattern of tape carriers. By combining improved photoetching technology with thinner copper foils and lower profile copper foil surfaces, the industry put to practical use in 2002 tape carriers having fine wiring with an inner lead pitch of 40 μm.

An IC chip and a tape carrier are generally joined together by inner lead bonding (ILB), based on gold bumps on aluminum electrodes and technology of thermal gang bonding with flying leads. The IC-mounting part of a tape carrier is therefore given an opening (a device hole) and the inner lead is made to stick out (a flying lead structure). However, ordinary flying lead type tape carriers have been reported to readily entail a deformed lead as they become finer. That prompted the industry to develop a chip on film (COF) package, where an IC is joined by a flip chip to a plated copper foil patterning on the polyimide film surface.

Figure 5.6 compares the method an IC chip is joined with ILB with the method with COF. The bonding of the electrodes on an IC and the inner lead on the base with COF is generally performed by alignment with the electrodes and leads made transparent from the tape surface side and by bonding. The method is therefore advantageous in that it is highly unlikely to let bonding tools get dirty by melting material. However, because of transparent alignment, polyimide tape is required to be transparent. It is therefore extremely important to ensure adhesiveness between smooth polyimide tape and copper foil.

Metallized tape is commonly used to ensure tape transparency. The method involves choosing the most transparent kind of polyimide, forming a thin layer of nickel-based alloy by sputtering it as a seed layer and thickening by electric copper plating. Other manufacturing processes include casting, where a copper foil is coated with polyimide precursor varnish and then heated to vaporize the solvent and heat

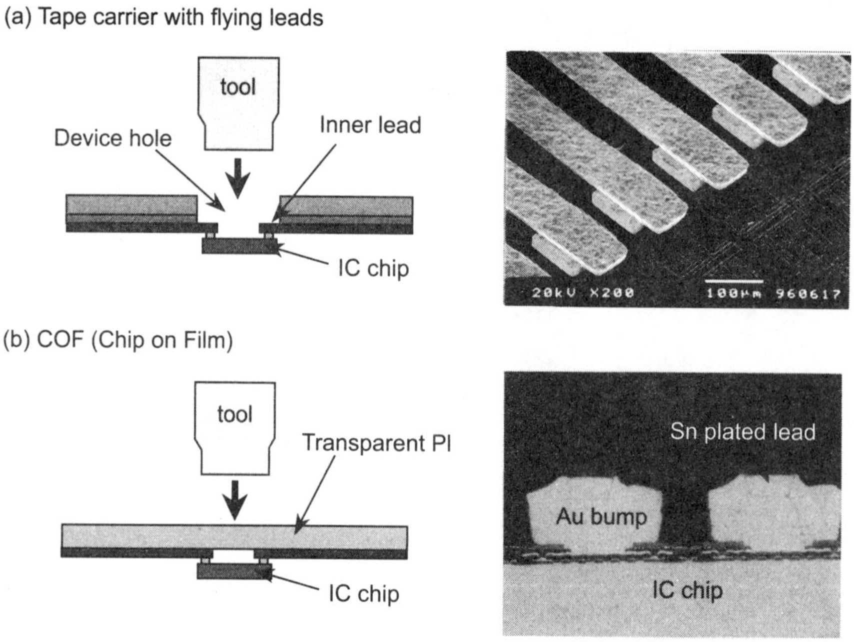

(a) Tape carrier with flying leads

(b) COF (Chip on Film)

Figure 5.6   Inner lead bonding by Au–Sn eutectic alloy forming.

curing it; and lamination, where a transparent polyimide film is attached to a copper foil with a transparent adhesive. The industry is also considering untransparent alignment junction based on a common untransparent tape material. Comprehensive consideration is given, involving manufacturers of tape materials, metal and organic materials, and IC devices.

The following chapters introduce a new tape carrier developed to make packages smaller in size and higher in functionality.

5.3   Double metal layer tape carrier

The operating frequencies of LSIs never stop growing year by year. The BGA structure is said to be shorter in wiring length than QFPs and more advantageous in frequency enhancement as well. Particularly, note that a double-metal layer tape carrier consisting of a signal layer separate from a gland layer can be enhanced greatly in electric characteristics. Tape BGAs (T-BGAs) based on such tape carriers are being put to practical use as a package best suited for high-frequency devices. This section describes double-metal layer tape material for T-BGAs developed in view of these circumstances. Figure 5.7 is an overview of the manufacturing process. This process is based on the formation of wiring by more efficient etching and uses as a material a piece of composite tape consisting of polyimide film laminated in advance on both sides with copper foils. The manufacturing process combines together many types of technologies, the connection of a via hole in the polyimide film with a double-sided copper foil by copper plating, the formation of a wiring

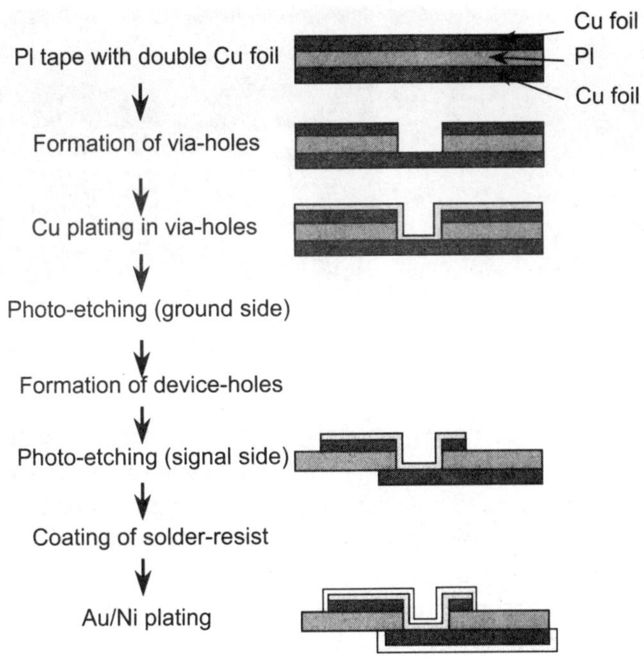

Figure 5.7   Process flow of double-metal layer tape carrier.

pattern on the gland layer and signal/power wiring layer by photo-etching, the formation of solder resist patterns in polyimide film openings of device holes and the land for forming solder balls, and multiple platings. In the present developed product, the laser drilling (hole opening) of polyimide films and the process of establishing a connect in it are characteristic as the results of establishing the new technology.

A laser was used to make holes in the polyimide films. The excimer laser has long been known as a means of fine processing polymers, but it is disadvantageous in that it is slow in processing. The industry has therefore recently applied a carbon dioxide gas layer used in the field of printed circuit boards (PCBs) as a technology for processing fine holes instead of conventional drilling.

The industry also applied the conformal masking method, which involves patterning the copper foil on the tape surface in the shape of a via opening by photo-etching and using it as a mask. The polyimide film is 50 μm. Hole-drilling technology with a special carbon dioxide gas processor is almost unprecedented, so that it has been important to determine the best conditions. Contamination with molten residues on the copper foil adhesive-side surface immediately after laser processing can be removed by chemical cleaning (it is called desmearing) and smoothens the surfaces. The limit on via processing is about 50 μm in diameter.

To connect the upper and lower patterns electrically via polyimide films, a new technology is applied to establish continuity in via holes. Metallization of double-metal layer tape carriers on polyimide vias has conventionally been conducted by electroless copper plating, similarly to PCBs. However, that method involves a complicated process, along with low plating speed and incapability of mass production,

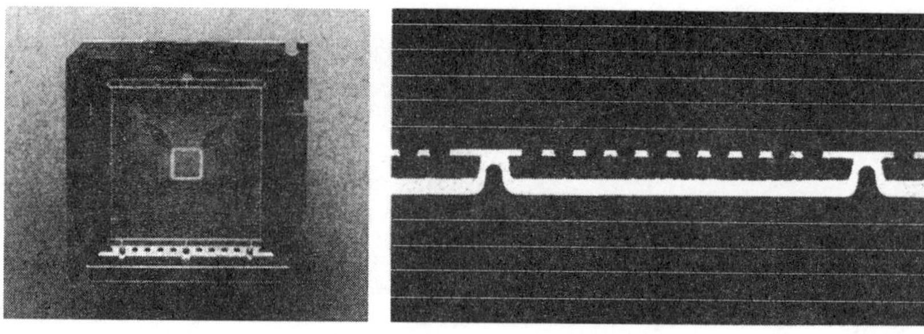

Figure 5.8   Appearance and cross section of double-metal layer tape carrier.

as well as difficulty in controlling plating agents and draining them. The industry therefore applied direct plating (DP) as an efficient plating process instead of electroless copper plating. DP is a through-hole plating process that involves forming a conductive thin film base on a through-hole wall of polyimide film and conducting electric copper plating directly on it.

Figure 5.8 illustrates an external view and a cross section of a typical double-metal layer tape carrier developed to solve the various technical challenges. Double-metal layer tape is smaller in its continuity part on the top and bottom of the tape than common single metal tape, therefore making it easier to form fine patterning. Figure 5.9 compares typical wiring designs of single metal tape and double-metal layer tape. Taking the example of CSP/BGA using a tape with a solder ball pitch of 0.5 mm and a minimum wire pitch of 44 μm, single-metal tapes can only be turned into three wiring patterns between lands for joining solder balls, while double-metal layer tapes can be turned into five wiring patterns between lands, because their continuity via holes is very small.

Moreover, double-metal layer tape has what is called a micro-strip structure, where signal wiring is opposed to a large-area gland layer with an insulating polyimide film in between. It therefore excels greatly in electric characteristics as well. Figure 5.10 compares the calculation results of electric characteristics of single-metal tape and double-metal layer tape. A simulation demonstrated that double-metal layer tape had a self-inductance reduced by a half, and a mutual inductance to one-eighth of that of single-metal tape. For that reason, when one considers the wiring design, double-metal layer tape can be provided with an LSI having a frequency as high as 2 GHz.

5.4   Electroless gold/nickel plating tape[1,2]

Joining an IC chip to a tape carrier usually requires a surface-treated copper circuit. The top surface is usually electrically plated with gold. This is due to high productivity, along with applicability to various methods of bonding and high connection reliability. However, forming a layer by electrical plating absolutely requires a power supply lead. However, when the wiring pitch is narrower, the power supply leads of all patterns become difficult to arrange. Furthermore, as the operating frequency of

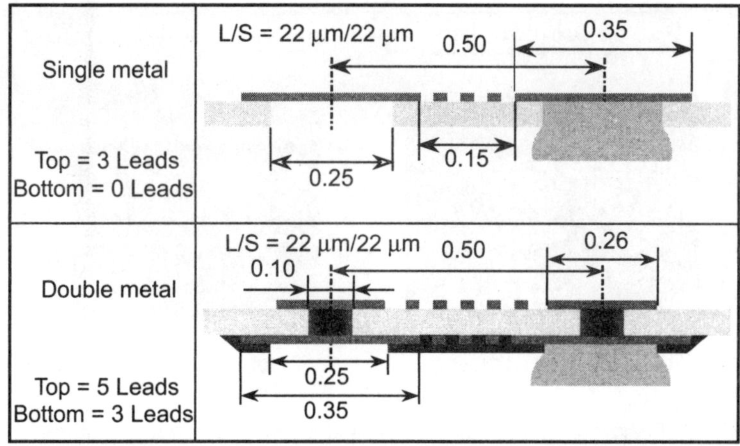

Figure 5.9 Comparison of patterning design of tapes.

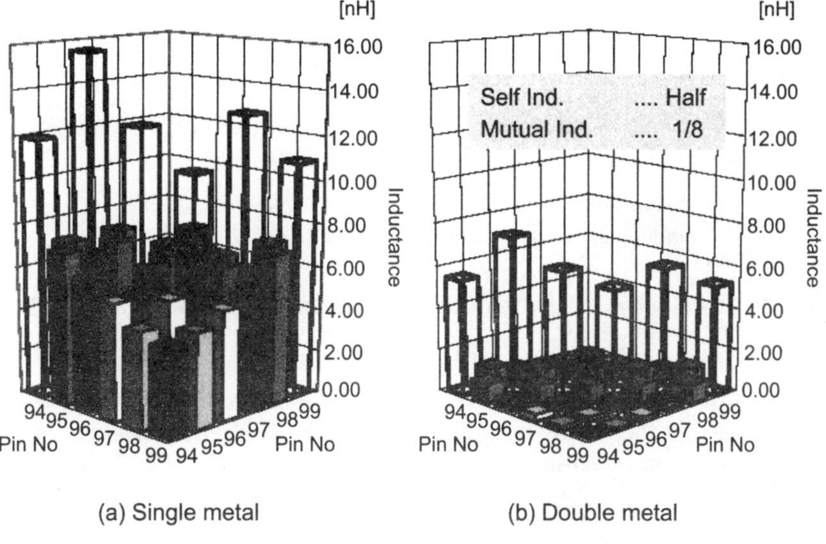

Figure 5.10 Comparison of electrical performance.

the IC increases, handheld electric products have recently begun to be concerned about the effects of noises due to the antenna effect of the residual lead. Demand then grew for electroless gold/nickel plating which requires no power supply lead.

Electroless gold plating has already been put to practical use in batch processing in sheet-type flexible printed circuits (FPCs) and printed circuit boards (PCBs). However, tape carriers can be turned into fine wiring as compared with FPCs. Further, tape carriers are also reel-shaped, so that continuous electroless gold plating ensures advantages in fine wiring and mass production. Figure 5.11 compares the pattern of electroless gold plating and the pattern of conventional electric gold plating. Electroless gold plating requires no power supply lead wiring. Therefore,

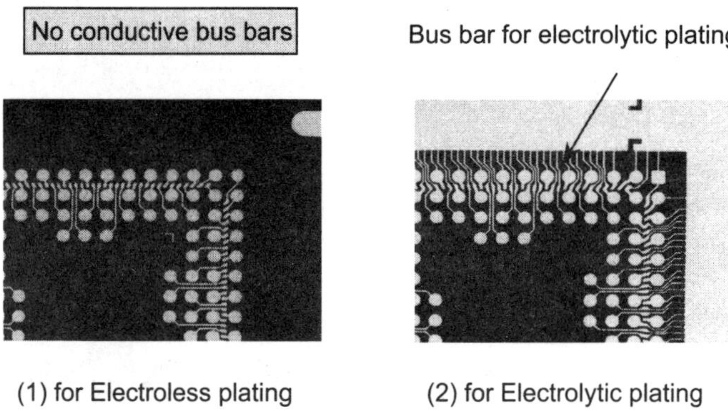

Figure 5.11   Comparison between electroless and electrolytic plating pattern.

Ni

Anodic reaction	$H_2PO_2^- + H_2O \rightarrow H_2PO_3^- + 2H^+ + 2e^-$
Cathodic reaction	$Ni^{2+} + 2e^- \rightarrow Ni$
	$2H^+ + 2e^- \rightarrow H_2$
	$H_2PO_2^- + H_{ad} \rightarrow P + OH^- + H_2O$

Au

Displacement	$Ni + 2Au^+ \rightarrow Ni^{2+} + 2Au$
Reduction	$Au^+ + Red \rightarrow Au + Red^+$

Figure 5.12   Deposition mechanism of electroless nickel and gold.

comparing patterns with identical arrangements of wire bonding pads and solder ball lands indicates that the wiring manufacturing pitch is extremely rough. The yield in etching is therefore high. However, phosphor generated by the decomposition of the reducing agent is co-deposited on the nickel layer as the base of gold plating, so that the layer characteristics are greatly different from those of the electric plating layer. This greatly affects the assembly conditions and reliability of the packages.

Figure 5.12 indicates the reaction mechanism of electroless nickel and gold plating. Common electroless nickel plating, which is provided as a base plating, uses phosphorus salt as a reducing agent. Phosphorus acid ions are hydrolyzed, resulting in phosphorus acid ions being generated and electrons emitted. This causes nickel ions to be deposited as nickel metal. However, some of the phosphinic acids react on adsorbed hydrogen (H_{nd}) and decompose, thus generating phosphorus. This phosphor is co-deposited into nickel. This hardens nickel, so that gold wire bonding in thermo-compression bonding based on ultrasonic waves involves difficulty in joining under conditions cultivated in electric gold plating. Solder joint involves a low-diffusion speed of phosphor when solder tin and nickel in the plating layer diffuse into each other. The result is the formation of a phosphor-rich layer near the joint interface, resulting in a decline in joint strength and interface peel off. Figure

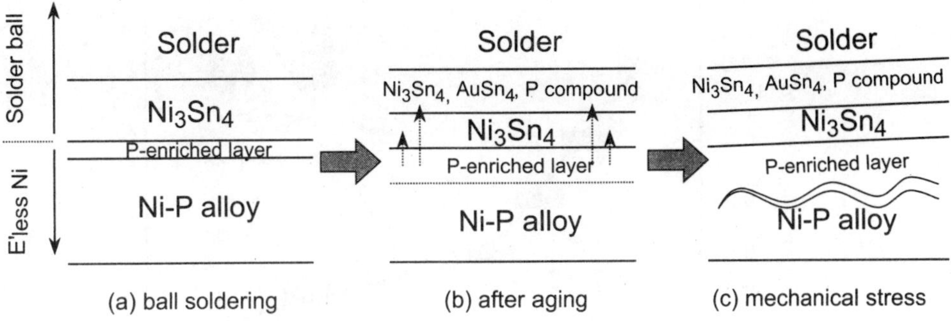

Figure 5.13 Solder ball inferior mechanism on electroless Au/Ni plating.

5.13 shows how solder joint strength declines due to the formation of a phosphor-rich layer, and how the interface peels off. It is particularly to be noted that, as gold-plating thickness, an inter-metallic compound layer of gold–tin–nickel is formed on the joint interface due to thermal aging. As a result, phosphor is easily left behind on the nickel-plating layer interface, so that the joint interface between the plating layer and solder balls becomes mechanically fragile. That is what has been observed.

Table 5.1 describes the advantages and disadvantages of electroless gold/nickel-plated tape carriers. Advantages include the non-necessity of lead wiring, resulting in high flexibility in wiring pattern arrangement; the absence of residual leads after packaging, resulting in good adhesiveness with the package-molding resin; and difficulty in catching noises generated by the antenna effect. On the other hand, particular consideration must be given to changes in the assembly characteristics of semiconductor packages due to the presence of phosphor in the nickel-plated layer.

Table 5.1 Merits and demerits of electroless Au/Ni-plating tape carrier

Merit	*Demerit*
1. High frequency of routing	Inclusion of P in nickel film
2. High resin adhesiveness around a package	↓
3. Lower noise without antenna effect	1. Reexamination of the wire-bonding condition
	2. Soldering inferior by forming of P-enriched layer after soldering

Electroless gold plating is a technology indispensable to even finer wiring and high-frequency applications. In adopting the technology, it is important to understand the layer characteristics. Future challenges include:

1. Considering nickel plating agents and plating conditions that produce layers not excessively hard and which do not easily produce phosphor-rich layers after soldering

2. Developing gold-plating processes where solutions do not easily decompose, temperatures are low and deposition speeds are high

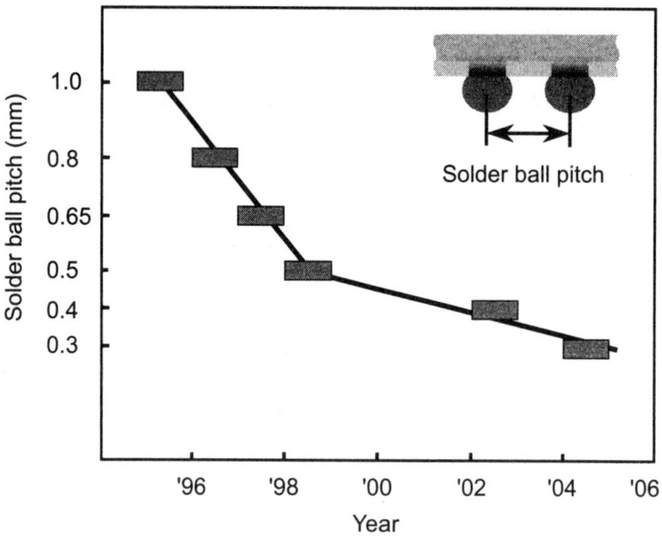

Figure 5.14 Transition of solder ball pitch of BGA package.

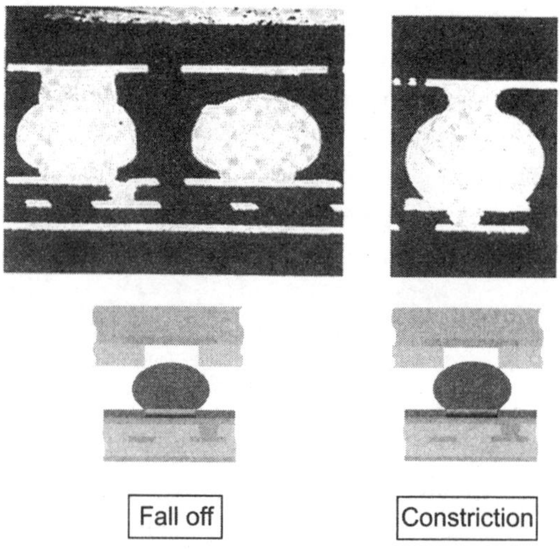

Figure 5.15 Lowering of solder ball reliability for fine pitch BGA package.[3]

5.5   Tape for filling up via holes with copper plating for joining solder balls[8]

BGA packages are already widely used as small, thin packages. Tape material is aggressively used to meet the smaller size, thinning, and multi-pin requirements of BGAs. The solder ball pitches of BGA packages made of tape material are already as small as 0.5 mm, and even smaller pitches are being developed (Figure 5.14). Incidentally, as ball pitches become smaller, the balls themselves become smaller (in capacity). Reports have indicated that, when a BGA package is mounted on a PCB,

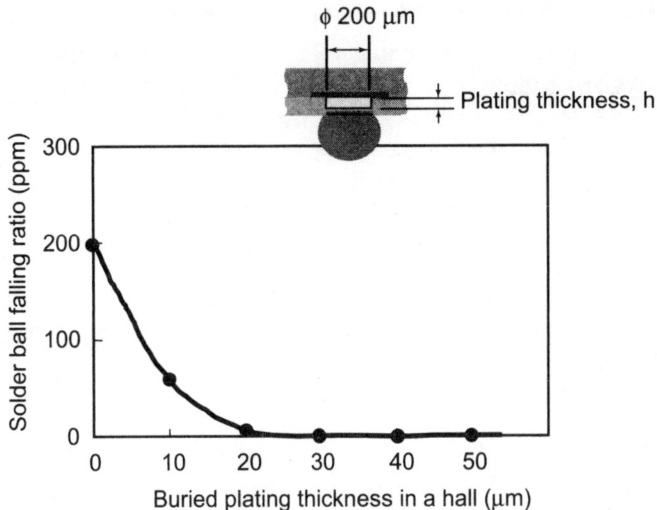

Figure 5.16 Relationship between buried plating thickness and solder ball falling ratio.

solder balls come off (falling) the package[3] (Figure 5.15). One solution conceived is based on a via bottom-up tape material, where the via holes in tape material for joining solder balls are filled up for more than a half with copper plating. Figure 5.16 shows the experimental results of the copper plating filling thickness (it is called buried plating) and the falling ratio of solder balls in the via holes in small-pitch BGA-use boards. It has been demonstrated that filling up via holes with a diameter of 200 μm and a depth of 62 μm with copper plating by at least 30 μm results in a solder ball come-off ratio of 0 ppm when a PCB is mounted. Users therefore required that mass production attempts be made for tape that fill up fine holes 200 to 250 μm in diameter with copper plating of about 40 μm.

As compared with conductive paste coating, plating is advantageous in that the filling of microscopic via holes is possible, air catching is less likely, and height (thickness) is easy to control. Various technical challenges had to be cleared in order to mass-produce tape highly homogeneous in surface shapes and plating thickness. Table 5.2 lists these challenges. The key words to development were three: fast deposition, uniformity of plating thickness, and continuous characteristic plating. For copper plating, the types of plating solutions, methods of controlling additives, methods of agitating solutions, and other elements, were considered in detail to establish the best conditions.[4,5]

Furthermore, it was particularly important to equalize the distribution of primary currents in the plating tank in order to equalize the filling thicknesses. Filling up fine openings provided in a tape carrier 70 mm wide with a reel-to-reel type copper plating device resulted in greater thickness at the end of the pattern in the widthwise direction of the tape. Then the distribution of currents between the anode and tape on a cross section of the plating tank with simulation software was analyzed in detail. As a result, they observed that the currents that came around from the circumference of the tank concentrated at the edges of the pattern. They then considered shelter

                                    Osamu Yoshioka and Akira Chinda

plates as a means to equalize the distribution of currents. They repeated simulations and found the best conditions in the size, interval, and other factors of the shelter plate that equalized the distribution.[6]

Table 5.2   Subjects of buried technique by copper plating.

Subject	Examination items
High speed plating	Selection of plating bath
	Agitation method of bath
High uniformity of plating height	High equalizing of current distribution
Continuous plating	Continuous plating machine

Consideration of these made it possible to deposit copper plating at high speeds with current densities of at least 20 A/dm^2 and to develop a continuous filling plating device. Figure 5.17 shows a typical appearance of how many microscopic vias 225 μm in diameter and 62 μm deep are recessed by about 40 μm thickness by copper plating. The tolerance in thickness is about ±5 μm with a target thickness of 40 μm. We are planning to stabilize film formation and increase production capacity as demand grows for thin packages.

The technology for filling up fine openings made in a tape automated bending (TAB) tape carrier continuously, fast, and equally by copper plating is being used in a wide range of applications, not only to increase the solder ball joint reliability of BGA packages but also to downsize packages even further. Figure 5.18 shows some applications. The left part of the same figure illustrates a board with a plating probe for burn-in inspection, which is a screening process where an IC package is inserted into a special-purpose inspection socket, and sample semiconductors are accelerated in degradation by heat and voltage to remove initial defects in advance. For BGA packages with solder ball pitches of up to 0.8 mm, conventional sockets with a metal

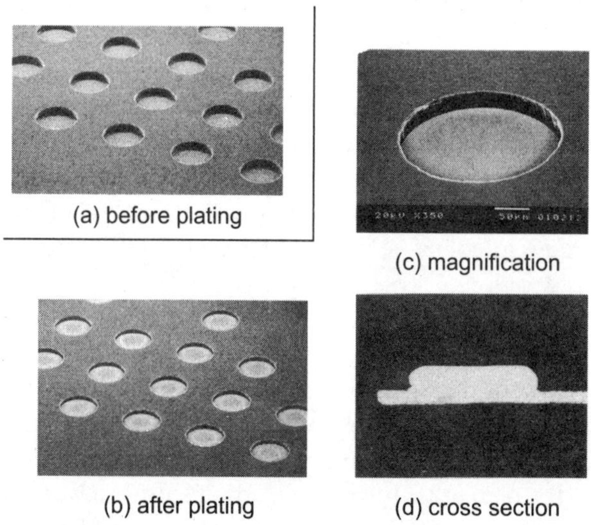

Figure 5.17   Appearance of buried copper plating via holes.

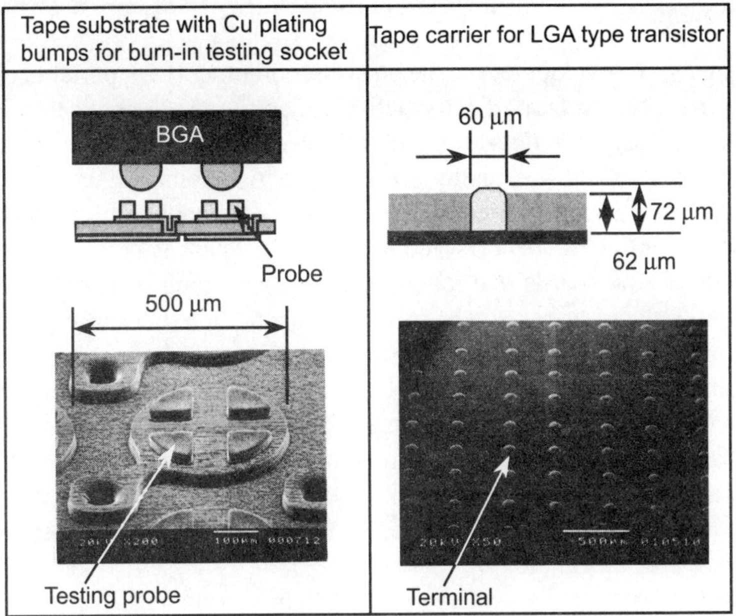

Figure 5.18 Application of buried copper plating technology.

probe pin in them are available on the market. However, as the ball pitch became no more than 0.5 mm, burn-in inspection was omitted because a probe was hard to erect and the burn-in board required an ultra-multi-layer board, so that no best sockets were available. Therefore a new sheet contact board [7] was developed which used a plating bump as a contact pin with the solder ball for double-sided wiring tape carriers that are highly flexible in wiring arrangement for burn-in inspection sockets for 0.5 mm-pitch BGAs. The contact terminal with the solder ball was made by laminating a thick dry film photoresist on the copper foil wiring land of the tape carrier; filling up copper plating in the fine openings in the resist formed by exposure and development, and eventually removing the photoresist, thus forming a copper bump.

The right part of Figure 5.18 illustrates a tape base made by filling completely with copper a blind via hole 60 μm in diameter and 62 μm deep with a high aspect ratio drilled with a carbon dioxide gas laser as an external terminal for boards for mounting small transistors, and by sticking it out by 10 μm from the top surface. The filling height is therefore 72 μm. It is significant that this downsizes the electronic components and obviates the need for joining solder balls. This technology will continue to be developed to make possible various products with filling plating technology for smaller electronic components.

There are many challenges in copper-plating technology for smaller electronic components including many difficulties such as the need to establish a plating technology for each required product and the need to make innovations. Many parameters remain unclear, such as the effect mechanism, and control methods for copper plating solution additives. Experience is therefore required. Also very important is close linkage between plating manufacturers and agent manufacturers.

5.6 Conclusions

With the explosive development of the personal computer (PC), portable electronics, and the Internet, the amount of information capacity rose dramatically. Efforts are active for developing circuits equipped with electronic components used for those purposes due to demand for smaller sizes, higher functions, higher electric characteristics, and other requirements. With the arrival of broadband networks, we have entered into an era of frequencies higher than 1 G. There is now a demand for the development of new boards that meet high-frequency requirements for electronics over 10 GHz. Polyimide-based tape carriers are spotlighted because of their advantages in terms of low dielectric constants and low dielectric dissipation factors. Under these circumstances, surface treatment has become a very important constituent technology for high-density mounting of electronics. Users' requirements in specifications never stop becoming stringent. It is therefore extremely important that users and processing manufacturers only adopt such technologies after fully understanding the deposition mechanism, properties of surface treatment layers, methods of processing, and control technology.

Needless to say, it is imperative to establish links with manufacturers of electronics devices, materials, processing chemicals, and other components, along with public institutions, colleges, and other establishments in order to apply new technologies of surface treatment. Plating technology should continue to create electronic components that excel in various characteristics.

References

1. A. Chinda, H. Akino, T. Yonekawa, *Surface Fin. Soc. Jpn*, **49**, 1291 (1998).
2. A. Chinda, O. Yoshioka, N. Miyamoto, *J. Jpn. Inst. Elec. Pack.*, **3**, 308 (2000)
3. S. Nogami, Y. Andou, F. Taniguchi, A. Takashima, K. Murakami, and N. Ito, *J. Jpn. Inst. Elec. Pack.*, **4**, 163 (2001).
4. A. Chinda, A. Matsuura, O. Yoshioka, and M. Mita, 50th ECTC Proceedings, p. 850 (2000).
5. A. Chinda and A. Matsuura, 104th Proceeding of Surf. Fin. Soc. Japan, p. 161 (2001).
6. A. Matsuura, A. Chinda, and K. Syzuki, *Hitachi Densen*, **21**, 53 (2002).
7. A. Chinda, A. Matsuura, K. Syzuki, Y. Wada, K. Morinaga, Y. Ohashi, and A. Soga, *J. Jpn. Inst. Elec. Pack.*, **4**, 4312 (2001).
8. O. Yoshioka and A. Chinda, *Electrochem. Soc. Jpn.*, **11** (70) (2002).

6 Flip-chip interconnection

Madhav Datta

6.1 Introduction

Miniaturization trends in integrated circuits have lead to the migration of inter-circuit wiring and connections from boards, cards, and modules to the chip itself. Multilayer wiring, involving fabrication of conductor and insulator configurations that were previously common in printed circuit boards (PCBs) and multilayer ceramic packages, are now being processed more effectively on chips. Increased performance is thus provided by the assembled microprocessor units. These advances have also put greater demands on the functionality and reliability of ever-increasing numbers of input–output (I/O) connections.[1] Chip–package interconnection technology will have to evolve accordingly in order to meet these requirements.

The basic function of the chip–package interconnection is to provide electrical paths to and from the substrate for power and signal distribution. In addition, the interconnection must provide mechanical support for the chip and help in dissipation of heat from the chip. Principal chip–package interconnection technologies currently used in the semiconductor industry include wire bonding, tape automated bonding (TAB), and flip-chip solder connection.

In wire bonding, electrical connections are created at the assembly stage by attaching a fine wire between each device I/O and its associated package pin. At best, serial wire bonding connects one or two rows of I/Os around the perimeter of the chip. A typical TAB process uses bumped chips and planar tapes with inner and outer leads that are bonded to chip and substrate, respectively. Similar to wire bonding, TAB interconnection is also peripheral but additionally involves bumping on the chip. Flip-chip interconnection utilizes solder bumps deposited on wettable metal layers on the chip and a matching footprint of solder bumps on wettable metal layers on the substrate.

Flip-chip interconnection is an area array configuration in which the entire surface of the chip can be covered with bumps for the highest possible I/O counts. The name flip-chip refers to the joining process that involves face-down soldering of chips to substrates.

Wire bonding has been used predominantly in microelectronic packaging for many years and is still being used in more than 90% of the assembled packages, primarily because the high speed wire bonders adequately meet most of the low- and medium-end interconnection needs of semiconductor devices.[2] However, flip-chip technology has witnessed an enormous growth during the past decade, and is

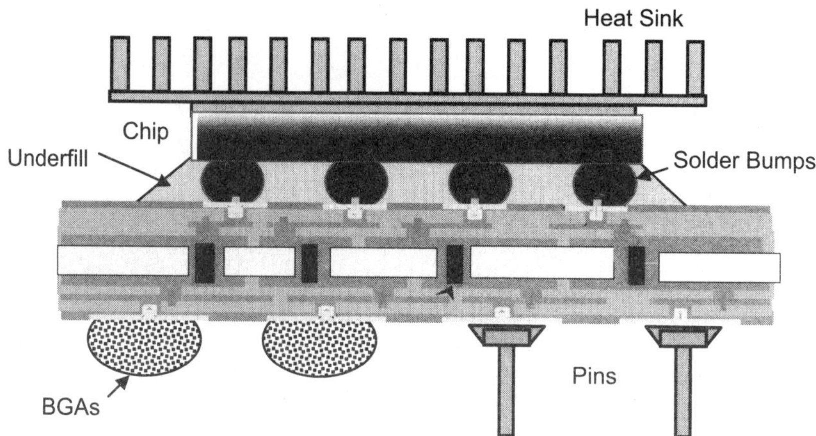

Figure 6.1   Schematic diagram of a chip–package assembly showing flip-chip interconnection including heat sink and BGA/pins for further connections.

expected to grow more than 20%/year over the next five years.[3] This growth is a result of the requirements of greater package density and higher performance. Compared to other methods of interconnection, flip-chip technology offers some distinct advantages.[2] On the one hand, area array bonding provides higher packaging density (high I/Os), and uniform power and heat distribution. On the other hand, shorter interconnect distances allow fast signal response and low inductance. These virtues of flip-chip technology make it an indispensable interconnect technology for advanced semiconductor packages.

Demands on performance, reliability, and cost have resulted in the development of a variety of flip-chip technologies using solder, high melting point metal bump and wire, conductive polymer, and anisotropic conductive adhesive interconnects. Among these materials, solders are predominantly used in flip-chip assemblies.[4]

A schematic diagram of a chip–package interconnect assembly is shown in Figure 6.1. Chips with solder bumps are flipped over and joined to the package substrate by reflow. An underfill is often used to improve reliability of the interconnect assembly. Also shown in Figure 6.1 are ball grid arrays (BGAs) and pins for further connection of the assembled package to boards, and a heat sink attached to the backside of the chip for efficient heat dissipation. Figure 6.1 thus serves to emphasize the importance of flip-chip interconnection in an assembled chip–package system. Besides providing an electrical connection and path for heat dissipation, solder joints provide the structural link between the chip and the substrate. The structural integrity of the solder joint affects both electrical and thermal performance of the assembled package; degradation in the structural integrity can be a reliability concern. Further connections involving reflow of BGAs and the like also influence the reliability. Therefore, the choice of solder material must take into consideration the temperature hierarchy of the complete packaging process so that temperature cycles do not adversely affect the structural integrity of the assembly. Some of these aspects will be discussed later in the chapter.

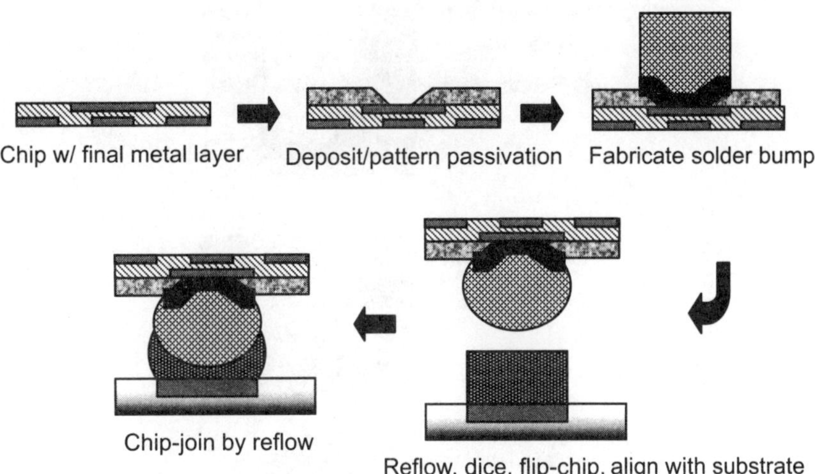

Figure 6.2 A typical process sequence for flip-chip interconnection.

6.2 Flip-chip interconnection: Process description

A typical process sequence for flip-chip interconnection is shown in Figure 6.2. The wafer, incorporating chips that have received their final metal/dielectric layers, is deposited and patterned with passivation layers consisting of inorganic (e.g., nitride, oxide), and/or organic (e.g., polyimide) thin films. An electrically conducting barrier layer is then deposited. This is required for adhesion to the passivation layer and to form the bond to the solder bumps. Depending on the choice of the bumping process, the layer may be deposited locally as in evaporated controlled collapse chip connections (C4s), or it may be a continuous film, as required in electroplating for passage of electric current through the wafer. As will be described later, the region of the barrier layer between the plated bumps needs to be subsequently etched to disconnect the bumps. The size and shape of a bump after reflow is determined by the size of the barrier layer under the bump. Therefore, the layer is also popularly known as the ball limiting metallurgy (BLM) and underbump metallurgy (UBM). The terms BLM and UBM are used interchangeably in this chapter. Following the metallization step, the wafer is deposited with the desired bumping material, generally a solder. Evaporation and electroplating are the two most commonly used bumping methods for high-end applications. Other methods include screen printing, solder jetting, and pick and place solder transfer process.[4] Evaporation and electroplating methods will be described in greater detail in the following sections. After the bump fabrication step, the wafer is reflowed (i.e., subjected to a thermal cycle in a protective environment to melt the solder bumps and form spherical balls). The wafer is then cut into individual chips by a saw operation. Individual chips are then placed face down and aligned to a substrate package, which generally contains another solder bump array. The chip is then joined to the package in a reflow module such that all joints are made simultaneously by reflowing the solder. The module thus joined by the flip-chip method is then filled with underfill and/or encapsulated.

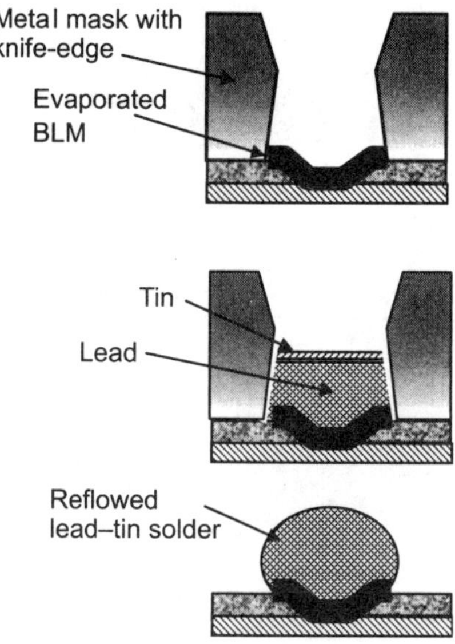

Figure 6.3   Key process steps involved in the fabrication of evaporated C4s.

6.3   Evaporated flip-chip interconnects

Flip-chip interconnect technology was first introduced by IBM in 1964 at which time the joining process was named C4 technology.[2] This terminology very well describes the preciseness and science behind the joining process: "A solder joint connecting a substrate and a flip-chip where the surface tension forces of the molten solder support the weight of the chip and control the height (collapse) of the joint."[5] The term C4 has now become widely popular throughout the semiconductor industry. However, flip-chip interconnection is a more generalized term for joining technology that does not always depend on the collapse nature of the molten solder, for example, as in the case of high-melting point, hard metal bump, and wire interconnect technologies.

IBM's original C4 technology involves evaporation of both BLM layers and the solder.[6] The evaporation process selectively deposits these materials through a molybdenum mask, which is aligned with the passivation openings in a wafer. Figure 6.3 shows the key process steps involved in the fabrication of evaporated C4s. The first process step is an argon ion sputter etch to remove oxidation and other residual products on the final metal pad and the passivation surface. This is to ensure low electrical contact resistance. The multilayered metals of the BLM are then deposited. These consist of Cr, phased Cr–Cu, Cu, and Au. This combination of the BLM structure provides an electrically conductive diffusion barrier, and establishes a good mechanical base for the solder bump. The next step is the evaporation of the solder. High lead solders, especially 95Pb/5Sn and 97Pb/3Sn, have been the most widely used with ceramic substrates because of their high melting point (315°C). Use of these alloys for chip connection allows other lower melting point solders to be used at other packaging levels without melting the chip C4s. During evaporation of solder, although lead and tin are usually in the same charge, Pb being the higher vapor pressure component evaporates and deposits first, followed by Sn on top of the Pb. In the un-re-flowed state, uniformly deposited bumps of consistent height across the wafer provide a good interface for probing and burn-in. The final step consists of bump reflow in a hydrogen furnace at around 350°C. This homogenizes the PbSn solder, brings the solder to a spherical shape, and allows Sn to form an intermetallic with the Cu in the BLM, thus providing necessary adhesion between the chip and the bump. With increasing C4 technology demands for a wide range of cost-performance products and with solder materials shifting from high-lead to Sn-rich solders, the limitations of the evaporation method have become obvious:

- Evaporation is an expensive process, involving large capital investments and operating costs. In addition to expensive evaporators, the mask and its low life and waste disposal contribute significantly to the evaporated C4 cost.

- Evaporation is typically less than 5% efficient. Indeed, more than 95% of the evaporated material ends up on the evaporator wall and on the metal mask, which need to be cleaned, and the Pb-containing waste has to be disposed of by an expensive, regulated procedure.

- Problems due to thermal mismatch between metal mask and wafer makes evaporated C4 processing on 300 mm wafers doubtful.

- Evaporated C4s are not extendible to finer bump pitch, a direction that the semiconductor industry is already heading towards. This limitation arises from the inability to fabricate molybdenum (moly) masks with finer pitch and diameter while maintaining the mask thickness.

- A very significant fundamental limitation of the evaporation process is its inability to evaporate Sn-rich lower melting point solders. The relatively low vapor pressure of Sn limits the effective rate at which it can be evaporated, thus leading to very long solder deposition times or solder melting on the wafer because of excessive tin source power level, both of which are unacceptable.

6.4 Electrochemically fabricated flip-chip interconnects

An alternative method to evaporation is the electrochemical fabrication of flip-chip solder bumps.[7,8] Some of the earliest applications of the electroplating method of bumping appeared in the 1970s for tape automated bonding.[9] Japanese Industries first introduced gold TAB bumps in the manufacture of low-cost consumer products such as watches and calculators.[10] Since then, several variations of the TAB bump process have led to the evolution of electrochemically fabricated advanced flip chip bumping technology for die–package interconnection.

Compared to evaporation, electrochemical fabrication of bumps is an extremely selective and efficient process. A key advantage of the process is that it relies on photolithographic means to define the solder bump. The currently available resist technologies are adequate to permit fabrication of extremely small bumps without

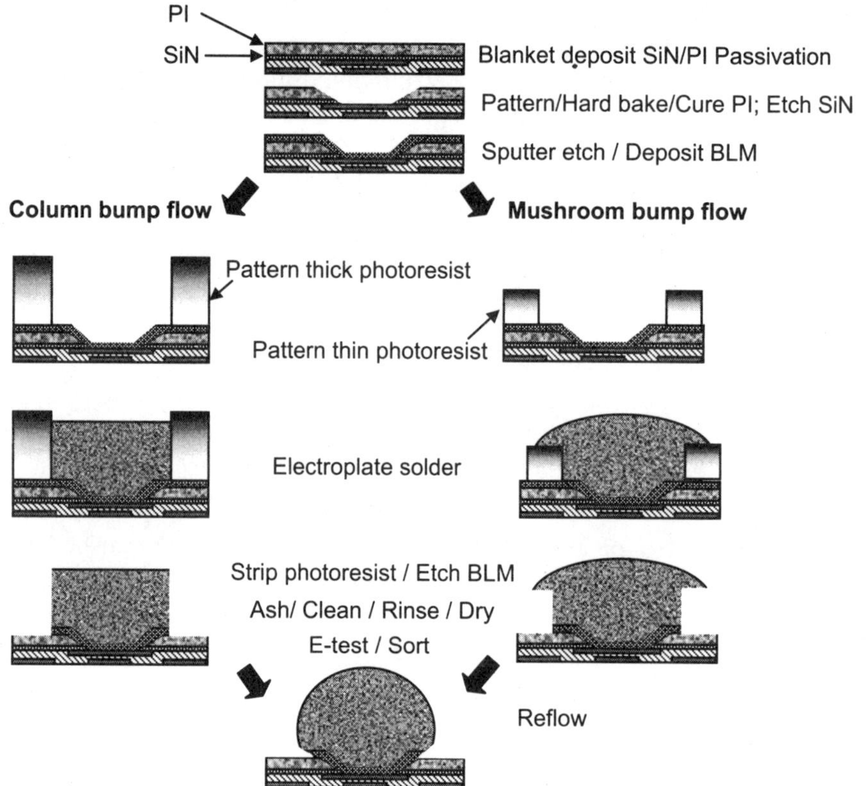

Figure 6.4   Process flow for electrochemical fabrication of column and mushroom-shaped
C4 interconnects.

any limitation to the practical minimum pitch. Furthermore, electroplating is not
limited to high-lead solder. A variety of lead–tin compositions can be plated from
commercially available baths. Electroplating is also adaptable to tin-rich and some
lead-free solder alloys. With the advent of 300 mm wafers, for which the misalignment issues of moly masks rules out evaporation, electroplating technology is
expected to be the technology of choice for wafer bumping.

The overall process flow for electrochemical fabrication of flip-chip bumps is
shown in Figure 6.4. Blanket layers of BLM are vacuum deposited on top of the final
metal bond pads and the passivation layer. A brief argon sputter etch is needed just
prior to the BLM layer deposition to improve adhesion between the BLM and the
passivation layers. A photoresist in the form of spin coat or a laminate is applied.
Depending on the required geometry of the bump, the thickness of the photoresist
may vary. Thin photoresists varying between 10 and 25 μm are used for mushroom
bumps, while for column bumps the resist thickness may vary between 50 and
100 μm. The photoresist is then patterned to the desired opening by careful exposure
and development.

After photoresist patterning, a brief plasma ash (descum) process is used to
ensure that no photoresist residue remains in the bottom of the resist openings.

Following the plasma ash descum step, the bumps are deposited in the photoresist openings using an electroplating process. For mushroom-shaped bumps, the electroplated deposit forms within the confines of the openings in the thin photoresist layer until the opening is completely filled; then the deposit grows out over the surface of the photoresist. For column-shaped bumps, the entire electrodeposition occurs within the confines of the thick photoresist openings.

The next step in the process flow is resist strip. Organic strippers such as *N*-methyl-pyrrolidone (NMP) are commonly used for stripping the resist. Following resist strip, the BLM layers are etched in the field areas between the bumps. Finally, the wafers are reflowed in an inert atmosphere of forming gas (a mixture of 90% nitrogen and 10% hydrogen). The reflow process converts the as-deposited mushroom or column bump solders into spherical solder balls (Figure 6.4). In case of high-temperature bump materials such as Au or Cu bumps, an annealing process in a diffusion furnace helps to stabilize the microstructure and to remove the entrapped volatile matters such as organic additives.

6.5 Barrier layer metallurgy

The performance and reliability of a solder bump is determined by the robustness of the barrier layer under the bumps. For electrochemical fabrication of bumps, a blanket layer of BLM is deposited on top of the final metal bond pads and the passivation layer. The BLM layer has dual functions: it provides an electrical current path for electrodeposition of bumps, and after etching, it becomes the ball-limiting metallurgy for the solder pad. The BLM must provide the following capabilities: excellent adhesion to the passivation layer and to the IC final metal pad; low resistance between the metal pad and the solder bump; protection of the IC final metal pad from the environment; effective solder diffusion barrier; and solder wettable surface. The BLM for flip-chip bumping is composed of at least two thin film layers (each on the order of 100 to 500 nm): an adhesion layer and a solderable layer. Stress in the adhesion layer must be low, the layer must be devoid of cracks and other defects, and it must act as a perfect barrier between the solder and the chip wiring. The solderable layer should have the ability to react with the solder upon reflow, thereby forming an intermetallic compound. However, the reactivity of the solderable layer with the solder must be low. An excessive reaction between these layers is undesirable because the formation of intermetallic compounds is accompanied by volume changes that induce undesirable stresses. Formation of thick brittle intermetallic layers must be avoided. If the adhesion layer and solderable layer are mutually soluble, an intermediate layer can be deposited between the two films to prevent adhesion failure.

Based on the above criteria of BLM selection, the combination of several metallurgies has been considered in the literature.[2,6–8] Films of Cr, Ti, TiW, TiN, TiW(N), and Ta adhere well to polymer as well to silicon dioxide. Cr, Ti, and TiW are most widely used as adhesion/barrier layers. Copper, cobalt, and nickel are suitable choices as solderable metal layers for commonly used lead–tin alloy solders.[11] For the Cr/Cu BLM, an intermediate layer containing a mixture of (phased) Cr and Cu is required to prevent adhesion failure, which occurs upon multiple reflow when the Cu is consumed to form Cu_3Sn intermetallic, which does not adhere well to Cr.[2]

With the change in conductor metallization from Al to Cu, more burdens have been put on the barrier layer integrity for the reliability of microprocessor assemblies using Sn-rich solders. Due to fast kinetics of the Sn–Cu reaction, diffusion of Sn through the BLM must be prevented. The refractory metals such as Ti, W, TiW, Mo, Ta, and Nb are applicable as a barrier layer for Cu because of their high melting temperature and low diffusivity. Ta as a barrier is of particular interest due to its low diffusivity (9×10^{-4} cm^2/sec in the temperature range of 400 to 700°C) and high melting temperature, 2996°C. The effectiveness of Ta as a barrier layer in the fabrication of solder bumps has been demonstrated.[12] Cracks, induced in the barrier layer during thermal cycling, and other defects may act as migration path for Sn, which may react with the Cu conductors thus creating electrical failure. Grain boundary diffusion in crystalline material is another possible mode of compromising barrier properties. Use of amorphous layers, such as TiW, helps alleviate such issues.

The BLM layers are generally vacuum deposited using evaporation or sputtering.[2,6–8] The vacuum deposition processes are optimized to obtain uniformity of film thickness over the wafer, low-film resistivity, and low-film stresses. The stresses in the deposited film are strongly dependent on the material, the deposition technique and conditions, and film thickness.[13] High-film stresses can be the cause of mechanical failure of bumps. A slightly compressive stress is generally preferred for good adhesion between the UBM and the passivation layer. Stresses in selected films deposited by evaporation and sputtering are compared in Table 6.1. For bump reliability, film stress is significant only for the overall BLM stack. The data presented in Table 6.1 indicate that the overall stress of sputtered Cr/CrCu/Cu stack is slightly compressive. This is in agreement with the experimental results where sputtered Cr/CrCu/Cu layers were found to have better adhesion than the evaporated layers.[8]

Table 6.1 Stresses in thin films of different BLM materials and thickness as deposited by sputtering and evaporation.[8]

Material	Thickness (Å)	Deposition technique	Stress (MPa)
Cr	1000	Magnetron sputtering	−1000
Cr	1000	Evaporation	800
Cu	1000	Magnetron sputtering	120
Cu	1000	Evaporation	60
Co	3000	Diode sputtering	−53
Co	2500	Evaporation	1000
Ni	1000	Diode sputtering	−160
Ni	1000	Evaporation	800
Cr/CrCu/Cu	1600/1400/2500	Evaporation	598
Cr/CrCu/Cu	500/1500/2500	Magnetron sputtering	−29

Electroless deposition can also be employed to deposit some selected BLM materials such as nickel, copper, and gold. Electroless nickel has attracted the attention of the industry due to its potential for being a lower cost alternative to other BLM approaches. Electroless nickel does not require the use of photomask and therefore eliminates processes such as patterning, developing, and stripping. Electroless nickel BLM can be used with conductive adhesives or various solder alloys. The process involves creating a BLM by building up nickel on the aluminum bond pad. Pd activation or zincating is typically needed to prepare the Al to receive nickel. Zincating is superior to Pd activation as far as adhesion is concerned. Zincating, however, involves the use of a strong alkaline zincate solution[14] that is corrosive to thin film Al. There are other alternative technologies which use special electroless nickel baths to eliminate the zincating step. Once the nickel is deposited, a layer of electroless gold is applied over the nickel for oxidation prevention.

Electroless nickel is generally deposited from an acidic nickel sulfate solution using sodium hypophosphite as the reducing agent. Nickel is deposited as an alloy (NiP) with 8 to 10% P. An attractive feature of the electroless Ni(P) BLM is its amorphous nature, which makes it an ideal diffusion barrier layer. Nickel BLM has the advantage that it also acts as the solderable layer with relatively low-dissolution rate in liquid solder. Electroless Ni, however, presents many challenges. The deposits are generally porous thus requiring a thick film of the order of 5 μm or more. Since electroless nickel only adheres to the bond pad and not to the passivation, corrosion can be an issue. In addition, the wafer must be compatible with the process. Except for the bond pads, all exposed metal must be passivated to avoid plating at undesirable sites.

6.5.1 BLM for Sn-rich solders

Until recently the majority of flip-chip technology used a high-lead solder (97Pb/3Sn or 95Pb/5Sn) with BLM systems mainly consisting of thin Cu or Ni solderable layers. However, with the growing interest in Pb-free solders and growing need for flip-chip bonding directly on to organic boards, Sn-rich, lower melting temperature solders are drawing the attention of the industry. This in turn requires a critical investigation of the BLM system.

Because of the limited content of tin in high-lead solders, the Cu–Sn intermetallic formation is limited and the thin Cu layer usually present in the BLM is adequate to ensure good metallurgical bonding over a desired lifetime without thermo-mechanical failure. For good bonding, the intermetallic reaction between Cu in the BLM and the Sn in the solder produces Cu_6Sn_5 that sticks well both to the solder and to the Cu in the BLM. To obtain good adhesion, it is imperative that all Cu is not consumed in intermetallic reaction because the Cu_6Sn_5 intermetallic does not adhere well to the underlying Cr, Ti, or TiW layer. With high tin content solder, complete consumption of the thin Cu in the BLM causes a loss of adhesion and a weak interface. Therefore, one of the accepted processes used for eutectic 63Sn/37Pb solder bumping is the use of a Cu stud as the BLM.

The copper stud is electroplated through a photoresist mask on top of a sputtered TiW/Cu BLM. The function of the Cu stud (or "minibump") is to provide an increased solder wettable area and to provide a solder diffusion barrier. However, these functions of the Cu stud have to be evaluated vis-à-vis its impact on the solder fatigue life. Companies that practice this process include Texas Instruments, Motorola, National Semiconductor, Aptos Semiconductor, and OKI Electric.[15] The Cu stud process is expected to be extendible to other Sn-rich alloys that are being considered as Pb-free solders for bumping.

The use of a nickel stud is another possible extended BLM for Sn-rich solders. Since reactivity of nickel with Sn is less than that of copper, a relatively thinner nickel stud may be used. However, the compatibility of a nickel stud with the underlying dielectric layer must be evaluated in view of its higher stiffness. In a recent study, Kang et al. demonstrated that a NiFe alloy is an effective BLM for Sn-rich alloys.[16] Electroplated NiFe films of different composition were compared with electroplated pure Ni film and electrolessly deposited Ni(P) films. A SnAg eutectic solder paste was placed on the nickel and on the nickel alloy surface and melted in a chip joining furnace at 250°C in forming gas for different times. Cross-sectional metallography revealed the extent of the interfacial reactions of Ni and NiFe films with molten SnAg solder. The 80Ni/20Fe behaved similar to 50Ni/50Fe in terms of dissolution of the NiFe film and of the intermetallic layer formed. The reaction rate of these layers was found to be significantly less than that of pure Ni, and 90Ni/10Fe. It was concluded that the composition range up to 80% Ni was effective as a barrier layer for Sn-rich solders.[16]

6.6 Flip-chip solder materials and their electrodeposition

A variety of solder materials have been considered and many of them have been investigated as possible candidates for die-package assembly and other levels of packaging.[16–20] Table 6.2 contains a list of selected bumping materials and their properties that are relevant to die–package joining. While the solder provides electrical and mechanical connection between the die and the substrate, it may also introduce a tremendous amount of stresses during assembly and thermal cycling leading to die–package integration challenges. Selection criteria for bumping material, therefore, involve a careful consideration of its electrical and thermomechanical properties. Solder joint fatigue is the principal failure mechanism of flip-chip devices. From the point of view of resistance to solder fatigue, a bumping material with high-yield strength, high-creep strength, and good ductility is desirable. Use of fragile ultra-low-k dielectric material in the die puts an additional burden on the bump selection criteria, which should aim to minimize inner lead bonding (ILD) stresses. These criteria include low melting point, low elastic modulus, low yield strength, and low creep resistance for increased ability to absorb deformation. In the present chapter the discussion on solder materials will be restricted to the materials that can be electrodeposited and will be divided into two categories: one on Pb-containing and the other on Pb-free solders.

Table 6.2 Thermo-mechanical properties of some selected solder/bumping materials.[17–19]

Metal/Alloy	Melting temp, °C	CTE, 10^{-6}/°C	Conductivity, $10^4\,\Omega^{-1}\,cm^{-1}$	Thermal cond., W/m K	Yield strength, MPa	Young's mod., GPa
Pb	327.5	28.9	4.8	35		
Sn	232	22	9.1	66		50
In	156.6	32.1		82		10.6
Bi	271.3	13.4	0.8	8.0		
Sb	630.5	11		24		
Ag	960.5	18.9	62.1	429		
Au	1063	14.3	45.5	317	308	65.5
Cu	1083	16.5	58.8	401	56.4	128.9
Ni	1455	13.4	14.3	91		224
97Pb/3Sn	310			35	11.6	21.6
63Sn/37Pb	183	24	7.0	50.9	26.88	36.05
Sn/0.75Cu	227–229				20.4	50.0
Sn/3.5Ag	221	22	9.0	56.15	39	51.00
Sn/5Sb	234–240					
Sn/3.6Ag/1.5Cu	225					51.00
Sn/20In/2.8Ag	178	28		53	37	32
In/48Sn	117	20	7.0			23.6
In/3Ag	141	20		73		
In/34Bi	110					
Bi/42Sn	139	17.5	3.0	19.56	48	28.5
Bi/32In	109.5					

6.6.1 *Electrodeposition of PbSn alloys*

Pb-containing solders are the principal joining materials used in the semiconductor packaging industry. High-lead solders, especially 97Pb/3Sn and 95Pb/5Sn, have been most widely used with alumina ceramic substrates because of their high melting point (310 to 315°C). Their use for the chip connection allows other, lower melting point solders to be used at chip to substrate and substrate to board packaging levels without remelting the chip solder bumps.

Another composition of PbSn solder that is very commonly used in both first level and second level packaging is the eutectic (63Sn37Pb), or near eutectic (60Sn/40Pb) composition. With a melting eutectic temperature of 183°C, the SnPb binary system allows soldering conditions that are compatible with most substrate materials and devices.

Standard cathodic reduction potentials of Pb and Sn from their ionic state in solution are separated by only 10 mV ($Pb^{2+} + 2e^- = Pb$: -0.126 V; $Sn^{2+} + 2e^- = Sn$: -0.136 V). The closeness of their potentials makes their codeposition one of the simplest alloy plating processes. Furthermore, because the reduction potential of hydrogen is slightly more noble (positive) than their potentials, and since both of them have high-hydrogen overpotential, their alloy can be electrodeposited from acid solutions with high-cathode efficiency close to 100%.

A wide variety of plating baths for electrodeposition of PbSn alloys has been reported in the literature.[21–27] The bulk of the earlier literature dealt with the development of fluoborate baths. These investigations focused, in particular, on the effects of additives used to improve throwing power and deposit properties. The kinetics of deposition from fluoborate baths and the effects of commonly employed peptone additives have been examined by Tam.[21] A fluoborate bath with stable additives, which allows a significant improvement in deposition rate, has been developed by Kohl.[22] However, the toxicity and corrosiveness of fluoborate baths have made them increasingly undesirable as environmental regulation became stricter. This led to the investigation and invention of a class of less toxic baths. Among them, sulfamate[7,23] and methane sulfonate[24–26] baths have been investigated because of their importance in microelectronic packaging applications.

6.6.1.1 Sulfamate baths

A systematic investigation of electrodeposition of PbSn alloys from sulfamate solution has been reported by Samel et al.[23] The study focused on selecting suitable additives for solution and examining the current density range for obtaining the desired plating. Polarization behavior of the solution and the effect of current density, lead–tin ratio in the bath and temperature on the deposited composition were investigated. The additives studied included peptone, resorcinol, polyethylene glycol, gelatin pyrogallol, and polyethylene oxide. Based on a Hull Cell study of the effectiveness of these additives in influencing leveling, an optimum combination of additives was proposed that was composed of 1.5 g/l each of peptone, gelatin, and pyrogallol. The effect of these additives could be recognized on the polarization curves, particularly on their influence in the limiting current. The effect is to inhibit hydrogen evolution and, as a consequence, a large range of potential becomes available in which metal deposition takes place without loss of efficiency. Sulfamate solutions, however, lack long-term stability; the stability diminishes when the solution is left idle for a period of time.

6.6.1.2 Methane sulfonate baths

Methane sulfonate based lead and tin proprietary electrolytes are now commonly used in the packaging industry. There is, however, very little information reported in the open literature on this family of electrolytes. Rosenstein[24] provides a general discussion on several types of baths based on methane sulfonic acid for lead, tin, and lead–tin plating. The role of different classes of additives on the cathodic polarization behavior in these systems was investigated by Goodenough and Whitlaw.[25] A variety of patented sulfonate electrolytes are now commercially available.[27] These are based on methane sulfonates of lead and tin together with different additives.

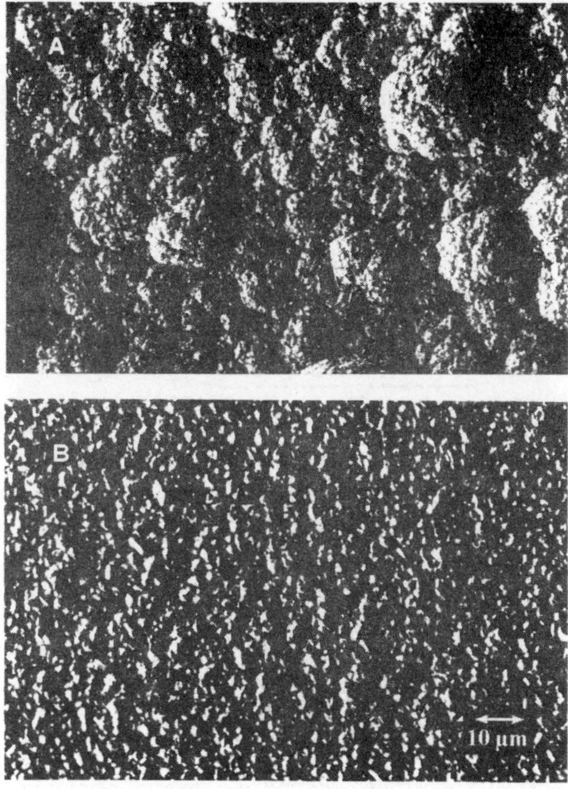

Figure 6.5 SEM photograph of PbSn deposits obtained from sulfonate bath at an average current density of 75 mA/cm^2. (top): DC plating with 5 ml/l Triton-X; (bottom): pulse plating with 0.5 ml/l, 10% duty cycle.

An understanding of the electrochemistry of the deposition processes of Pb and Sn in methane sulfonate baths has been obtained for electrodeposition of solder alloys varying from 63Sn/37Pb eutectic composition to the high melting point 97Pb/3Sn alloy.[26] The addition agents in electrolyte have been found to control both the deposition morphology and the alloy composition. In a recent study, Lin et al. investigated the influence of pulsating current on the morphology of lead–tin deposits.[28] Their study indicated that the use of pulsating current leads to fine-grained deposits and minimizes the need for additives as shown in Figure 6.5.[28]

The kinetics of Sn and Pb deposition from methane sulfonic acid are extremely fast, resembling the deposition of Sn and Pb from other acids. Metals deposited through such highly reversible reactions typically have poor morphology, tending to grow needles and large crystals. Organic addition agents can greatly impede the deposition reaction. They are universally used in Sn–Pb deposition to control the deposition properties and they are also important in controlling the deposit composition. Examples of typical additives in methanesulfonic acid (MSA) are given by Hurley et al.[29]

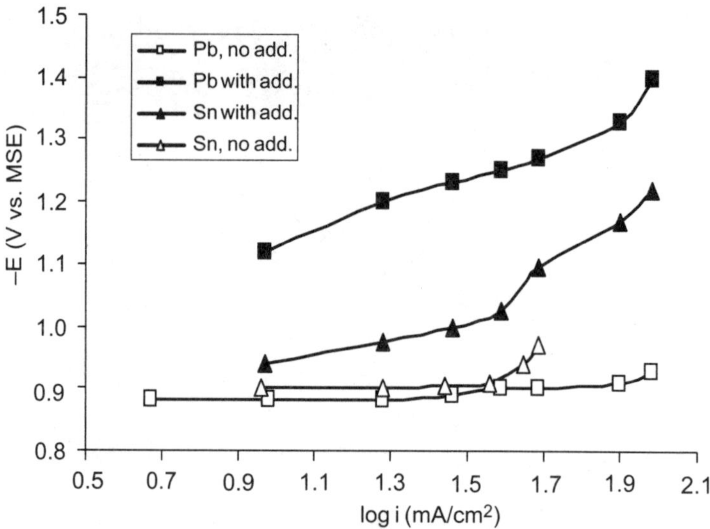

Figure 6.6   The effect of commercial additives on the cathodic polarization of Sn and Pb in methane sulfonic acid based bath containing 0.2 M Pb^{2+} and 0.2 M Sn^{2+}; additive is 10% by volume.[8,26]

   The polarization behavior of Sn and Pb deposition from MSA and the effects of a commercial plating bath additive are given in Figure 6.6, which shows the dependence of the partial currents i_{Sn} and i_{Pb} on potential. Alloy deposition was examined in a solution containing 0.2 M Pb^{2+} and 0.2 M Sn^{2+} without or with 10% by volume of commercially available additive. The curves shown in Figure 6.6 were derived from steady-state measurements of potential when a constant plating current was applied to a Pt rotating disk electrode (RDE). The deposits were stripped anodically from the RDE, and the stripping charge was used to calculate the metal deposition current, as distinct from the total applied current.[26]

   In the absence of additives, the electrochemical behaviors of Sn^{2+} and of Pb^{2+} are similar in acid solution; their deposition potentials are separated by only a few millivolts. Both metals are deposited with nearly 100% current efficiency at lower current densities. The current efficiency of Sn is less than 100% at high current densities, but Pb deposition is essentially 100% at all practical current densities because of the slow kinetics of hydrogen evolution on Pb. The very fast kinetics of both Sn and Pb deposition in pure MSA are reflected in the very steep current rise with potential. The commercial additive strongly polarizes both Sn and Pb deposition, especially at high current densities. The effect of the additive is very large for both metals, but Pb deposition kinetics is more strongly affected than that of Sn.[8,26]

   Figure 6.7 summarizes the influence of the additive and Sn concentration on the compositions of the deposits from an MSA solution with 0.2 M Pb^{2+}. In an additive-free solution, under conditions of fast reaction kinetics, the composition of PbSn alloy is difficult to control, as shown in Figure 6.7.[8,26] When the solution concentrations of Sn^{2+} and Pb^{2+} are equal, the deposit is primarily Pb, and at very high Sn^{2+}

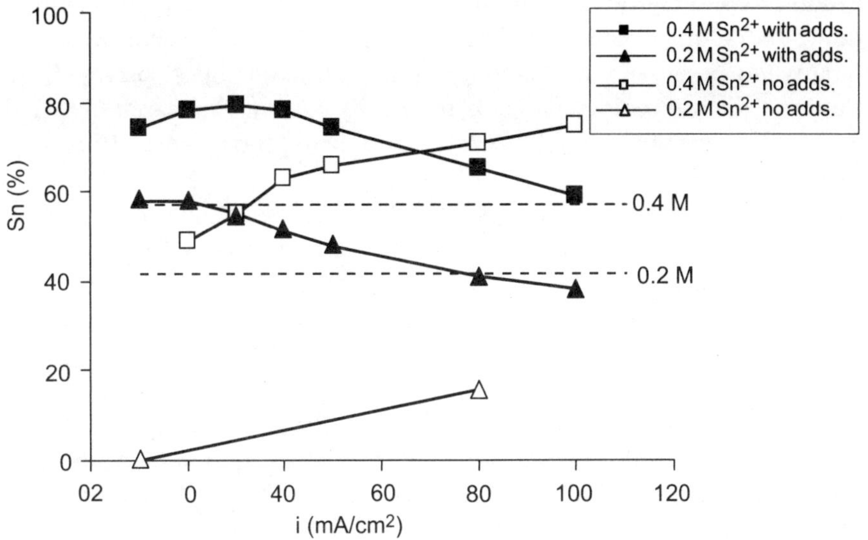

Figure 6.7   The effect of a commercial additive on the composition of Pb–Sn alloys deposited form methane sulfonic acid based bath.[8,26]

solution concentrations (not shown in Figure 6.7) the deposit is essentially pure Sn. At intermediate Sn^{2+}/Pb^{2+} ratios, alloys can be deposited but control of both morphology and composition is poor. The addition agent allows the plating of a deposit, which over an order of magnitude variation in plating current density, is of high quality and has a composition resembling the composition in the plating electrolyte. The methane sulfonic acid based plating system is compatible with the commonly employed under bump layers and photoresist.

6.6.1.3   PbSn plating bath monitoring

An essential part of implementing any manufacturing process is the understanding of how to control the process within the product specifications. The dependence of the overall process on each of the operating parameters must be known, the operating window must be established within which each parameter can vary, and a procedure of process control must be set up. Each component of the PbSn plating bath can be determined analytically.[30] Methane sulfonic acid concentration in the bath can be determined by neutralization titration using sodium hydroxide. Pb can be monitored by complexometric titration with ethylenediaminetetraacetic acid (EDTA) after oxidation and tin masking. Iodometric titration in the presence of an antioxidant can be used for determining Sn concentration.

One of the key concerns in controlling electroplating baths is the monitoring of organic additives. Voltammetry techniques can, in principle, be used to measure the concentrations of organic species that act as inhibitors in metal-plating baths, even when they are proprietary unknown chemicals. The monitoring method is based on the principle that additives act through their effects on the kinetics of the metal deposition reaction and thus reaction $M^{n+} + ne^- = M$ can be used as a probe to determine the additive concentration. The amount of metal deposited during a potential scan is a measure of the concentration of the inhibitor species. However, the heavy metal ions in solder plating baths readily poison probe electrodes, and thus electrochemical methods on solid electrodes are irreproducible and unreliable in these systems.

A dropping mercury electrode is a clean and reproducible surface for electrochemical measurements. In contrast to cyclic voltammetry on a solid electrode, square-wave voltammetry on a dropping mercury electrode (DME) has produced reproducible and reliable determination of the polarizing addition agent in Sn–Pb plating solutions.[31] The reduction of Pb^{2+} on the DME is used as the sensing reaction. Because Pb^{2+} reduction to the Hg amalgam is activation controlled in the presence of the strongly inhibiting addition agent, the size of the Pb peak in square wave voltammetry is indicative of the inhibitor's concentration. A peak height vs. addition volume relationship is determined when a Pb^{2+} stock solution is titrated with a plating solution. The end point of the titration is inversely proportional to the concentration of the additive in the plating bath. The measurements can be calibrated with known additive concentration in the solution.

6.6.2   *Electrodeposition of Pb-free bumps*

As noted in the previous section, lead–tin alloys are the most commonly used solder materials for microelectronic packaging. However, in view of the increasing health and environmental concerns, a serious effort to develop lead-free solders is underway. Driven by enforcement of lead-free regulations, Japanese and European companies have taken the lead in this direction and are already marketing some "green" microelectronic components. The U.S. microelectronic industry is now gearing up to the worldwide call for Pb-free solders.

Current microelectronic packages demand solders with enhanced metallurgical and thermo-mechanical properties. The lead-free solders are, therefore, expected to have melting temperatures close to that of lead–tin solders; better physical properties in terms of electrical conductivity, thermal conductivity, thermal expansion, and wettability; better mechanical properties such as elastic modulus, creep, and tensile strength; and better corrosion resistance and solderability. A relatively large number of lead-free solder alloys have been proposed in the literature.[16–20] The proposed solder alloys are binary, ternary, and even quaternary alloys. A large number of these alloys are based on Sn as the major component. Some other major elements include indium and bismuth. Alloying elements mainly include Cu, Ag, and Sb (Table 6.2).

After consideration of the criteria mentioned above, only a few lead-free alloys can be listed as serious contenders for flip-chip interconnection. These alloys are further narrowed down to only a handful if they are to be electroplated. The list of lead-free solders that can be plated includes Sn, Sn/0.7Cu, Sn/3.5Ag, SnBi, and In. Even in this list, only the first three alloys have been investigated to any extent by the microelectronics industry. A detailed description of electrodeposition of lead-free solder alloys for flip-chip interconnection is given in Chapter 8.

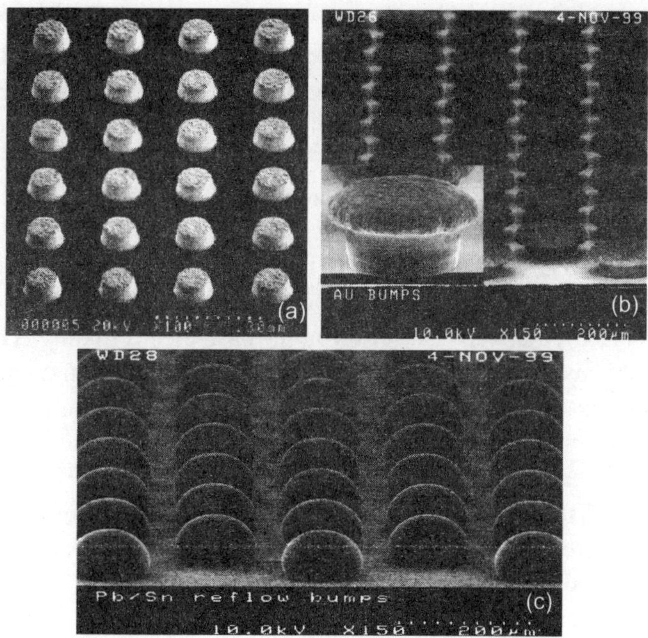

Figure 6.8   SEM photographs of (a) column, (b) mushroom, and (c) reflowed 97Pb/3Sn C4s.

6.7   Bump shape and uniformity

The trend in flip-chip interconnect pitch indicates that the bump pitch will shrink to 50 μm by 2012.[1] The driving force for this pitch reduction is to satisfy the requirements for high-performance silicon devices, and includes a dramatic increase in the number of I/Os due to increases in the number of signal lines and power requirements. Higher power devices require more signal and ground lines and, to limit point sources of heat, the power and ground interconnects should be spread evenly across the area array. In order to satisfy these requirements, a tighter control of bump shape and uniformity will be required.

Currently, two types of as-plated bump shapes are common in the microelectronics industry: mushroom bumps and column bumps (Figure 6.8). While mushroom bump technology has the advantage that it uses industry standard thin (up to 25 μm) photoresist capability, it may not be applicable to advanced products with narrower pitch bumps and high I/Os, due to possible bridging problems associated with these bumps. Furthermore, volume uniformity of mushroom bumps is difficult to control. On the other hand, the use of a thicker photoresist mask to restrict the bump-plating process within the photoresist feature permits fabrication of so-called column bumps. Column bumps are extendible to finer pitch and higher I/Os, and provide better volume uniformity control.

Uniformity of plating is an essential criterion to achieve high yield in chip joining. This includes uniformity of plated bump height and composition. Solder ball uniformity is critical because large variations between bumps could result in electrical open circuits for small balls and shorts for large balls. Recent advances in plating tool design have made it possible to achieve volume and compositional uniformities in the deposited alloys.[32] Providing the desired current distribution and mass-transport conditions at the electrode surface are the key design considerations. Two types of plating cells are now commercially available for precision plating of bumps: paddle cells and fountain cells. The paddle cell was developed in the early 1970s as a precision tool for the thin film magnetic recording head industry.[33] A paddle, which moves in close proximity to the workpiece, provides electrolyte agitation right at the surface where the agitation is needed. It also causes periodic reversal of flow, which is extremely critical for through-mask electroplating. The paddle cell, in its different forms, has now found applications worldwide in a variety of applications in microelectronics. The fountain plating system uses an electrochemical reactor where the wafer is held face down over an impinging flow of electrolyte.[34] The wafer may or may not be rotated during electrodeposition. The current distribution in these cells is modified by using auxiliary electrodes for current thieving or by inserting insulating elements within the electrolyte for current shielding. Details of electrodeposition tools are described in Chapter 16.

The choice of an appropriate plating current density is an important aspect of bump thickness and compositional uniformity. Typically, any non-uniformity in the current distribution (potential field) can be improved by increasing a dimensionless number, well known in electrochemistry as the Wagner number ($Wa = RT\kappa/\alpha FAiL$; with R = universal gas constant, T = electrolyte temperature, κ = electrolyte conductivity, α = cathodic transfer coefficient describing the electrode kinetics, F = Faraday constant, A = platable fraction of the chip area, i = average plating current density, and L = characteristic length, e.g., chip length). It follows that the non-uniformities within a chip can be improved by increasing bath temperature and conductivity, and by decreasing the chip size. Keeping all other parameters constant, plating at low-current density should provide more uniform current distribution. The choice of operating current density, therefore, involves a compromise between high degrees of uniformity vs. plating rate and is a strategic judgment to be made for a manufacturing process.

6.8   BLM etching

After completion of bump plating, the photoresist is stripped and the BLM layers between bumps are etched to isolate the bumps. During the etching process, the solder bumps remain exposed and act as masks for the underlying BLM layers. The etching process must be inert to the exposed metals, it must be compatible with the passivation layers that get exposed to the etchant after the BLM is removed, and it should yield a BLM edge profile that can sustain the stresses in the bump and metal films and the thermo-mechanical stresses that are generated during assembly and reliability testing. The seed layer etching process is, therefore, crucial to obtaining highly reliable electrochemically fabricated bumps.

Dry techniques for thin film etching include ion milling, sputter etching, reactive ion etching, and plasma etching. Dry etching techniques are used particularly in ultra-large-scale integration (ULSI) because of their ability to remove material anisotropically and to achieve high resolution. Some of the disadvantages of dry etching are its inability to provide sufficient selectivity, problems arising from re-deposition, and high equipment cost.[13] Datta et al. investigated the feasibility of ion milling to remove Cr/phased CrCu/Cu BLM layer during electrochemical fabrication of 97Pb/3Sn C4s.[8] The process was evaluated due to its directionality and its ability to produce oversized BLM pads with tapered edges that are similar to the so-called feather-edged BLM of evaporated C4s. The etch rate selectivity in ion milling is determined by the sputter yield of each material. The study concluded that the ion milling process, similar to other vacuum processes, redeposits the BLM material on the bump surface. This led to the formation of an intermetallic layer on top of the C4 bumps. Since the intermetallic layers have melting points higher than the PbSn C4s, formation of these surface layers prevented the solder from shaping into a ball during the reflow process.[8]

Wet etching processes are predominantly used for BLM etching during electro-chemical fabrication of die bumps. As indicated earlier, commonly employed BLM materials include Ti, TiW, Cr, phased CrCu, Cu, Co, and Ni. An understanding of the reactivity of these BLM materials in different etching baths is required to develop a viable selective wet etching process for removal of BLM without significant attack of the bump material.

6.8.1   *Ni, Cu, Cr, and phased CrCu etching*

Chemical etching. Thin films of nickel, copper, and chromium are easily etched in a wide variety of chemical etching baths that include nitric acid, hydrochloric acid, hydrogen peroxide, ammonium persulfate, and permanganate-based baths. However, selective etching of BLM layers in the presence of bump materials such as 97Pb/3Sn, 63Sn/37Pb or Sn-rich lead-free solders make the development of etching processes more challenging. Attack on the solder during BLM etching should be minimized in order to avoid: (1) loss of solder volume, (2) preferential dissolution of active phase of the solder alloy compared to other phases that may lead to changes in solder composition, and (3) formation of rough solder surface that may impede e-test and sort probing. Depending on the solder/BLM combination, therefore, a variety of etching baths have been developed in the microelectronics industry. For example, thin layers of nickel and nickel–vanadium are etched in hydrogen peroxide–sulfuric acid based etching baths. The exact composition of each component in the bath may vary depending on the solder composition.

Most of the technical know-how on Cu etching is derived from its application in the PCB industry.[35,36] A large number of formulations used for Cu etching include ferric chloride, chromic–sulfuric acids, ammonium persulfate, hydrogen peroxide–sulfuric acid, cupric chloride, and ammoniacal etchants. Cupric chloride and ammoniacal etchants are the most commonly used Cu etching chemistries in the PCB industry due to the possibility of automatic regeneration or replenishment that is needed to maintain steady-state etching conditions. However, these solutions also easily etch Pb–Sn alloys. Furthermore, wherever possible, chloride-based etchants should be avoided in order to develop a corrosion resistant die bumping process. Chloride ions, in principle, are extremely soluble so that final rinsing and cleaning operations are sufficient to completely remove these ions from the surface. On the other hand, penetrations of even trace amounts of chloride ions through defects such as grain boundaries and pinholes may be sufficient to cause corrosion or stress corrosion cracking of bumps leading to early failure during thermal cycling.

In the microelectronics industry, ammonium persulfate based processes are frequently employed for Cu etching. Since persulfate baths also attack PbSn bumps, development of a selective Cu etching process requires optimization of the bath components such that PbSn attack is retarded without significant influence on the Cu etch rate. Such an etching process for removal of Cu BLM during fabrication of 97Pb/3Sn C4s has been reported.[37] The bath consists of 12.5 g/l of ammonium persulfate, 5 cc/l of sulfuric acid, and 1 g/l of copper sulfate. The process gives an etch rate of 460 Å/min for immersion etching, 560 Å/min for submerged spray etching, and 610 Å/min for spray etching.[37] Spray etching eliminates problems of non-uniformity of etching caused by gas bubbles that are generated at the lead-rich bumps and tend to stick to the BLM surface thus preventing their removal at those locations. The process is compatible with PbSn bumps and other exposed materials.[37] The ammonium persulfate etchant is also applicable to the selective high speed etching of the Co BLM layer.[8]

Several etchant formulations in the acid and alkaline ranges are available in the literature for chemical etching of Cr.[13,37,38] The acidic etchants are generally based on hydrochloric acid or nitric acid. These formulations are not applicable in the C4 fabrication since they attack solder material and are expected to severely undercut the solderable layers (Cu, Co). A potassium permanganate based Cr etching process for C4 fabrication has been reported in the literature.[8,37] Since Cr is used as an adhesion layer between the passivation layer and the solderable layer, the compatibility of the Cr etching process with these layers is extremely critical. The temperature and the pH of the process is optimized to render an etching process that is completely resistant to polyimide or silicon dioxide attack. The etching bath consists of a mixture of 0.25 M potassium permanganate and 0.5 M sodium hydroxide, which operates at room temperature. The etching rate of Cr is estimated to be 75 Å/min for immersion etching, 100 Å/min for submerged spray etching, and 107 Å/min for spray etching.[37]

Chemical etching of phased CrCu is not straightforward. The layer involves a combination of heterogeneous phases of an active material (Cu) in contact with a passive material (Cr). The use of the Cu etching process described above does not attack the phased CrCu layer, indicating that the overall property of the phased layer is dictated by the passivating property of the Cr phase. The Cr etching process described above removes the phased layer by selective removal of Cr phases, which detaches from the Cu phases. Continuation of the process eventually also leads to removal of the Cu phase, but leaves behind islands of Cu phases that need to be removed by further processing. Electroetching is a more effective way of removing phased CrCu.

Electroetching. Electroetching is a promising alternative method of BLM

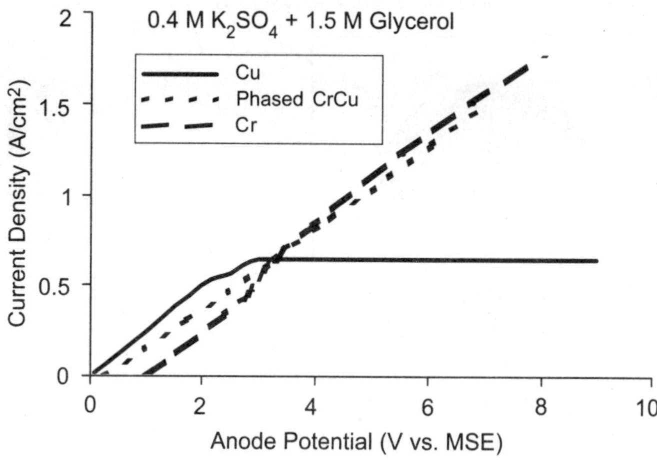

Figure 6.9   Anodic polarization curves for Cu, Cr, and phased CrCu (50/50) in potassium sulfate solution with glycerol. (Source: Ref. 43.)

etching. Since the driving force for the metal dissolution is derived from an external current source, the electrolyte in electroetching does not have to be aggressive. Indeed, most of the thin films of metals that are of interest in the electronics industry can be anodically dissolved in neutral salt solutions such as sodium nitrate, sulfate, or chloride.[39] In these non-toxic electrolytes, the dissolved metal ions form hydroxide precipitates which remain in suspended form in solution and can be easily filtered, thus significantly minimizing safety and waste disposal problems. Accumulation of reaction products in solution and depletion of bath components are of significantly less concern in electroetching of thin films. All of these reasons make electroetching a greener, safer, and simpler manufacturing process. On the other hand, frequently encountered loss of electrical contact during electroetching of thin films is one of the main reasons for slow acceptance of electrochemical metal removal as a microfabrication technique.[40,41]

Electroetching of Cu, Co, Ni, and Cr has been extensively studied and the available literature has been reviewed.[42] These studies demonstrate that these metals can be effectively etched in neutral salt solution and selective dissolution of these films can be performed by properly selecting the range of potential or voltage. In many cases, the anodic behavior of one metal with respect to another can be influenced by adjusting the proportion of passivating vs. non-passivating anions in the electrolyte. In an effort to develop an electroetching process for removal of Cr/phased CrCu/Cu, rotating disk electrode studies were performed to characterize the anodic dissolution behavior in neutral salt solutions of sodium nitrate, and sodium sulfate.[43] Figure 6.9 shows the potentiostatic anodic polarization curves for Cr, phased CrCu(50/50), and Cu in a mixture of 0.4 M potassium sulfate and 1.5 M glycerol. Thin films of each material were sputter deposited on Cu disks and were subjected to anodic polarization by potentiostatic scanning the potential from open circuit potential to very high anodic values at a scan rate of 100 mV/sec while rotating the disk at 1000 rpm. Cu shows a limiting current over a wide range of potential

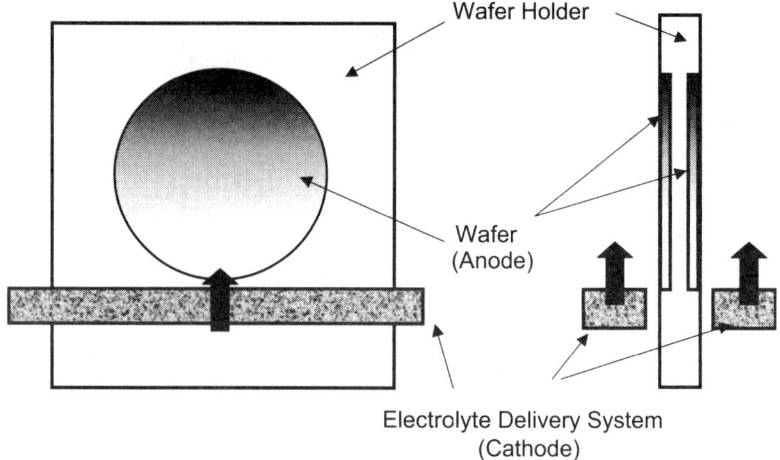

Figure 6.10 Schematic diagram of an electroetching tool for removal of BLM.[43]

beyond 3 V, whereas the current density continuously increases with increasing potential for both Cr and phased CrCu. This behavior allows one to choose optimum operating conditions at which all the layers can be etched without significant undercutting of the uppermost Cu layer. From a practical point of view, the data given in Figure 6.9 have only qualitative significance since in a manufacturing electroetching unit, the controlled parameter is expected to be cell voltage rather than electrode potential. However, current density vs. cell voltage data similar to that of Figure 6.9 can be easily established for optimization of the electroetching tool and process.

An electroetching tool for removal of Cr/phased CrCu/Cu has been patented and its applicability in the fabrication of PbSn C4s has been demonstrated.[43] Figure 6.10 shows a schematic diagram of the electroetching tool where two wafers are mounted vertically, back-to-back in a wafer holder. An alternative configuration of the tool, in which the wafer is mounted horizontally, has also been demonstrated.[43] An electrolyte delivery system in the form of multi-nozzle assembly is attached to a linear motion table and is scanned from one end of the wafer to the other end during electroetching. The wafer is connected as an anode. A stainless steel plate in the electrolyte delivery system acts as the cathode. The stainless steel plate is engraved with small nozzles through which an electrolyte impinges onto the wafer. The motion of the linear electrolyte impingement jet created by the multi-nozzle assembly is a single pass over the wafer surface, in an upward direction. The scanning rate is adjusted according to the dissolution rate of the material stack and the thickness of material to be removed. Using a mixture of 0.4 M potassium sulfate and 1.5 M glycerol as the electrolyte, a stack of Cr/phased CrCu/Cu has been successfully etched for fabrication of 97Pb/3Sn C4s with minimum undercutting of the Cu pad so that the desired C4 size after reflow could be achieved. The use of glycerol in the electrolyte helped in lowering the limiting current of Cu, thus allowing minimized Cu undercut during dissolution of the phased CrCu and Cr layers. For the TiW/phased CrCu/Cu BLM stack, the electroetching process was used to remove the top CrCu/Cu layers. The TiW layer remained completely passivated in the potential region where CrCu and Cu layers are anodically dissolved at high rates in the above electrolyte. This behavior served as an autostop for the electroetching process and the electroetching stopped at the TiW layer, which was removed in a separate wet etching process described below.

6.8.2 Ti, TiW, TiW(N) etching

Titanium, 10%Ti/90%W, and titanium–tungsten–nitride have the combined property of being an adhesion layer as well as a barrier layer. These layers are in direct contact with the final metallization layer and passivation layer. Etching processes for removal of these layers should have a sufficient etch rate for these materials while keeping the chip metallization, passivation, and solder materials practically intact. Hydrofluoric acid based etchants are popular in the microelectronics industry for etching titanium. These etchants have the advantage that they do not attack copper and aluminum, and the bath can be optimized to minimize attack on commonly used Pb- and Sn-rich solders.

The use of hydrogen peroxide as the etchant for TiW and TiW(N) BLM layers for C4 fabrication is well documented in the literature.[44–46] Van Oekel employed a buffered hydrogen peroxide solution to etch 10%Ti90%W films.[44] Acetic acid and ammonium acetate were used as a buffer in the pH range between 1 and 6 and citric acid and sodium hydroxide, in the alkaline pH range.

Datta et al. developed a chemical etching process based on hydrogen peroxide to selectively remove TiW in the presence of PbSn, CrCu, Cu, and Al.[45] The etching bath consisted of a mixture of the following constituents: (1) hydrogen peroxide acting as the etchant; (2) potassium sulfate acting as a passivating agent that forms protective layers over PbSn C4s; (3) potassium-EDTA acting as a stabilizer for hydrogen peroxide, a buffer and a complexant for the etched products. The bath is operated at 50°C. An end point detection method consisting of a contactless real-time *in situ* monitoring by impedance measurement[37] permits one to stop etching at a point that corresponds to complete removal of TiW between C4s, while providing a minimum undercut. Both hydrogen peroxide and EDTA in the TiW bath degrade with time leading to pH changes and degradation of etching performance. Analytical methods for monitoring, and a strategy for replenishing individual components in the bath are described by Cooper et al.[47]

Ramanathan et al. developed an etchant for selectively etching TiW(N) in the presence of electroplated 95Pb/5Sn solder[46]. The TiW(N) layer served as a glue layer to ensure excellent adhesion to overlying Cu layer and also as a barrier layer to prevent Cu diffusion through the barrier film. A hydrogen peroxide–ammonium hydroxide based etchant was developed by conducting a statistically designed set of experiments to establish the etchant composition, and process temperature that gave desired responses with respect to etch time, minimized undercut, and acceptable bump shape after reflow. An optimized etchant was thus developed that consisted of 6% by volume of 30% hydrogen peroxide and 0.75% by volume of 30% ammonium hydroxide. The bath operated at a temperature of 37°C.[46]

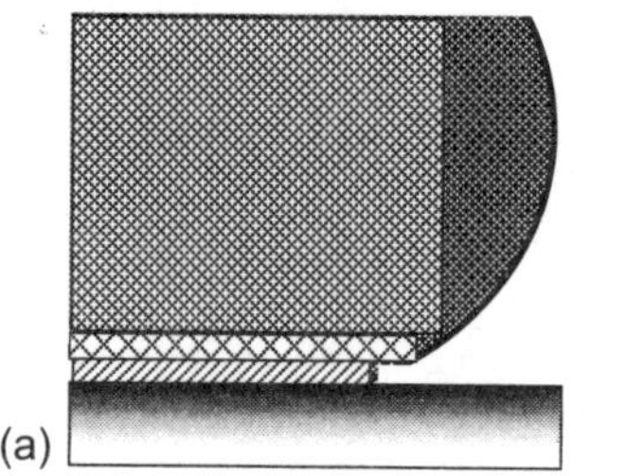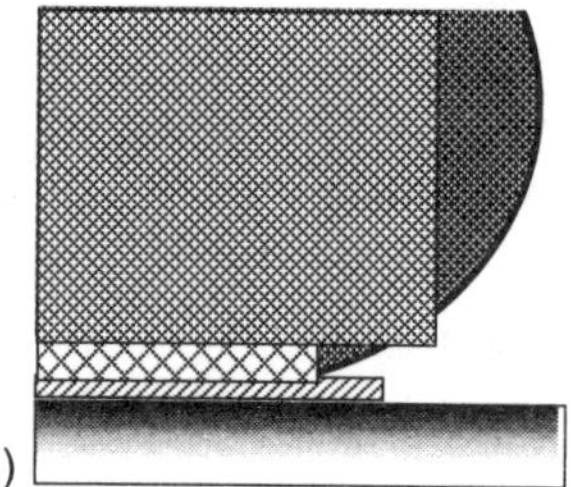

Figure 6.11   Schematics of BLM edge profiles: (a) after a two-step chemical etching, and (b) a stepped edge profile obtained by etch back of the top layer.

6.8.3   *Etched BLM profile and reliability concerns*

Wet etching of a BLM stack leads to undercut of the layers. Whereas these layers are generally sub-µm thick, the undercut may run into several micrometers, which can act as initiation points for failure during thermo-mechanical cycling.

The main causes of failure in reliability testing are delamination and/or under-bump cracking. Stresses in the film, stresses generated during intermetallic form-ation, and stresses due to thermal cycling are the principal causes of failure during reliability testing. Minimizing the stresses and distributing them properly along the BLM are key to the mechanical integrity of chip–package joining. Delamination is mainly caused by poor adhesion of the seed layer and is determined by the stresses in the film. Sputtered BLM layers are generally slightly compressive which is desirable for excellent adhesion of BLM to silicon nitride and polyimide. Indeed, sputtered BLM layers have been found to give better results than evaporated layers in terms of BLM delamination.[8]

Underbump cracking is believed to be related to stress distribution in and around the BLM. It has been shown that the edge effects play an important role in the incidence of insulator cracking at the metal terminal.[8,48] Specifically a coincident edge of the adhesion layer and the solderable layer leads to high stress levels at the abrupt edge and to cracking of the interlayer dielectric (ILD) layers. By giving the BLM edge a graded or stepped profile, stresses can be properly distributed to eliminate underbump cracking.

Mechanical robustness of evaporated C4s is due to a feathered or tapered edge of the BLM.[2,6] Electrochemically fabricated solder bumps that use wet etching for BLM removal, in principle, cannot provide a tapered BLM edge. A two-step etching process will lead to a negative step due to undercutting of the bottom layer as shown in Figure 6.11a. Such a negative step in the BLM is expected to cause mechanical failure during thermal cycling. The effect of BLM edge profile on reliability data is expected to be much more pronounced in the case of assembly processes where underfilling is not used. In order to alleviate such problems, wet etching processes have been optimized to minimize undercutting of BLM layers and an etchback method has been employed to produce a non-coincident stepped edge in the BLM as shown in Figure 6.11b.[8,48] The stepped edge profile of BLM reduced the geometrical stress concentration at the BLM edges, thus minimizing the probability of underbump cracking. The reliability test data demonstrated that through proper considerations of the BLM edge profile during wet etching, electrochemically fabricated C4s can have mechanical reliability comparable to that of evaporated C4s.

6.9 Copper bumps for flip-chip interconnection

A variety of bump materials in conjunction with different BLM materials are employed in flip-chip interconnection. Table 6.3 summarizes some of the selected bump/BLM combinations derived from the published literature.[7,8,45,49–57] Copper bumps represent a special case of bumping material because of extremely high melting point and relatively high stiffness compared to conventional solder materials.

Use of copper as a bumping material has been well documented in the literature. Copper as the bump material provides a superior electrical and thermal conduction path from the chip. The electrical and thermal resistance can be greatly influenced by the geometry, number, and location of bump interconnections. On the other hand, the stiffness of copper limits its use as a bumping material for applications where a highly compliant assembly system is required. Indeed, Heinen et al. demonstrated

Table 6.3 Selected electroplated die bumping processes.

BLM	Bump material	Number of plating steps	As plated bump shape	Reference
Ti/NiV	97Pb/3Sn	One	Mushroom	Shukla et al.;[49] Intel Corporation
Cr/CrCu/Cu	97Pb/3Sn	One	Column	Datta et al.;[8] IBM Corporation
TiW/CrCu/Cu	97Pb/3Sn	One	Column	Datta et al.;[45] IBM Corporation
Cr/CrCu/Cu	95Pb/5Sn	One	Column	Yung et al.;[7] MCNC
TiW(N)/Cu	Cu stud/95Pb/5Sn Cu stud/63Sn/37Pb	Two	Mushroom	Clegg et al.;[50] Motorola Inc.
Ti/Ni/Pd	63Sn/37Pb	One	Column	Takubo et al.;[51] Toshiba
Ti/Ni/Pd	Sn/3.5Ag	Two	Column	Ezawa et al.;[52] Toshiba/Ebara
TiW/Cu	Cu	One	Wire (WIT)	Moresco et al.;[55] Fujitsu
Ti/Cu	Cu (+ Sn/37Pb cap)	Two	Column	Yamada et al.;[54] Toshiba
Cr/Cu/Au/Cu	Cu (+ Pb/10Sn cap)	Two	Column	Heinen et al.;[53] Texas Inst.
TiW(N)/Cu; Cr/Cu	Cu stud/95Pb/5Sn	Three	Mushroom	Wolf et al.;[56] Univ. of Berlin
TiW/Au	Au	One	Column	Zakel et al.;[57] Univ. of Berlin
TiW/Au	80Au/20Sn	Two	Column	Zakel et al.;[57] Univ. of Berlin

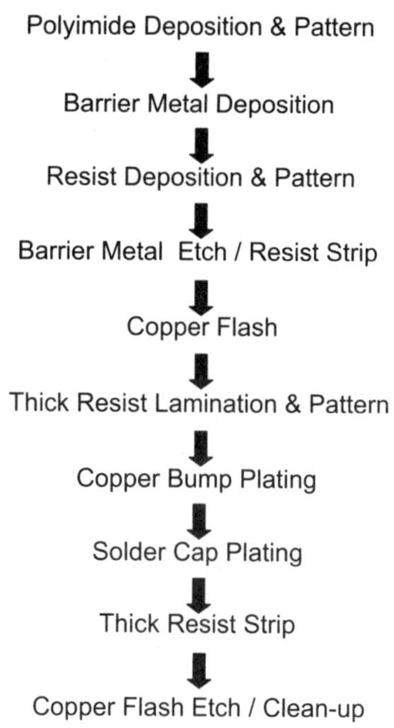

Figure 6.12 Copper bump plating process using two resist steps as described in Ref. 54.

that the use of copper bumps places restrictions on the type of passivation material.[53] Thermal stresses from copper bumps caused cracks on brittle silicon nitride or silicon oxide passivation layers.[53] However, the use of a more compliant polyimide passivation layer provided a stress buffer and was found to be compatible with copper bumps.

A copper bump based flip-chip interconnect technology for multichip modules has been described by Heinen et al.[53] The approach involves two lithography and two plating steps. The electroplated copper bumps are capped with sufficient solder to provide reflow bonding. The process flow for Cu bump formation is shown in Figure 6.12. The steps consist of deposition and patterning of a polyimide passivation layer, deposition of a blanket layer of BLM, deposition and patterning of a photoresist, etching of the barrier layer, removal of a photoresist, deposition of a blanket copper layer as the electroplating base, application and patterning of a thick photoresist, electroplating of copper bumps through a thick photoresist, electroplating of the solder cap on top of copper bumps, removal of a thick photoresist, and finally etching of a plating base copper layer to electrically disconnect the bumps.

The BLM layers were sputter-deposited and consisted of Cr/Cu/Au stack. An ammonium iodide/iodine mixture was used for chemical etching of the gold and copper layers and a caustic ferricyanide etchant was used for etching the chromium layer. Copper bumps, 100 μm in diameter with 100 μm space and 85 μm in height, were plated from an acid bath in a plating tool with a specially designed plating fixture that provided planarity of ±5 μm for bump height. The solder cap consisted of plating 20 μm of a 90Pb/10Sn layer from a fluoborate bath. The solder quality with respect to composition control and voids from this bath was superior to that from a sulfamate bath.

6.9.1   *Fine pitch high-aspect ratio Cu bumps*

Ordinarily, the bumps for flip-chip interconnections are around 75 μm high and 100 μm in diameter, and have a pitch varying between 150 and 200 μm. For high-speed optical devices operating at more than 10 GH and high density processing modules, smaller bumps with finer pitch are needed.[54] The bump height must be sufficiently high to reduce the strain caused by mismatch in the coefficient of thermal expansion (CTE) between the chip and the substrate. Such high aspect ratio bumps pose significant processing and reliability challenges. Yamada et al. developed a fine pitch, high aspect ratio bumping process for fabrication of an array of 5 μm diameter and 20 μm high, and pitch bumps of 10 μm for high reliability flip-chip interconnection.[54] The I/O pad geometry was modified and relocated at the center of the device to reduce the strain on the bumps using multilayer aluminum/polyimide wiring process. A precision resist patterning process was developed using a positive photoresist. The bumps were made of a copper base pillar capped with PbSn solder. Precision plating involved the use of an electroplating tool with recirculation of a well-controlled plating bath, and monitoring of bath components. The authors claim that the fine-pitch high aspect ratio bumps are expected to improve the thermal fatigue life four times longer than the conventional bumps.[54]

Another example of fine pitch and high aspect ratio flip-chip interconnection is Fujitsu's wire interconnect technology (WIT). WIT is a Cu metal post, approximately 10 μm in diameter and 50 μm high that is soldered to a micro-bump (approximately 25 μm diameter) on the substrate. WIT was developed to meet the need of >6000 I/O per chip on a minimum of 50 μm pitch. These interconnects required low electrical parasitic, high reliability, and low α-radiation flux.

An electroplating method of fabricating fine pitch Cu WITs has been described by Moresco et al.[55] A high aspect ratio via is patterned in a 50 μm thick layer of a positive photoresist. Electroplating into the deep high aspect ratio via requires special plating techniques. Activation of the surface is accomplished by chemical polishing or electropolishing, depending on the material under the WITs. A wetting agent in the activator is used to enhance the transport of etchant to the surface into the deep via. Sufficient solution agitation is essential to expel the rinse solution prior to electroplating and to replenish the copper ions during plating. Moresco et al.[55] used a high throwing power acid/copper sulfate type of electrolyte. The anode size, interelectrode distance, and electrolyte flow were optimized to minimize the variations in WIT height.[55] The electroplating process was optimized to yield a finely grained plated structure. Figure 6.13 shows an array of 10 μm diameter, 37 μm tall Cu WITs at a pitch of 40 μm. To allow multiple replacement of WIT attached chips, a thin layer of nickel is coated over the Cu WITs. The nickel layer acts as a diffusion barrier between the copper and the solder, thus allowing multiple solder attach and replacement cycles without degradation of the solder joint materials.

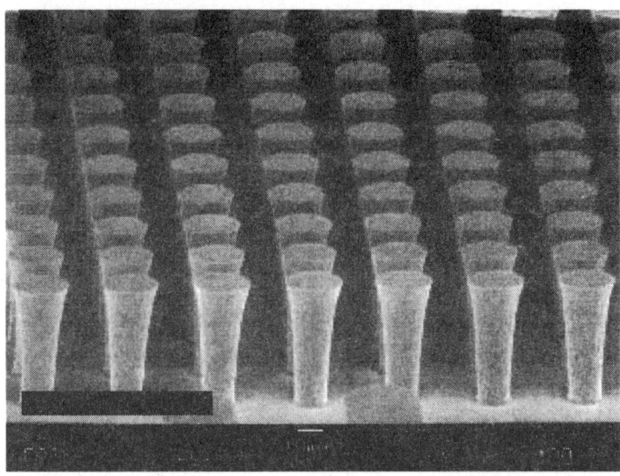

Figure 6.13 SEM photograph of an array of Cu wire interconnects (WIT) with 40 μm pitch. (Source: Moresco et al.;[55] photo provided by Fuzitsu Technologies Inc.)

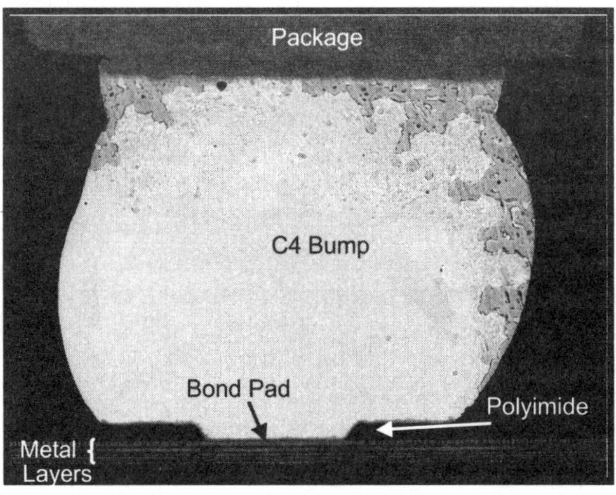

Figure 6.14 SEM photograph of a cross section of a chip–package assembly using flip-chip interconnect.

6.10   Assembly and reliability

6.10.1   Assembly process

Chip attach and encapsulation are the key assembly steps of a flip-chip interconnect system. Flip-chip attachment involves application of a flux on the substrate and alignment of the die bumps with the substrate solder, followed by heating the module to a specified temperature so that the solder melts and forms a metallurgical bond between the die and package solder bumps.

The shape and geometry of the solder interconnects depends on the chip attach process conditions. In the controlled collapse chip connection approach, surface tension of the molten solder determines the bump shape.[2] The gap height of the bonded module is determined by the balance of the surface tension force on the solder and the weight of the chip. Another method of controlling gap height is to pull the chip away from the substrate while the solder is molten, and then the connected module is cooled. This method of stretching a molten solder joint has been used to obtain larger standoff gaps in a high accuracy flip-chip interconnection.[58]

Following the flip-chip bonding process, the residual fluxes are cleaned. The module is then encapsulated by filling the gap between the chip and substrate with an epoxy-type underfill. Underfill protects the chip and interconnects during subsequent processes, and, more importantly, it helps in absorbing the thermo-mechanical stresses, hence improving the reliability of the interconnect. A cross section of a flip-chip package assembly is shown in Figure 6.14.

6.10.2 *Reliability issues*

Selection of a flip-chip interconnection system is determined by its reliability targets. The flip-chip solder joint must mechanically hold the chip to a substrate, but the solder must not impose significant strain to the underlying interconnect metallization and ILD layers or they could crack and fail. Thermal fatigue and corrosion of the solder joint can significantly affect the performance of a solder-based interconnect system during service. Electromigration and radiation-induced damage are some of the other reliability concerns that must be resolved. The kinetics of failure due to such phenomena is influenced by the solder composition, the substrate and underfill materials, the assembly process parameters, and the severity of the environment. Some of these aspects are discussed below.

Thermo-mechanical behavior of a solder joint is influenced by its composition. For example, high-lead solder (97Pb/3Sn; melting temperature of 310°C) interconnect exhibits three times the fatigue life of eutectic (63Sn/37Pb, melting temperature of 183°C) solder.[59] However, there is no direct correlation between the composition and the melting point of the solder and its fatigue resistance. Therefore, the selection of the solder composition for optimum thermo-mechanical integrity is most often empirical.

Thermal mismatch between different parts of an assembled unit impose relative displacements, which produce mechanical stresses in the solder joint. These stresses can lead to initiation and growth of cracks. The magnitude of the stress depends on the assembly stiffness and inelastic deformation properties of the solder joint. Propagation of cracks leads to structural and hence electrical failures.

Interaction of BLM with solder during assembly and other thermal cycling processes is extremely critical in determining the reliability of a flip-chip interconnection. In a recent investigation, Frear et al. investigated the solder microstructure and intermetallic reaction products for a variety of lead-containing and lead-free solder alloys for use as flip-chip interconnects.[20] During reflow, the intermetallic growth rate was found to be faster for the lead-free (predominantly Sn: ~96.0 to

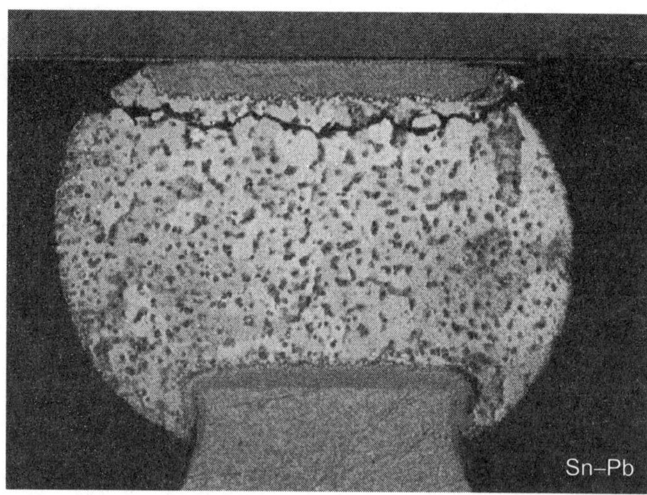

Figure 6.15 Cross section of an eutectic 63Sn/37Pb solder joint after thermal cycling showing formation and propagation of crack. (Source: Frear et al.[20])

99.3%) alloys, compared to eutectic tin–lead solder. Their reliability tests, which included shear strength and thermo-mechanical fatigue, indicated that failure occurred near the solder/intermetallic interface for all the alloys except Sn0.7 Cu which deformed by grain sliding and failed at the center of the joint. Figure 6.15 shows an example of the 63Sn/37Pb flip-chip interconnect fail by crack formation and propagation at the BLM/solder interface during thermal cycling.[20] The BLM/solder interface is the highest strain region in the joint where coarsening and failures typically occur. The cracks were observed to form on the outer edge of the bump and propagate to the center of the bump. The damage was localized to the heterogeneous coarsened band.[20]

Corrosion and stress corrosion cracking can also cause interconnect failures. Halide anions (e.g., from the flux residue or from the underfill) are known to cause corrosion of solder materials. High temperature and high humidity coupled with high thermo-mechanical stresses create favorable conditions for acceleration of corrosion reactions. Moisture has been found to be an important factor affecting thermal fatigue of eutectic SnPb solder and failure of indium solder flip chip interconnection. Hermetic packages and moisture-resistant underfill are essential to minimize corrosion related failures.

With an increasing trend towards smaller bump diameter and finer pitch, electromigration is becoming a critical reliability issue for flip-chip interconnection. According to current circuit design rules, some of the interconnect bumps are required to carry currents in the range of 0.2 to 0.4 A, which can result in a current density of the order of 10^4 A/cm^2. Although this current density is much lower than that found in Cu interconnects, atomic diffusivity of solders is high even at these current densities due to their lower melting temperature. An investigation of electro-migration phenomena in eutectic 63Sn/37Pb solder has been reported by Lee et al.,[60] who studied the effect of annealing at 120°C with 2×10^4 A/cm^2 over an extended period of time. Cross sections of tested samples showed formation of hillocks on the anode side of the solder ball and void formation on the cathode side. The anode was Pb rich, indicating that Pb is the dominant diffusing species. Selection of proper bumping material and modification of architectural design are some of the changes that must be incorporated to minimize electromigration issues associated with flip-chip interconnects.

Another reliability concern is the radiation-induced damage of devices, known as soft error rate (SER). High radiation counts may lead to damage of devices in high-end products due to mobile charge production in the silicon. The source of the radiation that causes the errors is attributed to trace amounts of radioactive materials in the package, particularly in the solder. The most significant emission of α-particles is from the lead in the solder interconnects. These radioactive sources are particularly damaging because the solder interconnects are located close to the silicon device, and there are no barriers to absorb the particles. Lead and tin–lead alloys (solder) contain radioactive elements of uranium and thorium that decay to form α-particles. The principle cause of α-emission is ^{210}Pb (an unstable isotope found in lead) that decays to form ^{210}Bi and then to the very unstable isotope of ^{210}Po, which subsequently decays to an α-particle. Cosmic radiation is another source of α-particles. However, no appreciable reduction in cosmic radiation is possible through material or process changes.

SER related reliability issues can become more predominant as gate size shrinks. Therefore there is a need for a source of lead with a low α-radiation rate to minimize the possibility of soft errors. Indeed, low α lead baths are currently used for 97Pb/3Sn C4 plating in leading chip making industries. Use of lead-free solders in flip-chip interconnects will further minimize the SER related reliability concerns.

6.11 Summary and concluding remarks

Flip-chip (C4) technology has long been recognized as a high density, high-performance area array interconnection. However, the original evaporated flip technology was mainly used in high-end applications that were restricted to high melting temperature lead-rich solders for joining chips to ceramic substrates. With the increasing demands for high density interconnects for organic packages, and the need for low melting temperature solders, the microelectronics industry moved to electrochemical processing of C4s. This led to a phenomenal increase in the applicability of the C4 technology. Indeed, electrochemical processing has enabled the fabrication of cost-effective, high-performance C4s that are extendible to finer pitch, larger wafers, and a variety of solder compositions including some lead-free alloys. These advantages have made the plated C4 technology a preferable chip–package interconnection for a variety of products.

The microelectronics industry is currently involved in the development of low temperature, lead-free solders. Due to environmental, and radiation induced damage related reliability concerns, there is a serious need to develop Pb-free flip-chip bumping technologies that will help assembly of advanced chips to packages with higher CTE mismatch. Development of Pb-free C4 materials requires a careful consideration of their electrical and thermo-mechanical properties. Fragile ultra low-k dielectric material in the chip puts additional burden to the C4 material selection criteria, which should aim at minimizing stresses imposed on the dielectric material by the C4 bumps. Electrochemical processing is expected to play a critical role in the industry's effort to develop reliable Pb-free flip-chip technologies for high performance microprocessors.

Acknowledgment

I would like to thank Professor Keith Sheppard of Stevens Institute of Technology, Hoboken, NJ, for a careful review of the manuscript.

References

1. The National Technology Roadmap for Semiconductors, SIA (1997).
2. P.A. Totta, S. Khadpe, N.G. Koopman, T.C. Reiley, and M.G. Sheaffer, Chip to Package Interconnections, in Microelectronic Packaging Handbook, Semiconductor Packaging Part II, R.R. Tummala, E.J. Rymaszewski, and A.G. Klopfenstein, eds., second edition, Kluwer Academic Publishers, Boston (1999).
3. S. Prasad, Technology Comparisons and the Economics of Flip Chip Packaging, *Adv. Packag.*, March (2001).
4. R. Sharma and R. Subrahmanyan, Solder Bumped Flip Chip Interconnection Technologies: Materials, Processes, Performance, and Reliability, in Flip Chip Technologies, J.H. Lau, ed., McGraw-Hill, New York (1996).
5. Microelectronic Packaging Handbook, Semiconductor Packaging Part II, R.R. Tummala, E.J. Rymaszewski, and A.G. Klopfenstein, eds., second edition, Kluwer Academic Publishers, Boston (1999).
6. P.A. Totta and R.P. Sopher, SLT Device Metallurgy, and its Monolithic Extension, *IBM J. Res. Dev.*, **5**, 226–238 (1969).
7. E.K. Yung and I. Turlik, Electroplated Solder Joints for Flip Chip Applications, *IEEE Trans. Components, Hybrids, Manuf. Technol.*, **14**, No. 3, pp. 549–559, September (1991).
8. M. Datta, R.V. Shenoy, C. Jahnes, P.C. Andricacos, J. Horkans, J.O. Dukovic, L.T. Romankiw, J. Roeder, H. Deligianni, H. Nye, B. Agarwala, H.M. Tong, and P. Totta, Electrochemical Fabrication of Mechanically Robust PbSn C4 Interconnection, *J. Electrochem. Soc.*, **142**, No. 11, 3779–3785 (1995).
9. A.D. Aird, Method of Manufacturing a Semiconductor Device Utilizing a Flexible Carrier, U.S. Patent 3,689,991 (1972).
10. T. Kamei and M. Nakamura, Hybrid IC Structures using Solder Reflow Technology, 28th Electronic Components Conference Proceedings, pp. 172–182 (1978).
11. J.F. Roeder, Proceedings of the 7th Electronic Materials Processing Congress, ASM International, 225 (1992).
12. K. L. Lin and Y.T. Liu, Manufacturing of Solder Bumps with Cu/Ta/Cu as Under Bump Metallurgy, *IEEE Trans. Adv. Packag.*, **22**, No. 4, November (1999).
13. J.L. Vossen and W. Kern, Thin Film Processes, Academic Press, New York (1978).
14. G.O. Mallory and J.B. Hajdu, eds., Electroless Plating: Fundamentals and Applications. American Electroplaters and Surface Finishers Soc., Orlando, FL (1990).
15. D.S. Patterson, P. Elenius, and J.A. Leal, Wafer Bumping Technologies – A Comparative Analysis of Solder Deposition Processes and Assembly Considerations, InterPACK '97, EEP-Vol. 19-1, June (1997).
16. S.K. Kang, J. Horkans, P.C. Andricacos, R.A Carruthers, J. Cotte, M. Datta, P. Gruber,

J.M.E. Harper, K. Kwietniak, C. Sambucetti, L. Shi, G. Brouillette, and D. Danovitch, Pb-free Solder Alloys for Flip-Chip Applications, 1999 Electronic Components and Technology Conference, IEEE, 283 (1999).

17. J.S. Hwang, Modern Solder Technology for Competitive Electronics Manufacturing, McGraw-Hill, New York (1996).

18. M. Abtew and G. Selvaduray, Lead-free solders in Microelectronics, *Mater. Sci. Eng.*, **27**, 95–141 (2000).

19. J. Glazer, Metallurgy of Low Temperature Pb-free Solders for Electronic Assembly, *Int. Mater. Rev.*, **40**, No. 2, 65–93 (1995).

20. D.R. Frear, J.W. Jang, J.K. Lin, and C. Zhang, Pb-free Solders for Flip-chip Interconnects, *JOM*, **53**, No. 6, 28 (2001).

21. T.H. Tam, *J. Electrochem. Soc.*, **133**, 1782 (1990).

22. P.A. Kohl, The High Speed Electrodeposition of Sn/Pb Alloys, *J. Electrochem. Soc.*, **129**, 1197 (1982).

23. M.A.F. Samel, D.R. Gabe, and D.R. Estham, Electrodeposition of PbSn Alloys from a Sulfamate Solution, *Trans. Inst. Met. Finish.*, **65**, 116 (1987).

24. C. Rosenstein, Methane Sulfonic Acid as an Electrolyte for Sn, Pb, and Sn–Pb Plating for Electronics, *Met. Finish.*, **88**, No. 1, 17 (1990).

25. M. Goodenough and K.J. Whitlaw, The Suppression of Sn, Pb, and Sn–Pb by organic Compounds, *Trans. Met. Finish*, **67**, 44 (1987).

26. J. Horkans, I.C. Hsu Chang, and P.C. Andricacos, A Rotating Ring-Disk Stripping Technique Used to Study Electroplating of Sn–Pb from Methane Sulfonic Acid Solutions, *IBM J. Res. Dev.*, **37**, 97 (1993).

27. U.S. Patent 4,565,609 (1986); 4,559,149 (1986); 4,617,097 (1986).

28. Q. Lin, K. Sheppard, M. Datta, and L.T. Romankiw, Pulse Plating of Lead–Tin Alloys from Methane Sulfonate Electrolytes, to be published.

29. M. Hurley, J. Lovie, and K. Miscioscio, Process control in Sn and Sn–Pb Plating, *Electron. Manuf.*, **36**, No. 1, 12 (1990).

30. H. Richter, K. Ruess, A. Gemmler, and W. Leonhard, Precision Tin/Lead Alloy Plating for Flip-Chip Mounting Technology, *Microelectron. Int.*, No. 42, 9 (1997).

31. J. Horkans, Polarographic Methods of Monitoring Addition Agents in the Electroplating of Sn–Pb Solders, *IBM J. Res. Develop.*, **42**, No. 5, 621 (1998).

32. S. Mehdizadeh, J. Dukovic, P.C. Andricacos, L.T. Romankiw, and H.Y. Cheh, *J.Electrochem. Soc.*, **137**, 110 (1990); **139**, 78 (1992).

33. L.T. Romankiw, I. Croll, and M. Hatzakis, *IEEE Trans. Magn.*, 6, 29 (1970); L.T. Romankiw, U.S. Patent 3,853,715, December 10 (1974).

34. T. Ritzdorf and B. Batz, Fountain Electroplating of Lead–Tin Solder for Semiconductor Flip Chip Applications, Int. Technical Conference Proceedings of SUR/FIN (1996), pp. 257–262.

35. M. Gurian, *Printed Circuit Fabrication*, **10**, No. 8, 12 (1987).

36. D. Ball and R. Campbell, *Printed Circuit Fabrication*, **10**, No. 8, 20 (1987).

37. S.G. Barbee, M. Datta, T.F. Heinz, L. Li, E.H. Ratzlaff, and R.V. Shenoy, Method and Apparatus for Contactless Real-Time *In-situ* Monitoring of a Chemical Etching Process, U.S. Patent 5,445,705, August 29 (1995).

38. P. Walker and W.H. Tam, Handbook of Metal Etchants, CRC Press, Boston (1991).

39. M. Datta, Microfabrication by Electrochemical Metal Removal, *IBM J. Res. Develop.*, **42**, No. 5, 655 (1998).

40. M. Datta, Electrochemical Micromachining: Opportunities and Challenges, *The Electrochem. Soc. Interface*, **4**, No. 2, 32 (1995).

41. R.V. Shenoy, M. Datta, and L.T. Romankiw, Investigation of Island Formation Problem during Through-Mask Electrochemical Micromachining, *J. Electrochem. Soc.*, **143**, 2305 (1996).

42. M. Datta, Anodic Dissolution of Metals at High Rates, *IBM J. Res. Dev.*, **37**, No. 2, 207 (1993).

43. M. Datta and R.V. Shenoy, Electroetching Method and Apparatus, U.S. Patent 5,543,032, August 6 (1996).
44. J. Van Oekel, U.S. Patent 4,814,293, March (1989).
45. M. Datta amd R.V. Shenoy, Selective Etching of TiW for C4 Fabrication, U.S. Patent 5,462,638, October 31 (1995).
46. L.N. Ramanathan and D. Mitchell, *IEEE Trans. Components Packag. Technol.*, **24**, No. 3, 425 (2001).
47. E.I. Cooper, M. Datta, T.E. Dinan, T.S. Kanarsky, M.B. Pike, and R.V. Shenoy, Methods for Monitoring Components in the TiW Etching Bath used in the Fabrication of C4s, U.S. Patent 6,238,589, May 29 (2001).
48. B.N. Agarwala, M. Datta, R. Gegenwarth, C.V. Jahnes, P.M. Miller, H.A. Nye, J. Roeder, and M.A. Russak, Etching Processes for Avoiding Stresses in Semiconductor Chip Solder Bump, U.S. Patent 5,268,072, December 7 (1993).
49. R. Shukla, V. Murali, and A. Bhansali, Flip Chip CPU Package Technology at Intel: A Technology and Manufacturing Overview, 1999 Electronic Components and Technology Conference, IEE, 945 (1999).
50. D. Clegg, R. Cole, J. Franka, D. Mitchell, and D. Wontor, C4 Makes Way for Electroplated Bumps, *Adv. Packag.*, March (2001).
51. C. Takubo, N. Hirano, K. Doi, H. Tazawa, E. Hosomi, and Y. Hiruta, Eutectic Solder Flip Chip Technology for Chip Scale Packaging, 1996 IEEE/CPMT Int. Electronics Manufacturing Technology Symposium, 488 (1996).
52. H. Ezawa, M. Miyata, S. Honma, H. Inoue, T. Tokuoka, J. Yoshioka, and M. Tsujimura, Eutectic Sn–Ag Solder Bump Process for ULSI Flip Chip Technology, 2000 Electronic Components and Technology Conference, 1095 (2000).
53. K.G. Heinen, W.H. Schroen, D.R. Edwards, A.M. Wilson, R.J. Stierman, and M.A. Lamson, Multichip Assembly with Flipped Integrated Circuits, Proceedings of 39th Electronic Component Conference, May 22–24, 672 (1989).
54. H. Yamada, Y. Kondoh, and M. Saito, A Fine Pitch High Aspect Ratio Bump Array for Flip-Chip Interconnection, 1992 IEEE/CHMT Intl. Electronics Manufacturing Technology Symposium, 288 (1992).
55. L. Moresco, D. Love, V. Holalkeri, P. Boucher, B. Chou, C. Grilletto, and C. Wong, Wire Interconnect Technology, a New Flip Chip Technique, Proceedings Flip Chip, BGA, TAB Adv. Packaging Symposium, Sunnyvale, CA, 947 (1996).
56. J. Wolf, G. Chimel, J. Simon, and H. Reicht, Solder Bumping – A Comparison of Different Technologies, 1994 ITAP & Flip Chip Proceedings, 105 (1994).
57. E. Zakel and H. Reichl, Flip Chip Assembly Using Gold, Gold–Tin, and Nickel–Gold Metallurgy, in Flip Chip Technologies, J. H. Lau, ed., p. 415, McGraw-Hill, New York (1996).
58. K. Moore, S. Machuga, S. Bosserman, and J. Stafford, Solder joint Reliability of Fine Pitch Solder Bumped Pad array Carriers, Proceedings of NEPCON West, Vol. 1, pp. 264–274 (1990).
59. H. Inoue, Y. Kurihara, and H. Hachino, Pb–Sn Solder for Die Bonding of Silicon Chips, IEEE Transactions, Components Hybrid Manufacturing Technology, CHMT-9(2), 190 (1986).
60. T.Y. Lee, K.N. Tu, S.M. Kuo, and D.R. Frear, Electromigration of Eutectic SnPb Solder Interconnects for Flip Chip Technology, *J. Appl. Physics*, **89**, No. 6, 3189 (2001).

7 Compliant interconnects

Paul A. Kohl

7.1 Introduction

The architectural construction of an electronic system from individual integrated circuits (ICs) and other components is dependent upon effective interconnection substrates and means of connecting to them. The essential function of electrical interconnects for ICs is to provide *communications*. Early ICs needed only a few input–output connections (I/O), such as that provided by the dual in-line package (DIP) shown in Figure 7.1. In addition to providing pathways for electrical functions, the package and pin structure provided mechanical stability and environmental protection. A key element of the mechanical structure is the coefficient of thermal expansion (CTE) mismatch between the silicon die and the other materials, namely, copper and thermoplastic polymers (e.g., epoxy).

The advancement in CMOS transistor technology, principally through the shrinking of device dimensions, has created a significant challenge for connecting the integrated circuit I/O to the module. It has also created a pressing need for substrates (or interconnection boards) to provide the IC with the services it requires. Today, the chip-to-module connection methodology and substrate technology are a critical part

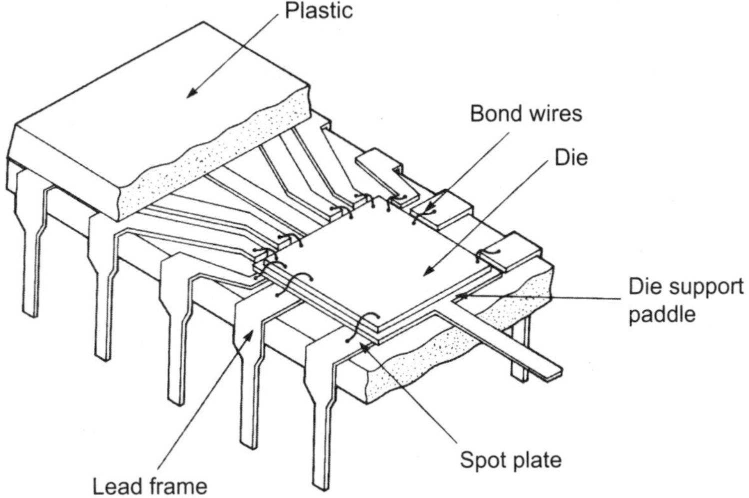

Figure 7.1   Sketch of dual in-line package.

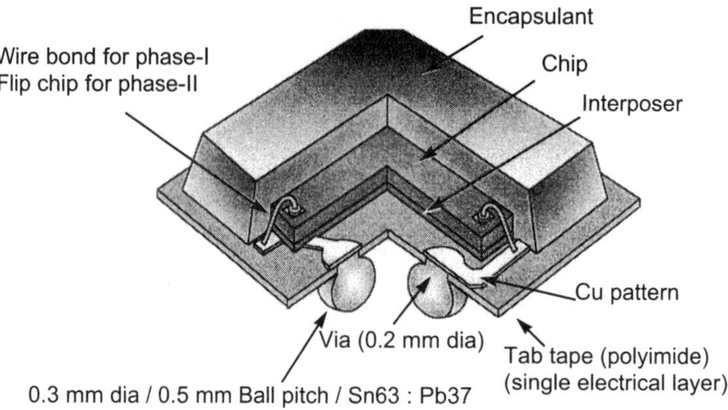

Figure 7.2   Sketch of ball grid array package.

of the IC design and partitioning. There is an ever greater risk that inadequate I/O connection and inferior substrate technologies will be the limiting factor in system performance and reduce the overall utility of the system. The progression of IC technology has demanded ever greater wiring and interconnection density and shorter interconnections. The ball grid array (BGA) is a contemporary version of the DIP. Figure 7.2 shows an IC mounted into a BGA package via short solder connections. The BGA would then be mounted onto a printed wiring board (PWB) via a larger solder ball. The example shows an increase in wiring density and I/O density as compared to the DIP. In this chapter, future trends in IC technology will be first presented so as to establish the need for higher interconnection density. The simultaneous need for denser I/O and shorter electrical path length leads into the discussion of *compliant interconnections.*

7.2   IC trends

The functions provided by the IC package and PWB can be divided into four classes:

1. electrical connections
2. structural and mechanical support
3. environmental and chemical protection
4. heat transfer

The International Technology Roadmap for Semiconductors (ITRS) provides guidance for GSI needs. Select values from the ITRS are presented in Table 7.1.[1]

7.2.1   *Electrical functions*

The electrical I/O provide signal, power and ground, and clock distribution for the IC. The critical parameter for signal and clock I/O is delay, as determined by the path

Table 7.1   ITRS (2000 update) goals for high-performance ICs.

Generation (node), mm	1.0	0.10	0.035
Clock speed, MHz	30	2000–3500	3600–13,500
MOS FET switching time, ps	20	5	2.5
Interconnect 'RC' delay, ps (length = 1 mm)	1	30	250
MOS FET switching energy, fJ	400	10	3
DC current, A	2.5	150	360
Maximum wiring length, m	100	5000	
Chip pin count	200	3000–4000	4000–4400

length, and electrical characteristics of the wiring (resistance, capacitance, and inductance). For power and ground, the cross-sectional area of the metal is critical because the current density cannot exceed about 1 MA/cm^2. Also, the voltage drop along power and ground lines as transistors switch is important because if the voltage is below a critical value, then transistor switching may not occur and swings in voltage appear as noise. The 35 nm technology node for high-performance CMOS calls for 360 A of DC current. It has been shown that the noise and voltage drop occurring within the DC power grid can be minimized by providing a large number of I/O dedicated to power and ground.[2]

The latency in interconnect and I/O is an emerging crisis for ICs. If a wire can be represented by series resistance (R) and capacitance (C), then the application of a potential step, V, to the RC circuit results in the flow of current, i, with respect to time, t, as shown in Equation 1.

$$i = \frac{V}{R}e^{-t/RC}$$ (1)

The exponentially decaying current decays to $1/e$ of its initial value (37%) in the time $\tau = RC$, and to 5% of its initial value at $t = 3RC$. At the 100 nm technology node, the RC delay for copper and low-k dielectric interconnects ($\varepsilon_r = 2.0$) is about 30 ps for a 1 mm long interconnect. This compares to only 5 ps required for changing the state of the transistor. The situation becomes more desperate as the technology advances. For the 35 nm node, the MOS FET switching delay lowers to 2.5 ps, whereas the RC delay for a 1 mm long global interconnect lengthens to 250 ps. As ICs become larger, as shown in Table 7.1, the use of global interconnects (long wires) becomes more important. While steps can be taken to reconfigure IC so as to mitigate some of the use of global wires (interconnect-centric ICs), global interconnects are essential parts of the modern IC.

There are several ways to reduce the penalty due to latency in RC wires. One approach is to fabricate and use better wires. In terms of physical dimensions, the time constant τ for an RC circuit can be represented in Equation 2,

$$\tau = [\rho\varepsilon][1/HT][L^2] \qquad (2)$$

where ρ is the resistivity of the metal, ε is the dielectric constant of the insulator, H is the metal height, T is the insulator thickness, and L is the length of the wire. In this case, one can easily see the advantages of using metals with lower resistivity (e.g., copper), and lower dielectric constant insulators. One can also see the desirability of using thicker metals (increase H), and thicker insulators (increase T), and keeping wires short (decrease L). However, the processes used to fabricate wires limit the dimensional shapes. Larger cross-sectional wires with thick insulation are more accurately represented by RLC or time-of-flight range of performance, where the electron velocity is that of the speed of light within a medium of dielectric constant, ε. The delay in a 1 mm long time-of-flight copper wire is about 4.6 ps, making it comparable with MOS FET switching times.[3] Such wires can more easily be fabricated on the PWB or within the package than on-chip. Thus, a large number of I/O will enable access to these high-performance wires. The second approach to reducing the latency of interconnect is to "keep wires short". By reducing the length of all interconnect and I/O, the penalty due to latency is reduced. Thus, the electrical characteristics of interconnects drive toward a very high density of I/O with minimum individual length.

7.2.2 *Mechanical and chemical aspects of I/O*

The mechanical requirements include the physical support for the packaged IC and protection from environmental agents. The susceptibility of IC to failure due to mechanical shock, atmospheric or residue induced corrosion, and failures due to thermal cycling (temperature excursions in-use) must be considered in designing I/O. For example, the mechanical impact of a device hitting a hard object after being dropped from several meters (dropping a component onto a hard floor) can produce deceleration values of 10,000 times that of gravity. Minimal mechanical and thermal shock protection needs to be built into the devices. When the interconnect and I/O are made short, layers of packaging will disappear bringing the IC and attachment metallization closer to atmospheric exposure. Hermetic packages using ceramic and metal enclosures are not very difficult for direct chip attachment to substrates.

The use of ICs results in thermal cycles where the ICs power-up from ambient temperatures to values close to 100°C, in the case of desktop microprocessors, and nearly 200°C in the case of automotive applications. Also during assembly, the ICs and packages are heated to over 200°C so as to reflow solder and cure polymer materials (passivation layers, underfill, etc.). Since the CTE of silicon is much less than that of copper and epoxy-based substrates (CTE (Si) = 3 ppm/°C, and CTE (PWB) = 16 to 22 ppm/°C), the CTE induced stress and resulting strain must be taken up within the I/O so as not to result in mechanical failure. For example, if a 2.5×2.5 cm IC were mounted on a PWB (CTE = 20 ppm/°C) and the temperature were raised 100°C, then the silicon would expand 7.5 μm, whereas the PWB would expand 50 μm. In Figure 7.1, the long leads of the DIP and wire bond connections to

Table 7.2 IC packaging historical trends.

Technology	DIP	QFP	BGA/CSP	DCA
Number I/O	64	500	1600	3600
Package area, cm^2	20	25	6.5	5.6
Lead pitch, μm	250	400	400	150
Heat transfer density, W/cm^2	0.025	0.3	4.5	20

the IC provide a means for mechanical deformation without failure. In the BGA shown in Figure 7.2, the solder connections provide some mechanical compliance, but additional support from underfill is often required. In the next section, a short review of other electronic packages will be presented which address this issue.

7.3   Packaging and I/O trends

The substrate and chip-to-module connection technologies (via the electronic package) are intimately tied together, and they evolved according to the needs of IC technology. The historical evolution of electronic packages and chip-to-module interconnections will be presented first followed by substrate trends. Table 7.2 summarizes typical parameters for different I/O packages.

7.3.1   *Perimeter attachment packages*

The DIP, and quad flat package (QFP) found decades of service. The low I/O count ICs had wire bond pads distributed around the perimeter of the die. The package size was up to 100 times the size of the IC. Typically, the die was attached to a

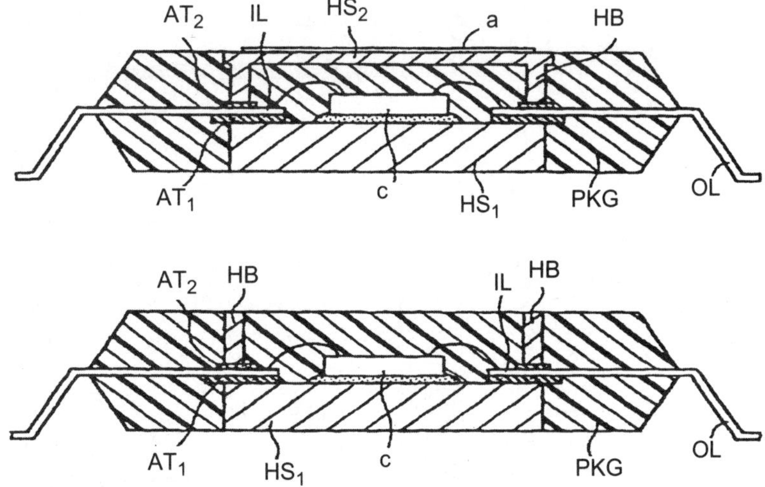

Figure 7.3   Sketch of surface mount package.

copper-based lead frame and the I/O pads were wire bonded to the lead frame. The die was molded onto the lead frame, and the metal leads trimmed so as to be inserted into a PWB (through-hole technology), or attached to the surface of the board (surface mount). In the DIP or QFP package, the wire bonds provided several important functions. First, the circumference of the die had sufficient space to provide adequate I/O for the IC. Second, the wire bond connections provided a low stress electrical connection for power and ground, and signal I/O. Wire bond automation reached amazing speed and reliability. Third, the copper lead frame or organic platform was mechanically matched to the PWB. That is, CTE of the board was adjusted to match that of copper by mixing different materials, such as low CTE filler in epoxy. Last, the pin-in-hole (DIP package) or surface mount attachment was robust against mechanical shock. The stress on the solder attachment was minimal. Figure 7.3 is a cross-sectional sketch of a surface mount device from U.S. Patent 5,629,561 showing the low stress wire bond connections to the IC and surface mount connections to the board.

7.3.2 *Area array packages*

Perimeter I/O and wire bonded packages (e.g., DIPs) did not have adequate chip-to-PWB I/O for high-performance ICs. As the area of the IC increased, the number of I/O that could be fitted on the perimeter was not adequate. That is, the need for I/O increased faster than the perimeter. Artificially increasing the area of the die so as to increase the number of I/O is not an acceptable option, and there is a physical limit as to the density of wire bonded pads. Second, the parasitic resistance, capacitance, and inductance of the wire bonds and their length became an important issue because the package became a limiting factor in system performance.

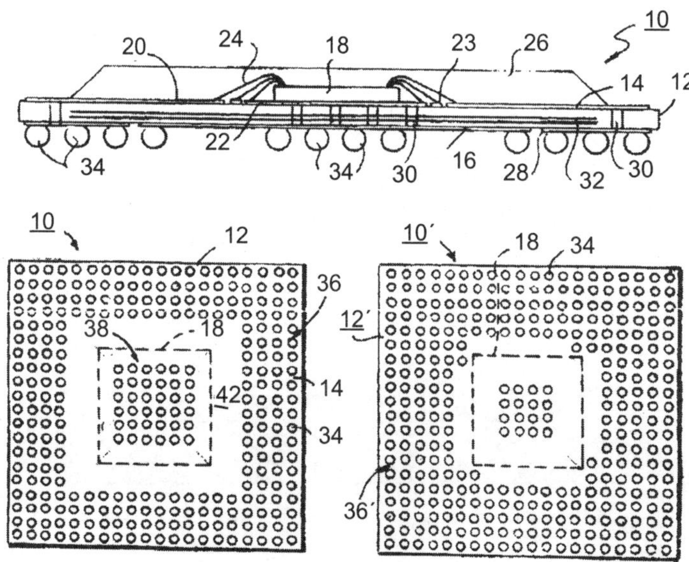

Figure 7.4   Sketch of BGA package, U.S. Patent 5,894,410.

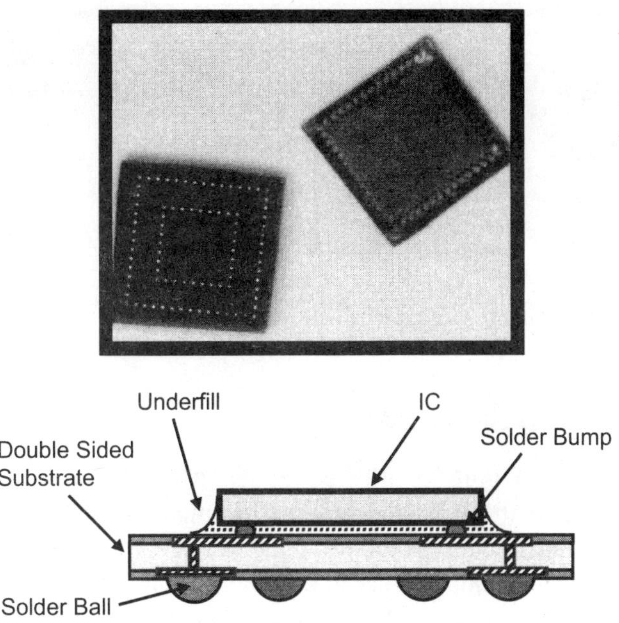

Figure 7.5   Ball grid array package used by Motorola.

The third and fourth columns of Table 7.2 show the evolution to area array devices as a means of increasing the package-to-board density. The pin grid array (PGA) and the BGA provided a high number of connections to the board. Implementation took place with first wire bonding of the IC to the package. Later, area array flip-chip attachment of the IC to the package lead to very dense packaging. It is instructive to examine specific technologies at the transition point to high density, compliant area array. Figure 7.4 shows a sketch of Intel Corporation U.S. patent 5,894,410. The IC (part 18) is wire bonded onto a multilayer printed circuit substrate (part 12). An area array of solder balls (part 34) is used to attach the package to the substrate. The multilayered package substrate provided a high density of fan-out to match the printed circuit board. The CTE mismatch between the silicon IC and the organic substrate (no solid copper lead frame is present like the simpler DIP) is addressed in this package. The area array of solder balls has an inner and outer array because a high number of failures were observed in solder balls when they were placed directly under the outer perimeter of the IC. The package also pays attention to the need for high heat transfer. The inner array of solder balls is available for heat transfer due to their close proximity to the PWB and low thermal resistance. The top of the electronic package is the primary heat transfer pathway in such packages. Further enhancements of the BGA have the IC flip-chip attached to the high-density package substrate, as shown in Figure 7.5.

Shrinkage of the package platform allows the footprint of the packaged die to approach that of the IC. This provides ever shorter chip-to-module interconnects, but the CTE mismatch issue becomes ever more pressing. Also, the reduced area of the package and the increase in IC power exacerbate the heat transfer issues.

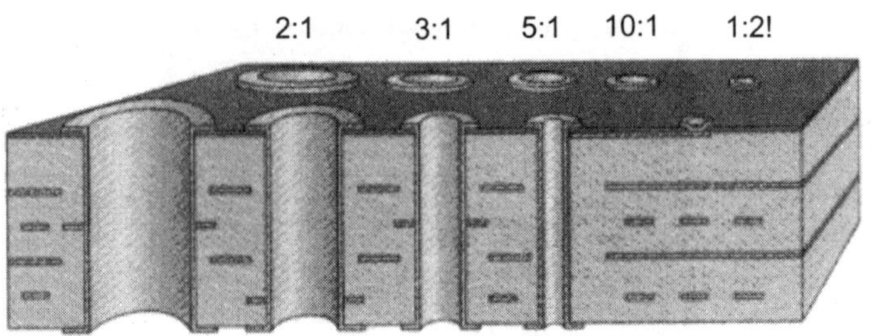

Figure 7.6   Generic printed wiring board with drilled and metallized through-holes.

7.3.3   *Chip scale packaging (CSP)*

The package shown in Figure 7.4 allowed for a large number of chip I/O to be distributed in an array where the packaged served as an intermediate between the fine pitch on the IC and the course pitch on the printed wiring board. As boards advanced to a finer pitch, it provided an opportunity to further shrink the IC package. High-density PWBs allowed the package size to approach that of the IC. High density PWBs, which are often called microvia boards, use fine-line circuitry usually built on a traditional PWB. Figure 7.6 shows a sketch of a traditional PWB where the layer-to-layer via connects are formed by mechanical drilling. A generic microvia PWB is shown is Figure 7.7. Multiple layers of fine-line wiring are fabricated on top of the

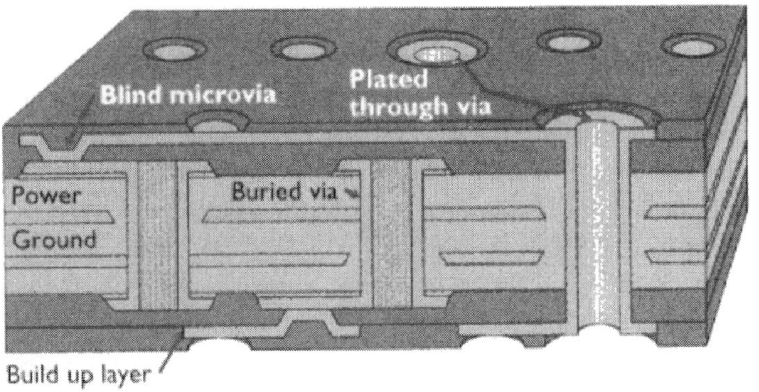

Figure 7.7   Generic microvia printed wiring board.

traditional board. The fine-line circuitry has dimensions that more closely match those of the IC and allow direct bonding of the IC to the board without the need for intermediate spatial fan-out of the I/O. The cost of the microvia buildup layers is modest because they are fabricated in parallel and not sequentially, that is, one drills the vias (holes) in the traditional PWB one after the other, but the microvia vias are printed in one lithography step using a photosensitive polymer or laser ablation.

Thus these chip-scale packages enabled a high silicon packing density to be achieved, but CTE mismatch between the silicon and the board needed to be

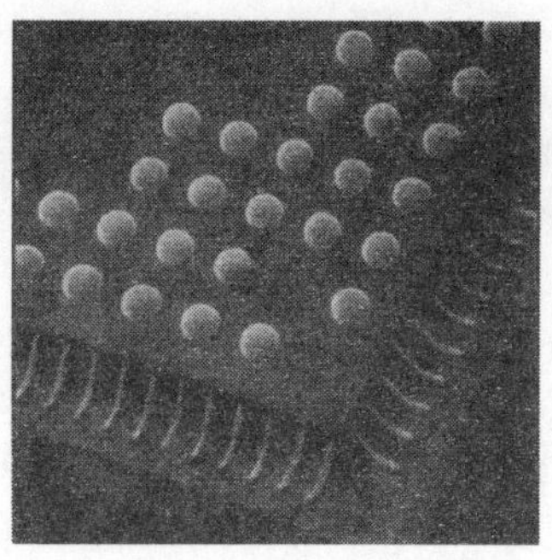

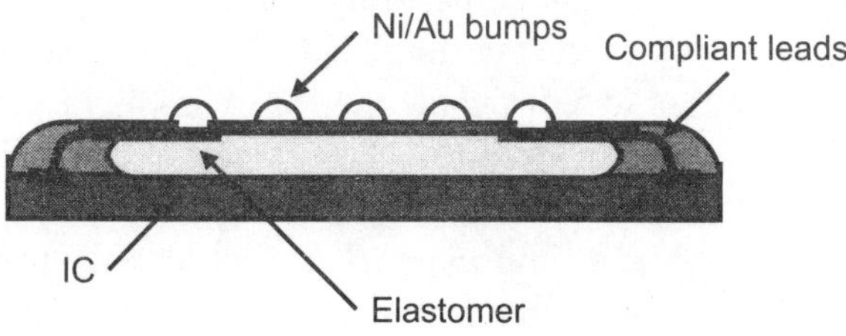

Figure 7.8　Tessera μBGA.

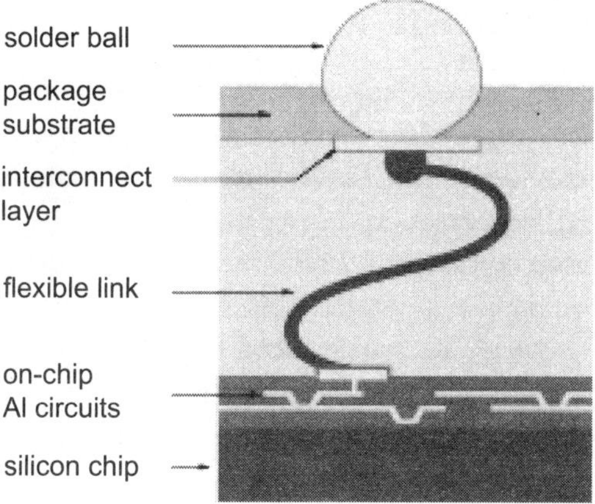

Figure 7.9　Tessera WAVE package.

addressed.[4,5] Numerous approaches have been taken to address the mechanical compliance. Pre-fabricated film or tape-based interconnections have been used to connect the ICs to PWBs. Amkor and Toshiba have used this style of interposer. The interposer (tape) is applied to the die followed by assembly. The IC can be wire bonded to the interposer followed by flip-chip attachment of the interposer-plus-IC to the PWB. A degree of compliance has been given to the interposer by use of three-dimensional (3D), spring-like structures. Tessera has made a compliant vertical link technology where the metal interconnection in the interposer lifts off the flexible foil by the injection of an encapsulant between the interposer and the IC. The device is then attached to the PWB by use of solder balls. Figure 7.8 shows the Tessera μBGA for perimeter I/O and Figure 7.9 shows the WAVE package for area array devices. The Tessera μBGA provides redistribution from perimeter I/O to area array so that the pitch on the PWB can be eased to coarser values. FormFactor has developed a micro-spring technology using nickel-plated gold wires, as shown in Figure 7.10. The micro-springs are formed on the wafer using a wire bonder. These examples have shown that adequate chip-to-module compliance can be achieved to compensate for CTE mismatches between polymer-based PWBs and silicon. They have further shown that higher density I/O and additional redistribution capabilities are necessary in many cases.

Redistribution of the I/O can be performed in wafer form by extension of the back-end-of-the-line processing. After normal IC fabrication, a final polymer metal buildup process can be used to redistribute the I/O. Flip-chip technologies fabricate copper redistribution traces in wafer form. The devices are flip-chip attached using solder paste, which is especially cost-effective compared to other solder deposition techniques. Compact electrical devices, such as cellular telephones, are primary applications. The use of an underfill material between the packaged IC and the PWB is necessary because the compliance of the interconnections is not adequate to take up the CTE mismatch between the polymer board and the silicon.

7.4 Wafer-level packaging (WLP)

Wafer-level batch packaging is an approach to forming chip-to-module interconnections that can address many of the current and future needs of ICs. Wafer level processing is where the traditional back-end-of-the-line IC interconnect processes (multi-layer metal buildup) are extended to include all packaging functions. This includes chip-to-module interconnection and full wafer-scale test and burn-in. Once the wafer-level batch packaging is completed, the ICs are separated and ready for assembly onto a PWB. The early uses have occurred primarily in lower pin-count devices and are driven by cost reductions. Since a traditional package is not required and the size of the PWB can be reduced (smaller IC footprint), the packaging costs

Figure 7.10 FormFactor MOST packaging technology.

are lower. Several approaches have been taken to address the need for compliance between the IC and PWB by fabricating metal posts in wafer form. A polymer then encapsulates the posts. The structure has a degree of compliance because the post can elastically deform and the polymer encapsulation can help distribute the stress. Additional compliance is provided by conventional solder bumps attached to ends of these posts. The Fujitsu SuperCSP is perhaps the best known wafer-level "copper post" technology.[4] Copper posts (350 μm diameter, 100 μm tall) are fabricated on top of an I/O redistribution layer. Special equipment has been developed to injection mold a polymer encapsulant around the posts followed by solder ball attachment. Oki and Casio have developed a similar structure for use in low pin count devices. Ibiden has developed a thin, flexible post in the second layer of polymer used in the redistribution buildup. These "post" technologies show the importance of in-plane (x–y direction) compliance in producing a reliable, wafer-level packaged device.

Many other issues remain to be addressed for wafer-level packaging (WLP). Among those are creation of an infrastructure for processing equipment, design tools for interconnection and I/O, wafer-scale test and burn-in equipment, and assembly tools that can handle different sizes of die, especially as die-shrinks occur. New technologies are often introduced as cost reductions on lower complexity parts. This can be seen in WLP where low I/O parts are directly attached to boards via solder bumps. Issues of assembly, reliability, and component layout and design are also challenging. The current issue of redistribution of I/O from perimeter I/O to area array exists primarily because the ICs were not originally intended for this form of packaging. More significant challenges come with WLP of high I/O count ICs and those operative at high clock speeds and power densities. These devices also bring the greatest payoffs due to the increased performance of the IC. The task of providing high I/O count is nontrivial due to the package I/O pitch limitations and CTE mismatch between the IC and PWB. Wafer level packages use large solder bumps with underfill support as their interconnection technology to withstand thermal and mechanical stresses. This limits the minimum pitch of the interconnects between the IC and the PWB.

7.4.1 Compliant WLP

Future IC packaging requirements and the performance of chip-to-module interconnections are determined by the expected chip performance. For example, the ITRS specifies that high performance chips at the 60 nm technology node will cover an area of 427 mm^2, have 4437 I/O pins, operate with an on-chip local clock speed of 6 GHz, and have a chip-to-board speed of 2.6 GHz.[1] These same chips will dissipate over 171 W of power and draw 190 A of DC current. The CTE mismatch between silicon (CTE $\approx 2.6 \times 10^{-6}$/°C) and typical printed wiring boards will only become more severe. The CTE of PWBs (currently about 18×10^{-6}/°C) may rise as the electrical performance becomes more critical. This translates into ~20 μm of lateral mismatch for a $\Delta T = 45$°C operation, at the 60 and 35 nm technology nodes. The chip-to-module interconnections must provide as much as 30 μm in vertical compliance (z-axis) to accommodate the non-planarity of the boards and to allow good

Figure 7.11 SEM micrograph showing portion of a SoL package with 12×10^3 x–y–z compliant chip I/O interconnects.

electrical contact between the test fixture and packaged IC. Other IC categories, such as handheld devices (including wireless), will present similar challenges defined by the I/O density, operating speed, power density, and heat dissipation. Also, there will be special demands arising from unusual form factors (e.g., 3-D chips) and mixed technologies (digital, analog, RF, optical, and micro-electro-mechanical devices). A final set of challenges (e.g., test fixture and assembly alignment and burn-in and attachment temperatures) will arise in the burn-in, testing, and assembly of these ICs.

The prospect of compliant WLP of a system-on-a-chip places enormous constraints on the chip-to-module interconnections. The package must enable high DC current with minimal ground bounce, and operate at high frequency with minimal signal degradation. Global interconnections, which can span a chip's corner-to-corner distance, pose special problems especially for global clock distribution.

7.4.1.1   *Sea of leads (SoL)*

Sea of leads (SoL) technology is proposed as an enabling technology for future chip-to-module interconnections. SoL WLP technology provides an ultra-high I/O density of x–y–z compliant leads ($>10^4$/cm^2) and can enhance the performance of a system on a chip (SoC) by routing critical on-chip global interconnects off-chip to reduce signal delay and thus increasing global clock frequency.[6]

Figure 7.11 is an SEM micrograph illustrating a portion of an SoL package with 12×10^3 x–y–z compliant leads per square centimeter.[7–9] The addition of embedded air gaps into SoL adds vertical compliance (z-axis) needed for wafer-level testing and mating to non-planar boards. Air gaps also serve to lower the dielectric constant of

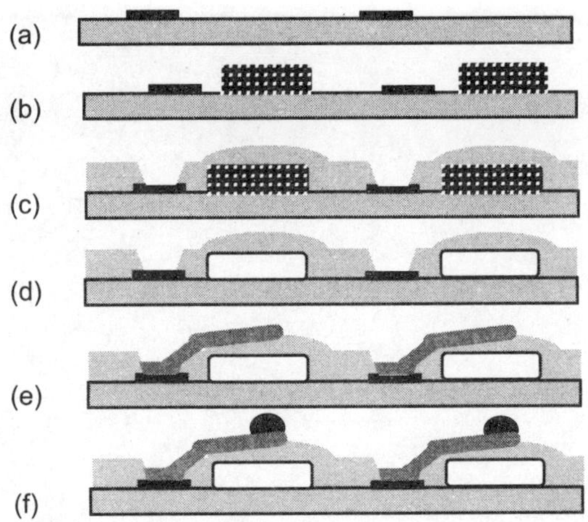

Figure 7.12 Build-up process for compliant wafer level interconnections.

the interconnect. Figure 7.12 shows a cross-section of process flow for an x–y–z compliant lead.[7,10] The use of a low modulus polymer encapsulating the air gap forms a structure capable of elastically deforming in all three dimensions. An IC with exposed bond pads is shown in Figure 7.12. A sacrificial material (e.g., Unity[TM], Promerus LLC) is deposited and patterned onto the wafer, Figure 7.12b. The sacrificial material serves as a placeholder and will later be removed to form an air cavity.[10] The sacrificial material is overcoated by a flexible material, and vias are opened to the bond pad in Figure 7.12c. When the overcoat elastomer is curing in Figure 7.12d, the sacrificial material simultaneously decomposes leaving a buried air cavity. The exact compliance of the final structure depends on the size and shape of the air cavity, and the elastic properties of the overcoat material. All these properties

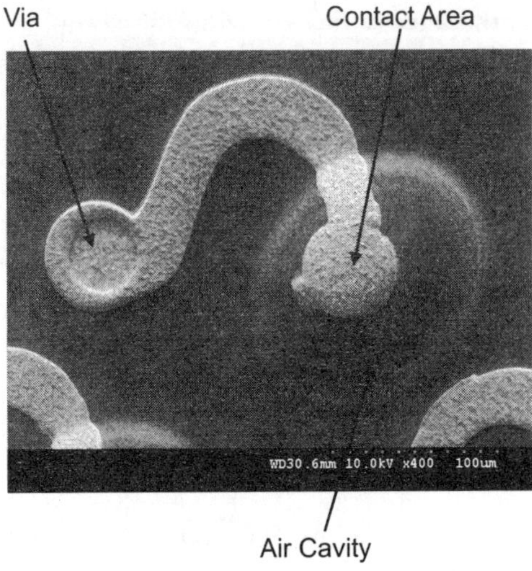

Figure 7.13 Components of compliant I/O interconnect.

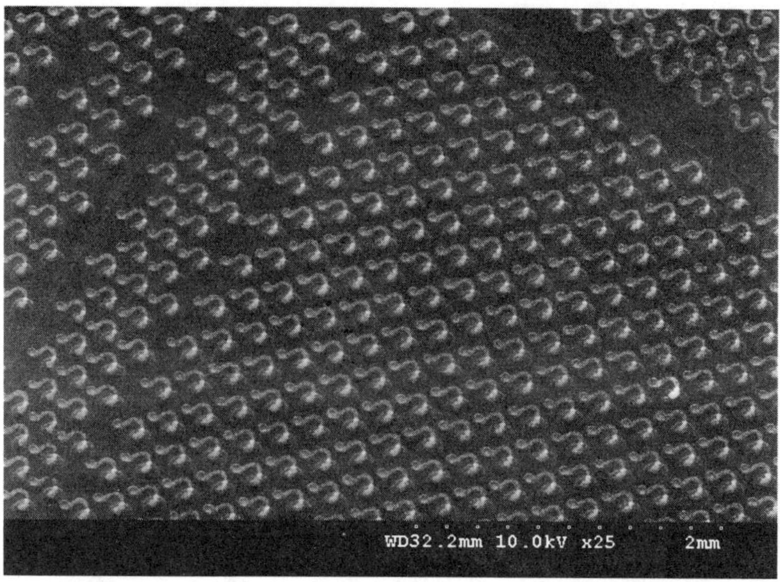

Figure 7.14 Compliant interconnect fabricated on buried air cavities.

are within the control of the designer. Finally, the metal leads are fabricated in Figure 7.12e, and attachment solder added in Figure 7.12f. Fine-line structures can be processed by use of photolithography, photosensitive polymers, and reactive ion etching. Courser features, such as for low I/O count ICs, can be processed by screen-printing and other similar processing techniques. It is important to note that the metal traces are made to not adhere to the surface of the overcoat polymer in Figure 7.12f, which can be accomplished in several ways, such as by using a polymer with poor adhesion. Figures 7.13 and 7.14 show metal leads fabricated on buried air cavities (the attachment solder has not yet been put in place). Figure 7.15 shows a cross-sectional slice through the air cavities showing the compliant region. This type of structure eliminates the need for chip underfill during package assembly of the chip to the PWB because of the x–y–z compliance of the leads.

The compliance of the membrane shown in Figures 7.13 and 7.14 can be chosen to meet system specifications. The length, width, and thickness of the membrane are adjustable parameters. Also, the material used in the elastic membrane can be chosen based on its elastic modulus. The deformation of a rectangular elastic plate is a reasonable approximation of the structure shown in Figure 7.13. The z-axis deflection is given by Equation 3.[11]

$$D = \alpha(Pa^2/Eh^3)(1 - \upsilon^2) \tag{3}$$

where D is the deflection of rectangular plate of dimensions a (width-shorter dimension) and b (length), P is the applied pressure, E is the elastic modulus, h is the thickness of the diaphragm, υ is Poisson's ratio, and α is a dimensionless coefficient.

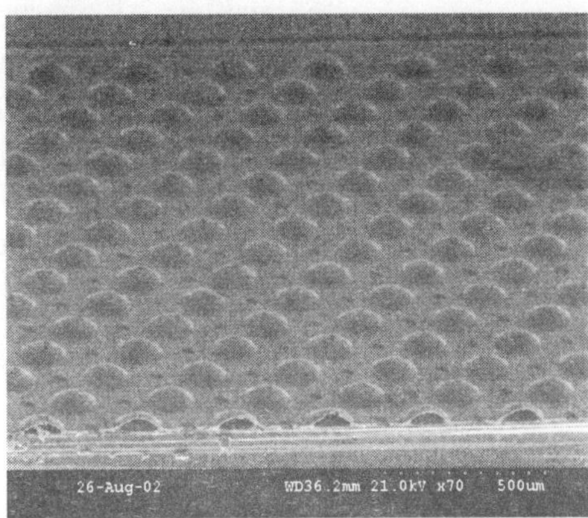 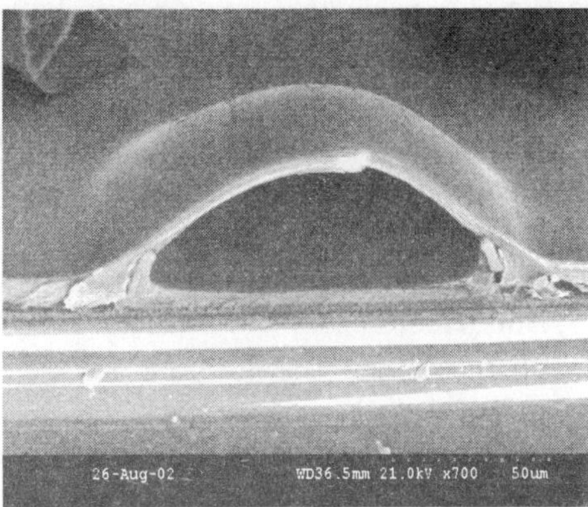

Figure 7.15 Sectioned compliant interconnect showing buried air cavities.

Values of α range from 0.0151 for $b/a = 1.0$ to 0.305 for $b/a = 2.0$. If b is much greater than a, then α approaches 0.0312.[11]

In-plane and out-of-plane compliance for the structures shown in Figure 7.13 were measured to be greater than 30 μm. While the leads are x–y–z axis compliant, they are short in length and thus exhibit minimal parasitics from DC to 45 GHz. The calculated resistance and inductance of the leads are less than 25 mΩ and 0.1 nH, respectively. Low electrical parasitics are desirable at both low and high frequencies for efficient conductive coupling of power, low power dissipation in the leads and thus low heat generation by the package. The microwave characteristics of SoL were measured at wafer-level using a two-port network analyzer with 150 μm coplanar ground-signal-ground (GSG) probes. To characterize the compliant interconnects, 15 μm thick Au leads were fabricated on a 15 μm thick polymer film. The return-loss and insertion-loss of the GSG lead interconnection were measured to be less than 20 and 0.2 dB, respectively, at 45 GHz.[9] In comparison, for example, insertion losses before and after the addition of underfill within a flip-chip package mounted on an alumina substrate with 75×150 μm bumps interconnected by 600 μm long 50-Ω coplanar waveguides, were found to be 0.6 dB and 1.8 dB, respectively, at 40 GHz.[10] The normalized impedance of the load as seen by the microwave probes was derived. At low frequencies (up to a few gigahertz), the leads appear as shorts (i.e., good metal interconnects). Another microwave measurement was made on a fully processed package where the leads were connected in pairs by 100 μm and 1 μm thick copper interconnects routed along the silicon wafer surface. The worst case return-loss and insertion-loss for a pair of interconnected leads were measured to be less than 12 and 1.2 dB, respectively. These measurements include the losses due to the Cu interconnects. Cross talk between adjacent parallel leads was measured to be less than 30 dB at 45 GHz, and cross talk between orthogonal leads was approximately 40 dB (i.e., 10 dB lower than the parallel leads). All interconnect structures were approximately 20 μm wide and 110 μm long.

Similar microwave measurements for other packaging technologies exist in the literature. It appears that the SoL measurements reported above compare very well to flip-chip packages, which are widely used in microwave applications. However, the ultra-high I/O density of SoL can provide an exceptionally high I/O bandwidth that would be difficult to match using alternate packaging technologies. For example, if 8,000 leads are assigned as signal I/Os and operated at only 5 GHz, the SoL package shown in Figure 7.11 can yield an aggregate electrical I/O bandwidth of 40 Tb/(cm^{2}s). Moreover, an SoL package avoids microwave performance degradation caused by underfill, as no underfill is required during assembly. This is in contrast to assembly requirements of flip-chip and BGA packages, where the presence of under-fill increases insertion-loss and shifts the return-loss frequency response to lower frequencies. [10,12]

An important aspect of SoL is the set of performance enhancements it offers a mixed-signal SoC. [9] SoL can enhance the DC power distribution as well as satisfy 3D structure I/O requirements. Because SoL processing requires low temperature (<250°C), it is expected that WLP processing will not damage previously fabricated structures during front-end-of-line and back-end-of-line processing. [7,13]

7.4.1.2   Compliant beams and MEM structures for I/O

Mechanically compliant beams can be used for the fabrication of chip-to-module I/O. A beam with preexisting curvature can form the basis for I/O connections which elastically deform under given loads. With proper design, the structures can be made to elastically deform in three dimensions.

As an example of the out-of-plane deflection of a beam, the mechanics of a stressed beam will be presented. A cantilever beam with dimensions of length, L (y-direction), width, w (x-direction), and thickness, b (z-direction) will have an initial deflection in the z-direction based on the intrinsic stress gradient. The beam is assumed to be fully clamped at one end and free at the other end, as shown in Figure 7.16. The stress in the top and bottom layer of the beam is assumed to be σ_1 and σ_2, respectively. It is assumed that the in-plane deformation of the beam is negligible compared to its out-of-plane deformation. Assuming a linear stress profile within the beam, the stress can be divided into two components, as given in Equation 4.

$$\text{Stress} = (\sigma_1 + \sigma_2)/2 + z(\sigma_1 - \sigma_2)/b \qquad (4)$$

The uniform stress, $[(\sigma_1 + \sigma_2)/2]$ accounts for the in-plane elongation of the beam whereas the $[(\sigma_1 - \sigma_2)/b](z)$ (the stress gradient component) causes out-of-plane

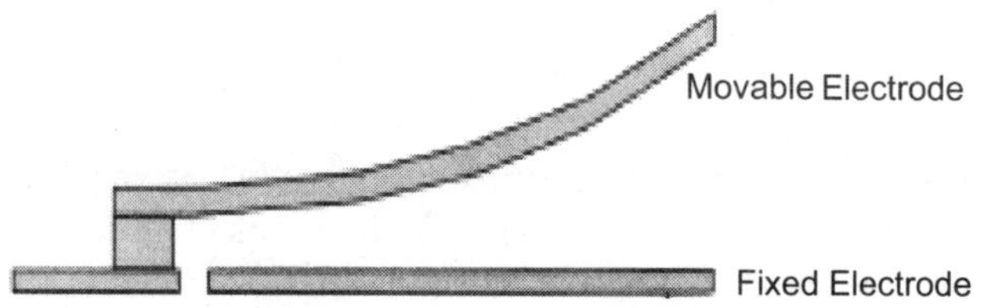

Figure 7.16 Cantilever beam with intrinsic stress gradient.

deflection. The net force, F, on the element of area $w(dz)$ is given by Equation 5.

$$dF = z((\sigma_1 - \sigma_2)/b)w(dz) \tag{5}$$

The moment about the axis through the center of the plane is given by Equation 6.

$$dM = (dF)z = z^2((\sigma_1 - \sigma_2)/b)wdz. \tag{6}$$

The net moment acting on the beam cross section can be obtained by integrating Equation 6 over the thickness of the beam, as shown by Equation 7.

$$\int_{-b/2}^{b/2} dM = \int_{-b/2}^{b/2} \left(\frac{\sigma_1 - \sigma_2}{b}\right)z^2 w(dz) \tag{7}$$

$$M = \left(\frac{\sigma_1 - \sigma_2}{12}\right)wb^2 \tag{8}$$

For a cantilever beam fixed at one end, with moment M acting on the other end, the governing equation for static deflection from continuum mechanics is given by Equation 9.

$$EI\frac{d^2 h}{dx^2} = M \tag{9}$$

where h is the deflection of the cantilever beam, E is elastic modulus of the material, and I is the moment of inertia ($wb^3/12$). For the beam fixed at one end, the boundary conditions used are $x = 0$, $h = 0$, and $x = 0$, $(dh/dx) = 0$. Solving the differential equation, Equation 9 with these boundary conditions, and solving for the deflection, z gives Equation 10.

$$h = \frac{Mx^2}{EI\,2} = \left(\frac{\sigma_1 - \sigma_2}{12}\right)\frac{b^2 wx^2}{2EI} \tag{10}$$

Bimetallic beams have been used to produce z-axis stress gradients in beams.[14,15] The CTE mismatch between the two metals in the beams produces a z-axis stress gradient and a curved beam with a single radius of curvature. The deflection of the beam will be a function of temperature, due to the increasing z-axis stress gradient with temperature.

Stress gradients can also be produced by changing the deposition conditions for a single-metal beam. Curved gold beams have been produced by first depositing

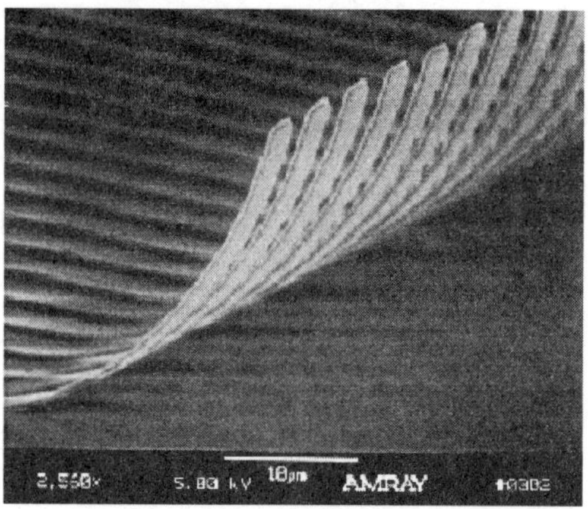

Figure 7.17 Picture of array of springs.

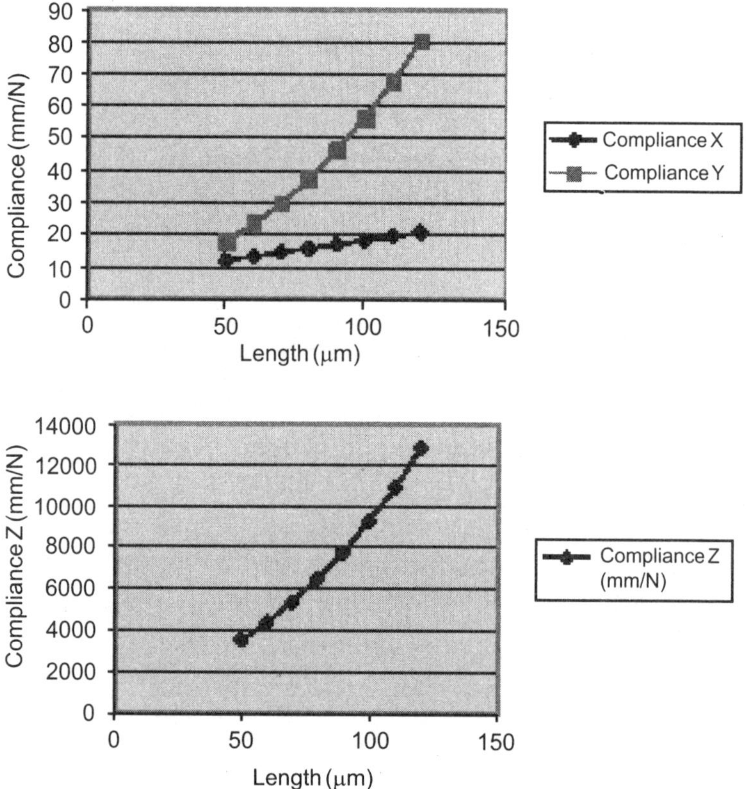

Figure 7.18 Compliance of spring structure shown in Figure 7.17.

stress-free gold followed by tensile stressed gold. The stress state was changed by altering the grain size of the deposited gold.[16,17] The curvature of the pure gold beams did not change with temperature, and the electrical conductivity was not

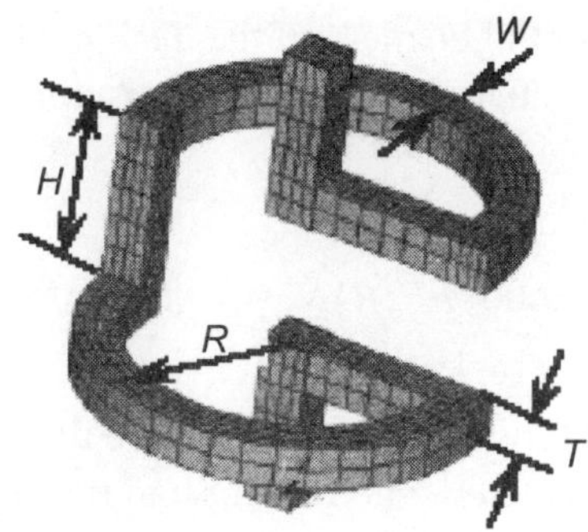

Figure 7.19 Metal helix structure for compliant I/O.

sacrificed by adding a second less conductive metal in order to induce a stress gradient. The curvature of the beam can be modified by inducing a stress gradient along the length of the beam (in addition to the existing through-plane stress gradient). This was accomplished by depositing a patterned layer of stressed gold. The stress gradient shown in Equation 4 can be altered to produce in-plane (y-axis) gradients resulting in beams with customized shapes.

Composite metal spring contacts have been fabricated using 80% Mo and 20% Cr (by weight). A picture of an array of compliant spring contacts are shown in Figure 7.17. The springs are 1 μm thick, 10 μm wide, and 100 μm long. The stress gradient in the z-direction was 2 GPa. Higher stress gradients resulted in greater curvatures. The x, y, and z compliance is shown in Figure 7.18.

Reliance on internal stress in metal beams for x, y, and z deflection can be problematic. Small spatial variations in stress state, thickness, or composition can cause significant changes or irregularities in beams. Further, striction forces need to be considered in the fabrication and lifting of the beams. Three-dimensional structures can be produced by using standard lithography and metal deposition.[18,19] Figure 7.19 shows a one turn helix structure produced by standard metallization and photolithographic techniques. The dimensions of the helix can be varied over wide ranges. Critical to the compliance is the radius of the arcuate beam, R. Figure 7.20 shows the compliance in millimeters per Newton. The electrical properties of the

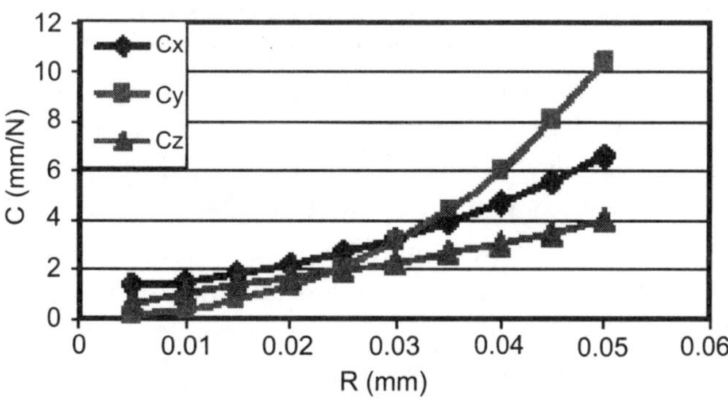

Figure 7.20 Effect of mean radius on directional compliance for beam shown in Figure 7.19.

compliant I/O are an essential element of the design. Of particular interest for helix structures is the self-inductance, which can be approximated for a helix of rectangular strip by Equation 11.

$$L_{\text{self}} = 0.001F(2R)N^2 - 0.004\pi RN(g + h) \tag{11}$$

where N is the number of turns of the helix, H is the distance between adjacent wires, R is the mean radius of the helix, F is the length of the beam/2R, and $(g + h)$ are geometric correction terms.[18,19] A wide range of compliance can be designed and built using these mechanical shapes; however, it comes at the expense of multiple lithographic and other processing steps.

7.4.1.3 *Other considerations for WLP*

Wafer level batch packaging presents special opportunities and challenges to the burn-in and testing portion of the IC manufacturing process. On the one hand, WLP allows a complete wafer containing fully packaged die to be taken through each stage of the burn-in and testing process. This is in contrast to burn-in and testing of a singulated die by inserting them into temporary test carriers and then removing them for final die-to-package integration and subsequent packaged part testing. Wafer-level testing sharply reduces part handling and die-to-test fixture alignment. It also eliminates the need of inserting a die into a temporary carrier for testing purposes. A major challenge to taking full advantage of wafer-level testing is to establish and maintain all the I/O connections during the burn-in and test process. The typical force applied to a single I/O by a test fixture can be from 5 to 10 mN; this translates into ~5 to 10 kN total force applied to a 300 mm wafer populated by high-performance chips at the ITRS 35 nm technology node. Fully powered, this same wafer would require 38 kA and dissipate 23 kW of heat. The test fixture and wafer must also accommodate the CTE expansion of the die during the burn-in and testing procedure (at 150°C an I/O at the edge of a 300 mm silicon wafer will be displaced by ~50 μm from its room temperature position). Testing algorithms need to be developed that speed up testing and allow the test equipment to efficiently examine a full wafer rather than be swamped by the bandwidth demands of distributing test vectors to and detecting responses from many dies rather than one at a time. Testing a mixed signal SoC will present many challenges. Integrating different cores such as DSP, RAM, microprocessor, RF, and analog on a single chip requires careful testing considerations early during the chip design cycle.[2] If 1,240 leads are allocated for only AC/DC testing, a mixed-signal SoC, direct access to the various cores may be possible, and great insight can be acquired into failure mechanisms. In addition, this high number of leads will provide very high bandwidth to perform faster testing.

Next-generation SoL technology will incorporate optical waveguide interconnection to permit global optical clock distribution within high-performance chips. Optical waveguide interconnection allows for planar packaging of a hybrid electrical/optical system in a manner conducive to future heat removal and power supply

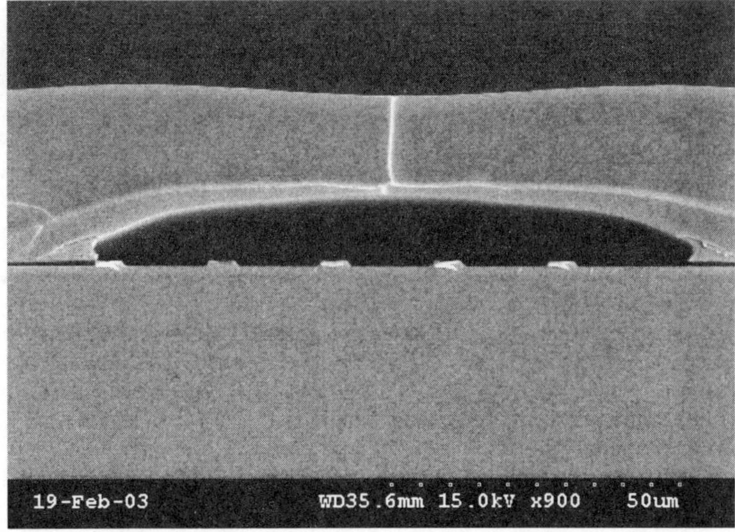

Figure 7.21 Cross section of air-encapsulated waveguides on silicon.

requirements. In addition, polymer waveguide technologies offer immediate and low-cost compatibility with wafer-level fabrication processes. Minimal redesign of the global distribution would be required with increases in clock frequency, as the design of the cross-sectional dimensions of an optical waveguide is independent of the clock frequency. Global propagation of the local clock frequency would be possible, thereby eliminating the need for global clock repeaters and allowing for the removal of cycle-to-cycle jitter from the global clock signal in its delivery to opto-electronic receivers for optical-to-electrical conversion. Finally, optical waveguide clock distribution would allow for enhanced predictability with respect to the arrival time of clock pulses at different optoelectronic receivers, thereby eliminating *unexpected* components of global distribution skew.[20] Embedded air gap cladding regions allow for a maximization in refractive index contrast, Δn, between waveguide core and cladding, and hence permit smaller bending radii and higher waveguide densities due to tighter confinement within the waveguide core. By placing optical waveguides within the package, via blockage concerns are eliminated with respect to waveguide routing, leaving only the electrical I/O interconnection to obstruct intra-chip waveguide routing. To create the embedded air gap cladding regions, a sacrificial photosensitive polymer composite (Unity[TM] 200) is applied and patterned over the waveguide. An overcoat polymer is then applied to embed the air gap and waveguide regions, where, upon thermal cure, the remaining polymer regions decompose to leave an embedded air gap. A cross section of five optical waveguides with air encapsulation on silicon is shown in Figure 7.21. By incorporating a buried air gap, a $\Delta n = 0.52$ is achieved between core and upper-cladding regions. Beam propagation methods reveal that, by incorporating an air gap cladding region, the minimum allowed edge-to-edge spacing between adjacent waveguides is reduced by over $3x$.[21]

In summary, WLP can be used to meet the performance and chip-to-module interconnection needs of future ICs. WLP with embedded air gaps may provide near-term, low-cost, course-pitch packaging as well as long-term advantages including improved electrical performance.

References

1. International Technology Roadmap for Semiconductors (ITRS), SIA/Sematech International (ITRS, http://public.itrs.net/).
2. P. Zarkesh-Ha and J.D. Meindl, "An integrated architecture for global interconnects in a gigascale system-on-a-chip (GSoC)", IEEE Symposium on VLSI Technology, June 2000.
3. J.D. Meindl, J.A. Davis, P. Zarkesh-Ha, C.S. Patel, K.P. Martin, and P.A. Kohl, "Interconnect opportunities for gigascale integration", *IBM J. Res. Dev.*, **46**, 245–265 (2002).
4. P. Garrou, "Wafer Level Chip Scale Packaging", Semi Chip Scale International '99, p. D-1 (1999).
5. *Fundamentals of Microsystems Packaging*, R. Tummala, Ed., McGraw-Hill, New York (2001).
6. A. Naeemi, P. Zarkesh-Ha, C. S. Patel, and J. D. Meindl, "Performance improvement using on-board wires for on-chip interconnects", IEEE 9th Topical meeting on Electrical Performance of Electronic Packaging, pp. 325–328, Oct. (2000).
7. M.S. Bakir, H.A. Reed, P.A. Kohl, K.P. Martin, and J.D. Meindl, "Sea of Leads ultra high-density compliant wafer level packaging technology", to be presented, *52nd Electronics and Components Technology Conf.*, San Diego, CA, May (2002).
8. C.S. Patel et al., "Low cost high density compliant wafer level package", *International Conf. on High-Density Interconnect and Systems Packaging*, Denver, CO, April 26–28, pp. 261–268 (2000).
9. M.S. Bakir, H.A. Reed, A.V. Mule', P.A. Kohl, K.P. Martin, and J.D. Meindl, "Sea of Leads characterization and design for compatibility for board level optical waveguide interconnection", *IEEE Custom Integrated Circuits Conference,* Orlando, FL, May (2002).
10. D. Bhusari, H. Reed, M. Wedlake, A. Padovani, S.A. Bidstrup-Allen, and P.A. Kohl, "Fabrication of Air-Channel Structures for Microfluidic, Microelectro-mechanical and Microelectronic Applications", *J. Microelectromechanical Syst.*, **10**, 400–409 (2001).
11. *Flat and Corrugated Diaphragm Design Handbook*, by Mario Di Giovanni, Ametek Controls Division, Fieasterville, PA, Marcel Dekker, New York (1982).
12. H. Kusamitsu, Y. Morishta, M. Ito, and K. Ohata, "The flip-chip bump interconnection for millimeter-wave GaAs MMIC", *IEEE Trans. Electron. Packag. Manuf.*, **22**, 23–28, Jan. (1999).
13. H.A. Reed, C.E. White, V. Rao, S.A. Allen, C.L. Henderson, and P.A. Kohl, "Fabrication of microchannels using polycarbonates as sacrificial materials", *J. Micromechanics Microeng.*, **11**, 733–737, Oct. (2001).
14. L. Ma, Q. Zhu, T. Hantschel, D.K. Fork, and S.K. Sitaraman, "Compliant cantilevered spring intersonnects for flip-chip packaging", Proceedings of ECTC (2002).
15. L. Ma, Q. Zhu, T. Hantschel, D.K. Fork, and S.K. Sitaraman, "J-Springs – innovative compliant interconnects for next-generation packaging", Proceedings of ECTC (2002).
16. A.K. Chinthakindi, D. Bhusari, B.P. Dusch, J. Musolf, B.A. Willemsen, E. Prophet, M. Roberson, and P.A. Kohl, "Electrostatic actuators with intrinsic stress gradient, Part I Materials and structures", *J. Electrochem. Soc.,* **149**, H139–H145 (2002).
17. A.K. Chinthakindi and P.A. Kohl, "Electrostatic actuators with intrinsic stress gradient, Part I Performance and modeling", *J. Electrochem. Soc.,* **149**, H146–H152 (2002).
18. Q. Zhu, L. Ma, and S.K. Sitaraman, "Design optimization of one-turn helix – a novel

compliant off-chip interconnect", Proceedings of Itherm 2002, International Conference on Thermal, Mechanics and Thermomechanical Phenomena in Electronic Systems, May 29 (2002).
19. Q. Zhu, L. Ma, and S.K. Sitaraman, "Mechanical and preliminary electrical design of a novel compliant one turn helix interconnect", Proceedings of InterPack '01, The Pacific Rim International, Intersociety, Electronic Packaging Technical/Business Conference & Exhibition, July 8 (2001).
20. P.J. Restle, T.G. Mcnamara, D.A. Webber, P.J. Camporese, K.F. Eng, K.A. Jenkins, D.H. Allen, M.J. Rohn, M.P. Quaranta, D.W. Boerstler, C.J. Alpert, C.A. Carter, R.N. Bailey, J.G. Petrovick, B.L. Krauter, and B.D. McCredie, "A clock distribution network for microprocessors", *IEEE J. Solid State Circuits*, **36**, 792–799, May (2001).
21. A.V. Mule', S. Schultz, E.N. Glytsis, T.K. Gaylord, and J.D. Meindl, "Input coupling and guided-wave distribution scheme for board-level intra-chip optical clock distribution network using volume grating coupler technology", *Proc. IEEE International Interconnect Technology Conference,* San Francisco, CA, pp. 128–130, June (2001).

8 Pb-free flip-chip technologies

Darrel R. Frear and W.H. Lytle

8.1 Introduction

Interconnects between the various levels of an electronic package form both an electrical and a mechanical joint. For flip-chip interconnects, this joint exists between the top metal layer of the Si device and the metallized pads on a substrate. The traditional material to create the solder interconnect has been an Sn–Pb alloy. Sn–Pb solder is used because it has a relatively low melting point, good wetting behavior, and good electrical conductivity; and can be used in hierarchical soldering, which is defined as the utilization of a solder that has a lower melting temperature than all other solder interconnects that precede it. The lower temperature solder must have a working temperature sufficiently low so that it does not melt a higher temperature interconnect. The melting temperature range for Sn–Pb solders is from 310°C for Sn–97Pb to 183°C for eutectic Sn–37Pb. The basic requirement of flip-chip solder interconnects is to form a reliable electrical and mechanical connection that retains integrity through subsequent manufacturing processes and service conditions. The joints are also required to have the capacity to dissipate strains generated as a result of coefficient of thermal expansion mismatches under service conditions over the lifetime of the assembly. As the number of joints increases, and the size decreases, the behavior of solder joints becomes problematical because they are more difficult to manufacture and reliability requirements become more difficult to satisfy. This is intensified for area-array applications because joints cannot easily be viewed or inspected after assembly.

The electronics industry extensively uses Pb–Sn solder alloys in flip-chip applications as well as in many other interconnects in the electronic package. However, medical studies have shown that Pb is a heavy metal toxin that can damage the kidneys, liver, blood, and central nervous system. Less than 1% per year of the global Pb consumption is used in solder alloys for electronic products, but electronics and electrical systems make up an increasingly larger fraction of landfills.[1] The issue of Pb leaching from landfills into the water table has raised alarm as a potential source of long-term contamination of soil and groundwater. Concerns about the presence of Pb in the environment and potential exposure scenarios that could result in the ingestion of Pb by humans and wildlife has prompted a concerted effort to limit the use of Pb in manufactured products (notably gasoline, plumbing solders, and paint). International laws have recently been proposed to expand Pb control laws to limit or ban the use of Pb in manufactured electronics products. The most aggressive and well known effort is the European Union's Waste in Electrical and Electronic Equipment (WEEE) directive that proposes a ban on Pb in electronics by 2006. The Japanese Environmental Agency has proposed that Pb-containing scrap must be disposed of in sealed landfills to prevent Pb leaching. The Japanese Ministry of International Trade and Industry and the Japan Automobile Industries Association called for a 50% reduction of Pb in vehicles (excluding batteries) by 2001 and a 2/3 reduction by 2003.[2] Electronics manufacturers have responded to these proposed bans in a variety of ways. Many companies have not taken a stance hoping that legislation will not be enacted. Other companies have aggressively pursued solutions to the proposed bans and are using Pb-free products as a "green" marketing strategy. Extensive research on Pb-free solders has been published. A comprehensive review of the status of Pb-free solders can be found in the literature that is primarily focused on carrier-to-board (surface mount and through-hole) interconnects.[3–15] A growing requirement is Pb-free solders for flip-chip interconnects.

One benefit of a Pb-free flip-chip interconnect is the reduction of Pb^{210} created alpha particle radiation. All mined Pb contains a small amount of radioactive Pb^{210} that decays and emits alpha particles. When an alpha particle enters an active element of the Si (such as a memory cell), it has sufficient energy to cause the stored charge to be released with the result of changing stored memory from a 1 to a 0 state. There is no permanent damage to the Si itself so this radiation induced fault is termed a "soft error." The alpha particles have a low energy that is dissipated over relatively short distances. However, due to their proximity to active elements, the flip-chip solder interconnects have sufficient levels of alpha particle radiation to induce soft errors in CMOS (complementary metal-oxide semiconductor) technology that become more critical as cell size on the die is reduced.[16,17] For Pb–Sn solders, one solution to alpha particle radiation is to use elemental Pb that was mined many, many years ago where the majority of Pb^{210} has decayed. The source of this Pb is typically found as the ballast of shipping vessels that sank almost 2000 years ago and is relatively expensive. The elemental constituents of Pb-free solders (Sn, Cu, Ag, Bi, In, Sb), however, do not radioactively decompose, and so alpha particle radiation is minimal.

Flip-chip interconnects are the electrical and mechanical connections between the semiconductor integrated circuit and the package (or board for direct chip attach (DCA)). These interconnects are formed on the periphery or in an area-array on the top surface of an active die. Flip-chip interconnects are formed by depositing solder onto a metallized Si wafer in the form of discrete balls, or solder paste, or by directly plating onto the pads on the wafer. The solder must wet and join to the pads on the Si devices, and so an underbump metallurgy (UBM) is typically deposited on the Al or Cu pads on the Si. The UBM typically consists of a barrier metal (e.g., Ti or W) followed by a solder wettable layer (e.g., Cu or Ni). The top layer of the Ni metallization is covered with a noble metal, such as Au, to prevent oxidation that inhibits solder wetting. The UBM also acts as a diffusion barrier between the Si and the solder and must be thick enough to withstand interactions (intermetallic formation) between the solder and UBM. Flip-chip interconnects are smaller (on the order of 100 μm diameter) than surface mount joints and are projected to have pitches that shrink below 150 μm. Flip-chip interconnects have a unique set of requirements. These joints must be able to withstand a potentially high level of strain mismatch between Si and an organic substrate. Flip-chip technology has moved from ceramic packaging with high-Pb solder (97.5Pb–2.5Sn) to an organic package that requires lower temperature reflow (<260°C). This can be accomplished by bumping the die with high-Pb solder and then joining it to an organic board with eutectic Sn–Pb but this is a cost adder that is eliminated with a monolithic solder. The joints must withstand board-level reflow environments compatible with joining to organic substrates that, again, have a maximum reflow temperature of 260°C. The Pb-free solder must meet these requirements and perform at, or above, the level of performance of the Sn–Pb solder it is intended to replace. The flip-chip solder alloy is typically deposited on Si wafers either as a solder paste stencil printed on the defined UBM pads or as direct plating on the UBMs. Figure 8.1 shows a schematic of the flip-chip bumping processes. The simplicity and low cost of plating the solder for flip-chip interconnects makes electrochemical deposition the most attractive choice for flip-chip bumping.

This chapter outlines plating strategies and solutions for baseline Pb-based alloys (eutectic Sn–Pb and 90 to 97.5Pb–Sn) and Sn-rich Pb-free alloys including an optimal Pb-free solder alloy and UBM for flip-chip interconnects. The physical metallurgy of these systems, including the reactions between the solders and UBM structures, are also presented. The mechanical and thermomechanical behavior of the joints is discussed, including reliability test results of the various plated and joined solder alloys.

8.2   Pb-free solder alloys and UBM structures

Pb-free solders for electronic applications are based on Sn-rich compounds that fall into a melting temperature range similar to the traditional eutectic Pb–Sn solder alloys (183°C). These include eutectic Sn–3.5Ag with alloying elements of Bi, Cu, Sb, In, or Zn. Other alloys based on the Sn–Cu, Sn–In, Sn–Sb, Sn–Bi, and Sn–Zn systems have also been proposed. A small two-phase region (temperature difference between liquidus and solidus) is desired because it prevents the joint from moving and becoming disturbed during solidification. Binary or ternary near-eutectic alloys are also desired because simpler alloys reduce the potential for compositional variations that affect the behavior of the solder joint. A chart of the melting temperatures of potential solder alloys is shown in Figure 8.2. The alloys Sn–0.7Cu, Sn–3.5Ag, and Sn–3.8Ag–0.7Cu are the most promising flip-chip solder alloys based on the criteria given above. The Sn–Sb alloy was deemed to have too large a two-phase region and the liquidus temperature of 240°C was too high for chip attachment to organic substrates (a process temperature of 245°C is desired). The Sn–Zn alloy had too many processing difficulties due to the rapid oxidation behavior of the Zn in the molten state and corrosion susceptibility of the alloy after solidification. The Sn–Ag–Bi alloy had too large a two-phase region that could result in a disturbed solder joint (very rough dull surface of the solder) that could affect the reliability, as the surface could act as crack initiation points. The Sn–Ag–Cu–Sb was not suitable for fear that in a flip-chip application the solder would damage the die rather than

                    Darrel R. Frear and W.H. Lytle

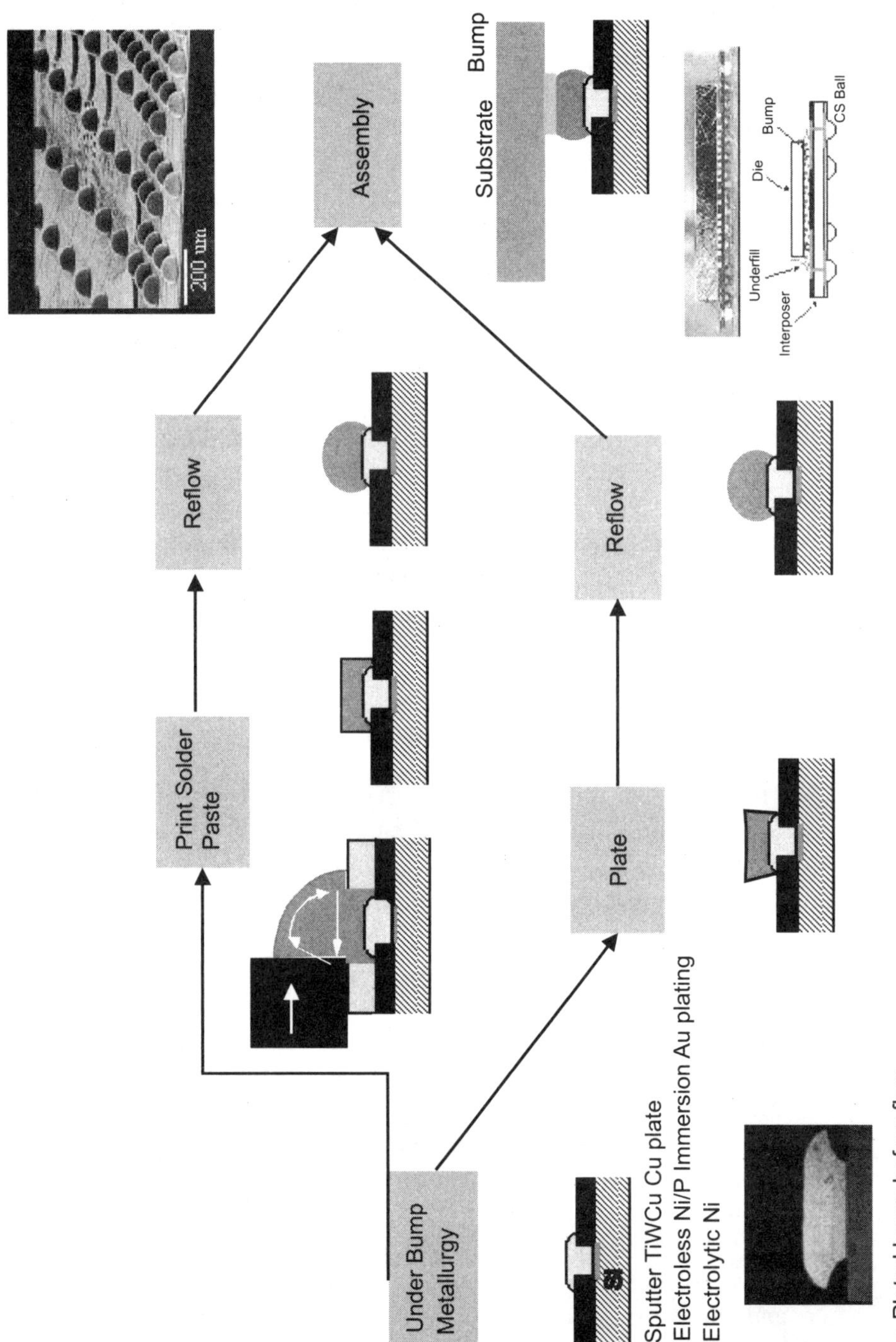

Figure 8.1 Schematic flow chart of the flip-chip bumping process by solder plating or solder paste deposition.

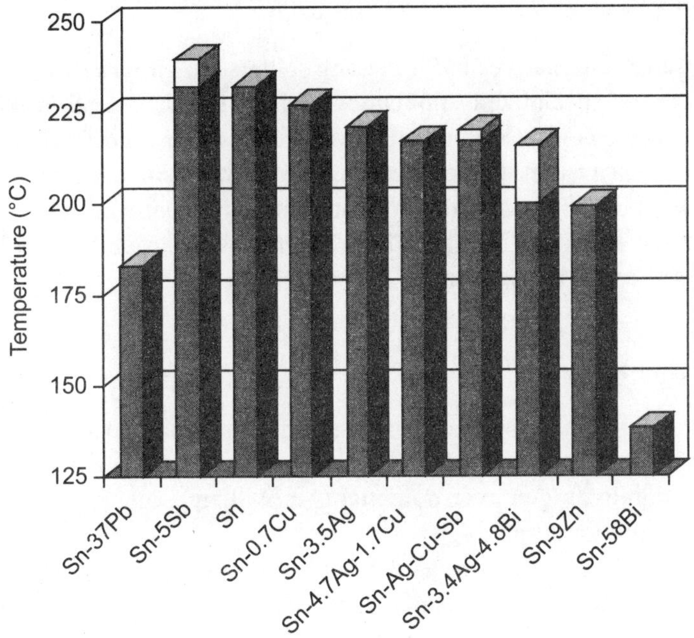

Figure 8.2   Chart of Pb-free solder temperatures (dark = solidus, light = liquidus).

deform it due to the high strength of the alloy,[18] and also because of the complexity of plating quaternary alloys. Flip-chip interconnects are part of a package that must undergo carrier-to-board reflow that occurs, at a minimum, at 220°C and the Sn–Bi eutectic alloy melts at far too low a temperature (138°C) to withstand this reflow. Limitations in what can be plated and result in a quality solder flip-chip interconnect will further limit the potential solder alloys as discussed later in this chapter.

Typical UBMs for flip-chip metallizations include sputtered TiW/plated Cu, electroless plated Ni–P/immersion Au, and electrolytic Ni (used only for solder plating). Ni reacts more slowly with Sn-based solders compared to Cu or Au. The TiW/Cu UBM requires sufficient Cu (on the order of 10 μm) to survive multiple reflows with the Sn-rich alloys.

The following sections describe the capabilities and processes for plating of flip-chip solders emphasizing the electrochemistry of Pb-free alloys. This is then followed by a brief discussion of the physical and mechanical metallurgy of the plated flip-chip solder interconnects.

8.3   Electrodeposition and plating processes for Pb-free solders

Electrodeposition is essentially the application of an electrochemical cell utilizing external electrical energy to cause a reduction reaction to occur at a specific location. Electrodeposition of the solder allows for a low cost, consistent deposition of solder for flip-chip interconnects. The following sections describe the electrochemical processes and procedures to create plated flip-chip interconnects.

8.3.1   *Electrodeposition of tin and Pb-free alloys*

Most alloys used as alternatives to Pb-bearing solders are Sn-based. The few non-tin alternatives are for specialized applications where standard lead-bearing solders would not be used, such as indium metal which has been used for flip-chip bump interconnections because of its high ductility at very low temperatures. Many alloys of Sn can be produced by sequentially plating the individual elemental components and reflowing the layers to form an alloy. Therefore, electroplating Sn is a natural point to discuss Pb-free alloy plating.

8.3.2   *Chemistry of Sn*

Tin is amphoteric, reacting with both bases and acids under the right conditions, but is relatively inert in neutral solutions. However, Sn-dissolution in acids and bases is slow due to a high hydrogen over-potential. An oxidizing agent must be present to depolarize the evolution of hydrogen.

Tin compounds usually exist as the Sn^{2+} ion in acidic solutions, whereas the higher oxidation state of Sn^{4+} usually does not exist except when complexed or hydrolyzed. In alkaline solutions, Sn^{4+} is the most stable species. Most tin compounds tend to hydrolyze in aqueous solution so excess alkali or acid must be maintained in the corresponding plating baths to maintain stability.

8.3.3   *Electrochemistry of tin (and lead)*

Tin and lead are similar not only in chemical reactivity and compound formation but also in electrochemical conditions. The standard reduction potentials for the two metals are very close with Pb slightly more noble.

$$Sn^{2+} + 2e^- \rightarrow Sn, \; E^0 = -0.136$$
$$Pb^{2+} + 2e^- \rightarrow Pb, \; E^0 = -0.125$$

Normal electrodeposition conditions differ significantly from standard or equilibrium conditions that are given as $[M^{x+}] = 1$ M, 298 K, and 1 atm. The Nernst equation gives the reduction potential (E) of an ionic species for any given temperature and concentration.

$$E = E^0 + \left(\frac{RT}{nF}\right)\ln a + C \qquad (1)$$

where E is reduction potential (volts), R is the gas constant, T is temperature expressed in degrees K, F is the Faraday constant, n is the number of electrons involved in the electrodeposition process, and a is the concentration or activity of the ion. C is a constant term that includes the other factors contributing to the electrodeposition process such as the effect of complexing agents, and other additives.

Pb and Sn can be deposited in nearly any alloy composition simply by varying the solution concentrations.

The Nernst equation can be used to determine the feasibility of co-deposition of metals for alloy preparation. The reduction potential for each metal ion in solution is calculated, and the smaller the differences are in potentials between the metals the more likely an alloy can be deposited. Co-deposition requirements can be met when E_1^0 and E_2^0 are similar, by adjusting the concentration of one or more species or by changing the C term. The C term is most commonly adjusted by using a complexing agent to modify the deposition potential. Pb and Sn can be plated in nearly any composition simply by varying the solution concentrations. This is not the case for alternatives to Pb. The concentration term is a logarithmic function and has minimal effect with the large differences of standard potentials. Even if simple adjustments of concentration favor co-deposition, the resulting deposit is not likely to be the desired composition. Therefore, manipulation of the C term must be used to plate the Pb-free alloys. Complexation offers a preferred avenue for deposition of Pb-free alloys. Aqueous complexing agents and their chemistry for the individual metals are fairly well known. However, for complex systems such as plating baths with many possible incompatibilities, determination of suitable complexing agents is difficult and requires extensive study and experimentation. Organic additives may be used in addition, or as an alternative, to complexation to inhibit the deposition of the more noble metal. In most cases, the trial and error approach seems to be the most common due to a lack of fundamental understanding of the role of additives during reduction and nucleation of ionic species at the cathode surface.

8.3.4 Considerations for the electroplating of tin and its alloys

Plated alloys can be formed by two methods: by sequence where the metals are individually deposited in layers that are alloyed during a reflow process or, by co-deposition where all metals are deposited in a single step. Before discussing specific Pb-free alloys deposition, some important fundamentals of selecting plating baths should be considered. Using tin as the primary example, the aspects of commercial electrodeposition are discussed first.

8.3.5 Plating bath selection (formulation)

In general, plating baths are composed of a metallic salt that supplies the cation to be plated, water, the electrolyte, and any additives. The metallic salts selected for use in electroplating should be very soluble, allowing a wide range of solution concentrations. This is even more important when considering alloy deposition. Ionic concentration affects a variety of dependant factors such as solution conductivity, throwing power, and deposition rate. The main purpose of the electrolyte is to provide a conductive medium for the exchange of electrons for reduction to take place at the cathode. The electrolyte also provides the proper pH in order to keep the metal ion in solution. As mentioned earlier, Sn hydrolyzes easily except at extreme pH (<1, >13). Additional criteria for selection of the electrolyte should also include cost, compatibility with the materials to be plated, and equipment construction and safety. Additives affect the plating processes and deposit characteristics and can be of a variety of forms: organic, metallic, ionic, and non-ionic. These are absorbed onto the plated surface and are often incorporated into the deposit. The ability of additives to affect the nature of the deposit seems out of proportion to their relative concentrations. In general, additives act as grain refiners and brighteners. Their actual mechanisms are not well understood but modern analytical techniques are beginning to shed light on their function at a molecular level.[19,20]

Additives also have several drawbacks because they are consumed either by incorporation in the deposit or they decompose in side reactions that usually occur at the anode. Additive concentration should be monitored analytically with periodic additions made to maintain the proper plating control. Loss of control can result in oxidation of Sn(II) ions or dendritic deposits which will affect uniformity. By-products of these side reactions can build up over time often becoming incorporated in the deposit, resulting in voids after reflow or decreasing the wettability of the solder deposit. Additives and their by-products are often difficult to monitor, requiring more sophisticated techniques not normally found in the plating facility such as high-pressure liquid chromatography (HPLC). Further discussion of additives aside from those selected for specific plating baths is beyond the scope of this chapter. The following is a discussion of some common commercial tin plating chemistries.

8.3.5.1 Alkaline tin chemistry

The first used, or perhaps first reported, tin plating baths that date back to the 1840s and 1850s were highly alkaline solutions based on stannite anions.[21,22] This type of bath chemistry was the basis of most deposition of tin for many years, but disproportionation of stannite(II) occurs yielding stannate(IV) ions and metallic Sn resulting in poor deposits. Stannate baths are an improvement over their predecessor and can be run without additives with high throwing power. The reduction of the stannate ion releases H^+, lowering pH which in turn reduces bath efficiency. Stannate baths are still used where cost and speed of deposition are important factors.

8.3.5.2 Acid tin chemistry

An obvious disadvantage with alkaline tin baths is the amount of current required to deposit tin from Sn(IV) solutions as opposed to Sn(II) found in acid solutions. Highly alkaline baths also degrade the photoresists used to define areas for solder plating on semiconductor wafers.[23,24] Simple acid baths tend to produce loosely adhering dendritic deposits with poor covering power. The Sn(II) ions are subject to both hydrolysis and oxidation. Additives are required for controlling acid tin plating baths in order to produce uniform deposits. Hydrolysis is controlled easily by maintenance of the pH below 1.0 or with complexation. A number of different plating bath chemistries have been used successfully to produce deposits of tin and its alloys for various industrial and electronic applications.[25–26] A few that have direct application for production of tin alloys used in the electronics industry are discussed below.

Sulfate. The stannous sulfate bath used primarily for bright tin coatings is very inexpensive to operate, especially where large tank volumes are required such as plating of large panels. Sulfuric acid is readily available and has low toxicity and waste treatment is easy to manage. Sulfate/sulfuric systems operate at 100% cathode efficiency with high throwing power, obviously advantageous to large-scale operations. These two factors are also of merit for flip-chip plating where the metal is deposited onto the wafer through small openings in a thick photoresist. High cathode efficiency eliminates concern over the formation of hydrogen bubbles that become trapped in the photoresist opening preventing further plating. High throwing power is important for producing uniform deposits in photo resist openings across a wafer. Unfortunately, this system has the major drawback of oxidation. At current densities over 30 amps/square foot (ASF), soluble tin anodes become passivated due to the formation of SnO_2. In the case of insoluble anodes, Sn(II) is oxidized to Sn(IV) at the anode in addition to natural oxidation from exposure to air. Sn(IV) hydrolyses to insoluble SnO_2. Another disadvantage of sulfate baths is that both Ag and Bi form insoluble compounds in sulfate-based chemistries, although sulfates of In, Zn, and Cu are very soluble.

Fluoborate. Fluoborate is among the oldest acid chemistries for tin plating. Tin(II) tetrafluoborate ($Sn(BF_4)_2$) is very soluble, allowing high concentrations Sn^{2+} in solution for higher current densities. Although Sn(II) does oxidize, the resulting $Sn(BF_4)_4$ remains in solution, meaning soluble anodes do not passivate and filters do not clog from the colloidal suspension of SnO_2 and other Sn(IV) compounds. Most metals have soluble tetrafluoborate salts including lead, bismuth, and silver. This fact coupled with the ability to plate at high current densities made the tetrafluoborate system the most ideal for high speed tin and solder plating applications.

Although Pb–Sn alloy plating is relatively easy, sequential plating can be used to plate tin over lead to protect it during subsequent processing. Fluoborate also undergoes hydrolysis in solution as:

$$BF_4^- + H_2O \rightarrow [HO-BF_3^-] + HF$$

Care must be taken to protect semiconductors from unwanted etching of silicon-based passivation by the resulting HF when plating in fluoborate solutions. The formation of HF can be reduced by the addition of boric acid. Technically, fluoborate would be one of the most ideal electrolytes for deposition of Pb-free alloys. However, concerns about the release of fluoride and borate ions to the environment and the high cost of waste treatment created a need for alternative electrolytes.[27]

Methane sulfonate. The steel industry is a major resource in the development of suitable plating baths for tin and its alloys. The high current density demands of large surface areas have been met with fluoride or phenol sulfonate chemistries. Fluoride chemistry is unacceptable in the electronics arena due to its corrosive effect on silicon compounds. Phenolsulfonic acid (PSA) has gained popularity over fluoride in the steel industry because it can act as its own antioxidant thereby improving bath maintenance. The drawback is environmental due to the release of the phenol group during processing. Methanesulfonic acid (MSA) can readily replace PSA, in theory, due to similar chemical properties. However, cost remains an issue in large tank operations because MSA is considerably more expensive than PSA. MSA has several advantages over the previously discussed chemistries. MSA is much less corrosive to electronic materials than fluoride, sulfate, or fluoborate chemistries. MSA is, by far, the most benign, as it is much safer to work with as well as less costly for waste treatment. Other advantages include high solubility of salts for all Pb-free candidate metals (an exception is Sb). Insoluble anodes are not attacked by MSA as with the aforementioned electrolytes. MSA is also reported to operate at current densities up to 1800 ASF.[28] MSA also appears to be the electrolyte of choice for Pb-free solder alloy deposition as evidenced by the growing number of patents that use this acid for plating of electronic materials.[29–35]

8.3.6 *Oxidation control*

Oxidation has been mentioned numerous times in the above sections and requires additional attention. With thermodynamic consideration of the reduction potentials in Table 8.1, the reaction of Sn(II) with dissolved O_2 to produce Sn(IV) in an acidic environment is quite favorable. The primary concern is twofold, the loss of Sn(II) from the electrolyte and the buildup of insoluble matter in the solution. The best case is that all tin remains soluble, although additional current is needed to further reduce Sn(IV) to tin metal negatively affecting plating rate. Monitoring Sn(IV) concentration and adjusting current and plating times is difficult and is usually not practiced. In most cases Sn(IV) ions generally hydrolyze into insoluble compounds (such as SnO_2) which turns the solution murky. The SnO_2 colloidal suspension can become incorporated into the deposit with detrimental effects; or, if the colloidal particles aggregate, they can clog filters. Antioxidants are predominantly used to reduce Sn(II) oxidation. Antioxidants can accomplish this by several mechanisms.[36,37] They can form stable complexes with Sn(II) reducing its reactivity with O_2, reducing the solubility of oxygen, and tying up or reacting with the dissolved oxygen. The latter methods are difficult to maintain since most baths are mechanically agitated, causing more oxygen dissolution. Most antioxidants used to complex Sn(II) are also reducing agents, such as pyrocatechol or hydroquinone.

8.3.7 *Anodes*

Anode composition must also be considered when selecting a plating bath for specific applications. There are two kinds of baths: soluble and insoluble. Soluble anodes are commonly used in acidic media, are made of the metal or alloy to be plated, and are consumable by design. They have large surface areas relative to the cathode area improving throwing power and current density, allowing a faster plating rate. Dissolution of the anode during plating assures that the concentration of metal ions will remain constant during plating. Soluble anodes work well for plating

Table 8.1   Standard reduction potentials.

Half reaction	Std. reduction potential (V)
$Zn^{2+} + 2e^- \rightarrow Zn,$	$E^0 = -0.7628$
$In^{3+} + 3e^- \rightarrow In$	$E^0 = -0.338$
$Sn^{2+} + 2e^- \rightarrow Sn$	$E^0 = -0.136$
$Pb^{2+} + 2e^- \rightarrow Pb$	$E^0 = -0.1263$
$2H^+ + 2e^- \rightarrow H_2$	$E^0 = 0.0000$
$Sn^{4+} + 4e^- \rightarrow Sn^{2+}$	$E^0 = 0.150$
$SbO^+ + 2H^+ + 3e^- \rightarrow Sb + H_2O$	$E^0 = 0.204$
$Bi^{3+} + 3e^- \rightarrow Bi$	$E^0 = 0.317$
$Cu^{2+} + 2e^- \rightarrow Cu$	$E^0 = 0.340$
$Ag^+ + e^- \rightarrow Ag$	$E^0 = 0.799$
$O_2 + 4H^+ + 4e^- \rightarrow 2H_2O$	$E^0 = 1.229$
$Pt^{2+} + 2e^- \rightarrow Pt$	$E^0 = 1.2$
$Au^{3+} + 3e^- \rightarrow Au$	$E^0 = 1.42$

pure metals, but can be problematic when used with alloy solutions. Differences in standard reduction potential can result in unwanted immersion or displacement of the more noble metal onto the anode when the power is off (Table 8.1). This will result in depleting the solution of one of the major components while increasing the concentration of the other. When the bath is restarted, the resulting concentration imbalance may lead to a deposit with an incorrect composition. A pure tin anode is quite susceptible to imbalance in Pb-free alloy baths containing ions of less reactive metals like Cu, Ag, or Bi. The displaced metals can also oxidize and passivate the anode thereby reducing the plating rate because oxidation takes place at the anode.

Insoluble anodes fabricated from a platinized titanium mesh are generally preferred for tin alloy baths. The platinum surface is at a potential too low to cause displacement of any of the more active metals, and so the anode can be left in the bath for greater convenience. Any exposed titanium is rendered inert by its very passive oxide. Insoluble anodes require more monitoring and replenishment of the metal content because the anode does not dissolve. Another disadvantage of the mesh-type anode is that the relative surface area of the insoluble anode is smaller than a conventional soluble anode. Thus a higher potential is needed to obtain comparable current densities. At higher potentials, the propensity for oxygen generation at the insoluble anode is greater leading to a drop in pH that must be monitored. At higher currents, O_2 gas is also evolved and bubble formation can be a problem in wafer plating for flip-chip interconnects. By minimizing these disadvantages, insoluble anodes have greater applicability for Pb-free alloy deposition.

8.3.8   *Pb-free alloy plating (basics)*

The development of electroplated Pb-free alloys requires consideration of the phase diagrams to determine the desired composition and redox potentials, the ease at which co-deposition can be accomplished, the chemical compatibility of the deposited metals in the plating baths, and the thermodynamic and kinetic considerations. Each of these are discussed in the following sections.

8.3.9 *Sequential plating*

The most direct way to produce Pb-free alloys is by sequential plating. Typically the Pb-free alloying element (e.g., Ag, Cu, Bi) is first plated onto the solderable metal surface (e.g., Cu or Ni/Au metallizations). The part is then rinsed and placed into the Sn bath for plating. The metals are then alloyed together by heating above the temperature required to melt the Sn. This sequence is preferred because the Pb alternative acts as a barrier reducing the metallurgical reaction between Sn and the solderable metal. In the case of zinc and indium, tin must be plated first due to corrosion of the more reactive metals by the Sn plating solution. Once the composition is determined, usually a eutectic, Faraday's law is used to calculate the amount of time at a given current to produce the quantity of each metal needed for the desired composition.

$$M = \frac{itA}{nF} \tag{2}$$

where M is the mass of the metal deposited in grams, I is the current in amperes, t is time expressed in seconds, F is the Faraday constant, n is the number of electrons involved in the electrodeposition process, and A is the atomic weight of the deposited metal.

The amount of metal deposited can be increased either by increasing time or current. The preferred choice is to increase current but limitations on the plating process may not allow this choice. Faraday's law also shows that the amount of metal deposited is inversely proportional to the number of electrons involved in the process. That is why acid Sn(II) baths are preferred over alkaline Sn(IV) chemistries.

8.3.10 *Co-deposition*

Single step alloy plating (co-deposition plating) is preferred given the cost of equipment, space, cycle time and waste treatment. Comparison of the standard potentials in Table 8.1 shows that unlike Pb, the potential differences between Sn and the alternative metals are sufficiently large to prevent co-deposition from a simple solution.

The Nernst equation shows that manipulation of ion concentration alone cannot change reduction potentials enough to allow co-deposition. Additives or complexing agents capable of modifying reduction potentials or electrode kinetics must be employed. Methods allowing the prediction of suitable additives and complexing agents are not fully developed, leading to time-consuming trial and error methods for determination of appropriate compounds. Furthermore, as in the case of cyanide, concerns for safety and the environment also limit the number of suitable chemistries for commercial Pb-free solder alloy plating.

8.3.11   Pb-free solder alloy creation

The two methods for electrochemically depositing flip-chip solder interconnects are sequential plating and alloy plating. The following sections describe specific processes available for producing alloys that are candidates for replacing Pb–Sn solders.

8.3.11.1   SnAg

Until recently, the eutectic alloy Sn–3.5Ag could only be produced by sequential plating. The most readily available silver electrodeposition processes use cyanide as the complexing agent. Such chemistries are well known and will not be discussed further. Alternatives to the highly toxic cyanide processes are becoming available. One of the first commercial non-cyanide Ag plating processes was the so-called Techni-silver Cy-less R product from Technic Inc. The chemistry is a lactate electrolyte with succinimide as the silver complexing agent. The bath is light sensitive and has a very narrow pH operating range around 8. This range is compatible with photoresist for semiconductor wafer processing. However, if the pH drops too far below that limit, metallic silver is deposited, and above 9 a complex silver salt precipitates. This range is incompatible for use in developing a tin–silver alloy bath. Sequential plating using this for bump manufacture has limitations. Free succinimide is also incorporated into the Ag deposit. Succinimide vaporizes at 287°C, and the vapor becomes trapped by the surface tension of the molten solder, forming voids. Another sequential SnAg bump plating process has been reported using separate alkane sulfonic acids and salt solutions for both the Sn and Ag depositions.[38] Recent patents indicate SnAg plating baths are being developed through the use of proprietary complexing agents that overcome the unfavorable thermodynamics. Ishihara Chemical Co. Ltd. has produced a commercial proprietary eutectic SnAg alloy process that is based on a heterocyclic ring compound containing nitrogen (similar to succinimide) to complex the Ag, and also a hydroxycarboxylic acid (i.e., citric, tartaric, lactic) for the electrolyte and Sn(II) salt.[39] Insoluble anodes are required because soluble Sn anodes cause displacement of Ag from the solution.

8.3.11.2   SnBi

Many bismuth salts are known to hydrolyze in aqueous solutions, and very few compounds are listed in general chemistry texts, leading one to suspect the choices for electroplating bismuth would be difficult. A variety of bismuth plating chemistries were evaluated in the late 1970s and early 1980s for development of extreme high and low temperature nuclear and aeronautical applications.[40,41] An In–Bi plating solution has also been investigated.[42] Sequential plating of Bi and Sn is straightforward but there are very few commercial Bi plating chemistries available. In spite of the large difference in reduction potentials, several high bismuth content tin alloys have been reported in patent literature. One patent reports Bi contents as high as 20% using soluble Sn anodes,[43] and another[44] claims eutectic compositions are possible using soluble Bi anodes. Both claim the use of organic sulfonates and a variety of complexing agents to decrease the difference in reduction potentials.

8.3.11.3   SnCu

Unlike other binary solder alloys, Cu does not appear to diffuse into Sn during reflow of sequentially plated layers to produce a eutectic alloy, at least in small geometries. Fortunately, the relatively small difference in standard reduction potentials allows Sn and Cu to be co-deposited rather easily. Similar to SnPb, the composition is variable using a variety of chemistries.[25,29,30] The presence of Cu(II) ions can catalytically cause oxidation of Sn(II) to Sn(IV) at a much faster rate than occurs through natural air oxidation in pure tin chemistries. The use of soluble Sn anodes promotes the displacement of the more noble Cu from solution. The non-adhering Cu metal on the anode tends to create a particulate problem and subjects the anode to corrosion as well as depleting the solution of Cu(II) ions. At least two eutectic SnCu alloy plating chemistries based on methane sulfonate are commercially available. The baths are quite stable with regards to oxidation and composition control even though Cu is only present at 0.7% need to produce the eutectic alloy.

8.3.11.4   SnIn and SnZn

Both In and Zn are considerably more chemically reactive than Sn as noted from the table of standard reduction potentials, Table 8.1. Sequential plating must be performed by plating Sn first because the oxidizing potential of the highly acid Sn baths will overcome the applied potential of the plating apparatus and dissolve the Zn and In deposits. The large differences in reduction potential make it highly unlikely that acid-based Sn alloy chemistries can be developed for these metals. SnZn alloys with as much as 25% Zn are routinely deposited from stannate baths containing zinc cyanide for use in corrosion protection of steel.

8.3.11.5   Ternary Pb-free solder alloys

Sequential plating of an alloy with a pure metal or another alloy appears to be an obvious method to produce ternary alloys with the least number of steps using the currently available chemistries. Several processes using this methodology have been reported in recent patents.[29,30,45] Until very recently the wide differences in reduction potentials and incompatibilities between solution chemistries of the candidate metals made achieving single ternary and quaternary plating baths seem a nearly impossible task. However, a eutectic SnAgCu plating process using thiourea as the complexing agent has been reported by researchers in Japan.[46] Thiourea has a great affinity for complexing Sn, Cu, and Ag making it possible to decrease their differences in reduction potentials. Although thiourea itself is very toxic, this new development promises the possibility of manufacturable solutions for this and other ternary and quaternary solder alloys in the future.

Table 8.2 is a list of solder alloys that can be plated for flip-chip applications. For flip-chip solder deposition, plating provides finer pitches and higher volume throughput than printing solder paste. However, solder plating has a limited number of alloy baths that are available particularly for ternary, and above, compositions,

Table 8.2 Solder alloys for plating flip-chip interconnects.

Alloy	Method	Availability	Plating comments	Metallurgical comments
Sn–Ag	Sequential	Commercial	Voids after plating; additional work needed to find more suitable complexing agent	More expensive due to Ag additions, stronger, shorter thermomechanical fatigue life
Sn–Ag	Alloy deposit	Commercial	Single step plating, no voids	As above
Sn–Bi	Sequential	Commercial	Eutectic-low melting	If joined with Pb-bearing solder will form 96°C low temperature melting phase, brittle at high strain rates
Sn–Bi	Alloy deposit	Commercial	Addition of 1% Bi minimizes whisker formation, Sn-coating	Low percentage Bi results in poorer fatigue life
Sn–Cu	Sequential	Not possible	Sequential plating possible, metallurgical interactions during reflow prevent formation of solder	High melting point SnCu intermetallics form instead of eutectic solder alloy
Sn–Cu	Alloy deposit	Commercial	Plating control accurate, simple plating chemistry	Lower cost, excellent joint compliance, best thermomechanical fatigue behavior; optimal choice for flip-chip interconnects
Sn–Ag–Cu	Sequential	Commercial by combining above alloy baths	Currently best option to manufacture this ternary alloy	More expensive due to Ag additions, same strength as Sn–Ag, better fatigue life than Sn–Ag but not as good as Sn–Cu
Sn–Ag–Cu	Alloy deposit	Reported in literature	Additional work needed to find more suitable complexing agent (reported complexing agent, thiourea, is toxic and possible carcinogen)	As above
Sn–In	Sequential	Commercial	Dissolves in acid Sn plating solutions and must be plated after Sn	Expensive due to In additions, very low strength, extensive oxidation during processing and as solidified, not a good flip-chip choice
Sn–In	Alloy deposit	Not reported	Possible using selected complexing agents	As above
Sn–Zn	Sequential	Commercial	Zn must be plated after Sn; Zn will dissolve in acid or alkaline Sn plating solutions	Low-cost alloy, extensive oxidation issues in processing and as solidified, strain rate sensitive
Sn–Zn	Alloy deposit	Commercial	High pH plating bath based on cyanide chemistry	As above

and the initial tool set and facilities for flip-chip bump processing are expensive. Printed solder paste allows virtually any alloy combination on an electroless Ni UBM, with a lower cost tool set. The binary alloys that can be best plated are Sn−Cu, Sn−Ag, and Sn−Pb. The SnCu alloy plating appears to be the easiest to control, with relatively simple chemistries commercially available in the U.S. and Japan. The SnAg plated alloy has a much more complex chemistry with possible voiding issues. The most promising ternary plated alloy is the near eutectic Sn−Ag−Cu solder.

8.4   Physical metallurgy of Pb-free solder alloys

The microstructure of the solder alloys that can be electrochemically deposited for flip-chip interconnects are shown in Figures 8.3 to 8.6. The Sn−Pb eutectic has a structure of Sn-rich and Pb-rich phases that form as lamella, Figure 8.3. Similarly oriented lamella form cells, or colonies, separated by slightly coarsened cell boundaries. This structure is known to be susceptible to heterogeneous micro-structural evolution that concentrates strain at the cell boundaries and causes them to further coarsen and, eventually, be the site of failure. In the fine scale of the flip-chip interconnect, there are only a few cells present in each bump but heterogeneous coarsening and failures are still observed.[47]

Figure 8.4 shows the Sn−0.7Cu microstructure. This solder is composed of large Sn-rich grains with a fine dispersion of Cu_6Sn_5 intermetallics. The solder grains form and grow out from the bond pad interfaces. The grains are large, on the order of 20 to 50 μm in size.

The Sn−3.5Ag microstructure consists of a fine structure of alternating Sn-rich/Ag_3Sn intermetallic lamella, Figure 8.5. Grain colonies also form in this microstructure but the boundaries are not coarsened. In addition to the fine Ag_3Sn intermetallics, large needles of Ag_3Sn are present and are typically attached to one of the bump pad interfaces. The Sn−3.8Ag−0.7Cu solder microstructure is shown in Figure 8.6 and is very similar to that of the Sn−3.5Ag with the addition of discrete Cu_6Sn_5 intermetallics throughout the bulk of the solder.

When the molten solders come into contact with the Ni or Cu surfaces, they wet and react to form interfacial intermetallics. The intermetallics grow out into the solder as rods, or plates, and continue to grow when the solder is in the solid state. Even though all the solders studied are Sn rich, the morphology and reaction kinetics differ between alloys.

On Cu, Sn−37Pb forms a two-phase intermetallic of Cu_6Sn_5 adjacent to the solder and Cu_3Sn adjacent to the Cu. The Cu_3Sn is planar with a columnar grain structure and the Cu_6Sn_5 consists of elongated nodules. On Ni, eutectic Sn−Pb forms irregularly shaped Ni_3Sn_4. The formation and growth of the interfacial intermetallics between Cu, Ni, and eutectic Sn−Pb solder are well known.[18,47,48] The intermetallics follow parabolic growth kinetics and do not extensively spall off into the solder.

The intermetallic formation for the Pb-free solders on Ni after two reflows is shown in Figure 8.7. Two reflows represent typical processing for flip-chip inter-connects; the first reflow represents ball formation on the wafer, the second represents flip-chip to substrate interconnect. Figure 8.7 shows the structure of the

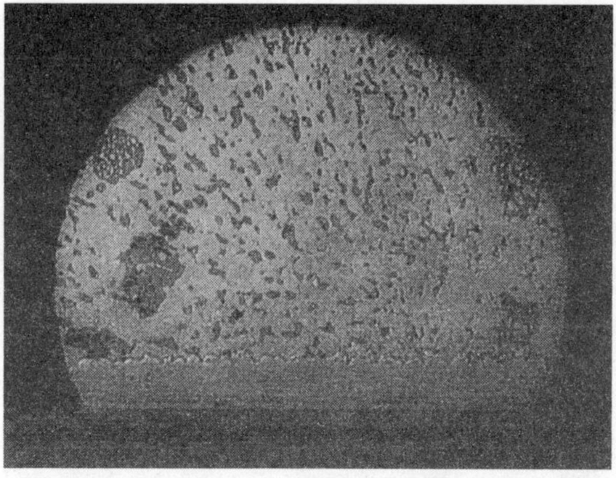

Figure 8.3   Optical micrograph of the microstructure of Sn–37Pb solder in a flip-chip bump on a Cu UBM.

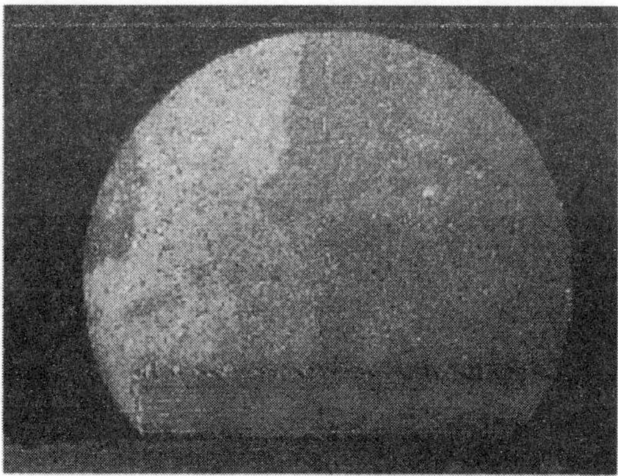

Figure 8.4   Optical micrograph of the microstructure of Sn–0.7Cu solder in a flip-chip bump on a Cu UBM.

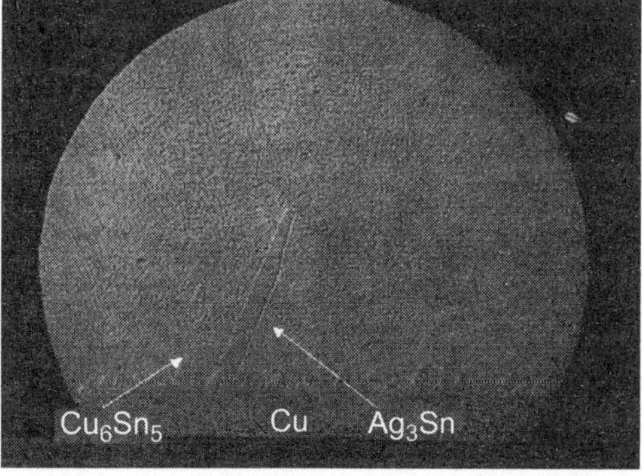

Figure 8.5   Optical micrograph of the microstructure of Sn–3.5Ag solder in a flip-chip bump on a Cu UBM.

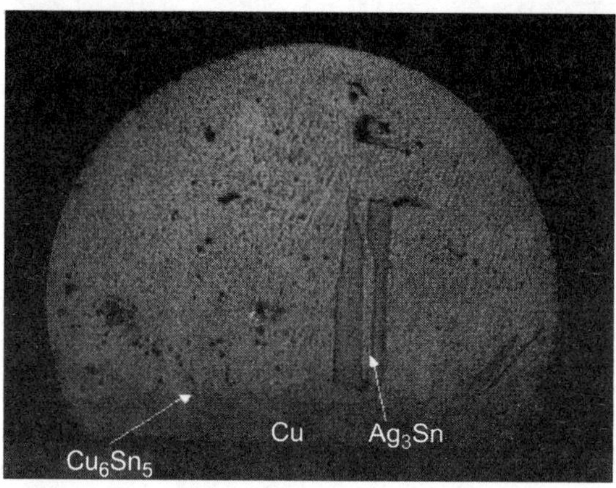

Figure 8.6   Optical micrograph of the microstructure of Sn–3.8Ag–0.7Cu solder in a flip-chip bump on a Cu UBM.

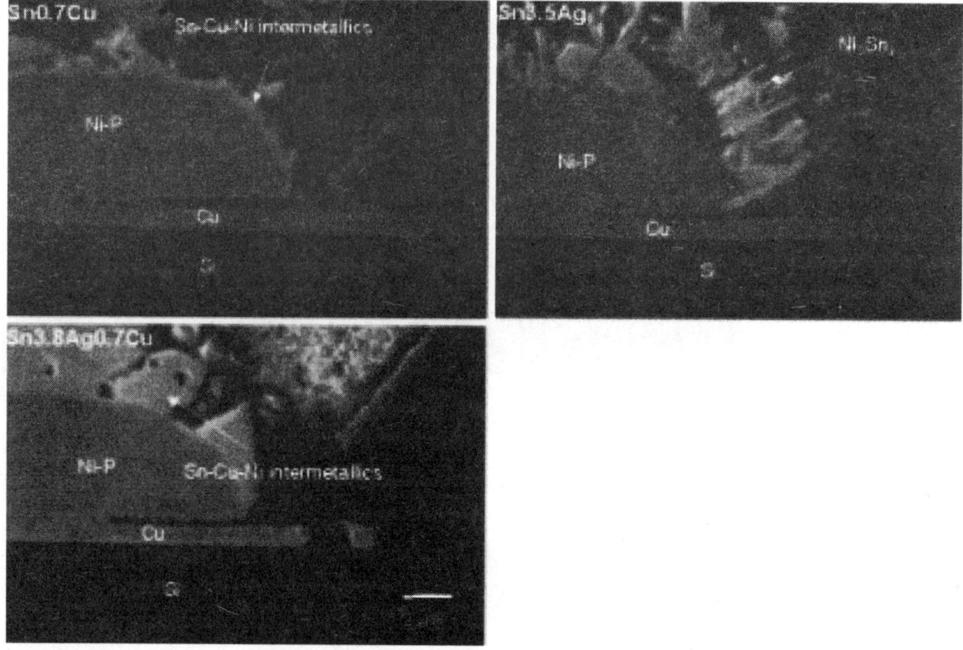

Figure 8.7   SEM micrographs of the Pb-free solders on an electroless Ni UBM showing the morphology of the interfacial intermetallics.

interface on electroless Ni–P, but the same observations were made for the electrolytic Ni UBM. The intermetallic that forms is Ni$_3$Sn$_4$. The Ni$_3$Sn$_4$ intermetallic between Sn–0.7Cu and Ni is thin and regular and is the most uniform of the Pb-free alloys. The Sn–3.5Ag solder on Ni has a different intermetallic morphology that consists of nodules and chunks of Ni$_3$Sn$_4$ that spall off into the solder. This morphology has been attributed to the lack of Cu in the solder. It is hypothesized that

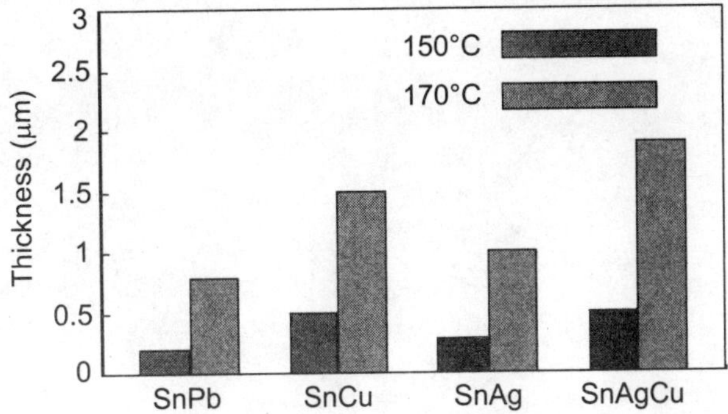

Figure 8.8   Thickness of consumed Cu for the solders on electroless Ni–P/Au UBM after 1000 hours aging at 150° and 170°C.

the Cu acts to saturate the solder with respect to the Ni and inhibits dissolution and spalling of the intermetallic into the solder.[49] However, this mechanism remains to be fully understood.

The growth of interfacial intermetallics while the solder is in the solid state is of concern for flip-chip interconnects. If the intermetallic layer coarsens significantly, it can consume the UBM and cause the joint to dewet at the layer beneath the UBM. Also, the intermetallic is brittle and if it becomes a significant fraction of the solder joint, it can act as a site for crack initiation and propagation when the joint is deformed.

The consumption of the Ni layer by the formation of Ni_3Sn_4 intermetallic during solid-state aging for each solder alloy is shown in Figure 8.8 for 1000 hours aging at 150 and 170°C. Figure 8.8 shows results on an electroless Ni–P/Au UBM (similar results were observed on the electrolytic Ni UBM). For these solders, less than 2 µm transformed into Ni_3Sn_4. The Ni reacts slowly with Sn-based solders and is the reason why it is often preferred for Sn-rich UBM structures. The Pb-free solders consume more Ni and form more intermetallic than Sn–Pb eutectic solder but this increase is relatively small. The Ag_3Sn intermetallic plates are attached to the Ni_3Sn_4 interfacial intermetallics, similar to those observed on the Cu UBM.

Figure 8.9 shows scanning electron microscope (SEM) micrographs of the solders on a Cu UBM after two reflows. The intermetallic that forms is Cu_6Sn_5; no Cu_3Sn was found but it may have been too thin to be observed. The interfacial intermetallic has the same morphology for all the Pb-free alloys as for Sn–Pb eutectic. The intermetallic consists of regularly spaced nodules of Cu_6Sn_5. The Ag-containing solders all have large Ag_3Sn intermetallics that are attached to the Cu_6Sn_5 interface (Figures 8.5, 8.6, 8.9). All three Pb-free alloys also have small discrete particles of Cu_6Sn_5 present in the bulk of the solder. For Sn–0.7Cu, the Cu is present in the solder before joining to the UBM. For the Sn–3.5Ag solder the Cu_6Sn_5 is present due to the dissolution of some of the UBM into the solder. Spalling of the interfacial intermetallics into the molten solder was not observed and this is believed to be due to the presence of Cu in each of the solders during reflow. The Cu inhibits growth and spalling of the intermetallic because it saturates the solder.[58] Of the three Pb-free

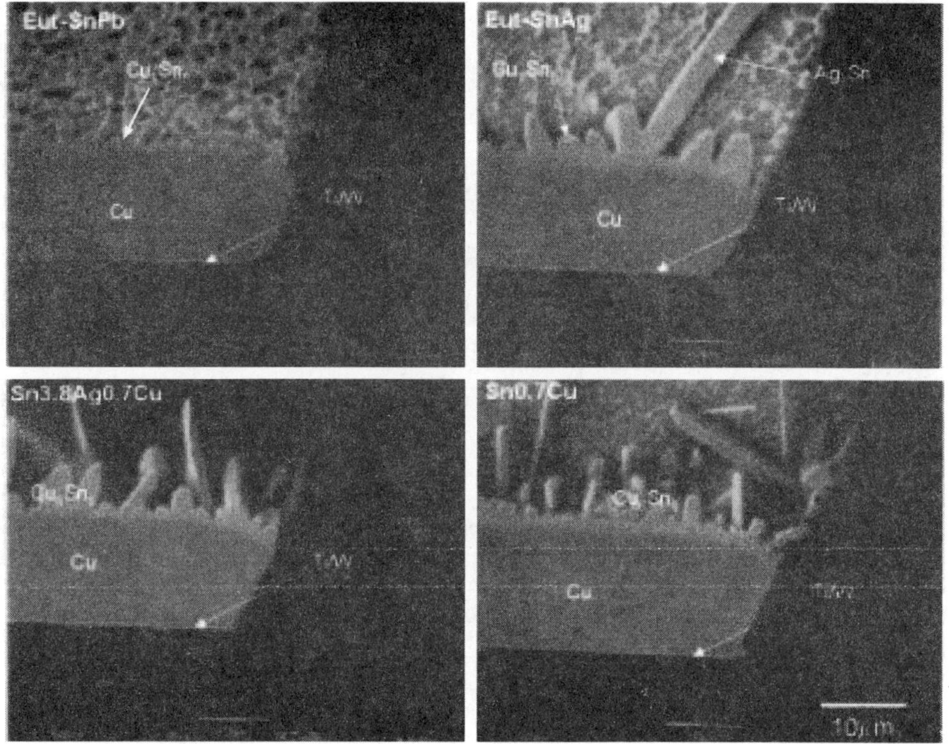

Figure 8.9   SEM micrographs of the Pb-free solders on Cu UBM.

alloys on Cu, the Sn−0.7Cu solder structure is the most uniform, and has the thinnest intermetallic structure.

One of the concerns of using Sn-rich Pb-free solders is the reaction of the Cu with the solders, feared to be so fast that the UBM will dissolve during reflow. A plot of the Cu consumed after two reflows for Sn−Pband the Pb-free solders is shown in Figure 8.10. The Pb-free solders consume only 10 to 20% more Cu than Sn−Pb and this is less than 2 μm after two reflows. A plot of consumed Cu during solid-state aging is shown in Figure 8.11 for the solders on Cu at 150°C for 500 and 1000 hours. In the solid state, the Cu was consumed at a slower rate in Pb-free solders than for eutectic Sn−Pb. The Pb appears to play a role in enhancing intermetallic growth, perhaps by enhancing Sn diffusion to the intermetallic/solder interface. Sn−0.7Cu had the slowest consumption rate of the Cu UBM.

8.5   Mechanical metallurgy of Pb-free solder alloys

The flip-chip solder joint must mechanically hold the chip to a substrate but the solder cannot impose significant strain to the semiconductor device or the device could crack and fail. The shear strength of the solder alloys is shown in Figure 8.12. The strength of the solder alloys is a measure of the ability of the flip-chip interconnect to be compliant to the imposition of strain. Sn−37Pb and Sn−0.7Cu solder bumps

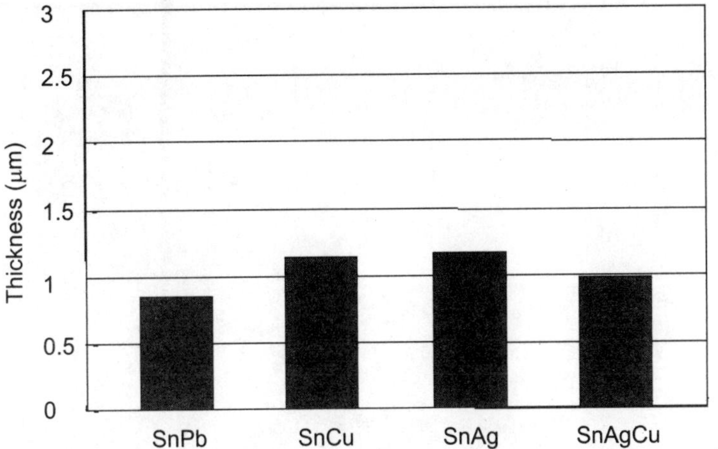

Figure 8.10 Consumed Cu thickness after two reflows for the solders on a Cu UBM.

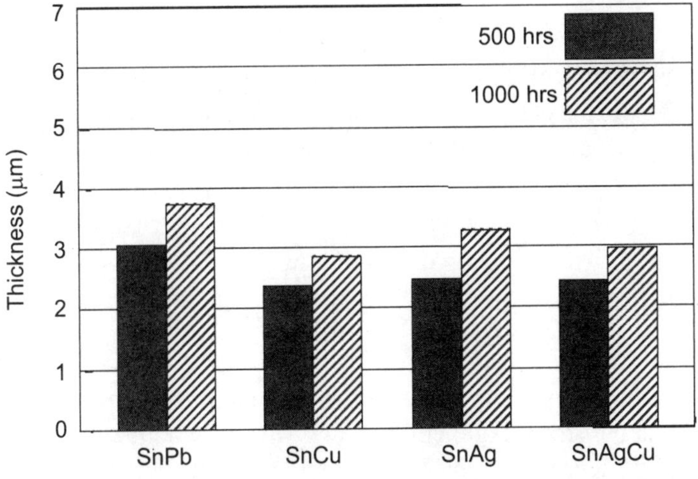

Figure 8.11 Bar chart of the consumed Cu thickness of the solders on a Cu UBM after aging for 500 and 1000 hours at 150°C.

have similar values of shear strength. The Sn–3.5Ag solder alloy has the greatest shear strength at 25% greater than Sn–0.7Cu, with the Sn–3.8Ag–0.7Cu having a slightly lower strength than Sn–3.5Ag. The shear failure for all joints tested occurred solely through the solder. The solder joint strength and failure mode were similar to, and independent of, the UBM type. Eutectic Sn–3.5Ag has shown susceptibility to brittle interfacial delamination in tensile[18] and shear tests[51] for surface mount interconnects. The failure occurs at the intermetallic/solder interface where one side of the fracture shows intermetallic and the other side reveals an impression of where the intermetallic grew into the solder.

Creep behavior is important for flip-chip interconnects because the solders deform to relax stress over time when held at a constant strain. The creep rate of a

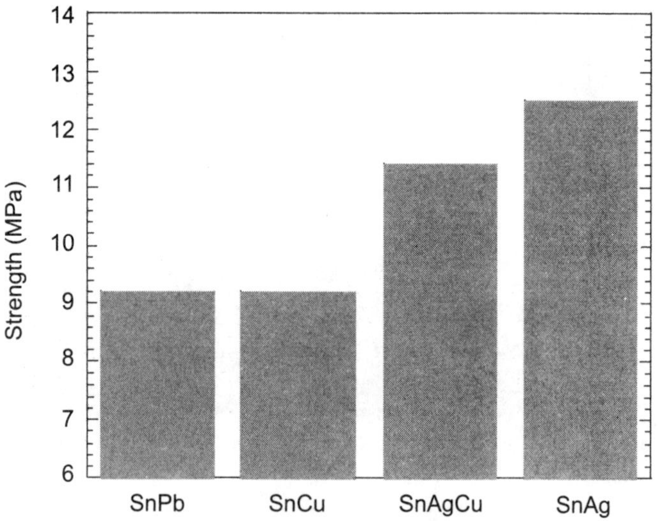

Figure 8.12 Shear strength of the solder flip-chip bumps.

solder must be sufficiently fast so that the strain is minimized in joined bulk components. However, the creep rate must not be so fast that the components move over time. The creep behavior of solders can be summarized empirically using one of two equations:

$$\mathrm{d}\gamma/\mathrm{d}t = A\,\sigma^n e^{-Q/RT} \tag{3a}$$

$$\mathrm{d}\gamma/\mathrm{d}t = A\,\sin h(\alpha\sigma)^n e^{-Q/RT} \tag{3b}$$

where $\mathrm{d}\gamma/\mathrm{d}t$ is the creep rate in shear, A is a constant, α is the stress constant, σ is the flow stress, n is the stress exponent, and Q is the creep activation energy. Equation 3a works well for creep mechanisms that remain constant over all test temperatures. Equation 3b is the Garofalo, or sinh, creep relation that captures up to two different creep mechanisms in a single formulation. Creep test results are found in the published literature.[52–55] Table 8.3 represents the fitted results of creep tests on Pb-free solders using Equations 3a and 3b with the Garofalo (equation 3b) having the best fit.[56] The creep rate of high Sn-content solders is slower than Sn–Pb.

Table 8.3 Constitutive creep relations.

Alloy	$A\ (s^{-1})$	α	n	$Q\ (kJ/mol)$	Ref.
Sn–40Pb	2.48×10^4	0.0793	3.04	56.9	52
Sn–40Pb	1.1×10^{-12}		6.3	20	53
Sn–3.5Ag	9.3×10^{-5}		6.05	61.2	54
Sn–3.8Ag–0.7Cu	2.6×10^{-5}		3.69	36	54
Sn–1Cu	1.41×10^{-8}		8.1	79.4	55

Eutectic Sn–3.5Ag has greater strength and a higher creep resistance than eutectic Sn–Pb.[56] The creep behavior Sn–3.5Ag has been found to have one of the highest levels of creep resistance of the Pb-free alloys.[57] The steady-state creep behavior for Sn–Cu has been found to be faster than eutectic Sn–Pb. The faster creep rate of Sn–Cu is desired for flip-chip applications because damage can be accommodated by the solder, rather than the more brittle joined components. Furthermore, a faster creep rate often translates into a longer thermomechanical fatigue lifetime.

Temperature variations encountered during use conditions, combined with the materials of differing coefficients of thermal expansion in the electronic package, result in cyclic temperature and strain on the solder joints. Glazer presented a summary of work on Pb-free solders during 1994 and concluded that thermomechanical fatigue data for low melting point solders are scarce and appear contradictory.[3,4] The published data have primarily focused on surface mount interconnects, and thermo-mechanical fatigue data for a flip-chip configuration are still needed. The thermal fatigue behavior of the solder in flip-chip interconnects is described in the following.

A diagram of thermal fatigue life vs. strain range was obtained, as shown in Figure 8.13, using the characteristic life extracted from Weibull plots of the thermomechanical fatigue data and thermal strain range calculated from Equation 1 both for the 0 to 100°C and –40 to 125°C tests. These data were collected on flip-chip packages with no underfill. By eliminating the underfill, this study evaluated the relative thermal fatigue performance of different Pb-free solders and characterized the failure mechanisms. Although the strain evaluated here tends to be much higher than that of packages with underfill encapsulation, the failure mode remains valid. The thermal fatigue failure data follow a power law function. In Figure 8.13, the thermal fatigue data from Sn–0.7Cu flip-chip joints on both NiP and TiW/Cu UBMs fall onto the same straight line. The thermal fatigue performance of Sn–0.7Cu is UBM independent and is confirmed by the failure analysis discussed below where failures occurred solely through the solder joint. The cross sections of the failed joints shown in Figure 8.14 are on the TiW/Cu UBM, but similar failure behavior was exhibited for the electrolytic Ni and electroless Ni/Au UBM structures. Sn–3.5Ag has the shortest characteristic thermal fatigue lifetime.

The Sn–37Pb flip-chip interconnects fail by crack formation and propagation through heterogeneous coarsened bands near the UBM/solder interface, as shown in Figure 8.14. The heterogeneous coarsened region forms as a result of the strain concentrating at the eutectic cell boundaries.[58–60] The boundaries are slightly coarsened and are the weakest regions of the joint. Damage accumulates at the cell boundaries and anneals at the high temperature portion of the thermal cycle, resulting in coarsening. Coarsening continues until the Sn and Pb grains in the coarsened bands are so large that they can no longer slide and rotate to accommodate the strain, resulting in grain boundary crack initiation and propagation. The UBM/solder interface is the highest strain region in the joint and is where coarsening and failures typically occur for eutectic Sn–Pb flip-chip interconnects. The cracks were observed to form on the outer edge of the bump and propagate to the center of the bump. The surface of the solder joint remains smooth after thermal cycling indicating that the damage was localized to the heterogeneous coarsened band.

The eutectic Sn–0.7Cu solder exhibited a failure mode that differed from the

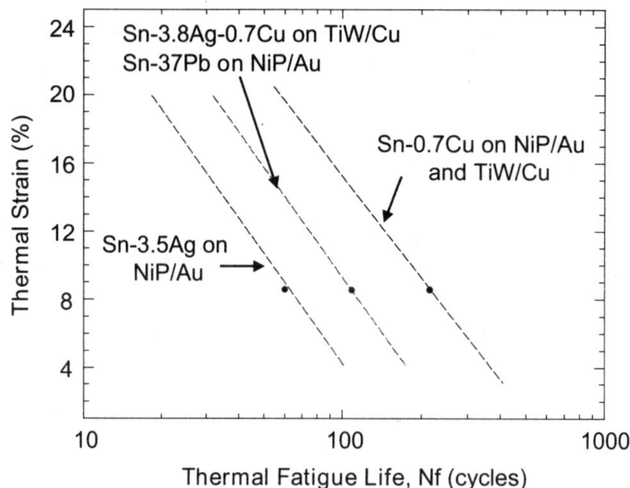

Figure 8.13 Plot of thermal fatigue life as a function of applied thermal strain for the Sn–3.5Ag, Sn–37Pb, Sn–3.8Ag–0.7Cu, and Sn–0.7Cu.

other solder alloys studied. The initiation and propagation of fatigue cracks is through the grain boundaries in the Sn–0.7Cu solder, as shown in Figure 8.14. The fatigue cracks initiated on the highest point of strain on the bump, closer to the edge of die, and then propagated across the middle of the bump toward the center of the die. After thermal cycling, the surface of the solder bumps is no longer smooth. The Sn–0.7Cu solder deforms by grain boundary sliding. The cracks were observed to propagate at the grain boundaries significantly removed from the UBM/bump interface, near the center of the joint. This solder is the most compliant in thermal fatigue and undergoes massive deformation before failing by crack propagation.

The failure behavior of Sn–3.5Ag solder is shown in Figure 8.14. Similar to the Sn–3.8Ag–0.7Cu alloy, the thermal fatigue cracks in the Sn–3.5Ag solder cracks initiate and propagate through the intermetallics and at the intermetallics/solder interface. The surface of the solder bump also exhibited no deformation with the damage concentrated at the solder/intermetallic interface. The Sn–3.5Ag solder has a shorter thermal fatigue lifetime than the other Pb-free alloys. It is hypothesized that large Ag_3Sn intermetallic plates strengthen the joint and further reduce the compliance of the solder thereby shortening the thermal fatigue life.

For the solders examined in this study, the shear strength has a direct correlation to fatigue performance. The weaker, more compliant, and faster steady-state creep rate solder (Sn–0.7Cu) has a longer fatigue lifetime. Eutectic Sn–Pb has the same low shear strength as Sn–0.7Cu but has a shorter thermal fatigue lifetime due to heterogeneous coarsening. It is well established that heterogeneous coarsening concentrates strain in the joint to the coarsened band and accelerates crack formation and propagation.[58–60]

The Sn–0.7Cu alloy undergoes massive deformation during thermal cycling. This deformation protects the semiconductor device from damage due to imposed strain. Interestingly, even with significant surface deformation, the crack formation

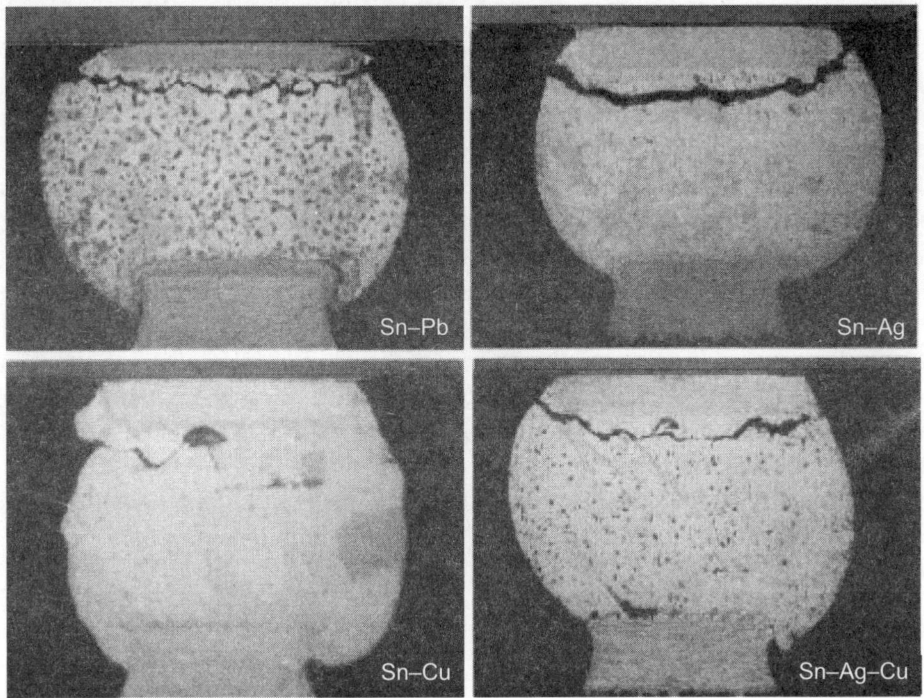

Figure 8.14 Optical micrographs of solder joint cross sections on Cu UBM/Cu pads on organic substrates after thermal cycling 0 to 100°C.

takes longer in Sn−0.7Cu than the other solders that had no surface deformation. The Sn−0.7Cu accommodates the thermal strain by grain boundary sliding and rotation. The self-diffusion of Sn to accommodate the sliding and rotation is sufficient to delay the formation of cracks. An additional positive aspect of the Sn−0.7Cu solder is that failure is only observed in the solder joint away from the brittle intermetallic/solder interface.

8.6 Summary

Pb-free solder interconnects may become a necessity for flip-chip interconnects to satisfy environmental legislative bans of Pb in electronics. The low cost of plating Pb-free solder flip-chip makes electrochemical deposition attractive. Solder alloys can be plated either by sequential deposition of alternate layers of pure elements (the composition is determined by each layer thickness) or by alloy plating. Alloy plating is preferred for simplicity. There are a number of two-component plating baths that are commercially available for Pb-free solders including Sn−Bi, Sn−Ag, and Sn−Cu. It is also possible to plate Sn−In and Sn−Zn but this involves either development of new complexing agents (for Sn−In) or hazardous cyanide complexing chemistry (Sn−Zn). Ternary Sn−Ag−Cu solder alloy plating has been reported, but commercially this involves the less than preferred method of sequential plating.

From a metallurgical standpoint, the optimal Pb-free solder alloy for flip-chip interconnects is Sn–0.7Cu. The alloy contains no Ag and so is the lowest cost alternative of the Pb-free alloys. Sn–0.7Cu has the highest melting temperature of the Pb-free alloys proposed for flip-chip applications and thus provides some melting temperature hierarchy for subsequent package solder processing. The intermetallic structure of the Pb-free alloys is similar to that of eutectic Sn–Pb. The interfacial intermetallics grow at a faster rate during reflow of the Pb-free alloys but during solid-state aging the Pb-free solders have a slower intermetallic growth rate. Sn–0.7Cu has superior thermal fatigue performance of the Pb-free alloys surpassing that of the low melting temperature standard of eutectic Sn–Pb. The Sn–0.7Cu joint is compliant and undergoes massive deformation before cracks initiate or propagate in thermal fatigue. When the cracks do form, they occur near the center of the joint. Crack initiation and propagation in the other Pb-free solders occur at the more brittle solder/intermetallic interface.

For lower cost, ease of processing, and superior thermal fatigue performance, Sn–0.7Cu is the optimal Pb-free solder alloy for flip-chip interconnects.

References

1. N.C. Lee, "Pb-free Soldering – Where the World is Going", *Advancing Microelectronics*, Sept./Oct. (1999), p. 29.
2. A. Grusd, "Integrity of Solder Joints from Pb-Free Solder Paste", Proc. NEPCON West'99 (1999).
3. J. Glazer, "Metallurgy of Low Temperature Pb-Free Solder for Electronic Assembly", *Int. Mater. Rev.*, 40, No. 2, pp. 65–93 (1995).
4. J. Glazer, "Microstructure and Mechanical Properties of Pb-Free Solder Alloys for Low-Cost Electronic Assembly: A Review", *J. Electron. Mater.*, **23**, No. 8, pp. 693–700 (1994).
5. G. Whitten, "Pb-free Solder Implementation for Automotive Electronics", Proc. 50th Electron. Comp. Tech. Conf. (2000), pp. 1410–1415.
6. E. Bradley, III, and J. Hramisavljevic, "Characterization of the melting and wetting of Sn–Ag–X Solders", Proc. 50th Electron. Comp. Tech. Conf. (2000), pp. 1443–1448.
7. W.K. Choi and H.M. Lee, "Effect of Soldering and Aging Time on Interfacial Microstructure and Growth of Intermetallics between Sn–Ag Solder and Cu substrates", *J. Electron. Mater.*, **29**, No. 10 (2000), pp. 1207–1213.
8. F. Guo, S. Choi, J.P. Lucas, and K.N. Subramanian, "Effects of Reflow of Wettability, Microstructure, and Mechanical Properties in Pb-free Solders", *J. Electron. Mater.*, **29**, No. 10 (2000), pp. 1241–1248.
9. T.S. Choi, K.N. Subramanian, and J.P. Lucas, "Thermomechanical Fatigue Behavior in Sn–Ag Solder Joints", *J. Electron. Mater.*, **29**, No. 10 (2000), pp. 1249–1257.
10. Y. Miyazawa and T. Ariga, "Microstructural Change and Hardness of Pb-free Solder Alloys", Proc. 1st Int. Symp. on Environ. Conscious Design (1999), pp. 616–619.
11. M. Abtew and G. Selvardery, "Pb-free Solder in Microelectronics", *Mater. Sci. Eng.*, **27** (2000), pp. 95–141.
12. H.K. Seelig and D. Suraski, "The Status of Pb-free Solder Alloys", Proc. 50th Electron. Comp. Tech. Conf. (2000), pp. 1405–1409.
13. K.G. Snowden, C.G. Tanner, and J.R. Thompson, "Pb-free Soldering Interconnects: Current Status and Future Developments", Proc. 50th Electron. Comp. Tech. Conf. (2000), pp. 1416–1419.
14. T.M. Korhonen, P. Su, S.J. Hong, and M.A. Korhonen, "Reactions of Pb-free Solders

with CuNi Metallizations", *J. Electron. Mater.*, **29**, No. 10 (2000), pp. 1194–1199.

15. J.C. Foley, A. Gickler, F.H. Leprevost, and D. Brown, "Analysis of Ring and Plug Shear Strengths for comparison of Pb-free Solders", *J. Electron. Mater.*, **29**, No. 10, 2000, pp. 1258–1263.

16. M.W. Roberson, P.A Deane, S. Bonafede, A. Huffman, and S. Nangalia, "Conversion between Standard and Low-alpha Pb in Solder Bumping Production Lines", *J. Electron. Mater.*, **29**, No.10 (2000), pp. 1274–1277.

17. Z. Hasnain and A. Ditali, "Building-in Reliability: Soft Errors-a Case Study", 30th Annual Proceedings. Reliability Physics (1992), pp. 276–280.

18. D.R. Frear and P.T. Vianco, "Intermetallic Growth Behavior of Low and High Melting Temperature Solder Alloys", *Metall. Trans.*, **25A** (1994), pp. 1509–1523.

19. T. C. Franklin, "Some Mechanisms of Action of Additives in Electrodeposition Processes", *Surface Coatings Technol.*, **30**, pp. 415–428 (1987).

20. L. Oniciu and L. Muresan, "Some Fundamental Aspects of Leveling and Brightening in Metal Electrodeposition", *J. Appl. Electrochem.*, **21**, pp. 565–575 (1991).

21. A. Smee, *Elements of Electrometallurgy*, London (1843).

22. G. Gore, *Theory and Practice of Electrodeposition*, London (1856).

23. E.K. Yung and I. Turlik, "Electroplated Solder Joints for Flip-Chip Applications", *IEEE CHMT Trans.*, **14**, pp. 549–559 (1991).

24. K.K. Yu and F. Tung, "Solder Bump Fabrication by Electroplating for Flip Chip Applications," 1993 IEEE/CHMT International Electronics Manufacturing Symposium, pp. 277–281 (1993).

25. J.W. Price, *Tin and Tin-Alloy Plating*, Electrochemical Publications Lmt., Ayr, Scotland (1983).

26. R.E. Horn, "An Overview of Tin and Tin Alloy Plating Particularly as Applied to the Electronics Industry", 12th Capacitor and Resistor Technology Symposium. CARTS '92 pp. 66–75 (1992).

27. M. Datta, J.M. Fenton, and E.W. Brooman, "Developments for More Environmentally Benign Electrodeposition Technology", Proceedings of the Symposium on Environmental Aspects of Electrochemical Technology: Applications in Electronics, pp. 177–187 (1997).

28. M. Schelsinger and M. Paunovic (eds.), *Modern Electroplating*, 4th Edition, John Wiley & Sons, New York (2000), p. 261.

29. R.A. Schetty III et al., European Patent 1091023, April 11, 2001.

30. A. Egli et al., European Patent 167582, January 2, 2002.

31. F. Nobel et al., U.S. Patent 4617097, October 14, 1986.

32. M.P. Toben et al., U.S. Patent 4880507, November 14, 1989.

33. J. L. Little et al., U.S. Patent 5110423, May 5, 1992.

34. G. Federman et al., U.S. Patent 5174887, December 29, 1992.

35. M. Wild and D. Crosby, U.S. Patent 52528112, November 2, 1993.

36. R. Moshohoritou, I. Tsangaraki, and C. Kotira, *Plating Surface Finishing*, **81**, pp. 53–58 (1994).

37. S. Meibuhr and P.R. Carter, *Electrochem. Technol.*, **2**, p. 267 (1964).

38. H. Ezawa, M. Miyata, S. Honma, H. Inoue, T. Tokuoka, J. Yoshioka, and M. Tsujimura, *IEEE Trans. Electron. Packag. Manuf.*, **24**, pp. 275–381 (2001).

39. S. Masaki et al., Japanese Patent JP11021693, January 26, 1999.

40. R. Walker and S. J Snook, "The Electrodeposition of Bismuth", *Met. Finish.*, June, pp. 79–84 (1980).

41. I.M. Kenawy, M.A. Hafez, M.M.A. Gouda, and M.M. Elsemongy, "New Baths for Cathodic Deposition of Bismuth", *Indian J. Technol.*, **24**, pp. 205–208 (1986).

42. Y.N. Sadana and Z.Z. Wang, "Bismuth–Indium Alloy Deposition", *Met. Finish.*, December, pp. 23–30 (1985).

43. S. Masaki et al., European Patent 255558, February 10, 1988.

44. World Patent WO9004048, April 19, 1990.

45. P. Andricacos et al., U.S. Patent 6224690, May 5, 2001.
46. M. Fukuda, K. Imayoshi, and Y. Matsumoto, "Effects of Thiourea and Polyoxyethylene Lauryl Ether on Electrodeposition of Sn–Ag–Cu Alloy as a Pb-Free Solder", *J. Electrochem. Soc.*, **149**, pp. C244–C249 (2002).
47. D. R. Frear et al. (eds.), *Solder Mechanics: A State of the Art Assessment*, TMS, Warrendale, PA (1990).
48. W.J. Boettinger et al., The Mechanics of Solder Alloy Wetting & Spreading, Van Nostrand Reinhold, New York (1993), Ch. 4.
49. H.K. Kim, K.N. Tu, and P.A. Totta, "Ripening-Assisted Asymmetric Spalling of Cu–Sn Compound Spheroids in Solder Joints On Si Wafers", *Appl. Phys. Lett.*, **68**, No.16, April (1996), pp. 2204–2206.
50. S. Chada, R.A. Fournelle, W. Laub, and D. Shangguan, "Cu Substrate Dissolution in Eutectic Sn–Ag Solder and Its Effect on Microstructure", *J. Electron. Mater.*, **29**, No. 10 (2000), pp. 1214–1221.
51. B. Roesner, X. Baraton, K. Guttmann, and C. Samin, "Thermal Fatigue of Solder Flip-Chip Assemblies", Proc. 48th Electron. Comp. Tech. Conf. (1998), pp. 872–877.
52. J.J. Stephens and D.R. Frear, "Time Dependent Deformation Behavior of Near Eutectic 60Sn–40Pb Solder", *Metall. Trans*, **30A** (1999), pp. 1301–1313.
53. Z. Mei, J.W. Morris, "Characterization of Eutectic Sn–Bi Solder Joints", *J. Electron. Mater.*, **21** (1992), pp. 599–607.
54. D.R. Frear, "Constitutive Behavior of Pb-Free Solder Alloys" Sandia National Labs Report SAND96-0037 (1997).
55. J. Liang et al., "Creep Study for Fatigue Life Assessment of Two Pb-Free High Temperature Solder Alloys", *Mat. Res. Soc. Symp. Proc.*, **445** (1997), pp. 307–312.
56. A. Grusd, "Integrity of Solder Joints from Pb-Free Solder Paste", Proc. NEPCON West'99.
57. A. Grusd, "Lead Free Solders in Electronics", Proc. Surface Mount Int. Conf. (1998), p. 648–661.
58. D.R. Frear, D. Grivas, and J. W. Morris, Jr., "Microstructural Study of the Thermal Fatigue Failures in 60Sn–40Pb Solder Joints", *J. Electron. Mater.*, **17** (1988), 171.
59. D.R. Frear, D. Grivas, and J.W. Morris, Jr., "Parameters Affecting Thermal Fatigue Behavior of 60Sn–40Pb Solder Joints", *J. Electron. Mater.*, **18** (1989), 671–680.
60. D.R. Frear, "Microstructural Evolution During the Thermomechanical Fatigue of Solder Joints", *The Metal Science of Joining*, M.J. Cieslak, M.E. Glicksman, S. Kang, and J.H. Perepezko (eds.), TMS, Warrendale, PA (1992), 191–200.

Part IV

Packages and PC boards

9 Materials overview in organic packaging

Saikumar Jayaraman, John Tang, and Vijay S. Wakharkar

9.1 Introduction

Organic packaging serves many key functions. These functions include the interconnection between the silicon to the motherboard, in which the tight pitch of the bond pads on the integrated circuit (IC) are fanned out to the loose pitch of the input/output (I/O) pads on the motherboard; protection of the IC and its connections from the environment; and serving as a thermal path for the heat to be transferred out of the IC.

In the last few decades, there have been rapid changes in substrate and packaging technologies. These changes are driven by the continued evolution of the traditional computer, communication, and consumer electronic market. In addition, newly emerging areas like consumer digital and mobile devices also are driving key technology improvements and changes. Packaging technology has had continuously to scale with IC process technology and the ever-shrinking size of transistors, as predicted by Moore's law.

9.2 Organic packaging and substrate technology

9.2.1 Plastic molded packages

In the 1970s and 1980s, typical high volume organic packages and substrate technologies are limited to injection-molded packages with metal lead frame substrate. Due to the low cost and vast infrastructure base, injection-molded plastic package technology is still being used today. The substrate technology on which this package is based is a simple, single level metal lead frame. The lead frame substrate is the foundation of the molded plastic package. The two most common lead frame metals employed by the industry are nickel–iron alloy and copper alloy. The process of lead frame manufacturing is relatively simple; it starts with a metal sheet, the sheet is formed into metal interconnect by either stamping or chemical etching. Stamping is a lower cost/high-volume process, whereas etching provides a tighter design rule with higher routing density. In the etching process, photolithography is used to transfer design image to lead frame interconnect. The die paddle and bond fingers are spot plated with silver to form a bondable surface for IC bonding and wire bonding. The IC is bonded with a die attach adhesive onto the die paddle; gold wire bonding is used to make the connections between the IC and the lead frame. The injection-molding process is used to encapsulate the IC and bond wires. The outer leads are solder plated for board mount.

There are a wide variety of form factors of molded plastic packages; the older packages are plastic dual in-line package (PDIP), single in-line package (PSIP), and others for mounting on through-hole print circuit motherboard and socket. The advent of surface mount technology brought in the surface mounted packages like small outline package (SOP), plastic leaded chip carrier (PLCC), plastic quad flat pac (PQFP), etc. Surface mount components have the advantages of increased lead count, hence increased routing density on both the motherboard and the components. Molded packages are still widely used today in areas like memory packages.

9.2.2 Organic multi-layers laminate packages, PBGA

In the late 1980s and early 1990s, continued scaling of ICs and change in market requirements demanded new package and substrate technologies. A microprocessor unit (MPU) like the 486-processor family and its compatible were packaged in lead frame substrate and ceramic substrate. For low density products like memory, lead frame substrates such as thin SOP (TSOP) continued to meet the requirements, but due to its limitations on feature size and routing density, lead frame packages were less than desirable for the production of high-performance central processing units (CPUs) or application specific ICs (ASICs). In addition, the arrangement of peripheral I/O leads arrangement of a high pin count PQFP is not scalable beyond the approximately 300+ leads maximum, when it runs into major yield problems. It is very difficult to maintain lead co-planarity and to avoid solder bridging during board assembly at very high lead count.

In the late 1980s, organic plastic ball grid array (PBGA) package and organic substrate technology emerged and became the technology of choice for IC packaging. It provides numerous advantages over the lead frame based organic package. It is based on the widely used and matured printed wire board (PWB) manufacturing. It is low cost compared to alternative packages with high wiring density, it is capable of multi-level metal interconnect and high routing density, it uses plated copper interconnect for high electrical performance, it is reliable, and it has a wide mature supplier infrastructure. The key design advantage of PBGA is that it provides the scalability to much higher I/O count, due to the array structure of the solder balls arrangement.

In order to make possible the PBGA substrate, new incremental process improvements were made on the conventional PWB manufacturing processes. These process improvements included high resolution photoresist for fine line patterning, high resolution expose technology, smaller diameters mechanical hole drilling technology, improved layer to layer alignment technology, new roughening and inter-layer adhesion technology, and many others. New dielectric materials were also developed to increase the reliability and performance.

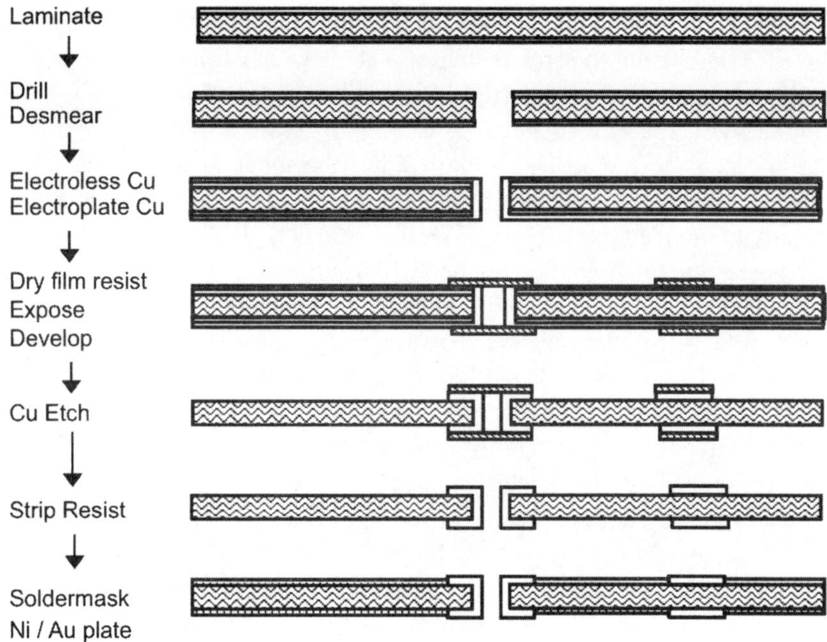

Figure 9.1   PBGA substrate subtractive process flow.

The PBGA substrate patterning process is based on the subtractive patterning process. It starts with a laminate sheet (bis-maleimide triazine resin (BT) with laminated copper foil). Dry film photoresist is laminated onto the surface of the copper foil, the photo mask is aligned, the pattern is exposed and transferred onto the resist, and the sheet is processed through the developer and etcher. The copper that is not protected by the photoresist pattern is etching away. This process transfers the design on the photo mask onto the laminate sheet as copper interconnects. A typical subtractive process flow is illustrated in Figure 9.1.

In recent years, miniaturization of consumer electronic devices has continued to drive IC packages to an ever smaller form factor. The dimensions are reduced on both the X–Y direction (body size) and Z-direction (thickness). Chip scale package and stack package were developed to increase the packing density per unit area. Substrate suppliers and package subcontractors develop innovative package solutions to continue to push the technology envelope.

9.2.3   Tape carrier packages (TAB packages)

Tape automated bonding (TAB) packages emerged in the early 1980s as one of the alternative packaging technologies for high lead count IC devices. It was developed by many major semiconductor companies to meet their requirements, and is capable of high routing density. The tape substrate manufacturing process is highly automatic and economical, it employs the reel-to-reel process, wheel and sprockets through the manufacturing process drive the tape, and it is similar to the processing of photographic film. This enables a high throughput, fully automated process line to be achieved. The tape width is up to 70 mm. The dielectric film for the tape substrate is typically polyimide with copper foil laminated on the surface.

For one layer tape substrate, the patterning process is a subtractive process. The substrate has exposed metal lead in both sides, the inner lead bonding (ILB) side where the leads are bonded onto the IC, and the outer lead bonding (OLB) side where the I/O leads are bonded to the board. The ILB side is typically finished with Ni–Au plating, whereas the OLB lead is finished with tin or solder plating. The substrate is then processed through the ILB TAB bonding. The subsequent encapsulation process covers and protects the chip joints. The package is excised at this point from the reel into a single unit; the OLB leads are formed and mounted on board.

One of the key advantages of the TAB substrate is its thin profile. That led to the use of the tape carrier package for mobile CPU by Intel in the early 1990s. The ultra low profile package is perfect for the emerging laptop market, where minimization of package size is key. The reduction in package Z height is achieved by the combination of the following factors: the reduction of substrate thickness, elimination of wire bond loop height, elimination of mold cap (or mold body) height, and the reduction of OLB height. TAB substrate is being used today on state-of-the-art packages.

9.2.4 *PPGA/PLGA packages (multi-bond tiers cavity packages)*

In the mid-1990s, the mainstream CPU volume migrated from a ceramic package to an organic package called the plastic pin grid array (PPGA) and plastic land grid array (PLGA). The substrate technology is a multi-layer laminate process based on an improved printed circuit technology. The package is unique in that it is an organic substrate with multiple bond tiers, and it has a cavity in the middle of the substrate. A nickel-plated copper heat slug is bonded to the bottom. This provides the platform for the CPU IC to be die bonded to, and a direct thermal path for heat removal. The bond tiers and staggered pitch result in very high interconnect density, and wires are bonded at different loop heights for the different tiers to minimize wire sweep shorting issue. The bonded chip is then encapsulated with an epoxy to protect the IC from the environment.

The manufacturing process for PPGA and PLGA substrate is based on a complex subtractive laminated process. It starts with a sheet of BT laminated of reinforcing glass fiber with bonded copper foils. The sheet is of a standard size for the printed circuit board (PCB) industry, which accommodates PPGA or PLGA substrates. A cavity is routed out of the center of each package. The trace patterning is done by a conventional subtractive process. Multiple sheets of patterned layers are then stacked and laminated into a composite panel. The top and bottom layers with no cavity are used to cap off the inner opening and thus protect it from additional processing. The panel is then drilled and copper plated; a final litho process and etch is performed to complete the patterning process on the external layers. Solder mask is applied to the surface. The cavity is opened through a routing process, and all the I/Os are finished with Ni–Au plating. The panel is routed into individual singulated

Figure 9.2 PPGA, PLGA packages.

units. A heat slug is bonded onto the substrate at the bottom and chip caps are placed on the top side. Figure 9.2 illustrates a typical PPGA or PLGA package.

9.2.5 OLGA and FCPGA (organic land grid array and flip-chip pin grid array)

In the late 1990s, Moore's law continued to drive toward higher transistor counts and smaller transistors. The requirement of tighter feature sizes and better electrical performance led Intel to convert from wire bonding to flip-chip packaging. Flip-chip packaging had been used by IBM for decades on mainframe computers with ceramic packages, and it had a proven performance and reliability record. The major challenge was to implement it economically into organic packages for the mass market, where it could be deployed in desktop and laptop computers and other digital devices. Due to the significant difference in the coefficient of thermal expansion (CTE) between silicon and organic substrates, new processes and materials for chip joint and underfill were developed to improve reliability and performance.

For the organic substrate, feature sizes needed to be reduced substantially to meet the routing requirement for flip-chip packages. The reduction in feature sizes included the reduction of trace width, spacing between traces, micro-via (i.e., vias between buildup layers) diameter, plated through-hole (vias through all layers of the core) diameter, solder mask opening diameter, solder mask alignment accuracy, and layer-to-layer alignment accuracy. Figure 9.3 illustrates a typical organic flip-chip package developed by Intel.

In order to improve the routing density as stated above, the leading suppliers in the industry are now using a process known as semi-additive buildup. (See Figure 9.4 for process flow.) This is a sequential process instead of a parallel laminate process as in the PBGA substrate process. The flow of the semi-additive process starts with a core for power and ground planes, which is fabricated by a procedure very similar to the PBGA process. Dielectric layers are applied onto the core and via holes are laser drilled to form interlayer connections. Laser drilling is employed to minimize

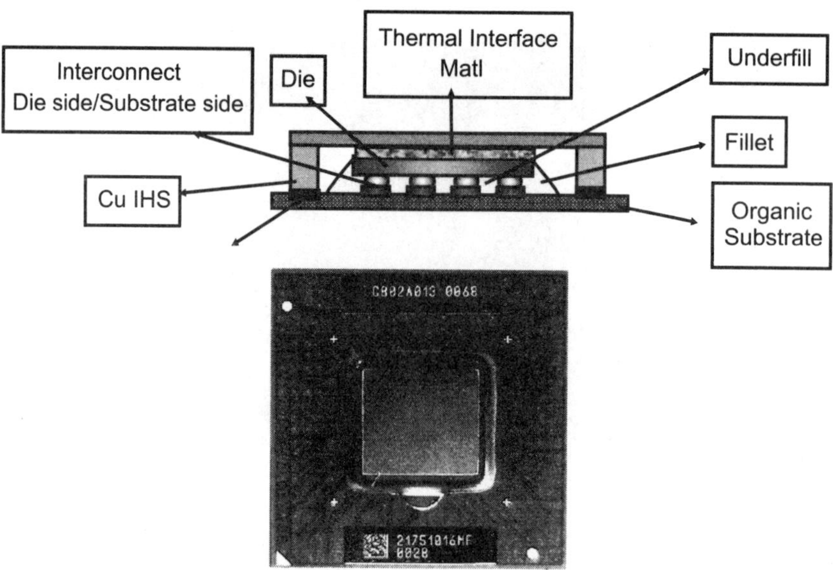

Figure 9.3　Organic flip-chip package.

the via diameter and to increase the via count. A very thin copper seed layer is deposited onto the dielectric; photoresist is applied, exposed, and developed to transfer the design pattern onto the resist, leaving openings with exposed copper seed. Electrolytic plating is then used to build up the copper thickness in the resist openings. To complete the patterning, the resist is stripped and a quick etch removes the seed copper between the plated-up areas. Additional interconnect layers can be added by repeating the above process flow. To complete the package substrate, solder resist is applied and patterned, exposing the flip-chip pads and the pads on the opposite side for board connection. A protective layer is applied on the Cu pads.

In contrast to the subtractive process, in which the copper pattern is protected by photoresist and unwanted copper is etched away, the semi-additive process plates copper interconnect in the area where it is needed. The trace sidewall is defined by the photoresist profile, and therefore the semi-additive copper trace typically has straighter sidewalls and higher resolution. On the other hand, the subtractive etching process is isotropic, and therefore it always has an etch angle which limits its resolution. Sequential processing also improves layer-to-layer alignment. Each layer can be aligned to the previous layer. In a typical subtractive/laminate substrate all the layers are aligned in one step with alignment pins.

9.3　Electroplating in the organic substrate manufacturing processes

This section provides an introduction and a discussion of the copper electroplating process in the organic substrate manufacturing process. In both subtractive and semi-additive technologies, copper electroplating is one of the most important processes in the manufacture of printed circuit substrates. The plating process defines the

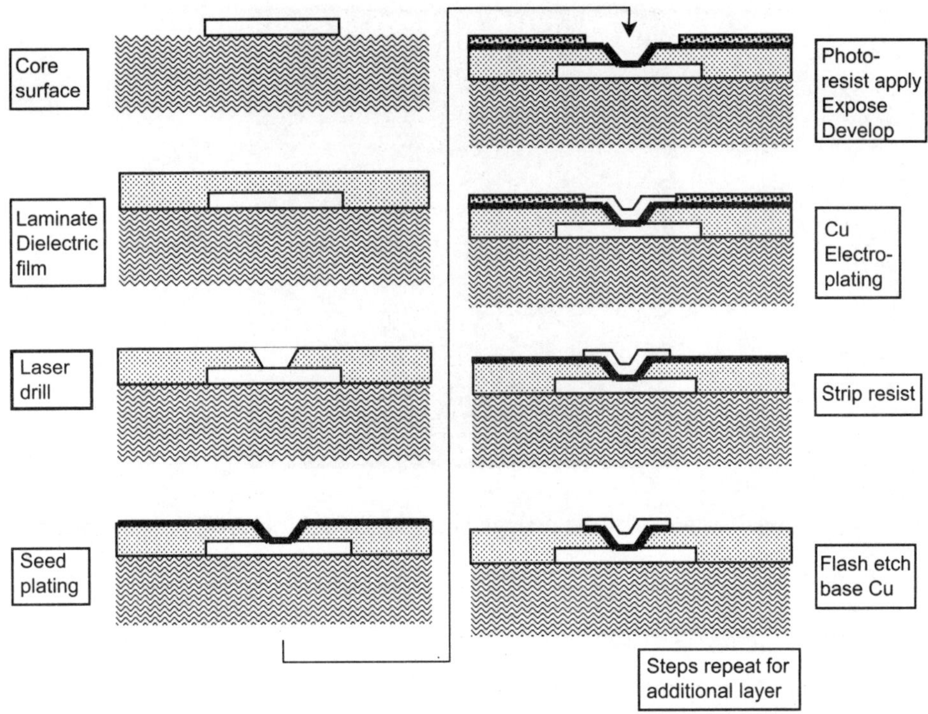

Figure 9.4   Semi-additive process flow.

properties of the conductors that form the interconnects; as a result, it determines the electrical performance of the package. As mentioned earlier, the capabilities of the lithography and electroplating processes define the design rules and routing density of the interconnects. The mechanical properties of the conductors also directly affect the reliability of the interconnects: copper plating determines the ductility of the interconnect traces. This is especially important as line widths continue to shrink and plated through-hole aspect ratios continue to increase. Trace cracking and barrel cracking can result from poor quality copper deposition.

9.3.1   *Requirements of an electroplating process for organic substrate manufacturing*

The basic requirements for a substrate plating process are unique and differ to those of plating finishes for other industries. The key attributes are as follow:

- Plating distribution uniformity across the surface of the panel
- Uniform plating thickness for high aspect ratio through hole
- Uniform plating thickness for step coverage of micro-vias
- Plated copper with high ductility and tensile strength
- High plating rate

Figure 9.5   Cross section of plated through hole (PTH), center of PTH has poor plating.

9.3.2   Surface plating uniformity

It is important to maintain uniform plating thickness across the whole panel area. Thickness variation of the conductor affects the electrical performance; it also can reduce the yield of the process. Local areas on a panel can be under-plated whereas other areas are over-plated. In a semi-additive process, conductors that are over-plated cause mushrooming over the photoresist and result in process defects. In the case of a subtractive process, non-uniform plating thickness results in over-etching in areas where there is thin plating and under-etching where there is thick plating; this also results in process defects. For a high volume manufacturing process, economies of scale drive larger panel sizes and a maximum number of parts per panel. The larger the panel, the more challenging it is to control the uniformity of the plating.

9.3.3   Plated through-hole uniformity

Advance laminate substrates have tighter design rules to accommodate high lead counts and small form factor packaging requirements. Tighter design rules translate to smaller plated through holes (PTH) and more copper layers. The aspect ratio, defined as the ratio of length to the diameter of the PTH, has been increasing with each generation of packages. The local potential field distribution at the opening of the PTH tends to drive thicker plating at the rim and thinner plating on the inside wall of the through hole. This thickness variation worsens with higher aspect ratio. Figure 9.5 shows a cross section of a PTH with low plating thickness in the center. Commercial plating bath suppliers develop proprietary additives to improve the plating uniformity of PTHs. In addition, the plating bath is typically arranged such that the plating solution flows through the PTH, improving transport of copper ions to the center.

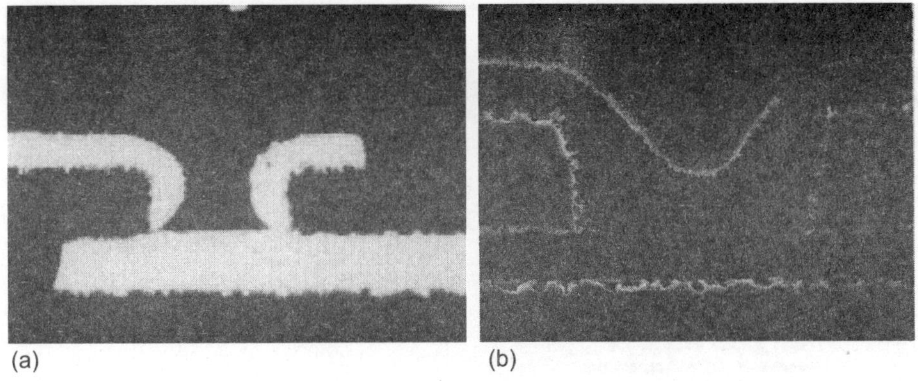

Figure 9.6   (a) Via with thin plating at bottom; (b) Via with good plating coverage.

9.3.4   *Micro-vias plating coverage*

In a semi-additive buildup process, micro-vias are laser-drilled to form the interlayer connection. These layers have thinner dielectric thickness, and therefore micro-vias typically have a lower aspect ratio. The main plating issue for micro-vias is the blind via (dead end) structure. At the bottom corner of the via, the solution exchange is limited and inefficient, and results in a thicker diffusion layer. Diffusion of Cu^{++} ions across this boundary layer can often be the rate-limiting step, limiting the plating rate in this location. In addition, the geometric effect at the bottom corner results in a much lower potential gradient and thus causes lower plating current density and lower thickness. Figure 9.6a shows a defect of low plating thickness at the via bottom; Figure 9.6b shows a uniform plated via. One way to improve solution exchange at the via bottom is by the use of jet plating. Jet nozzles are aligned in the plating cell to direct a high-pressure stream of electrolyte onto the cathode surface, which facilitates solution exchange and reduces the diffusion layer thickness. Another method to improve plating via plating thickness is by adding an organic leveling agent to the plating bath to promote more throwing power. This use of leveling agents is analogous to the approach used for wafer-level Damascene processes, but is typically less effective for package level plating due to the larger dimensions of the recessed structures.

9.3.5   *Primary current distribution and plating cell design*

For a typical electro-plating cell, an external voltage is applied between two electrodes in a conductive electrolyte. The external potential is needed to drive the deposition of Cu on the cathode and the dissolution of Cu on the anode.

The primary current distribution is determined by the geometry of the plating cell and the potential field established between the two electrodes. Other polarization effects on the electrode surface are neglected. Primary distribution is important to the understanding of plating cell layout design and optimization, and also the geometric effect of anode /cathode arrangement in order to maintain plating uniformity.

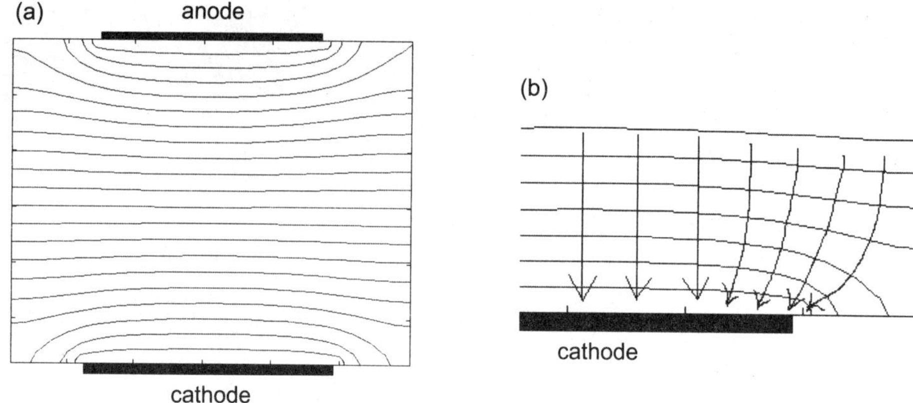

Figure 9.7   (a) Potential field of a typical plating half cell; (b) Current flow and edge effect at lower right-hand corner.

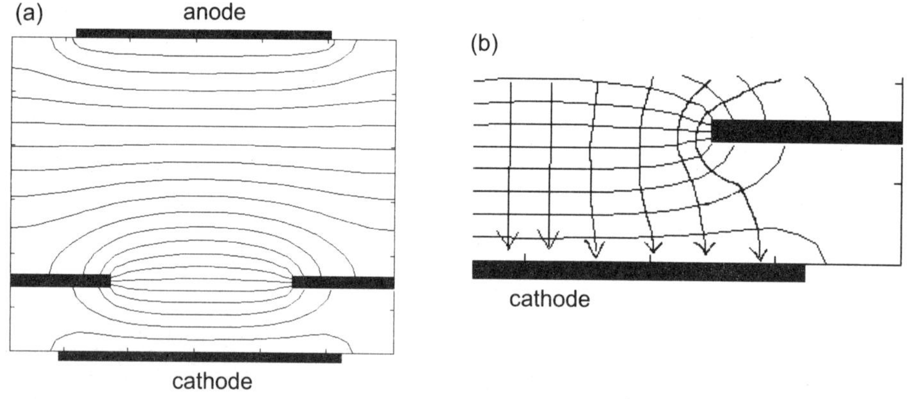

Figure 9.8   (a) Potential field of a shielded cathode; (b) Current flow and edge effect.

Primary current density distribution is a function of the potential gradient close to the cathode surface. The localized current density is highest when the potential gradient is the greatest. Therefore, it is essential to design a plating cell which maintains uniform potential gradient at the cathode surface. An ideal plating cell is infinite parallel plates electrodes. The key factor that makes these arrangements ideal is in the elimination of the edge effect, which results in a uniform potential gradient on the cathode surface and uniform plating thickness. These arrangements are theoretically desirable but impossible to implement in practice.

In a typical plating tank, all dimensions are finite; and gaps must exist between the electrodes and tank sidewall for panels agitation, panels transport, and solution flow. Figures 9.7a and 9.7b show a contour plot of potential field and the resulting edge effect on current flow of a typical plating cell. (Note that this is a half cell representation; a typical substrate plating cell has anodes on both sides, and line of symmetry at the cathode.) As illustrated clearly in the contour plot, the potential gradient is the steepest at the edges of the cathode, since the current density is directly related to the gradient of the potential field, and the deposition rate is the highest at these locations.

The general practice to optimize the geometry of a plating cell is by adding baffles and shields. Since the shield is made of insulator material, the potential gradient on the surface of a shield is zero. Therefore, there is no deposition on the shield. The shield helps rearrange the potential field and redirects current flow to even out plating distribution. Figures 9.8a and 9.8b illustrate the effect on a typical plating cell. Another practice to reduce edge effect and improve uniformity is by the addition of current thieves. Rack design with conductive frame is extended from the critical areas to take up the high current density area due to the edge effect.

9.3.6 *Other factors that impact the current density distribution*

There are several simultaneous transport phenomena and reactions that are occurring on the surface of the electrodes that hinder the deposition process. The result of these effects is defined as secondary current distribution. These activities include charge transfer, diffusion, reaction, and crystallization.[1]

9.3.7 *Mass transport and its effect on secondary current distribution*

In a production facility in the substrate industry, it is essential to maintain high throughput in the plating process for economical reasons. As a result, a typical high volume plating process needs to be optimized for plating rate, thickness uniformity, and plating coverage.

In the limiting case, when the surface concentration of the depositing ion is zero, the deposition rate is diffusion controlled. This means that the deposition rate is limited by the rate of diffusion of the depositing ions on the surface. The current density in this condition is defined as the limiting current density. When the plating rate of a process exceeds that of that of the limiting current density, the copper deposit becomes dendritic. Such a deposit has a surface with a burnt appearance, it has poor mechanical strength and poor adhesion to the substrate, and in the extreme case, the deposit has a powdery appearance and can be easily brushed off. Therefore, the plating process operates below the limiting current density. One way to increase the limiting current density of a plating cell is to increase the force convection on the depositing surface. Since limiting current density is inversely proportional to the diffusion layer thickness, it is important to minimize the diffusion layer thickness.

Modern electroplating processes are designed to have high forced convection on the plated surface. Plating processes employing rack agitation, air agitation, and high pressure jet impingement to reduce the diffusion layer are thus common practices.

9.3.8 *Commercial additives and current distribution*

One of the key factors influencing the properties of the depositing metal and the uniformity of plating is the addition of plating additives. Commercial plating baths

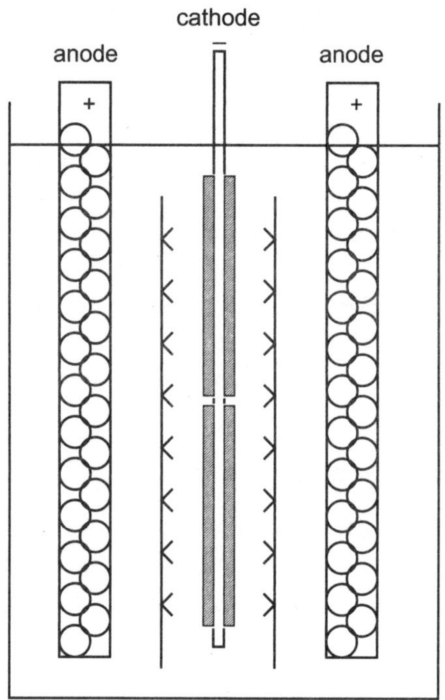

Figure 9.9   A typical vertical plating tank.

include complex proprietary, multi-component chemical systems. Typical classes of components include brightening and leveling agents. A brightener functions to refine the grain size of the deposited copper. It is the key ingredient in determining the mechanical properties, including ductility and tensile strength, which affect the reliability of the interconnects.

A leveler is needed to provide uniform plating thickness across all features and topography on the surface of the cathode. It is also essential for plated through-hole sidewall plating uniformity and micro-via coverage.

The leveler and the brightener typically come in a two-part system, and are added into the plating tank through an automatic dosing schedule. The schedule is based on the amp-hour of the plater, and the consumption rates of the additives are replenished accordingly.

9.3.9   Commercial plating line and design

9.3.9.1   Vertical plating line

The conventional plating process employs a vertical design, as shown in Figure 9.9. Multiple panels are mounted on a large plating rack. The rack has clips that attach to the panel where the voltage is applied. Multiple anode baskets are typically arranged on both sides of the cathode. Filter curtains are placed to separate the anodes to the cathode to eliminate contamination. The solution is pulled through from the bottom of the tank, passes through a particle filter, and returns into the tank through a series of nozzles. Air agitation is also employed to increase convection.

The advantage of such a system is high throughput, ease of maintenance, modular design, product mix flexibility, and programmability. A line can comprise many plating tanks, and an individual plating tank can be optimized to run a specific product. Therefore many part numbers and product types can be run in the same line simultaneously. All these plating tanks share the same pre-clean and conditioning baths. An individual tank can be serviced even when the line is running, and so downtime can be minimized. One of the drawbacks of the vertical plater is its edge effect, and this is especially pronounced in a large-scale, large-rack plater.

9.3.9.2   Horizontal electroplating line

In recent years, the implementation of horizontal plating lines has improved the plating quality and efficiency over standard vertical plating lines. Key advantages of horizontal plating lines include plating distribution uniformity and ease of material handling automation. As described in earlier sections, the edge effect due to the geometry of the tank design, anode to cathode dimensions, and location are key causes of plating thickness variation. In horizontal plating, the panel travels through the full length of the plating cell, and each panel sees the time average of the whole anode basket arrangement. As a result, the variation of thickness between the leading edge to the trailing edge of each panel is minimized, and the variation of average thickness between panel to panel is also minimized.

As opposed to the vertical plating process, where extra space between anode and cathode is needed for hoist transport and panel agitation, the horizontal plater can significantly reduce the anode to cathode distance due to its fixed and linear panel travel path. This tight spacing between the anode and cathode results in a more uniform potential field and current density distribution. The drawback of the horizontal line is that it is not as flexible as the vertical line.

In order to improve solution flow and force convection, the horizontal plating cell is typically designed with high pressure solution jet impingement, which enables high speed plating (see Figure 9.10[2]).

9.4   Flip-chip interconnect materials

Packaging has mainly acted as a protection for chips and as a mechanical carrier to bring power and signals in and out of chips. Although there are several ways one can bring power in and out of the chip, the controlled collapse chip connection (C4) has been the most preferred way of attaching the chip to the mechanical carrier such as the substrate.

C4, widely known as flip-chip technology,[3–5] is an interconnection technology that was first invented by IBM over 35 years ago for attaching chips to ceramic carriers. Intel was the first to extend this technology toward organic packages, thus driving flip-chip technology as the most dominant packaging solution for current

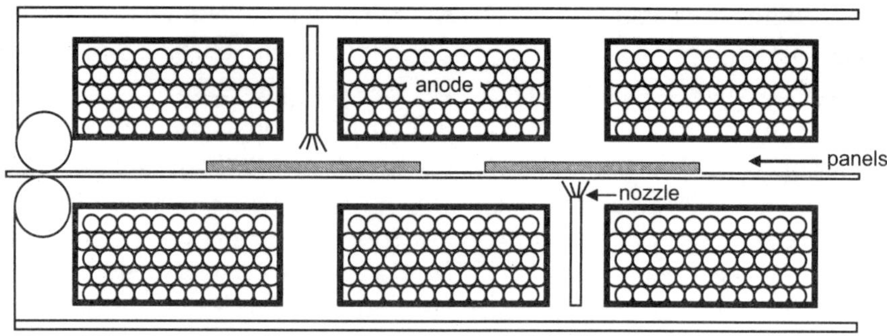

Figure 9.10 A cross section of a horizontal plater.[2]

high-performance Pentium[®] chips.[6] In the C4 assembly process the active chip surface is mounted facing toward the organic or ceramic substrate by a variety of interconnect materials, using a reflow process. Materials for interconnects in today's packages are Pb–Sn-based solders. The interconnects from package to motherboard are also Pb–Sn-based solder. Eutectic Pb–Sn is used in most of the applications except chip bumping, where high lead bumps (Pb–3 wt. % Sn) are used. A detailed review of Pb–Sn solder materials and their characteristic properties have been provided by Wassink.[7] Flip-chip technology not only provides a path to a potentially less expensive interconnection/ assembly process, but also offers many more advantages over the traditional face-up wire bonding technology. These include high I/O density, shorter interconnection length, better electrical performance, improved reliability performance, and finally better manufacturability resulting in high process yields. However, during subsequent manufacturing steps, where the package is subjected to cycles of elevated and lowered temperatures, high shear stress is placed on the solder joints during the temperature cycling of the device. This shear stress is partially a result of the difference in the CTE of the flip-chip and the mounting substrate. Die materials such as silicon and gallium arsenide have CTEs in a range from about 3 to about 6 ppm/°C. Mounting substrates (especially organic) are usually composites of organic resin-impregnated fiberglass composites with metallic circuitry. These organic substrates have CTEs in the range from about 15 to about 25 ppm/°C. Consequently, a mismatch in the CTEs exists between the flip-chip and the mounting substrate. The presence of high shear force on the interconnect bumps CTE mismatch is a major cause of interconnect joint fatigue failures for an area array package, especially during temperature cycling. In order to reduce interconnect/solder joint failures due to stress during thermal cycling, the solder joints are reinforced by filling the space between the flip chip and the mounting substrate, and around the solder joints, with an underfill encapsulant.

9.5 Underfill materials

By tightly adhering to the chip, solder joints, and substrate, the adhesive material of the underfill redistributes the stresses and strains from the CTE mismatch and

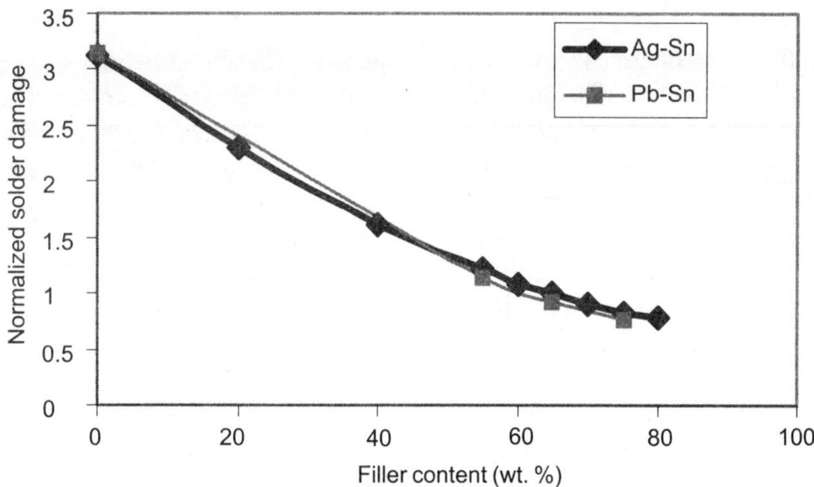

Figure 9.11 Impact of filler content of an underfill on Pb–Sn- and Sn–Ag-based solders joint fatigue/temperature cycle stress.

mechanical shock over the entire chip area. As a result, crack initiation and growth in the solder joint are either prevented or drastically reduced, thus extending the life of the flip-chip devices significantly.[8,9]

Figure 9.11 shows the impact of underfill CTE (controlled by the filler amount in the underfill) on solder damage per temperature cycle stress. Irrespective of the solder or interconnect material, it can be clearly seen that as the filler amount is increased and hence the CTE of the underfill is decreased, the effective solder damage/cycle decreases in an exponential manner.

Thus underfills act as one of the key materials in providing protection to the solder interconnect, improving mechanical integrity to the package and thereby the overall reliability of the package significantly. The secondary benefit from underfilling is partial protection against moisture ingress, and other forms of contamination. Consequently, the quality and reliability of the underfill material itself will greatly impact the component reliability.

In general, an underfill material when cured is a high modulus, low CTE composite made up of a cross-linked organic resin and low CTE inorganic particles (up to 75 wt. %). A typical underfill formulation before cure is usually formulated with a liquid resin such as epoxies, a hardener such as anhydride or amines, an elastomer for toughening, a catalyst for promoting cross-linking, and other additives for flow modification and adhesion. In the development of an underfill material, a wide number of properties have to be optimized. A few of the key properties apart from CTE (as a function of temperature) are high modulus, low viscosity for good flow at the dispense temperature, high glass transition temperature (T_g), high toughness, good adhesion to the substrate, die passivation layer, and low moisture absorbance. Excellent flow characteristics between the die and the substrate are very critical for capillary underfill process.

9.5.1 *Underfill process*

Underfilling is considered a slow process in the flip-chip assembly line. The process of underfilling between the die and the substrate is primarily governed by the capillary flow of the material under the bumped die. In a typical process, underfill is dispensed along one side of the die in-line pattern. The filling action between the die and substrate is completely driven by the capillary action. In order to increase the speed of the underfilling process, packages are typically heated from the substrate side, in order to reduce the viscosity of the material. In the post-under-filling process, the packages are cured in a oven to complete the underfilling process. Figure 9.12 shows a typical underfill process used in the flip-chip assembly line.

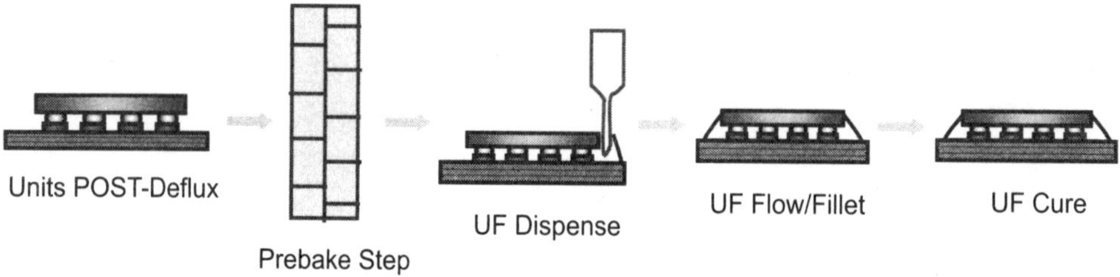

Figure 9.12 Typical underfill process: steps involve prebaking of package for moisture removal, underfill dispensing at elevated temperature, followed by underfill flowing under the die, and finally curing at elevated temperature of underfill coupling the die and the substrate.

9.5.2 *Fundamental parameters affecting capillary flow*

If we assume the underfill to be a Newtonian liquid, the time it takes to flow the length of the die can be calculated using the following two equations 1:[10]

$$\Delta p = \frac{2\gamma\cos\theta}{h} \quad \text{Equation 1: Capillary driving force} \tag{1}$$

where Δp is the drop in pressure between the liquid–vapor interfaces, γ is the surface tension of the liquid underfill–vapor interface, θ is the contact angle between the underfill and the interface it touches (i.e., solder resist/die passivation material), and finally h is the gap height. between the substrate and the die. Overall, the larger the drop in pressure, the higher is the capillary force for underfilling the die and the shorter is the underfill process. The underfill flow time under the die is driven by the following Equation 2:

$$t = \frac{3\mu L^2}{h\Delta P} \quad \text{Equation 2: Underfill flow time} \tag{2}$$

Thus, the flow time is directly proportional to the length of the die (L) and the viscosity of the underfill (μ), and inversely proportional to the capillary pressure and the gap height between the die and the substrate. The incomplete underfilling process results in the formation of voids under the die, resulting in lower reliability performance of the package.

As a result the current capillary underfill process faces tremendous challenges. This arises because of the inherent difficulty of highly filled underfill resins to flow via capillary action into a very narrow gap height between the die and the substrate within a tight bump pitch. Thus for effective underfilling process via capillary flow, the underfill material is expected to have low viscosity at dispense temperature, low contact angle for excellent wetting to the interfaces, and high surface tension for quick capillary underfilling process. Further, the underfill is expected to have uniform flow front, so that no flow voids are created. The materials are also expected to self-fillet, so that no corner structure (die/substrate level) failures are observed. Other parameters that affect underfill flow are bump pattern, bump pitch, substrate temperature, dispense pattern, amount of underfill dispensed, and wetting characteristics of the underfill to the substrate and die passivation surfaces. The impact of both material and package properties affecting capillary flow are explained in the following section.

9.5.3 *Underfill material properties*

Critical properties of the underfill materials that impacts capillary flow are

1. Viscosity of the underfill material
2. Surface tension of the underfill material at the dispense temperature
3. Contact angle of the underfill material to both solder resist and die passivation materials
4. Cure kinetics

9.5.3.1 *Viscosity*

One of the key properties for good capillary flow of a material is viscosity. Figure 9.13 shows the performance of two different underfills, dispensed at the same temperature and on a package configuration (die size, bump pitch, gap height, etc), but varying very widely on room temperature viscosity, and having almost the same viscosity at 110°C, the dispense temperature. This clearly demonstrates that, with respect to viscosity of the material, it is not the viscosity of the material at room temperature that matters, but what is important is the viscosity at the dispense temperature, and rate of change in viscosity over time at the same dispense temperature. Lower viscosity at the dispense temperature is expected to have a huge impact on the ability to underfill at very high line output, traditionally measured by units per hour (UPH). The rate of change in viscosity of the underfill at the dispense temperature also affects the ability of the material to underfill effectively. This becomes much

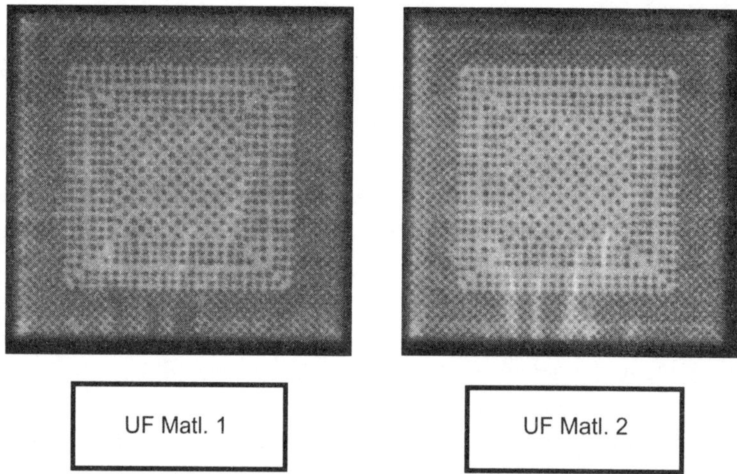

Figure 9.13 Effect of underfill (UF) material viscosity at room temperature on underfilling performance, under the same processing conditions. UF matl. 1 had a room temperature viscosity of 39 Pa·s, while UF matl. 2 had a room temperature viscosity of 180 Pa·s.

more apparent, when the material is used to underfill large die packages, or packages with extremely tight bump pitches. Thus the viscosity of the underfill at the dispense temperature and the change in viscosity as a function of time are two very critical parameters for the performance of an underfill material. This is explained in detail in the following section.

9.5.3.2   *Factors impacting viscosity of the underfill material*

Viscosity of the underfill depends on several factors:

1.   Choice of underfill chemistry
2.   Base epoxy resin used
3.   Amount of filler
4.   Type of filler and its distribution
5.   Presence and amounts of additives such as coupling agents
6.   Type and amount of elastomer in the formulation

Although other types of materials such as cynate esters, silicones, or hybrid materials have been tried as underfills, Epoxy-based materials typically dominate the underfill market. Key drivers for epoxy-based materials are ease of formulation, excellent adhesion to the substrate of choice, and lower cost. Typically there are three types of epoxy-based materials depending upon the choice of hardener used in the formulation. Homopolymerization of epoxy using imidazole-type catalysts leads to polyether-type materials, while use of amine- or anhydride-based curing agents leads

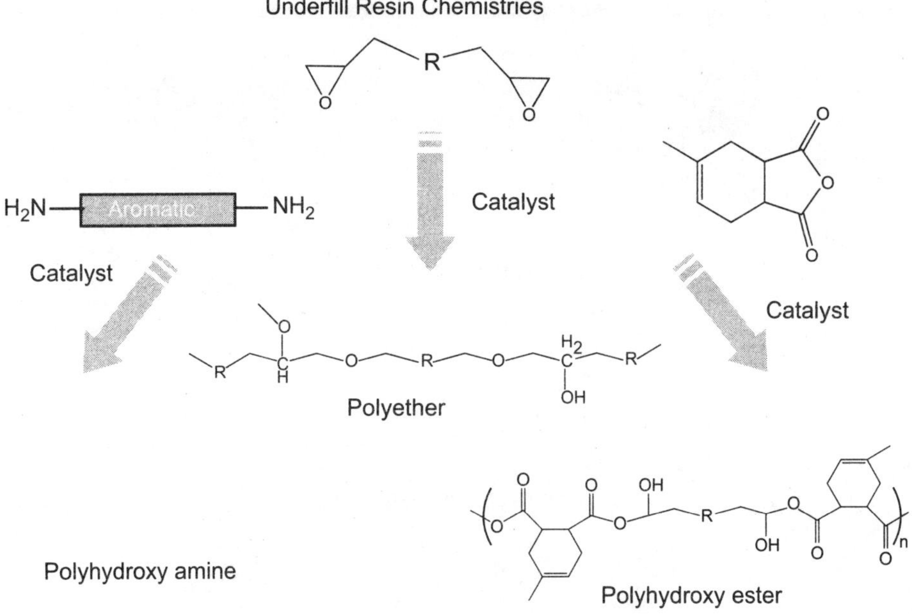

Figure 9.14 Typical methods of curing epoxy-based underfill materials.

to polyhydroxyl amine- or polyhydroxyl ester-based linkages. Figure 9.14 shows the various ways of curing epoxy-based systems. In general, due to the very low viscosity and the dilution effects of the anhydride hardener, underfill materials formulated with anhydride curing agents have much lower room temperature viscosity compared with epoxy systems with no hardener or amine-based hardener systems, with all other parameters being equal.

The type of epoxy used in the formulation also plays a very key role in defining the viscosity of the underfill material. Underfills, irrespective of the type of cure, have lower viscosities both at dispense and room temperature, with use of liquid-type epoxies rather than solid epoxies, with use of cycloaliphatic epoxies rather than aromatic epoxies, and with use of lower molecular weight epoxies rather than higher molecular weight epoxy systems. Addition of diluents such as mono-epoxides play a very important role in reducing the viscosity of the underfills. Details on the various epoxy systems and the impact of diluents on the overall viscosity of the system can be obtained from an epoxy handbook.[11] Fillers play a very important role in underfill formulations.[12] Fillers are typically used to reduce the CTE of the underfill resins. Reducing the CTE of the underfill material via addition of filler particles is very critical to the reliability of the device, as the underfill material is expected to compensate the difference in CTE between the substrate and the chip. In order to reduce solder joint fatigue and extend solder joint life, the CTE of the underfill material is tuned to be in the range of about 20 to 40 ppm/°C at temperatures below its glass transition temperature (T_g). Addition of fillers typically increases the viscosity of the underfill formulations, due to the interaction of the epoxy resin with the high surface area of the fillers, especially silica-based fillers. Figure 9.15 shows the effect of silica

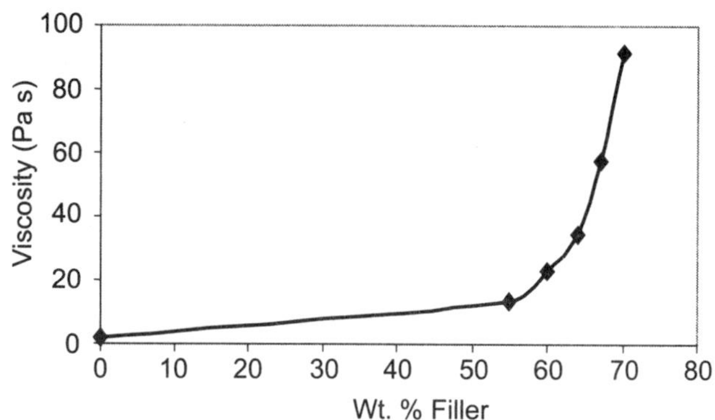

Figure 9.15 Impact of filler amount on underfill viscosity at room temperature.

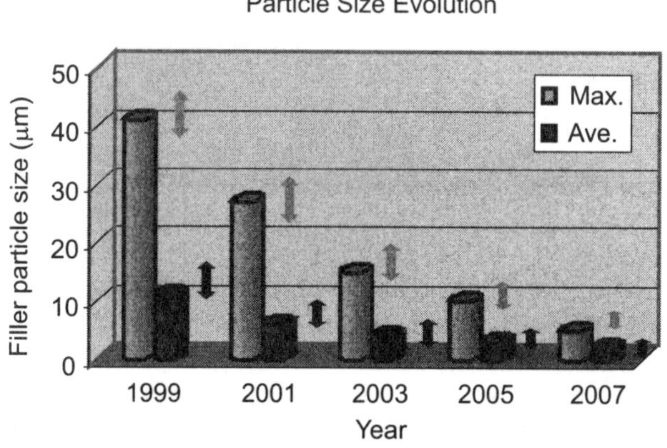

Figure 9.16 Expected trend in filler mean and maximum particle size in the years to come, with ever-decreasing chip gap height and bump pitch.

filler amount and its impact on viscosity of the underfill system. In general, as the filler amount is increased, the viscosity of the underfill is observed to increase, with an exponential rise being observed as the filler amount is increased to greater than 70 wt. %.

Filler size and filler distribution also play a very key role in controlling the viscosity of the underfill material. In general, as the gap height between the chip and substrate becomes smaller, and bump pitch decreases, underfill materials with better flow performance become essential. Due to the reduced gap height, smaller and smaller filler sizes are needed for effective gap filling of the underfill. Figure 9.16 shows the decreasing trend in filler size, with the decrease in graph and pitch from generation to generation expected in years to come.

As the particle size of the underfill is reduced, the surface area of the filler is observed to increase, resulting in exponential increase in viscosity of the underfill. Figure 9.17 shows the effect of three different filler types with decreasing mean and maximum particle size, and its impact on underfill viscosity and performance.

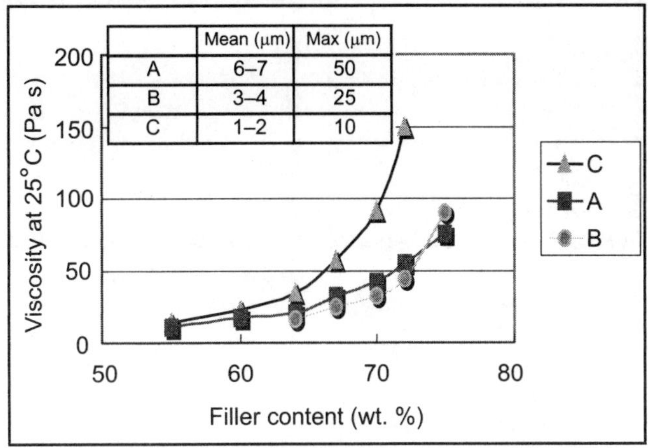

Figure 9.17 Impact of filler distribution on viscosity of underfill formulation.

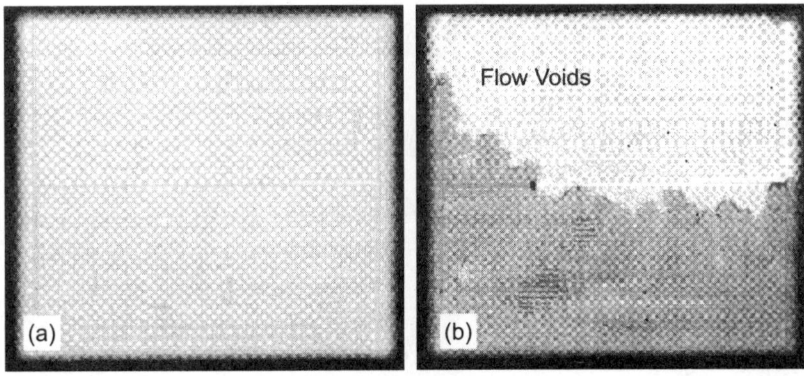

Figure 9.18 Impact of fine filler on flow performance of same underfill formulation. (a) Underfill was 65 wt. % filled with fine filler, while (b) was 72 to 73 wt. % filled with fine filler. Both materials processed using same process conditions.

The impact of underfill performance due to the use of finer filler is shown in Figure 9.18. As the filler amount in the underfill formulation increases from about 65 to 73 wt. %, using finer filler (mean 1 to 2 µm), the viscosity is observed to triple at room temperature, leading to higher viscosity at the dispense temperature. This increase in viscosity decreases the flow of the material under the die, resulting in huge flow voids.

One way to improve the performance of the underfill, especially using finer filler particles, is via use of right particle size distribution. It is well known that by use of right particle distribution in an underfill, one can optimize the particle packing density. This is very critical in the overall flow performance of any particulate/fluid systems. Detailed research in the area of particle packing has been to develop algorithms for calculating packing density and porosity as a function of filler shape, filler roughness, etc. so one can predict the performance of particulate/fluid

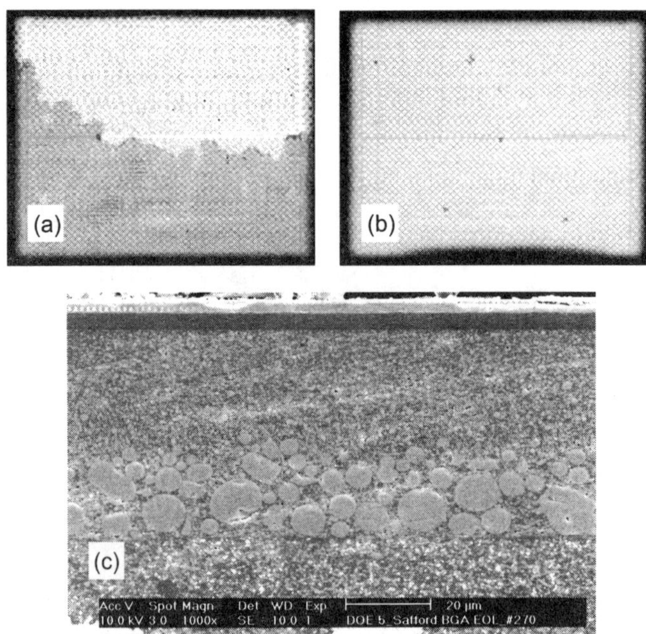

Figure 9.19 Demonstrates the impact of using the right filler distribution on flow performance of an underfill system. (a) Underfill formulation contains 100% of finer
filler, resulting in flow voids, while (b) is underfill formulation using 50/50
mixture of finer and coarse filler distribution, resulting in complete flow, at the
same process conditions. (c) Cross section of unit (b), showing good filler
distribution on the entire graph of the die.

systems.[13,14] Figure 9.19 demonstrates the impact of using the right filler distribution on flow performance of an underfill system. As discussed earlier, use of a fine
filler alone was observed to have very poor flow performance, for highly filled
system. By mixing a fine filler with coarse filler distribution, as at a certain optimized
level, the flow performance of the same underfill is improved tremendously,
resulting in complete underfilling of the die at the same process conditions. The use
of right filler distribution also is expected to modulate the underfill flow front. Flow
front control is very important to have uniform flow to the end of the die without
voiding. The use of the right filler size and distribution plays a very critical role in
controlling the flow front of the material.[15]

Viscosity of the underfill can be measured via several methods, but the two most
common methodologies are the (1) Brookfield viscometer at controlled revolutions
per minute, for room temperature viscosity measurements; and (2) rheological dynamic analyzer, using a parallel cup and plate geometry. In an RDA experiment, using
a temperature sweep experiment on a thawed underfill material, the storage modulus
G', the loss modulus G'', and the viscosity of the material are recorded, as a function
of temperature. Figure 9.20 shows a typical dynamic viscosity curve as a function of
temperature, measured on two different underfill materials measured at constant
strain rate.

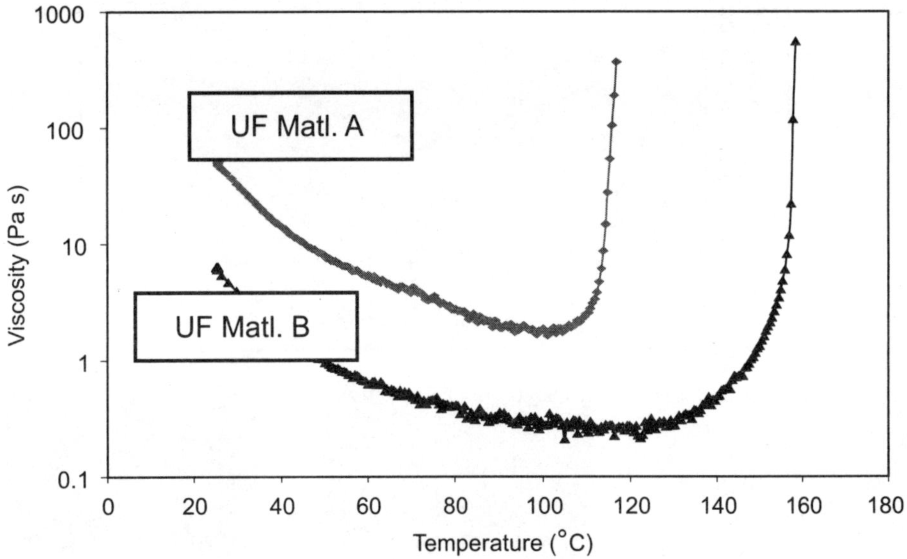

Figure 9.20 UF material A shows typical dynamic viscosity vs. temperature cure for two different underfill materials. UF material B shows typical isothermal viscosity measured at 110°C for two different underfill materials.

It is clear from the curve that UF material A has a higher room temperature viscosity compared to UF material B. As the temperature is increased, the viscosity of the underfill materials starts to drop in viscosity, reaching around 100°C for material A and 120°C for material B. Typically, underfill dispense temperature is chosen very close to the temperature where the viscosity of the material reaches its lowest, in a dynamic viscosity vs. temperature curve. Processing the material close to the lowest viscosity temperature is expected to decrease flow time, and increase module throughput. As the temperature increases further, the cross-linking reaction starts, resulting in increase in viscosity of the material. This increase in viscosity of the material occurs for material A around 110°C, while for material B this occurs around 140°C. Thus, material B is observed to have a wider process window compared to material A. For an ease of processing, material with a wider processing window is preferred over a material with a shorter process window.

Rheological analysis, as a function of time, but at a constant temperature, is also a very critical property to ascertain the robustness of a material from the processing point of view. In order to have a very robust process, very little change in viscosity of the material is expected to occur as a function of time during the processing operation. Figure 9.21 shows the rheology characterization (in poise) curves as a function of time at the testing temperature of 110°C after the underfill material has been exposed at 40°C for various lengths of time. The 40°C temperature was chosen to simulate the temperature experienced by the syringe in the dispenser. As seen from Figure 9.21, the viscosity and gel time of the material at 110°C changes with exposure time at 40°C. This indicates that the material starts to react (probably oligomerize) at temperatures as low as 40°C, thereby leading to a shorter process window

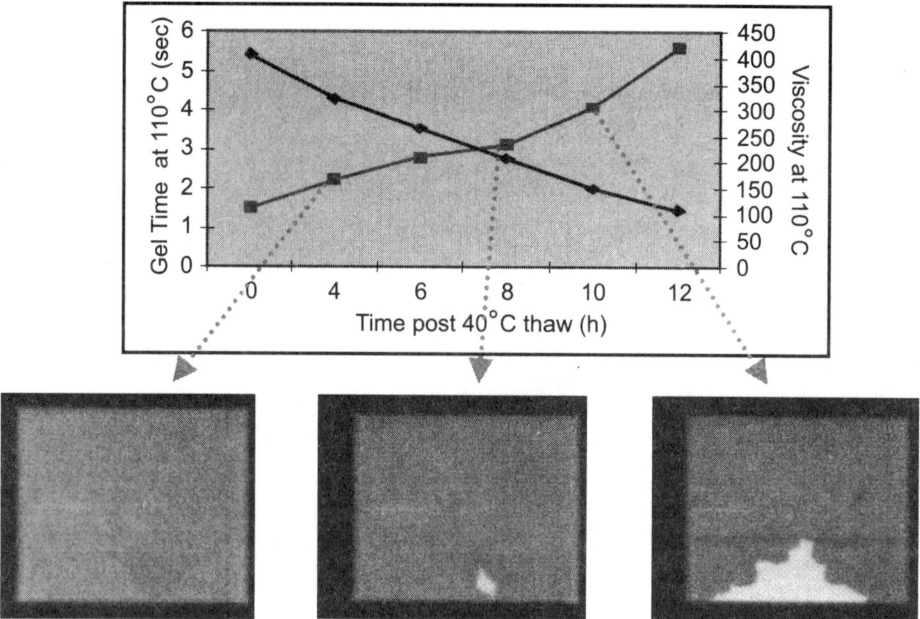

Figure 9.21 Viscosity and gel time vs. time measured at 110°C post-exposure at 40°C, for 0, 4, 6, 8, 10, and 12 h and corresponding flow performance of the material. Flow voiding observed post 8 h of exposure to 40°C, indicating poor pot life performance.

for capillary flow. Experimental builds with this material indicate that capillary flow is not an issue up to 6 h. After 8 h the flow of the material drops significantly due to the rapid increase in viscosity, with corresponding decrease in gel time, resulting in flow voiding at the opposite end of dispense. This performance was observed to be worse for large dies, as time to flow the entire die is expected to be larger.

9.5.3.3 *Impact of underfill gel time on performance*

An important parameter of an underfill material is its gel time, which is the time required for the underfill polymer to form a cross-linked network at a constant temperature. When gel point is reached, the molecular chains in the underfill become linked into a network of infinite molecular weight. Consequently, the viscosity and modulus of the underfill material increase drastically, and the underfill loses its ability to flow, which renders it non-processable. The kinetics of the curing or cross-linking reaction is very important in controlling the gel time of the underfill material. [16–18]

There exists a strong correlation between the underfilling process window and the gel time of the underfill material. Specifically, for the current C4 process line configuration and with complexities in package configuration, the underfill process margin can be very tight, and small variations in underfill gel time would result in void formation in the cured underfill material. Experimental results as shown in

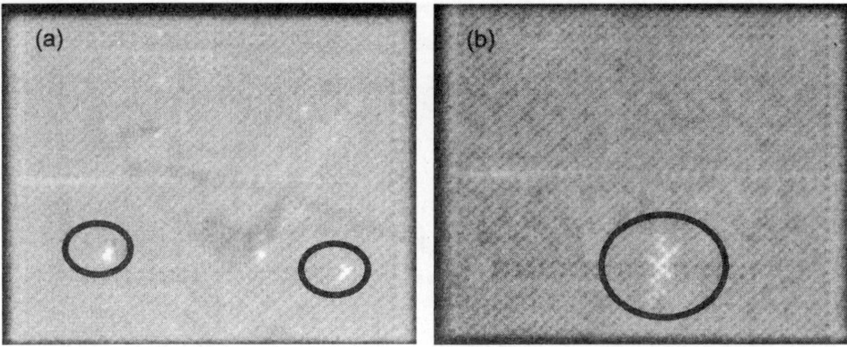

Figure 9.22 (a) Small voids formed due to entrapment of volatiles in an underfill material having short gel time; (b) an example of squiggle voids formed in an underfill material having large gel time.

Figure 9.22 indicate that voiding and the type of voiding can be mitigated by the underfill gel time (i.e., shorter gel time of an underfill epoxy can lead to small voids). On the other hand, squiggly voids would form if the gel time is too long. Therefore, underfill gel time is a key parameter for material screening.

Several techniques exist in measuring the gel time of an underfill material. The first technique is based on differential scanning calorimetry (DSC), which monitors endothermic and exothermic reactions.[19] The second technique is based on rheological characterization, which detects any changes in underfill viscosity as a function of time at a selected isothermal temperature.[20]

In the first method, isothermal DSC technique is used to detect the onset of the curing reaction, which is correlated to the onset of gelation. DSC has been widely used to study thermoset curing reactions. The basic assumption for the application of DSC is that the measured heat flow, dH/dt, is proportional to the reaction rate, $d\alpha/dt$. During the underfill curing process, heat is released from the underfill material. Therefore, an exothermic peak develops in the DSC curve. The heat-releasing rate is directly proportional to the chemical reaction rate, or the curing rate. In an isothermal DSC experiment, the time to reach the onset of the exothermic peak is taken as the DSC gel time of the underfill.

Figure 9.23 shows an isothermal DSC trace (open circles) of an underfill material as a function of time at 145°C. Also plotted in Figure 9.23 is the derivative of heat flow curve (dashed line), which is used to accurately determine the DSC gel time. The ramping time from 25 to 145°C, which is 191.25 sec, is included in the time. From Figure 9.23, it is clear that the curing reaction rate is quite small at $t \leq 310$ sec, which is about 2 minutes after the isothermal temperature is reached. As time increases, the curing reaction rate also increases. At $t = 424$ sec, a peak appears in the heat flow curve, indicating that the reaction rate, $d\alpha/dt$, reaches its maximum. At $t = 411$ sec, a downward peak develops in the derivative of heat flow curve, indicating that at that time, the change in the reaction rate, $d^2\alpha/dt^2$, reaches a maximum. To calculate the gel time, a tangent line is drawn from a point on the DSC curve before substantial curing reaction has taken place, as shown in Figure 9.23. A second tangent

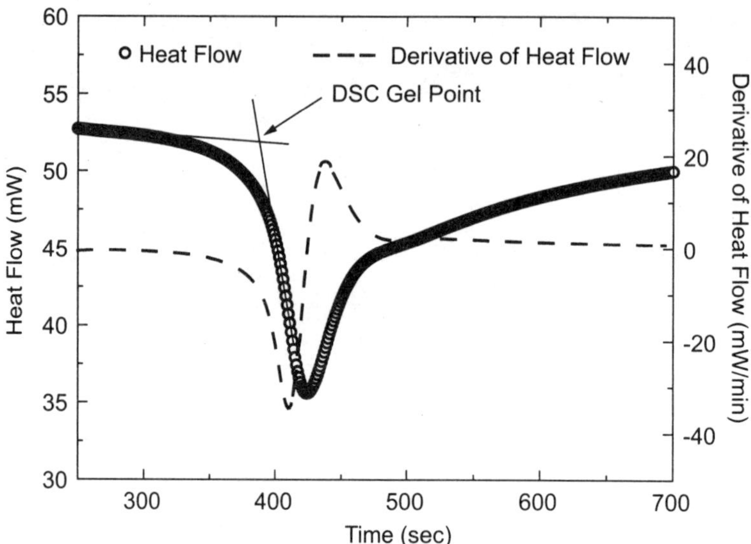

Figure 9.23 Isothermal DSC trace of an underfill as a function of time at 145°C. The derivative of heat flow with respect to time, shown in dashed line, is also plotted. The arrow indicates the DSC gel point.

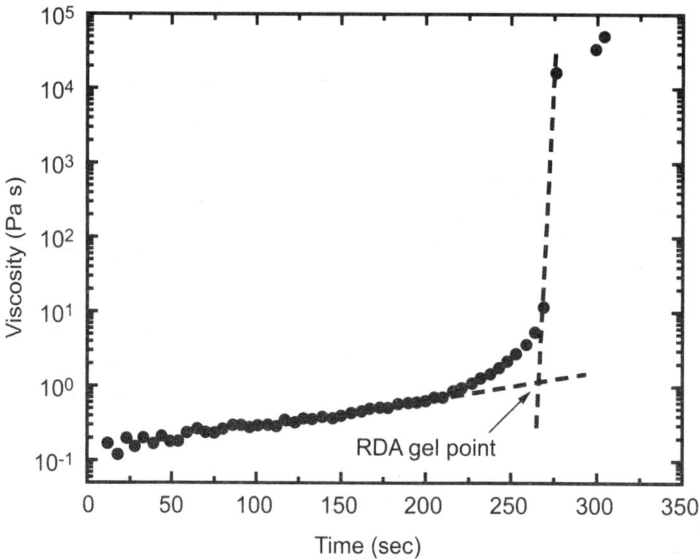

Figure 9.24 Viscosity as a function of time at 145°C determined by RDA. The arrow indicates the gel point.

line is drawn from the DSC curve, which corresponds to the downward peak in the derivative vs. time curve. The intersection of these two lines is the DSC gel point. In the second technique, the rheological or flow characteristics of the underfill epoxy are measured at a constant temperature. As gel point is approached, the underfill

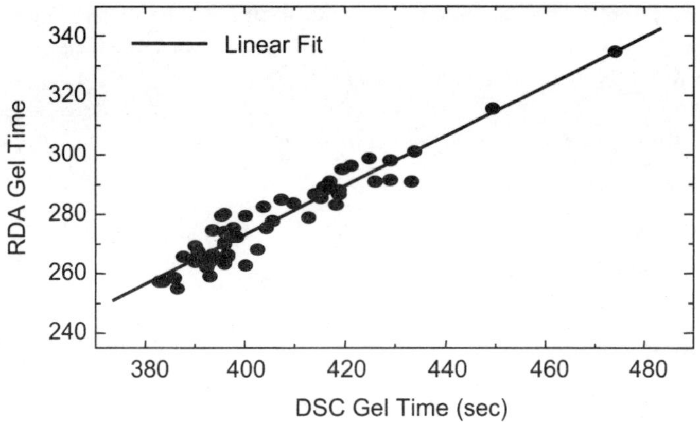

Figure 9.25 Correlation analysis of DSC gel time and RDA gel time of underfill material.

transforms from a viscous liquid to an elastic gel network, resulting in a sudden increase in viscosity and modulus. The onset of this rheological change is taken as the underfill gel time.

Figure 9.24 is an example of the RDA curve, showing the viscosity η as a function of time at an isothermal temperature of 145°C. In the beginning, the viscosity of the underfill is very low, on the order of 0.2 Pa·s.

As time increases, viscosity also increases. This is an indication that the curing reaction is taking place slowly. Between $0 \leq t \leq 200$ sec, η shows an exponential dependence on time, or $\log \eta \propto A + Bt$. As time increases further, $\log \eta$ begins to deviate from the linear time dependence, and η increases faster than the exponential. When the gel point is reached, material viscosity increases very rapidly, as shown in Figure 9.24. Indeed, η increases by more than three orders of magnitude between $t = 269$ and $t = 276$ sec. At gel point, a three-dimensional polymer network begins to form, and the molecular weight reaches infinity.

9.5.3.4   Correlation between DSC and RDA on gel time

DSC and RDA monitor different material properties of the underfill material. One measures the heat-releasing rate during the curing reaction, while the other monitors the mechanical response of the material. Therefore, it is expected that the DSC gel time is not the same as the RDA gel time. However, DSC and RDA results are correlated, as shown in Figure 9.25. Statistical analysis revealed that there is a linear correlation between the RDA gel time and the DSC gel time, and the correlation can be expressed as:

$$t_{RDA} = -59.944 + 0.8325\, t_{DSC} \tag{3}$$

as shown in Figure 9.25. The R-square for this fitting is better than 90%, indicating that the correlation is very strong. Thus one can use potentially RDA or DSC based

Table 9.1   Effect of underfill resin chemistry on underfill to solder resist contact angle.

Substrate solder resist type	Contact angle of underfill (degrees)			
	Epoxy anhydride 1	Epoxy anhydride 2	Epoxy amine	Epoxy homopolymer
SR 1	22.8 ± 1.87	25.23 ± 3.75	29.40 ± 2.23	30.67 ± 2.38

on the sensitivity of the material to either of the analysis techniques for quantifying gel time.[21]

9.5.4   Impact of underfill contact angle on flow

In order for the underfill to flow under the die both quickly and effectively, the underfill is expected to have excellent wetting characteristics to the interfaces in contact. The wetting characteristics of an underfill to the interface depends not only on the UF formulation, but also on the dispense temperature and the polarity of the substrate/passivation surfaces. From an underfill formulation perspective, more polar the resin matrix, the better is the wetting characteristics to both the solder resist and polyimide passivation that is of interest to Intel. This has been clearly demonstrated in Table 9.1, where a comparison between epoxy anhydride, epoxy amine, and epoxy homopolymer systems. Contact angle measured as a function of epoxy material type, at constant temperature and on one specific solder resist A, clearly indicates epoxy anhydride systems to have a much lower contact angle, compared to epoxy amine systems, or epoxy homopolymer systems. Epoxy homopolymer is the most hydrophobic, translating to the lowest contact angle.

The wettability of the underfill material also depends on the type of the solder resist used on the substrate. The contact angle of the underfill to the solder resist can be modulated by the choice of solder resist used on the substrate. This directly translates itself into flow performance, as evident in the underfill flow rate in between the substrate/glass sandwich of the substrate in Figure 9.26.

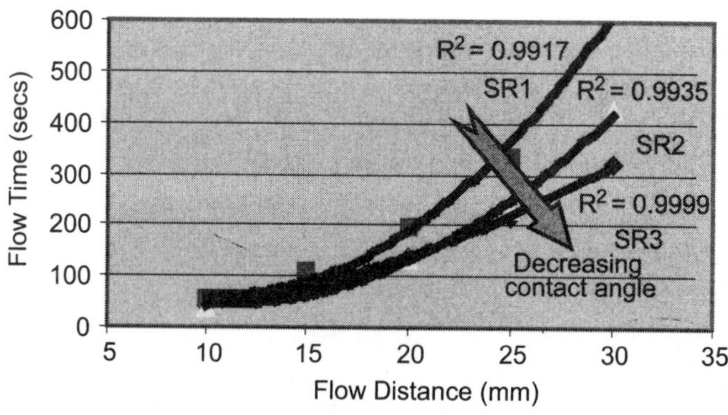

Figure 9.26 Impact of underfill contact angle to substrate solder resist on flow time.

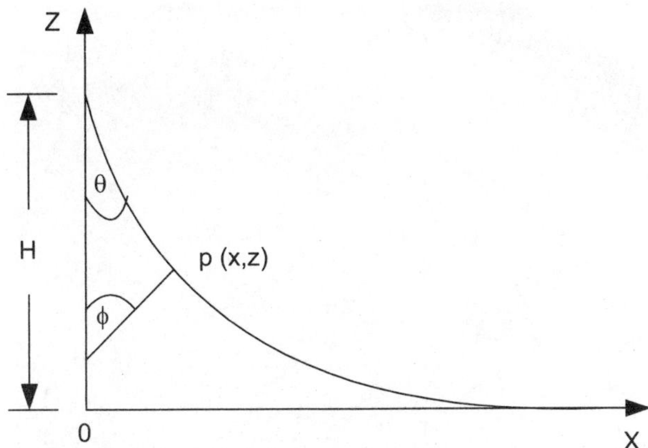

Figure 9.27 Capillary rises against a vertical wall.

The ability of the underfill to form a self-fillet is another key property that is modulated by the wettability of the underfill material to the sides of the silicon die. This is very critical to the underfill flow process. It usually requires that the underfill form a fillet with a height of half of the die thickness. The fillet is similar to a meniscus against a plane wall, as shown in Figure 9.27.

The capillary rising has the following relationship with the contact angle and surface tension

$$H = \sqrt{\frac{2\gamma(1 - \sin\theta)}{\rho g}} \tag{4}$$

where θ is the contact angle, γ is the surface tension, and H is the capillary height. Equation 4 shows that the capillary height is proportional to the square root of the surface tension and $(1 - \sin\theta)$.

Contact angle is one the key factors affecting fillet ability. Poor wettablity of the underfill to the substrate results in incomplete fillet formation, resulting in incomplete die edge coverage (IDEC) resulting in high yield loss. As shown in Table 9.2, underfill A has a lower contact angle, or better wetting characteristics, to substrate 1, compared to substrate 2. This also shows up in the surface energy calcu-

Table 9.2   Contact angle and surface energy of substrate A and B.

Substrate	*Substrate 1*	*Substrate 2*
H_2O contact angle (degree)	96	114
CH_2I_2 contact angle (degree)	56	81
Underfill contact angle (degree)	23	41
Surface energy by geometric mean method (dyn/cm)	31	17
Surface energy by harmonic mean method (dyn/cm)	34	21

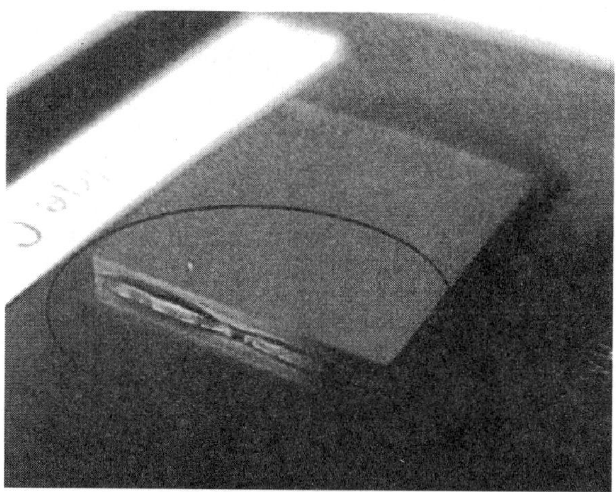

Figure 9.28 Incomplete die edge coverage as indicated in circle.

lations. Figure 9.28 clearly shows the effect of incomplete filleting, especially at the opposite side of the dispense.

9.5.5   Reliability performance of underfill material

As described earlier, the main function of the underfill is to protect the interconnect from solder fatigue failure, arising from the CTE mismatch between the silicon die, the solder interconnect, and the substrate. This is especially true in the case of an organic package where the CTE mismatch between the die and the substrate of choice are much higher. Thus, underfills are mainly used to extend the service life of flip-chip packages significantly. In this section, typical failure modes of flip-chip packages and the fundamental underfill material properties mitigating these failures, will be discussed. A typical organic package is expected to withstand temperature cycling and humidity stresses such as TCX and HAST. The reliability stressing is used to assess the materials performance in an accelerated fashion to extrapolate the materials behavior under normal operating conditions. Typical failure modes attributed to underfill material on an organic flip-chip package when exposed to temperature cycling or humidity stressing are (1) solder bump fatigue, (2) thin film cracking and delamination on the die, (3) delamination of the underfill to die or underfill to substrate, and (4) bump bridging due to solder migration.

9.5.5.1   Thin-film cracking and delamination

Thin-film cracking is the cracking of silicon nitride passivation layer on the die edge corners in a packaged unit post-reliability testing. Thin film cracking is hypothesized to be caused by shear stresses, applied by the package material to the active die surface. Post-package removal, thin-film cracking are recognized by the nitride cracks oriented at 45 degree angles relative to the line edge. On the other hand,

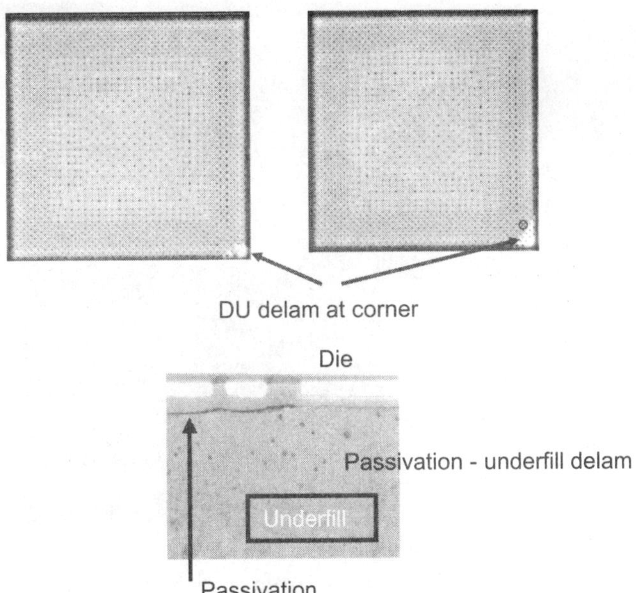

Figure 9.29 Package cross section showing underfill to passivation delamination resulting
due to die edge delamination of underfill.

thin-film delamination refers to the debonding or peeling at the polyimide/nitride
interface initiated along the die edges. Thin-film delamination is believed to be driven
by high residual stresses, both normal and shear forces along the die periphery. The
residual stresses originate from assembly processing and from accumulated reliability stressing. Although thin-film delamination originates at the die edges, it then is
expected to propagate into the interconnect layers underneath the passivation toward
the chip center, leading to interconnect opens, as shown in Figure 9.29. Another mode
of failure is the delamination/crack propagating in the bulk of the fillet material,
toward the substrate, leading to cracking of the copper traces underneath the solder
resist. Figure 9.30 clearly illustrates a typical fillet crack fail leading to Cu trace
cracking on the substrate. Apart from improving adhesion at the interfaces of
concern, arresting the crack growth in the fillet and in the bulk of the underfill material has been observed to prevent thin-film cracking and thin-film delamination fails.

One of the key underfill properties mitigating fillet crack performance of an
underfill material is bulk fracture toughness of the underfill. From an underfill formulation perspective, several key ingredients impact fracture toughness of an underfill.
These include underfill resin chemistry, epoxy molecular weight, addition of elastomer, and use of right filler distribution in the formulation. Figure 9.31a shows the
impact of use of anhydride-based formulations over epoxy homopolymers with
respect to fracture toughness. Anhydride formulations typically provide much higher
fracture toughness, due to inherent hydrogen bonding between the hydroxyl groups
and the carbonyl functionality of the cured resin matrix. Also use of higher molecular
weight epoxy resin reduces the cross-link density in the cured underfill materials also
results in higher fracture toughness, as shown in Figure 9.31b.

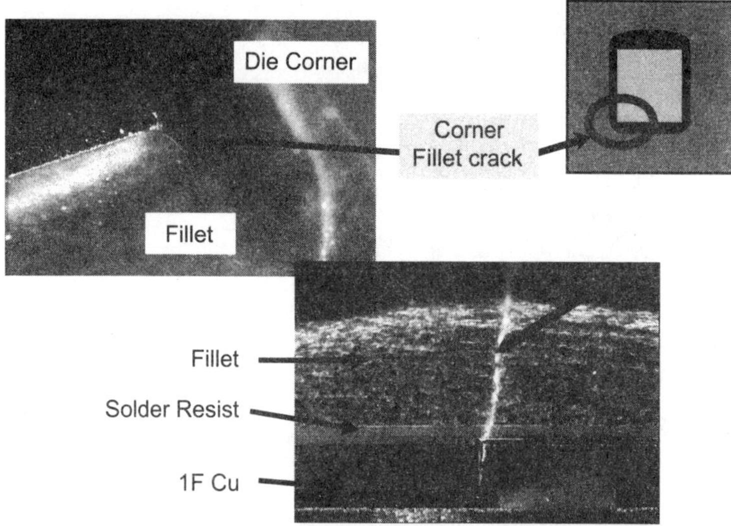

Figure 9.30 A typical fillet crack fail observed in an organic flip-chip package. Fillet crack is observed to arise at the die edge, propagating in the filleting resulting in cracking the substrate Cu trace.

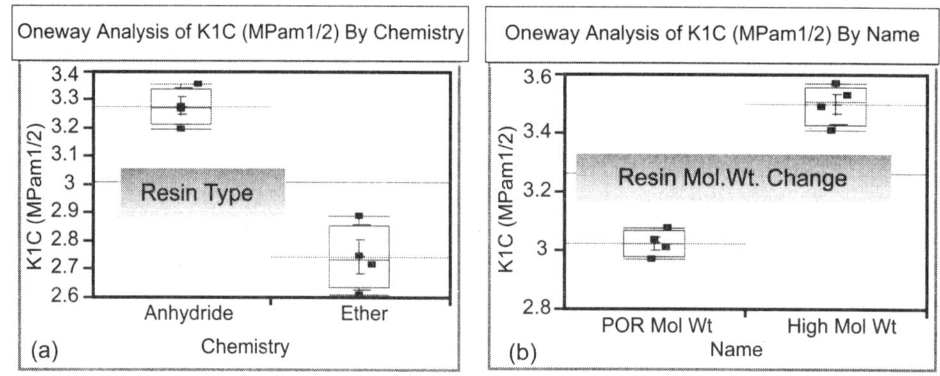

Figure 9.31 (a) Shows the impact of underfill chemistry on fracture toughness, while (b) shows the impact of resin molecular weight on fracture toughness of cured underfill.

It is very well known that addition of elastomers improves the fracture toughness of cured resins through phase separation of the elastomers. This has been observed to be true even for underfill materials. Butadiene-based elastomers are one of the most preferred elastomers for improvement of fracture toughness in underfills. The butadiene-based elastomers could be both non-functionalized or functionalized with appropriate groups, thus reacting with the epoxy matrix during cure. In general, elastomers that react into the epoxy matrix are expected to provide better fracture toughness performance when appropriate molecular weight of the elastomer is chosen. One surprise toward improvement of fracture toughness is the presence of

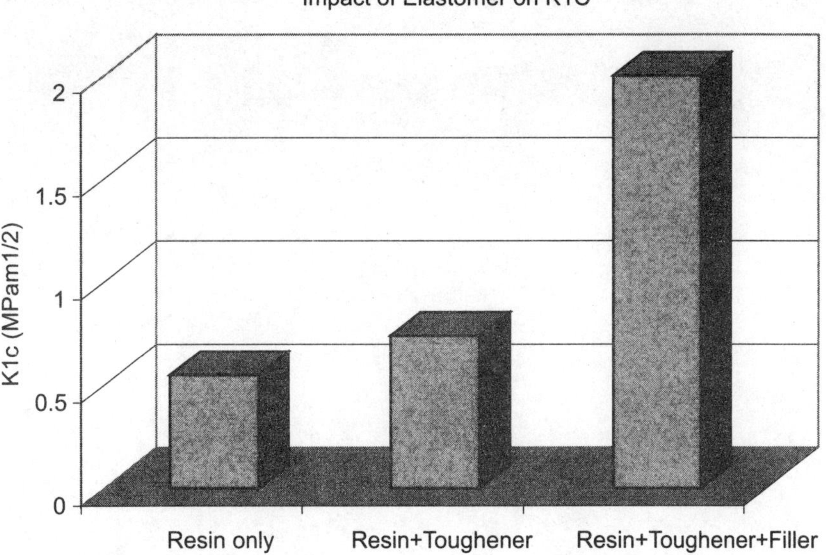

Figure 9.32 Impact of elastomer and elastomer + filler on fracture toughness of an underfill material.

right filler distribution. Figure 9.32 shows the impact of fracture toughness improvement via addition of an elastomer alone and in combination with the same elastomer and right filler distribution. Clearly, the impact of filler is much larger over that of even the presence of an elastomer. A similar phenomenon has also been observed by Hashemi et al. and Mitsui et al. for the polystyrene system, with various particulate fillers.[22,23]

9.5.5.2 Bump crack

Bump crack is another mode that has been observed as a failure. The initial hypothesis suggested that delamination of polyimide–underfill interface propagated and caused solder bump to crack. The delamination of the interface was suspected to be caused by CTE mismatch, that was caused by a resin-rich layer (RRL) in the underfill as a result of filler settling (Figure 9.33). Both material formulation changes and processing have been used to address this issue.

9.5.5.3 Underfill delamination at die or substrate interface

Delamination of underfill at the die or the substrate interface can occur both during the temperature cycling and under humidity stress conditioning. Delamination of the underfill during the temperature cycling is observed due to CTE mismatch between the die and the underfill, and to both nominal and shear stresses. These stresses have been observed to be very high especially at the die edges, resulting in delamination at these areas.

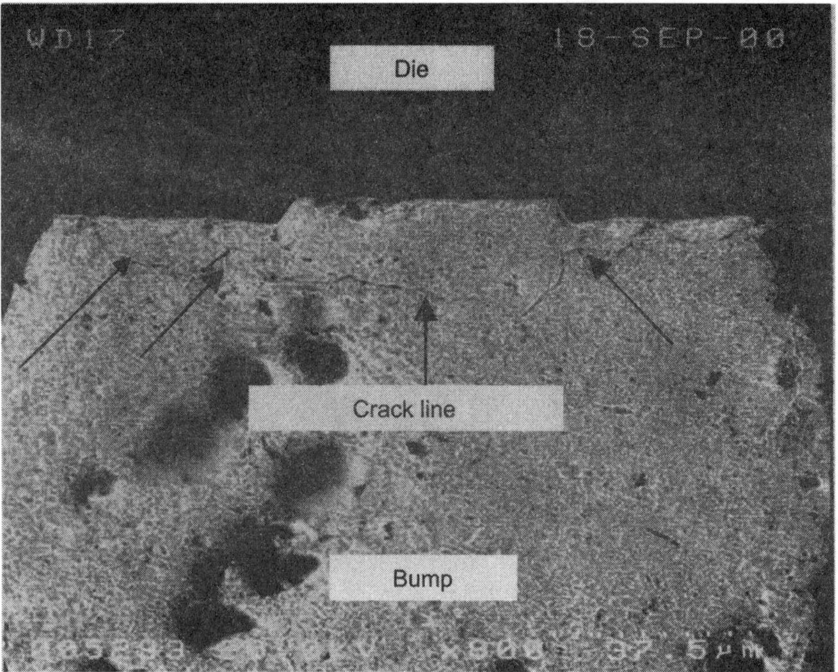

Figure 9.33 (a) Resin rich layer and (b) solder bump crack.

Delamination of underfill under moisture conditions can occur either at the underfill to die passivation interface or underfill to solder resist interface. Delamination of underfill can lead to solder bump fracture resulting in opens, or metal migration causing shorts under biased moisture conditions. Typically, delamination under moisture stress occurs due to poor adhesion of the underfill at the interface of concern. Adhesion of the underfill to the interface depends not only on the underfill formulation but also on the presence of reactive sites at the interface and also absence of any impurities at the substrate or passivation surface. Adhesion of the underfill depends once again on the right choice of under-material chemistry, and on the ability of the underfill to pick up moisture. Figure 9.34 shows the performance of two underfill formulations; one is anhydride based and the other is epoxy homopolymer based. As can been seen, although the anhydride-based underfill has very high adhesion post-cure to the solder resist interface and post-moisture exposure, the anhydride formulations are expected to drop in adhesion much more sharply when compared to epoxy homopolymer-based underfills. This phenomenon is fundamentally due to the ability of the anhydride-based formulations to pick up moisture, resulting in degradation of adhesion at the interface.

One way to improve the adhesion of the underfill to solder resist or passivation of interest is via addition of coupling agents to underfill formulations.[24] In general, coupling agents are silyl ethers with reactive groups on the other end. Several coupling agents have been explored in order to improve the adhesion of the underfills to various substrates.

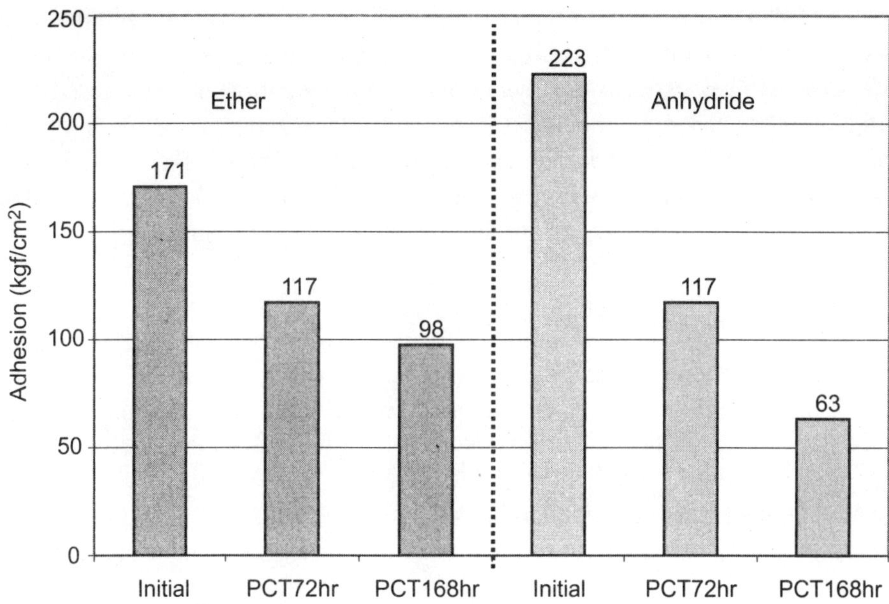

Figure 9.34 Button shear adhesion test indicating the impact of underfill material type of adhesion to solder resist post cure and post exposure to moisture stress.

9.6 Thermal interface materials and heat spreaders

In the context of heat removal for microprocessor packaging we can identify two generic architectures: (a) Architecture I typically dealing with low-power (<30 W) microprocessors or microprocessors in height-constrained applications where the die is directly attached to the heat sink or a heat pipe and (b) Architecture II, typically dealing with medium-to-high-power processors (>30 W) where an integral heat spreader (IHS) is used to spread the heat. Figure 9.35 shows the basic implementation of these two architectures.

In either case, successful thermal management requires the development of a thermal interface material (TIM) that comes in contact with the die and heat sink (as shown in Figure 35a) or the die and heat spreader (as shown in Figure 9.35b).

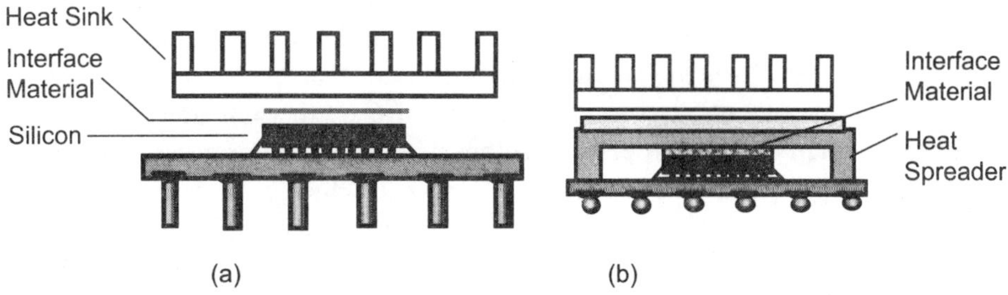

Figure 9.35 Basic implementation of two architectures for thermal management: (a) Architecture I, (b) Architecture II.

Typically TIMs are made of a polymer matrix in combination with highly thermally conductive fillers (metal or ceramic) and can be classified as phase change materials (PCM), thermal greases, gels, etc. Performance considerations as well as cost and manufacturability concerns, inevitably result in trade-offs that are made during the selection of an interface material for a given application. The attributes that are considered include (1) choice of polymer matrix, (2) choice of fillers, (3) design considerations, (4) manufacturability, (5) reliability, and (6) total thermal solution cost.[25]

9.6.1 Thermal materials technologies for Architecture I

This architecture encompasses packaging solutions for low- and medium-power microprocessors, predominantly encountered in the value and mobile processor market segments. As indicated earlier, the technology involves interfacing the heat sink to the die through a compliant interface material. Development of thermal materials technologies for this application has evolved over time and includes materials technologies such as elastomeric thermal pads, thermal greases, and phase change films (PCFs).

9.6.1.1 Elastomeric thermal pads

These are used to improve heat dissipation across large gaps, by establishing a conductive heat-transfer path between the mating surfaces. Thermal pads are typically 200 to 1000 μm thick and are popular for cooling low-power devices, such as chip sets and mobile processors. The pad consists of a filled elastomer, with filler materials ranging from ceramic to boron nitride for varying thermal performance. Typical failure mechanisms observed in this application are increased thermal resistance due to inadequate pressure or loss of contact at one or more surfaces. Nevertheless, these constraints significantly limit the thermal performance achievable with this class of materials.

9.6.1.2 Thermal greases

Thermal greases offer several advantages over pads, including the ability to conform to the interfaces. They require no post-dispense processing (e.g., no cure) and they have higher effective thermal conductivity compared to other classes of materials. Greases have been used very successfully in combination with various packaging form factors and have shown excellent performance. However, certain design and environmental considerations can preclude the use of thermal greases. Thermo-mechanical jeopardy to the processor has been observed under certain conditions. Greases are subject to loss in thermal performance under extended conditions of power cycling due to a phenomenon called "pump out." Under cyclic loading, extensive thermo-mechanical stresses exerted at the interface because of the relative motion (flexure) between the die and the base of a heat sink lead to loss of grease material from the interface. Also under high-temperature bake, the formulation chemistries utilized in typical thermal greases result in separation of the polymer and filler matrix due to the migration of the polymer component. The separation and loss of polymeric material could result in poor wettability at the interfaces, resulting in an increase in thermal resistance, also known as "dry out." The failure mechanisms encountered are a strong function of the thermal grease operating temperature and the number of on/off cycles that the processor assembly has been subjected to. The rate of thermal degradation is also dependent on the surface finish of the mating surfaces (heat spreader surface vs. backside of silicon). The pump-out mechanism and phase separation mechanisms have an exponential dependence on temperature, with a twofold increase in degradation for every 10°C increase in average operating temperature of the interface material. Another area of concern is related to retention of a heavy heat-sink mass interfaced to the die through a compliant material such as grease. Data collected on mechanical shock and vibration of heat-sink masses between 200 to 250 g indicate that the retention of the heat sink to the processor is critical to prevent die damage. During shock testing, relative motion between the heat sink and die could lead to mechanical damage to the die surface, with the corners of the die being highly susceptible to damage.

This fail mechanism is strongly dependent on the design of the heat-sink retention feature as well as the mass of the heat sink used. Typically the heat-sink mass (or volume) is proportional to the power dissipation of the processor; a heavier heat sink is required to cool higher processor power. Data collected indicate that in the case of lighter die loading, die damage due to mechanical shock is not a concern. In summary, thermal grease-based materials are recommended for applications at lower operational temperatures (to alleviate phase separation), lower die loading (to alleviate mechanical damage), and lower power cycling requirements (to alleviate pump out). However, the limitations in the use of thermal grease triggered the development of an alternate material described in the following section.

9.6.1.3 Phase-change films

PCFs are a class of materials that undergoes a transition from a solid to semi-solid phase with the application of heat; the material is in a liquid phase under die-operating conditions. This class of materials offers several advantages including the ability to conform to profiles of the mating surfaces, no post-dispense processing (e.g., no cure), and ease of handling and processing due to its availability in a film format. However, from a formulation perspective, the polymers and filler combinations that can be utilized impose limitations on the thermal performance of these materials. PCFs are typically a polymer/carrier filled with a thermally conductive filler, which change from a solid to a high-viscosity liquid (or semi-solid) state at a certain transition temperature. The choice of materials is tailored such that the transition occurs below the operating temperature of the die. Key advantages of PCFs are related to their ability to conform to surfaces and their wetting properties, which significantly reduces the contact resistance at the different interfaces. These materials usually are reinforced with a fiberglass mesh, which acts as a core, providing mechanical rigidity. Due to this composite structure, PCF materials are able to withstand mechanical forces during shock and vibration, protecting the die from mechanical damage. The semi-solid state of these materials at elevated temperatures resolves issues related to "pump out" under thermo-mechanical flexure. Typically, dispense processes required for thermal greases are throughput limiters. The manufacturing throughput of the assembly line is greatly improved since PCFs can be pre-attached to the base of a heat sink or heat spreader using a pick-and-place operation. In general it has been observed that the thermal performance improves over the course of reliability stressing. Since the material is in a softened state during these stresses, the compressive loading at this interface resulted in a decrease in thickness as well as improved wettability (conformance) with the surface irregularities.

9.6.2   *Thermal materials for packaging Architecture II*

This architecture is designed to be scalable to meet the demands of medium- to high-performance (power) processors. The integration of the heat spreader on the die required the development of a thermally conductive polymer, a heat spreader, and an adhesive sealant. The heat spreader helps in spreading of the heat flux from the smaller die area to a much larger surface area. This in turn translates to improved thermal performance of the heat sink. The heat spreaders utilized in this technology is a nickel-plated copper heat spreader. The TIM material comes in contact with the nickel-plated surface and the heat spreader is bonded to the organic substrate using an adhesive, as shown in Figure 9.35b. In addition to heat removal the heat spreader provides mechanical protection to the die and is also used to provide critical information, which is either laser or ink, marked on the surface of the spreader. A variety of plating processes (e.g., electrolytic, electroless, etc.) were investigated to determine the optimum plating for the heat spreader – key factors that were considered included (1) integrated thermal performance (i.e., plating characteristics that minimized the thermal contact resistance with the TIM), (2) mechanical integrity of the plating as well as adhesion of the plating to the sealant polymer, (3) manufacturability, and (4) cost.

From a thermal material perspective, Architecture II precluded the use of thermal greases due to the thermo-mechanical failure mechanisms discussed in detail in Ref. 23. PCMs were precluded since they did not meet the stringent thermal-performance requirements. In addition, the need for a positive compressive load at the interface imposes limitations on the design of this packaging architecture. In this packaging architecture the TIM has two key functions: (1) to dissipate heat to allow higher processing speeds and (2) to absorb stresses resulting from the mismatch of CTE of the die, substrate, and IHS. A significant amount of theoretical understanding of the thermal resistance has been applied to the development of TIM formulations. Thermal models have been developed based on electrical resistance theory and percolation theory. The key TIM formulation development issues identified from these models include bulk thermal conductivity (BTC), thermal interfacial resistance, and bond line thickness for package level considerations. Filler conductivity, filler loading, filler particle size and distribution, and silicone resin oil viscosity have also been identified to play key roles. By manipulating these key parameters, significant improvement in TIM performance has been achieved. As a result, a new thermally

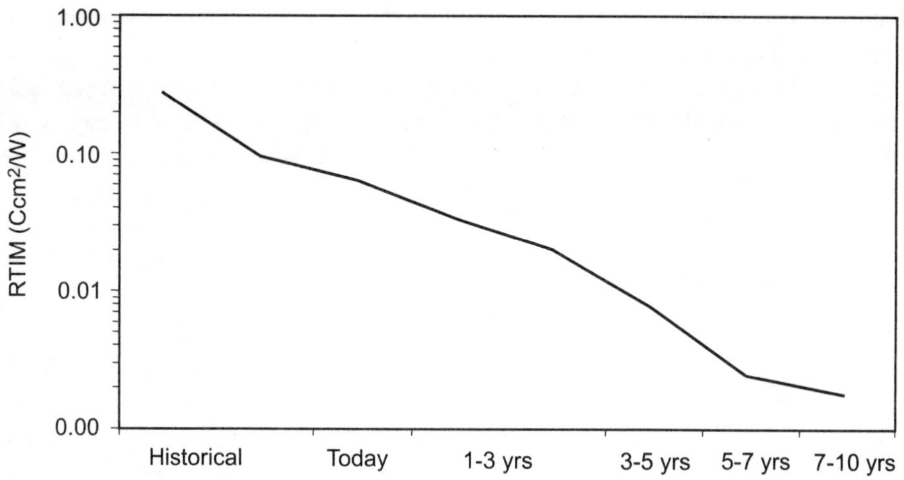

Figure 9.36 An illustration of the challenges in developing high performance thermal materials.

conductive gel TIM was developed. The gel is typically a metal or ceramic particle-filled silicone polymer and it combines the properties of both a grease and a cross-linked polymer. This is shown in Figure 9.36 where we continue to make approximately a 20% improvement in the thermal performance of the package.

9.7 Future trends

The major driving forces for changing materials in interconnects are (1) environmental concerns – Pb-free; (2) miniaturization of micro-electronic packages, and (3) cost constraints. European legislation and Japan marketing pressure have led intensive research in Pb-free alternatives for interconnect materials. The force driving toward Pb-free ones has led to an increase in using new Pb-free solders and other Pb-free materials as interconnect materials to meet environmental concerns. The new materials do not have the large empirical databases as those of Sn–Pb solders. Therefore we cannot automatically assume to be able to use the same life prediction model as Sn–Pb solders. Table 9.3 shows the expected trends in bump

Table 9.3 Driver trends: bump metallurgy, ILD, reflow.

Present and future trend	Bump (die)	Pre-solder (substrate)	SMT peak reflow temp (°)
1998	C4	Sn/Pb	220
2000–2005	C4	Sn/Pb	220
2005–2008	Pb-free	Pb-free	240–260
2008–2010	Pb-free	Pb-free	240–260
Beyond 2010	Pb-free	Pb-free	240–260

metallurgy and potential reflow temperatures the packages have to withstand for providing a reliable interconnect.

In the application of BGA and attachment, eutectic Sn–Ag–Cu has been recommended by the industry as a general use solder. Sn–Ag–Cu solders are generally stiffer than the current Pb–Sn eutectic solders. While fatigue resistance of the eutectic Sn–Ag–Cu is better than eutectic Pb–Sn solder, the early failure mechanism may transfer to other areas of the package due to the stiffness of the solder joints, for example, package substrates or PCBs. It is a challenge to design mechanically compatible materials set to achieve overall package reliability targets. Metallurgical compatibility of the eutectic Sn–Ag–Cu solder with soldering surfaces, such as PCB metal finishes and eutectic Sn–Ag–Cu BGA balls with various attachments solder pastes; for example, eutectic Pb–Sn paste is not fully understood. New reliability life prediction models for eutectic Sn–Ag–Cu have to be generated to obtain accurate predictions of interconnect reliability. Many process and material issues arise, as a result of higher reflow temperature (peak temperature increases from 210–230°C to 240–260°C).

With the move toward lead-free interconnect systems in the next generation packages, underfills are now expected to be compatible to higher reflow temperatures such as 260°C. Adhesion to various interfaces, and fracture toughness of the material are expected to be key modulating properties for an underfill to be compatible to 260°C reflow temperature. Because ball grid area package must experience a reflow during assembly, adhesion, and moisture uptake become key properties. They must be optimized for compatibility with 260°C reflow and JEDEC level 3 preconditioning.

There are several technology options that are under consideration to satisfy future underfill technology needs.[26] Capillary underfill is the current technology of choice. The continued shrinking of gap height and bump pitch will eventually pose limitations on the capillary flow. From a material perspective, new material systems are under investigation for potential future underfill needs. Alternate material systems include cyanate ester, polybenzoxazine, siloxarane, bismaleimide, and silicon-based underfill systems. These new material systems possess many desirable physical properties such as lower polymer CTE, lower moisture uptake, higher glass transition temperature (T_g), and higher toughness and mechanical strength. The successful development of these materials will largely depend on understanding the overall material characteristics/performance and successful integration of these materials into reliable package performance.

From a process simplification point of view, the vacuum assisting capillary technique (AUF) and no-flow underfill[27,28] (NUF) are under evaluation. Wafer-level underfill technology at present is more in the early development stage. The main advantage of the vacuum process is that it eliminates most of the barriers for capillary by providing a negative pressure drop to flow. The use of the vacuum process provides new capability to underfill highly viscous materials and also can provide huge gains in UPH, especially for large dies. This technique might also provide for a relaxation in bump design rules, opening a pathway to test alternate designs. Further, this can extend the life of flip-chip technology to a few more generations of products.

One potential alternative to vacuum underfill is the no-flow underfill technology. In this case, the underfill is pre-dispensed on the substrate and the gap is simultaneously filled with underfill as the C4 attach process is performed. No-flow, self-fluxing underfills provide a means of simplifying the SMT process by combining die attaching and underfilling in one step. This process would be gap height and die size independence and therefore has the potential to be extended for multiple generations. It also provides more flexibility for the package design group to scale the bump.

The challenges for no-flow material development are balancing fluxing capability, with suitable curing profile, and mechanical properties. The built-in fluxing capability is critical for joint formation and for a high yield of interconnection. Depending on the joint metallurgy, the developed no-flow material must possess fluxing capability to wet Sn–Pb or Pb-free solder. The curing behavior of the underfill needs to be optimized by adjusting the amount and latency of the catalyst to have minimum gelling below and at solder melting temperature, and maximum curing at higher temperature. No flow underfills typically have any silica settling problems. However, the CTE of the material is normally higher than capillary style underfills. Hence, it demands excellent adhesion and toughness, and low moisture absorption rate to satisfy the reliability requirement.

A more revolutionary approach is the concept of wafer-level underfill.[29] Here underfill is pre-applied to the die after bumping or redistribution. Like no-flow underfills, wafer-applied underfill technology is also gap height- and die size-independent. Like the no-flow underfills, there is potential to extend this wafer-level underfill to multiple processor generations. Current development of wafer-applied underfill has been focusing on the low-density flip-chip device application. At this time, the process and materials are not mature enough for high-density flip-chip package applications. Innovative processes and materials are needed to make this technology capable for Intel high-density flip-chip applications.

The challenges for a successful wafer-level underfill include the alignment and dicing of the wafer after underfill, fluxing capability of the pre-applied underfill, fillet formation during underfill process, and finally the reliability of the package.

Tablet or molded underfill is a technology in which epoxy underfill is forced into the gap by a transfer-molding process. The material development has been focused on the application of a mixed non-CPU package by stacking a wire-bonding die and a flip-chip. The required underfills for flip-chip and overmold for wire bonding can be accomplished by one-step molding with this moldable underfill material. Tablet underfill technology offers a great simplicity of the process and significant cost reduction. For CPU, the equipment cost is still very high and infrastructure change is also a big concern for deploying this technology. Some process issues need to be solved including epoxy on die, warpage, and stress in the package after molding.

Anisotropic conducting film (ACF) is another choice for underfilling the flip-chip package. There are two kinds of ACF: ACF for interconnection of bumpless die and substrate, and ACF for the interconnection of bumped die and substrate. The former interconnection is through a metal wire conductor in the film and the latter is through a mechanical contact of the conductive filler in the film. Both films have been considered for the non-CPU flip-chip application. The main issues for the development of ACF is the very high bonding force for chip attachment and high electrical resistivity of the joints.

It is expected that in order to keep up with the power requirements it will be necessary to make significant breakthroughs in thermal material technologies. Heat dissipation through polymer TIMs occurs through the phenomenon of percolation. One can also consider providing thermal solutions for heat dissipation entirely through conduction. However, such technologies can pose an entirely new set of challenges from a stress, cost, and infrastructure perspective. An ideal case would be to develop a composite TIM material that provides enhanced heat dissipation while balancing the mechanical properties to minimize package stress. Developing these composite TIMs can be achieved in a variety of ways including utilizing recent developments in nanomaterial technologies. It is expected that one can develop materials with bulk thermal conductivities five to ten times those obtained using conventional approaches.[30]

Acknowledgments

The authors would like to acknowledge the support of Intel assembly and packaging division senior management for their support in allowing us to publish this chapter, specifically Nasser Grayeli and Mostafa Aghazadeh. Gratitude is expressed to Hamid Azimi for his valuable inputs in the substrate section. Special thanks go to Rahul Manepalli, Vassou Leboneur, Sabina Houle, and Yi He for pulling together some of the data presented in the chapter. This work would not have been possible without the support of a lot of people inside Intel, specifically module and quality and reliability personnel.

References

1. M. Paunovic and M. Schlesinger, Fundamentals of Electrochemical Deposition, John Wiley & Sons, New York, 1998, pp. 73–74.
2. M. Vogelaerek, V. Sommer, H. Springborn, and U. Michelsen-Mohammadein, "High-Speed Plating for Electronic Application", *Electrochim. Acta*, **47**, 2001, pp. 109–116.
3. D.O. Powell and A.K. Trivedi, *IEEE Trans. Components, Hybrids Manuf. Technol.*, **16**, No. 8, pp. 182–186, Dec. 1993.
4. F. Nakano, T. Soga, and A. Amagi, The Proceedings of the 1987 International Symposium on Microelectronics, ISHM, 1987, pp. 536–541.
5. D. Suryanarayana, R. Hsiao, T. P. Gall, and J.M. McCreary, *IEEE Trans. Components, Hybrids and Manuf. Technol.*, **14**, No. 1, pp. 218–223, March 1991
6. Pentium® Chips are trademark of Intel Corporation.
7. R.J. Wassink, "Solder Alloys", Soldering in Electronics, 2nd ed., Electrochemical Publications Ltd, Ayr, Scotland, 1989, p. 135.
8. K. Norris, "Reliability of Controlled Collapse Interconnections", *IBM J. Res. Dev.*, May 1969, p. 266.
9. V. Gektin, A. Bar-Cohen, and S. Witzman, "Coffin–Manson Based Fatigue Analysis on Underfilled DCA", EEP-Vol 19-2, *Adv. Electron. Packag.*, ASME, 1997.
10. S. Han and K.K. Wang, IEEE Trans. Components, Packag. Manuf. Technol. – Part B, Vol.

20, No. 4, Nov. 1997, pp. 424–433.
11. Epoxy Resins: Chemistry and Technology, Clayton A. May, Ed., Marcel Dekker, New York, 1988.
12. G. Wypych, Handbook of Fillers, 2nd ed., Chem Tec Publishing, Toronto, Canada, 1999, 291.
13. D.R. Dinger and J.E. Funk, *MRS Bull.*, Dec. 1997, pp. 19–23.
14. L.Y. Sadler and K.G. Sim, Chem. Eng. Process, March, 1991, pp. 68–71.
15. L. Nguyen, C. Quentin, P. Fine, B. Cobb, S. Bayyuk, H. Yang, and S.A. Bidstrup-Allen, *IEEE Trans. Components Packag. Technol.*, 22, No. 2, June 1999, pp. 168–176.
16. D.S. Kim and S.C. Kim, "Rubber Modified Epoxy Resin. I: Cure Kinetics and Chemorheology", *Polym. Eng. Sci.*, 34, 1994, pp. 625–631.
17. I. Sbarski and S.N. Bhattacharya, "Simplified Viscosity Model of Rubber Filled Polymeric System", *Plastic, Rubber Composites Process. Appl.*, 27, 1998, pp. 266–271.
18. J.J. Cai and R. Salovey, "Model Filled Rubber. VI: Dynamic Property Dependence on Filler Particle Size of Rubber Compounds During Curing", *Polym. Eng. Sci.,* 40, 2000, pp. 118–126.
19. G.W.H. Hohne, W. Hemminger, and H.J. Flammersheim, Differential Scanning Calorimetry: An Introduction For Practitioners, Springer, NewYork, 1996.
20. R.B. Prime, "Thermosets" in Thermal Characterization of Polymeric Materials, Vol. 2, 2nd ed., E.A. Turi, Ed., Academic Press, Boston, 1997, Chapter 6, pp. 1370–1431.
21. Y. He, "Physical Characteristics of FC-PGA Underfill Candidates" Intel Report, AMCL, 1999.
22. S. Hashemi, K.J. Din, and P. Low, *Polym. Eng. Sci.*, 36, No. 13, 1996, pp. 1807–1820.
23. S. Mitsui, H. Kihara, S. Yoshima, and Y. Okamoto, *Polym. Eng. Sci.*, 36, No. 7, 1996, pp. 2241–2246.
24. Q. Yao, J. Qu, J. Wu, and C.P. Wong, IEEE –ECTC, 1999, pp. 1079–1082.
25. R. Wiswanath, V. Wakharkar, A. Watwe, and V. LeBonheur, "Thermal Performance Challenges from Silicon to Systems", Intel Technol. J., Q3, 2000.
26. L. Wang and C.P. Wong, IEEE –ECTC, pp. 224–230, 1999.
27. S. Shi, G. Jefferson, and C.P. Wong, The Proceedings of the 3rd International Symposium and Exhibition on Advanced Packaging Materials, 1997, p. 42.
28. R. Thorpe, D. F. Baldwin, and L.P. McGovern, IEEE –ECTC, 1999, pp. 419–425.
29. Q.K. Tong, B. Ma, E. Zhang, A. Savoca, L. Nguyen, C. Quentin, S. Luo, H. Li, L. Fan, and C. P. Wong, "Recent Advances on Wafer Level Flip Chip Packaging Process, Proc. 50th Electronic Components and Tech. Conf, 2000, p. 101.
30. S. Jayaraman, P. Koning, F. Hua, T. Chen, C. Gettinger, and G. Clemmons, IATTJ, 4, *Int. Techn. J.*, Intel Corporation, 2001.

10 Glass–ceramic packages

Kazuhiro Ikuina, Yuzo Shimada, and Kazuaki Utsumi

10.1 Introduction

Electronic circuit miniaturization is continuing for integrated circuit improvements. High-speed computer systems and optical communication systems require new packaing technology with high propagation speed and high-wiring density.

For high-speed computer systems, the use of high density packaging is one of the key factors in reducing machine cycle time, which determines computer system performance.[1] The reduction in the machine cycle time can be achieved by reduction in propagation delays for signals between large-scale integrations (LSIs). The following formula can be represented as a performance factor Tm for the wiring delay in a given substrate.[2]

$$Tm = \tau \sqrt{S/G} \tag{1}$$

where τ is the signal delay per unit wiring length, S is the substrate area, and G is the number of gates mounted on the substrate. Taking $\varepsilon\gamma$ as the dielectric constant for a dielectric material, and c as the speed of light, yields the following formula:

$$\tau = \sqrt{\varepsilon r}/c \tag{2}$$

Lower dielectric constant materials and higher gate density on the substrate become the key points for packaging.

High speed and small size are required for optical communication systems. In order to reduce inductance and capacitance, which might cause electronic cross-talk at high speed, it is necessary to use new packaging technologies with high-wiring density.

The main requirement for high-speed computer and communication systems lies in how to reduce delay due to wiring. The following requirements must be met in the packaging substrate in order to improve system capabilities. In terms of construction, multi-chip packaging (MCP) or multi-chip module (MCM) technology with high density wiring and short wiring length are needed, while, in terms of materials, there are two important factors. One is to decrease the dielectric constant for dielectric material near a signal line. The other is to minimize electrical resistance for conductors to prevent undesirable voltage drops.

On the other hand, high-speed electrical equipment systems and ultra-large-scale integration (ULSI) packaging development demand new packaging technology to meet the high propagation speed, high-wiring density, and high frequency pulse requirement for a multi-chip mounted substrate.

For logical modules in high-speed computer systems and communication systems, as very large-scale integration (VLSI) performance and density increases, the propagation delays resulting from the length and capacitance of the interconnection on the packaging substrate can contribute significantly to the total propagation delay. The reduction in the machine cycle time that determines the performance of the computer system can be achieved by reduction in propagation delays for signals between VLSIs. Lower dielectric constant material for substrate and higher gate density on the substrate are key points for logic modules.

Table 10.1   Properties of glass–ceramic materials.

Materials	Dielectric constant (1 MHz)	Dissipation factor (1 MHz)	Flexural strength (1 MPa)	Coefficient of thermal expansion (ppm/°C)
1. LB(45)-A(55)	7.8	0.003	300	4.2
2. B(50)-A(50)	7.1	0.003	300	5.5
3. B(60)-A(40)	6	0.001	220	5
4. B(65)-C(35)	5	0.003	150	4.5
5. B(65)-(15)-C(20)	4.4	0.002	120	3.2
6. B(65)-(15)-C(20) (isolated pore structure)	3.4	0.002	85	3.2

For high-speed communication systems, the essential features are high speed, small size, and high performance. To meet these requirements, it is necessary to use new packaging technologies with high-wiring density.

From the viewpoint of developing multilayer substrates for high-speed and high-frequency multi-chip modules and packages, the following properties should be satisfied to achieve identical ceramic materials.

1.   The dielectric constant for the insulator materials should be reduced.

2.   It is necessary to use low electrical resistivity conductors to minimize line resistance and prevent any undesired voltage drop.

3.   The coefficient of thermal expansion (CTE) for the substrate should be matched with that for silicon semiconductor chips to avoid damage by thermal stress.

4.   Mechanical strength for the substrate should be tough enough to endure many kinds of stress (e.g., interface stress with metal in fabrication and assembly process).

5.   It is necessary to achieve high-wiring density and three-dimensional (3D) packaging for reduction of delay time and miniaturization.

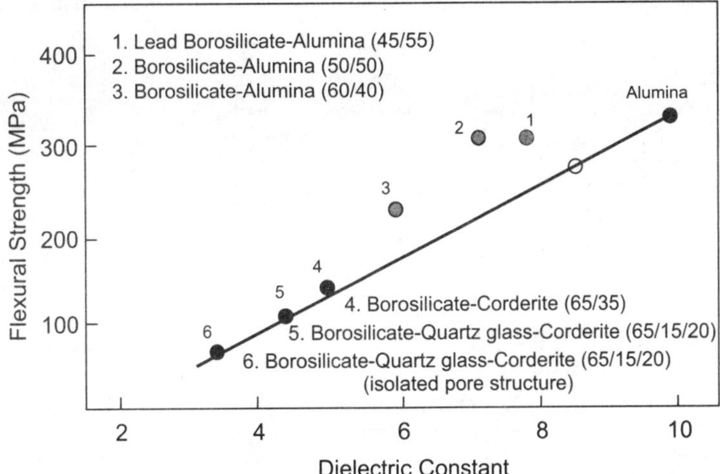

Figure 10.1 Relation between flexural strength and dielectric constant for dielectric glass–ceramic materials.

To meet these requirements, a new glass–ceramic material, a new organic insulator material, and a new substrate structure have been developed for advanced packages and modules.

This chapter describes dielectric materials with low dielectric constant; microstructure of glass–ceramic bodies; fabrication processes; new glass–ceramic multilayer substrates for high-performance packages and multi-chip modules, and their applications to high-speed computer packages, optical communication modules, microwave multi-chip modules, high-speed RISC modules, chip carrier for chip size packages, and 3D stack memory modules.

10.2 Glass–ceramic packages

10.2.1 Materials

To meet the high propagation speed requirement, the authors have developed several low dielectric constant glass–ceramic materials. Many kinds of glass–ceramic materials for the multilayer wiring substrate have been developed, as shown in Table 10.1. These glass–ceramic materials can be sintered around 900°C, so it is possible to use silver, copper, gold, palladium, and alloys of these metals as a conductor material. Figure 10.1 shows the relation between flexural strength and dielectric constant for glass–ceramic materials. It is necessary to use a glass–ceramic material that has a low dielectric constant, low dissipation factor, and high flexural strength to meet multilayer substrate for high-speed and high-frequency multi-chip modules. Therefore, we mainly use the glass–ceramic material whose composition is 50 wt. % borosilicate glass and 50 wt. % alumina in this chapter (material number 2 in Figure 10.1 and Table 10.1).

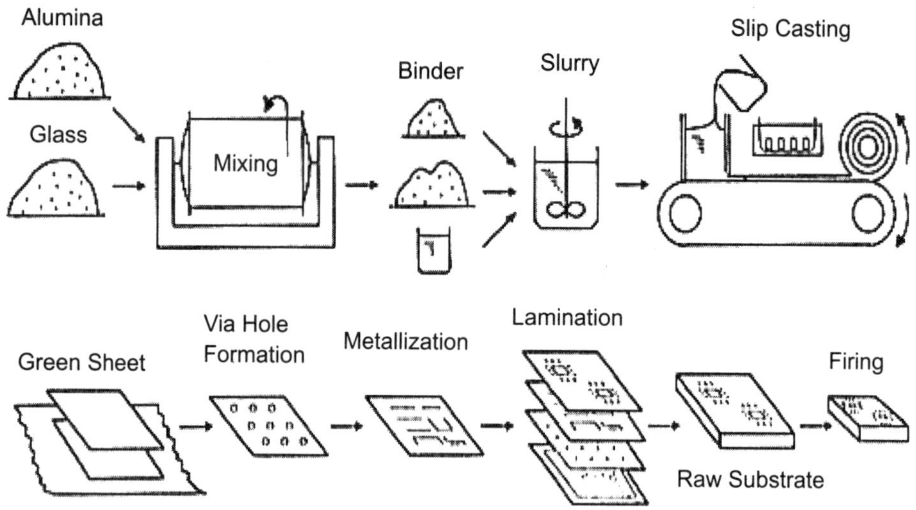

Figure 10.2 MGC package fabrication process.

Further, low dielectric permittivity glass–ceramic structure (material number 6, $\varepsilon\gamma = 3.4$) was realized, owing to closed pores in the glass–ceramic body.[3] Alumina–lead borosilicate glass system (material number 1) and alumina borosilicate glass system (material number 2) were applied to packaging substrates for computer and communication systems.

10.2.2 Fabrication process

A flow chart for the co-fired multilayer glass–ceramic package fabrication process is shown in Figure 10.2. The dielectric material is a mixture of alumina and lead borosilicate glass powder, which has controlled particle size, owing to electrical and mechanical properties and shrinkage control. The composition of the mixture is 55 wt. % alumina and 45 wt. % lead borosilicate glass. The slurry used consists of mixed dielectric powders and organic liquid vehicles. The slurry is cast into about 200 µm thick ceramic green sheets by a slip casting process. Via holes are formed through the green sheet by a punching machine with punches and dies. A green sheet may have more than 10,000 via holes in a 250 mm square area. Low resistivity conductor paste is applied onto the punched sheets. In this process, via holes are filled with paste to form contacts between signal lines. Low electrical resistivity for conductors reduces voltage drop along the line, and prevents a rounded pulse waveform. Since the dielectric material can be sintered at about 900°C, it is possible to use low electrical resistivity conductors with a low melting point, such as silver–palladium, and gold, instead of molybdenum or tungsten refractory metals in the alumina co-fired multilayer substrate. When sintered at 900°C, the resistivities are around 5 µΩ cm for 95 wt. % silver–5 wt. % palladium, and around 3 µΩ cm for gold. These conductors with low electrical resistivity reduce voltage drop as well as achieving high-wiring density.

The metallized green sheets are precisely stacked in a pressing die in sequence by the stacking machine. These sheets are then laminated together by hot pressing. Density heterogeneities in the laminated samples influence any shrinkage in the sintered substrate. Therefore, this lamination process should be homogeneously carried out by means of correct dimensional die and punch with flat surfaces. After burn-out processing at about 500°C, sintering is carried out in an electric furnace at about 900°C in air. Finally, LSI chips, capacitor chips, resistor chips and so on are assembled onto the multilayer glass–ceramic (MGC) substrate.

10.2.3 Substrate properties

The typical properties of the MGC substrate, using the alumina–lead borosilicate glass system, are summarized in Table 10.2.

Table 10.2   Typical MGC substrate properties.

Substrate	Material	*Alumina 55 wt. %* *Lead borosilicate glass 45 wt. %*
	Flexural strength	300 MPa
	Dielectric constant	7.8 (1 MHz)
	Dissipation factor	<0.3% (1 MHz)
	Thermal expansion coefficient	42×10^{-7} deg^{-1}
	Thermal conductivity	3.6 W/mk
	Linear shrinkage	13.0%
	Shrinkage tolerance	<0.2%

This substrate has high flexural strength, more than 300 MPa. The camber is extremely small, though the substrate size is large. The dielectric constant is 7.8, and the dissipation factor is less than 0.3% at 1 MHz. The CTE is one of the important factors, which must be considered for the packaging substrate. The CTE is 42×10^{-7} deg^{-1} and is close to that of silicon chips. Thermal conductivity is not as high. Substrate linear shrinkage and shrinkage tolerance are exactly controlled by the high-level control technologies.

10.2.4 Microstructure of glass–ceramic body

When sintered at about 900°C, crystallization takes place in the sintered ceramic body on alumina–lead borosilicate glass system. This reaction leads to high mechanical strength for the MGC substrate. This lead borosilicate glass mainly contains elements of Si, Pb, B, and Ca, and has a softening temperature at around 800°C.

In the case of coarse glass powder, crystallization takes place in the glass, and consequently cristobalite crystal is formed in it, while, in the case of fine glass powder, the microcrystalline structure grows as a result of chemical reaction of the alumina with the lead borosilicate glass. Figure 10.3 shows a transmission electron

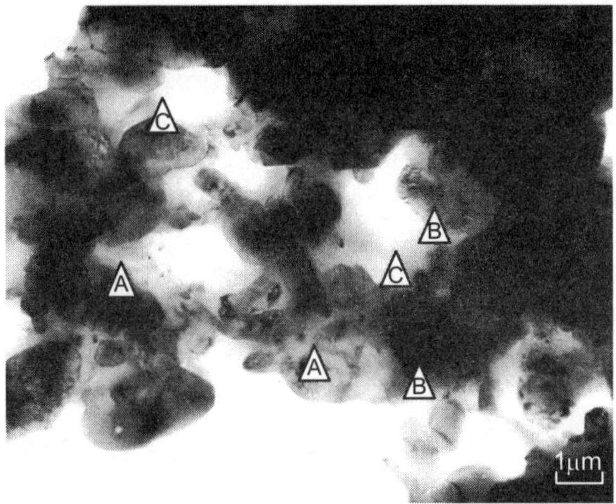

A: Anorthite   B: Alumina   C: Glass

Figure 10.3 TEM image of the glass–ceramic body in the multiplayer substrate.

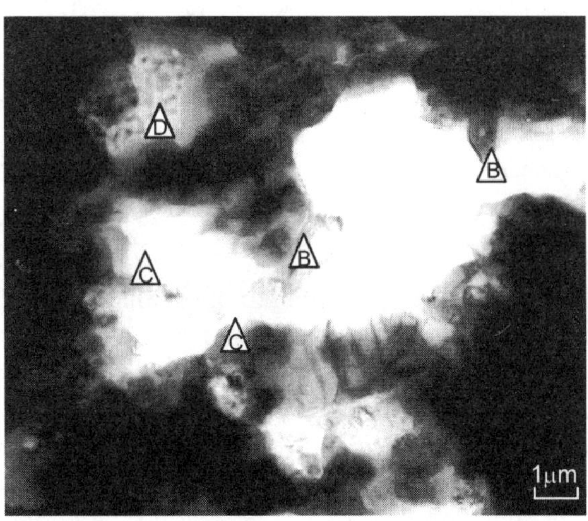

B: Alumina   C: Glass   D: Separated Phase

Figure 10.4 TEM image of the glass–ceramic sintered body at the first stage of sintering process (900°C).

microscope (TEM) image of the glass–ceramic body in the multilayer substrate. Alumina particles and the anorthite crystal phase, which is formed by a reaction between alumina and glass, exist in the vitreous network. This vitreous phase has the element of Al dissolved in the lead–borosilicate glass. Anorthite is formulated as $CaAl_2Si_2O_8$, and is formed because of the reaction of alumina with Si and Ca in the glass.

In the first stage of the sintering process, a glass softening reaction takes place at about 800°C, and alumina dissolves in the glass. Anorthite is then crystallized partially from the glass phase with alumina at more than 800°C. If the crystallization is insufficient, a heterogeneous phase remains in the glass phase. Figure 10.4 shows the microstructure of a glass–ceramic body sintered under conditions of high-speed sintering with short holding time at 900°C. The heterogeneous phase is observed in the sintered body and was identified using energy-dispersive X-ray (EDX) analysis. This heterogeneous phase consists of aluminum–calcium–silicon oxide, and changes into the anorthite crystal phase when crystallization is sufficient.

Anorthite crystal plays the role of reinforcement in the glass–ceramic body. As a result, the MGC substrate, which is made from the alumina–lead borosilicate glass system, has a high flexural strength of more than 300 MPa.

10.2.5　Shrinkage control technologies

It is very important to control shrinkage between raw substrate and sintered substrate for fine-line formation on the substrate. Since shrinkage is delicately influenced by any change in the raw materials, it is necessary to establish shrinkage control technologies. Highly accurate shrinkage can be attained by strict control at every process stage. The shrinkage is greatly influenced by differences between green sheet and raw substrate densities, and between raw substrate and sintered substrate densities.

Shrinkage control factors are mainly packaging density and glass–ceramic chemical reaction. The parameters, which influence these forms, are as follow:[4]

1. Particle-size powder package
2. Pressure on lamination raw substrate density
3. Binder content in slurry
4. Powder composition
5. Additive for crystallization
6. Temperature profile at sintering

Particle size and pressure can be specially altered for shrinkage control, because these parameters can suitably control raw substrate density.

10.2.6　Application

10.2.6.1　High-speed computer packaging

The packaging hierarchy for the high-speed computer system consists of 3 levels.[5] The first level is concerned with the connection form for LSI chips. The second level is a multi-chip package level containing high density flipped tape automated bonding (TAB) carriers (FTC), and is most important for a high-speed packaging. The third is a board level with zero insertion force (ZIF) connections. The MGC substrate is applied to the multi-chip package in the second level.[6]

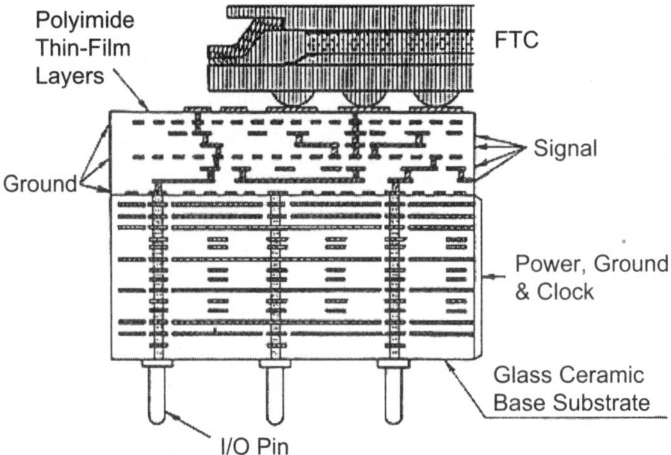

Figure 10.5 High-speed computer use multiplayer substrate cross-section view.

Figure 10.6 Cut-away view showing printed green sheets in the raw MGC substrate.

Figure 10.5 shows a cross section of the multilayer substrate (MLS), consisting of MGC base substrate and polyimide thin-film layers, for high-speed computer use. There are clock line, power, ground, and via hole layers in the MGC substrate. Via holes for electrical contact between layers are around 200 μm in diameter. This substrate has 40 glass–ceramic dielectric layers, of which 14 are conductive layers with 6 clock layers.

Figure 10.6 shows a cut-away view of printed green sheets in the raw MGC substrate. The sintered substrate size is 225×225 mm. The substrate thickness is 5.5 mm. The MLS, which has polyimide layers on the MGC substrate, is pictured in Figure 10.7. Polyimide dielectric layers have eight conductive layers with four signal layers. The signal wiring pitch is 75 μm, and its width is 25 μm. The cure temperature for polyimide is about 400°C, which is very useful for the combination of fine signal

Figure 10.7 MLS which has polyimide layers on the MGC substrate.

Figure 10.8 Cross-sectional photograph of MGC substrate.

layers with higher pin-counts and higher power distribution on the MGC substrate. The adoption of many polyimide layers brings about mechanical stress due to polyimide cure, and differences in thermal expansion between polyimide and MGC substrate. Therefore, an MGC substrate has to maintain high flexural strength.

Figure 10.8 shows an MGC substrate sectional photograph with input/output (I/O) pins. By using silver 95 wt. %–palladium 5 wt. %, the sintered metallization electrical resistivity is 5 $\mu\Omega$ cm using the thick film printing process, while resistivity for gold is 3 $\mu\Omega$ cm. I/O pins are attached to sputtered bonding pads by using 80 wt. % Au–20 wt. % Sn solder, with 1.8 mm pitch in the lattice. The I/O pin number on the substrate is 11,540. These pins are made of copper alloy, which has low electrical resistivity. The bonding strength hardly changes during a 40 heat cycle test (25°C→ 220°C→ 100°C→ 25°C), and is sufficient, more than 3 kg per pin. The multi-chip package can be used to mount up to 100 LSI chips.

The reliability tests on electrical migration for the silver–palladium conductor were carried out under high temperature and high humidity conditions. Consequent-

Table 10.3   Typical MGC substrate specifications for MLS.

Substrate	Size	225×225 mm
	Thickness	5.5 mm
	Line width	200 µm
	Via hole diameter	200 µm
	Number layer	40
	Conductive layers	14
	Conductor; material (resistivity)	Ag–Pd (5 µΩ cm)
		Au (3 µΩ cm)
I/O pin	Number	11540
	Pitch	1.8 mm
	Bonding strength	>3 kg

ly, high reliability for the MGC package was confirmed. Typical specifi- cations for the MGC substrate, which is used in the MLS for high-speed computer systems, are shown in Table 10.3.

10.2.6.2   *Fiber-optic transmitter/receiver*

Optical fiber local area network (LAN) systems have been attracting much attention due to the features of optical fiber (i.e., wide bandwidth and no electromagnetic noise interference). Recently, there has been an increasing requirement for high-speed transmission of various information services, such as voice, data, graphics, database, and file transfer. Fiber-optic transmitters/receivers are applied to such LAN systems for electrical-to-optical signal conversation.[7] The MGC substrate is applied to the transmitter/receiver modules to accomplish miniaturization.[8] This part describes the receiver module in particular.

Figure 10.9 shows an image diagram for a fiber-optic receiver module. Resistor network chip, capacitor network chip, LSI, and discrete chips are mounted on the MGC substrate. This substrate has power supply and solid reference layers, to reduce cross-talk coupling noise between the front-end output stages. This substrate has seven conductor layers and six insulator layers. Three of the seven conductor layers were power supply and solid reference planes.

The capacitor network chip is used with a low firing temperature (about 900°C) ceramic material in the ternary system $Pb(Fe_{2/3} \cdot W_{1/3})O_3 - Pb(Fe_{1/2} \cdot Nb_{1/2})O_3 - Pb(Zn_{1/3} \cdot Nb_{2/3})O_3$, and has 15 capacitor elements from 12 to 117 nF in it. The dielectric constant for this material is around 14,000 at room temperature. The

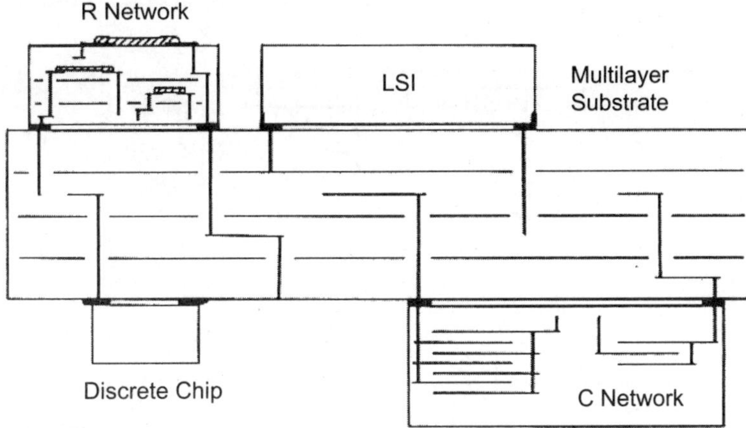

Figure 10.9 Image diagram for fiber-optic receiver.

Figure 10.10 Fiber-optic receivers with multiplayer ceramic components (left) and conventional components (right).

fabrication process uses green sheet lamination technology, and is basically the same as that for the MGC substrate.

The resistor network chip has seven resistor elements with a range of from 22 Ω to 3.6 kΩ in the insulation layers. The resistor material is mainly RuO_2, and is co-fired with glass–ceramic insulation material and conductor at about 900°C. The fabrication process is also similar to that for the MGC substrate.

Fiber-optic receivers which use the new multilayer ceramic components (MGC substrate, and capacitor and resistor network chips) and conventional components are shown in Figure 10.10. The dimensions of the new developed fiber-optic receiver are reduced one-third compared with those for a conventional receiver. The MGC substrate size is $21.8 \times 9.8 \times 0.8$ mm. The capacitor network chip size is $7 \times 5 \times 1.5$ mm,

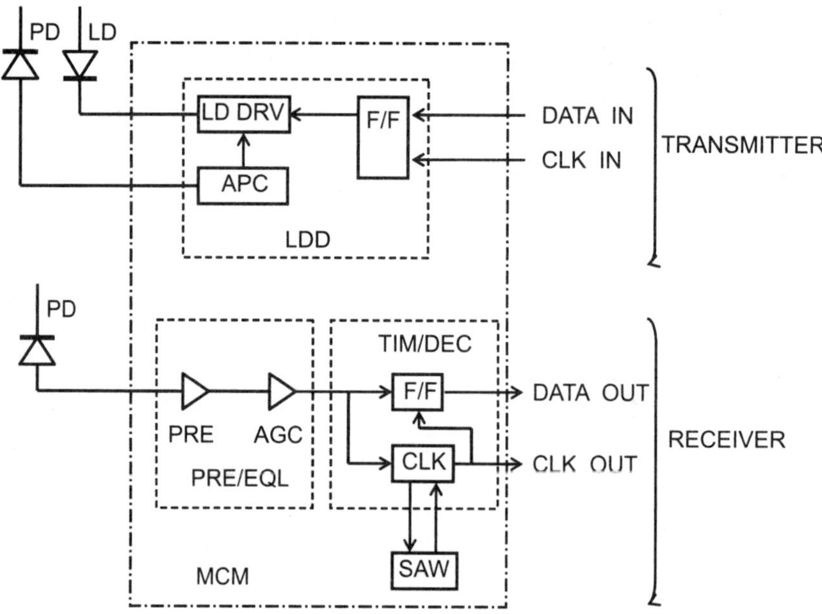

Figure 10.11   156 Mbps optical interface block diagram.

and the resistor network chip size is $4 \times 5 \times 0.8$ mm. The multilayer resistor network chip, LSI, and transistors are mounted with solder on one face of the MGC substrate, while the multilayer capacitor network chip is mounted on the other face of the MGC substrate. LSI terminals are connected to the multilayer capacitor network chip through via holes in the MGC substrate. The passive component number for the newly developed fiber-optic receiver decreases to 6 chips in comparison to a conventional receiver (26 chips). It has become possible to achieve small size. Bit error rate characteristics of 230 Mbps for the fiber-optic receiver are as good as those for a conventional receiver operating at 200 Mbps.

10.2.6.3   *Optical transmission module*[9,10]

Development of an optical interface, based upon the network node interface (NNI) on the CCITT counsel, is being actively carried out. High performance, miniaturization, and low consumption of electric power are required for optical interface modules, accompanied by miniaturization of the communication system. In order to meet these requirements it is necessary to increase packaging density in the module. However, high-packaging density brings about problems of chip radiation and increased wiring layers in the substrate. An MCM is one of the best techniques which can be used to solve these problems. The MGC substrate is applied to MCM for the 156 Mbps optical interface used in optical communication systems.

Figure 10.11 shows the 156 Mbps optical interface block diagram. The transmitter part consists of a 1.3 μm Fabry Perot laser diode (LD) and an LD driver LSI with automatic power control (APC) function. The receiver part consists of an

Figure 10.12   Multi-chip module (MCM) for 156 Mbps optical interface.

InGaAs-PIN-type photo diode (PD), a reshaping/amplification LSI, a timing/ decision LSI, and a surface acoustic wave (SAW) filter for timing extraction.

MCM structure is a six-layer glass–ceramic substrate 30×25 mm in size. All components, except for LD and PD, are mounted on the sides of the substrate. By using the bare-chip packaging technique, an extremely small size MCM, 10 cc, volume has been accomplished. Figure 10.12 shows the MCM for a 156 Mbps optical interface module.

Table 10.4   Glass–ceramic material characteristics and multi-layer substrate specifications.

Glass–ceramic material specification	
Dielectric constant	7.8 (at 1 MHz)
Thermal conductivity	3.6 W/m·K
Thermal expansion	4–5 ppm/°C
Insulation resistance	$>10^{14}\,\Omega$ cm
Bending strength	2800 kg/cm^2
Multi-layer substrate specification	
Conductor material	Silver/palladium
Minimum line width/pitch	0.2 mm/0.2 mm
Via-hole dimensions	0.2 mm ϕ
Minimum via-hole pitch	0.6 mm

Table 10.4 shows glass–ceramic material characteristics and multi-layer substrate specifications. The main features of the glass–ceramic substrate are a low dielectric constant and a low expansion coefficient close to that for silicon, compared with general substrates. Moreover, this substrate is appropriate to bare-chip mounting, because of better thermal conductivity than that of organic substrates.

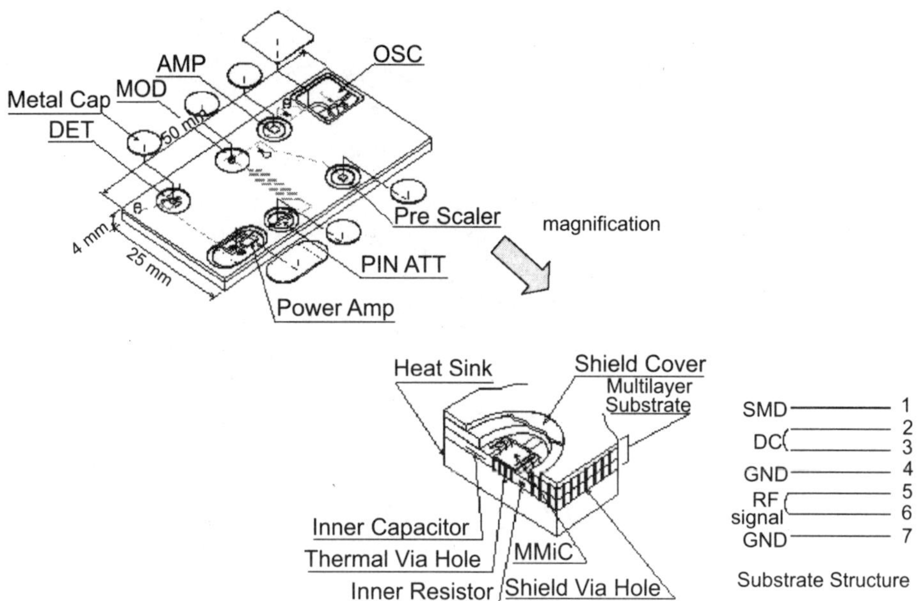

Figure 10.13   Overview of the newly developed microwave multi-chip module (MMCM).

10.2.6.4   *Microwave multi-chip module (MMCM)*[11]

Microwave to milliwave frequency range circuits are mainly used in the fields of national defense and public communications. These circuits are generally expensive and comparatively bulky. On the other hand, the microwave frequency devices used in the fields of DBS and mobile communication are steadily becoming less expensive and more compact, having a great effect on other fields. The microwave frequency devices used in such new areas as wireless LANs and vehicle guidance systems need to be reduced in both size and weight in the same way. Such reductions have been achieved throughout the history of functional progress in electric systems, especially computer systems. The driving force behind the reductions was the continuous evolution of LSI reduction technology, large scale circuit design technology, and the progress made in inspection technology.

It is difficult to achieve large-scale integration to a significant degree even in monolithic microwave ICs (MMICs). Attempts to do so have only lengthened the development period, decreased yield, and increased costs. No effective reduction technologies such as digital LSIs have yet been introduced into the microwave frequency area.

The new packaging technology using a glass–ceramic substrate within the microwave/milliwave frequency effectively solves these problems. It overcomes the defects of MMICs which make it difficult to achieve large-scale integration, produces high-shield performance in the glass–ceramic multilayer substrate itself, and makes it possible to design and produce almost any kind of microwave frequency circuit in a short time.

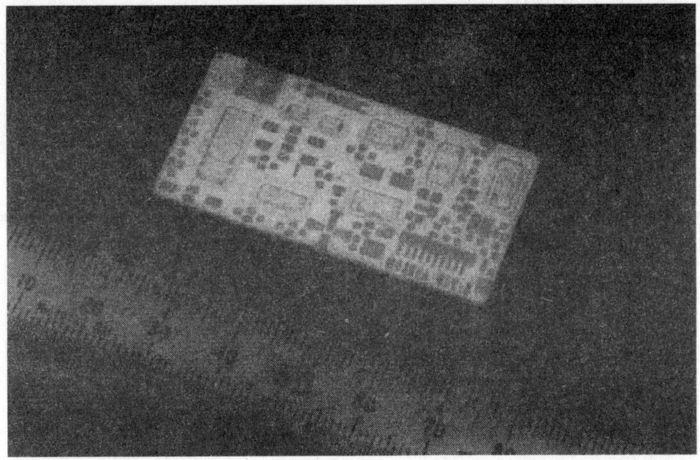

Figure 10.14   Photograph of newly developed MMCM substrate.

Figure 10.13 shows an overview of the newly developed microwave multi-chip module (MMCM). VCO, amplifier, modulator, and several other microwave circuits are set at the bottom of the chip cavities. Furthermore, MMIC chips are directly attached to the Ag electrode at the bottom of the cavity, and are electrically connected to an outer circuit using wire-bonding technology. A number of thermal via holes are set under a few MMIC chips that dissipate a large amount of power. Heat radiation is done via these thermal via holes to effectively reduce heat. The cavities are covered and sealed by metal caps with solder.

In this glass–ceramic substrate, many shield via holes are arranged at regular intervals in all areas except the parts where circuits are set. These via holes shield each circuit from the other circuits in this module and from the external electric field. There are seven conductive layers, the first three of which house the surface mount device (SMD) area and the DC circuit area. Layers five and six house RF signal planes and layers four and seven ground planes. All electric components, which are SMD, are attached to the substrate by using SMT technology.

A photograph of a newly developed MMCM substrate is shown in Figure 10.14. There are several special features in this MMCM. The first is the electric field shielding structure. Shielding in the electric field, which is very important for MMCM is done perfectly on the GND layer in Z direction and is also done in X–Y direction making 2,000 shield via holes 200 μm in diameter with a 500 μm pitch on a 25×50 mm substrate.

The second is the high-wiring density inner layer, especially the RF signal layer. High integration of wiring is effected by a minimum line width of 50 ± 5 μm and minimum line gap 40 μm.

The third is the thermal radiation structure. Advanced integrated GaAs chips dissipate large power and deliver heat up to 30 W/cm^2. In this substrate 500 μm in diameter with 800 μm pitch thermal via holes for the thermal flux path are manufactured directly under the bare chip to solve this thermal radiation problem. As a result, MMCM achieves less than 10°/W thermal resistance. The value is as small as one-fifth of the substrate with no thermal via hole. Table 10.5 shows the typical performance of this MGC substrate.

Table 10.5   Typical performance of MGC substrate for microwave multi-chip module.

Dielectric constant (1 MHz, 10 GHz)	7.1, 7.1
Dissipation factor (1 MHz, 10 GHz)	0.002, 0.005
Insulation resistance (Ω cm, 50 V dc)	$>10^{14}$
Coefficient of thermal expansion ($\times 10^{-7}/°C$)	5.0
Flexural strength (MPa)	280
Linear shrinkage (%)	13.0
Conductor resistance ($\mu\Omega$ cm)	2.5
Inner resistance (7pieces) (Ω)	$50\pm20\%$
Minimum line width/pitch (μm)	50/40
Via-hole dimension (shield via, thermal via) (μm)	Φ 200, ϕ 500
Minimum via-hole pitch (shield via, thermal via) (μm)	500, 800
Thermal resistance (thermal via-hole part) (°C/W)	10
He gas leak (cc/sec)	$<1.0\times10^{-8}$

10.2.6.5   3-Dimensional memory (3-DM) module[12–14,15]

A new stack memory module, which is convenient for high memory density and low cost, has been developed using the glass–ceramic memory chip carrier for the chip size package (CSP). Figure 10.15 shows the structure of the new 3-dimensional memory (3-DM) module.

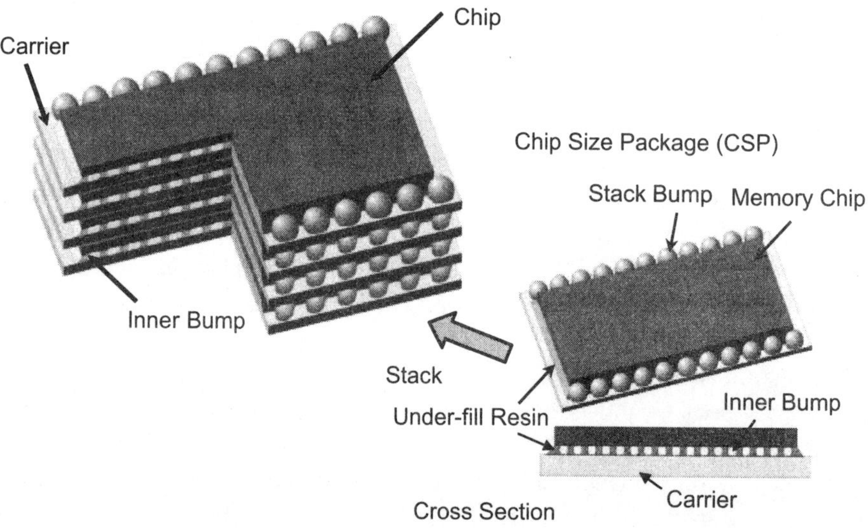

Figure 10.15   Structure of the new 3-dimensional memory (3-DM) module.

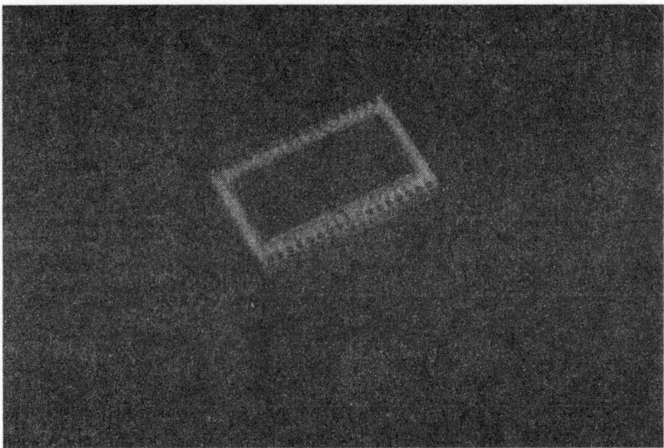

Figure 10.16 Stacked 3-DM module.

The module consists of four 64 Mb DRAM elementary modules connected to the carrier by solder bumps. Each 64 Mb DRAM elementary module consists of a 64 M DRAM and a thin carrier made of the glass–ceramic material.

The glass–ceramic chip carrier has a circuit pattern, interconnection pads and outer-connection pads on both sides. The 64 Mb DRAM is connected to the carrier by a flip chip bonding technique. Epoxy resin is filled into the 0.05 mm gap between the 64 Mb DRAM and the carrier. The upper surface of the bare chip mounted on the carrier is polished. After polishing, the thickness of the mini-module is 0.4 mm. Four mini-modules are stacked by solder bumps. Figure 10.16 shows the stacked 3-DM module. The 64 Mb DRAM module is almost the same size as the current 256 Mb DRAM package (thin small on-line package — TSOP). Therefore, it is possible to make the next generation memory device using current memory devices by this 3-DM module technology.

10.2.6.6 *Glass–ceramic CSP*[16]

Since the late 1990s, the flip-chip bonding technique has been used in many electrical systems to miniaturize the module size. Around the same time, the growing demand for miniaturization of electronic equipment led to the introduction of CSPs, and these packages are now widely used in the production of portable applications, such as camcorders. We have applied glass–ceramic multi-layer wiring substrate technologies, originally developed for high-speed computer systems, to several types of CSPs, and have hereby achieved high packaging density. In a 3-DM module we have developed the glass–ceramic substrate as an interposer, and also includes inner- and outer-connection pads, as shown in Figure 10.17.[14,17] The memory chip is connected to the inner-connection pads via Au bumps by using flip-chip bonding techniques, and the substrate is mounted on the motherboard via solder bumps formed on the outer-connection pads. In this package structure, the CTE mismatch between the memory chip and the motherboard affects package reliability. The CTE of the

Figure 10.17 3-DM module.

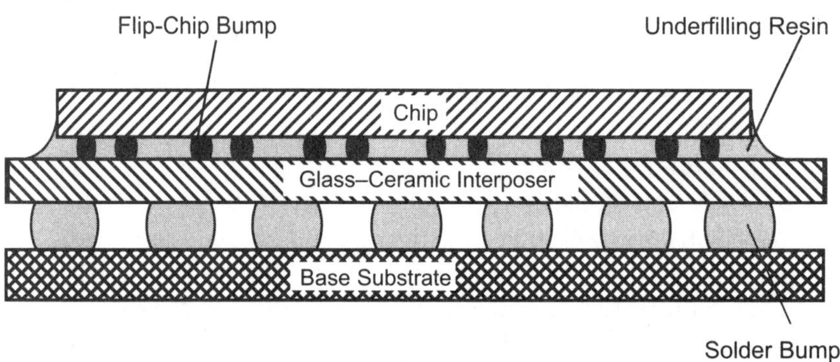

Figure 10.18 Structure of glass–ceramic CSP.

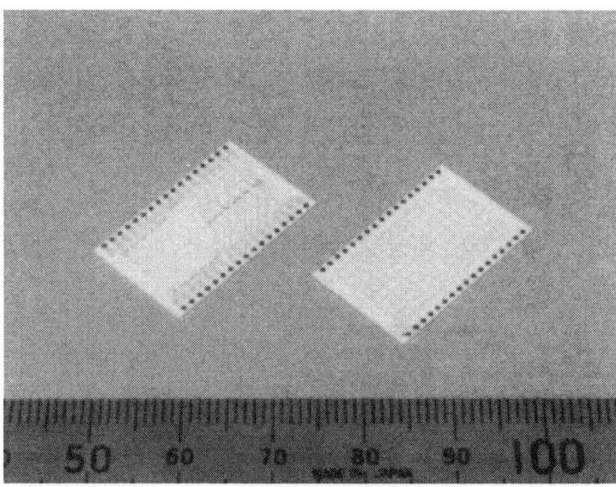

Figure 10.19 Photograph of glass–ceramic substrate.

memory chip and the CTE of the FR-4 are approximately 3 and 16 ppm/°C, respectively, and the developed glass–ceramic has an intermediate CTE of 5 ppm/°C. Accordingly, it is suggested that the glass–ceramic interposer functions as a stress buffer. In order to identify a suitable material for the interposer, we evaluated glass–ceramic CSPs and developed an advanced glass–ceramic CSP. This chapter describes the processing technology, thermal stress simulation, results of thermal cycle testing, and micro-structural analysis of the advanced glass–ceramic CSPs.

Structure of glass–ceramic CSP

The CSP consists of a 64 Mb DRAM chip and a substrate, as shown in Figure 10.18. The 64 Mb DRAM chip is about 19×9 mm, the memory chip is connected to the substrate via Au bumps by using flip-chip bonding technology, and the gap between the memory chip and the substrate is filled with epoxy resin. The substrate is made of glass–ceramics and has a circuit pattern, inner-connection pads and outer-connection pads, as shown in Figure 10.19. Typical properties of the glass–ceramics are shown in Table 10.6.

Table 10.6   Typical properties of glass–ceramics.

Dielectric constant (1 MHz)	7.1
Dissipation factor (1 MHz)	0.002
Insulation resistance	$>10^{14}\,\Omega$ cm
Coefficient of thermal expansion	5 ppm/°C
Flexural strength	280 MPa
Linear strength	$13.0\pm0.3\%$
Conductor resistivity	$2.5-12\,\mu\Omega$ cm

Fabrication process

First, bumps are formed on the inner-connection pads of the 64 Mb DRAM chip. Gold is used for the bumps, which are formed by the ball bump method using gold wire. The chip and the substrate are connected under pressure at 350 to 400°C. Figure 10.20 shows a cross-sectional view of the interconnection between the memory chip and the glass–ceramic substrate. Good flatness of the glass–ceramic substrate is

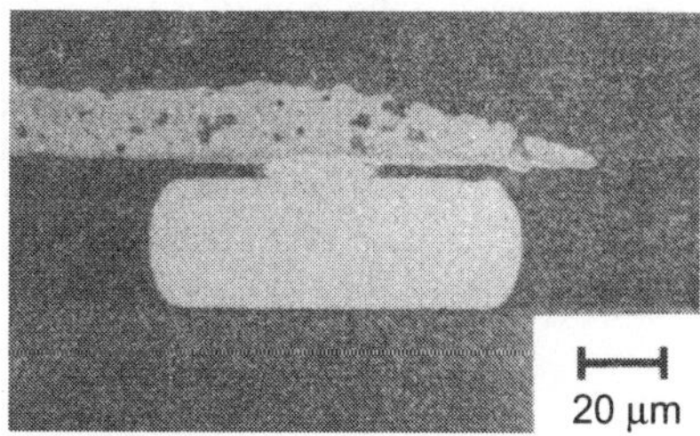

Figure 10.20   Cross-sectional SEM image of a flip-chip bump.

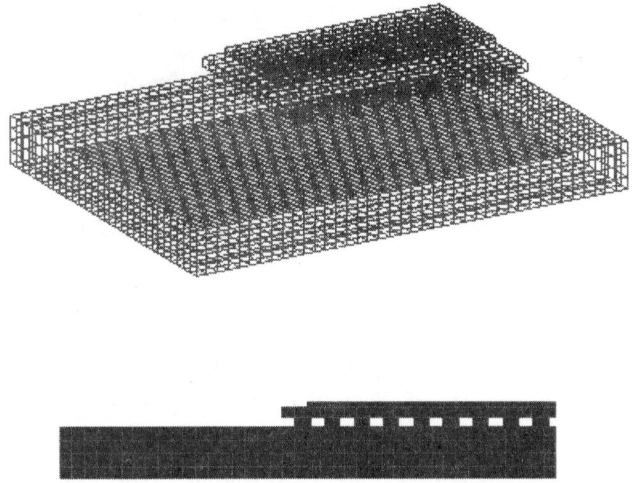

Figure 10.21 Simulation model.

essential for high reliability of the interconnections between the gold bumps and the gold pads on the substrate. The camber of the glass–ceramic substrate is less than 10 μm. After the chip bonding, epoxy resin is used to fill the gap between the memory chip and the substrate. Next, solder bumps are formed by the solder ball method. The solder ball material is Pb/Sn alloy, whose composition is Pb/Sn = 37/63 wt. %, and the glass–ceramic CSP is mounted on a printed wiring board (PWB) substrate at 200 and 240°C.

Thermal stress simulation

In order to estimate the thermal stress on the CSP, several analyses were conducted using the finite element method. Figure 10.21 shows the one-fourth model of the CSP mounted on a PWB base substrate. In this model, the chip size is 19×9 mm and the thickness is 0.2 mm. The substrate size is 23×14 mm and the thickness is 0.4 or 0.8 mm. The material properties used for this simulation are shown in Table 10.7.

Table 10.7 Material table.

	Material	*Young's modulus GPa*	*Poisson's ratio*	*CTE ppm/°C*
Chip	Silicon	130	0.28	3
Under-filling resin	Epoxy	9.2	0.3	25
Interposer				
Glass–ceramics	Glass–ceramics	80	0.22	5.5
Alumina	Alumina	340	0.25	8.7
Solder bump	Solder	20	0.39	25
Base substrate	PWB	18	0.19	16

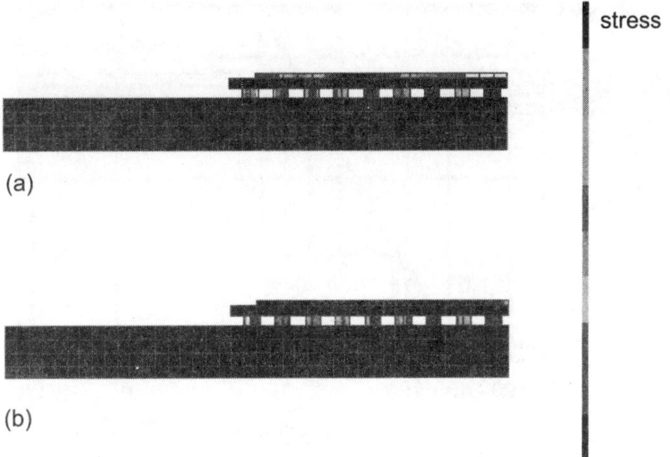

Figure 10.22 Thermal stress on alumina CSP and glass–ceramic CSP.

Figure 10.23 Thermal stress on thin and thick glass–ceramic CSPs.

First, we compared the glass–ceramic CSP and an identically structured alumina CSP. Figure 10.22 shows the results of the thermal stress simulation when both CSPs are cooled to a low temperature from a high temperature. The greatest stress occurred at the corner bump. As compared with the alumina CSP, less stress occurred on the glass–ceramic CSP. The CTE mismatch between the glass–ceramic substrate and PWB is greater than the one between alumina and PWB. The reliability of this structured package seems to be related not only to the close match of the CSP CTE to that of the chip or PWB, but also to the rigidity of the CSP. Next, the substrate thickness was evaluated. Figure 10.23 compared the thickness of glass–ceramic CSPs. The greatest stress occurred at the corner bump. Less stress occurred on the thin CSP. The flexibility of the glass–ceramic thin substrate seems to be effective as a stress buffer. These results indicate the high reliability of thin glass–ceramic CSP.

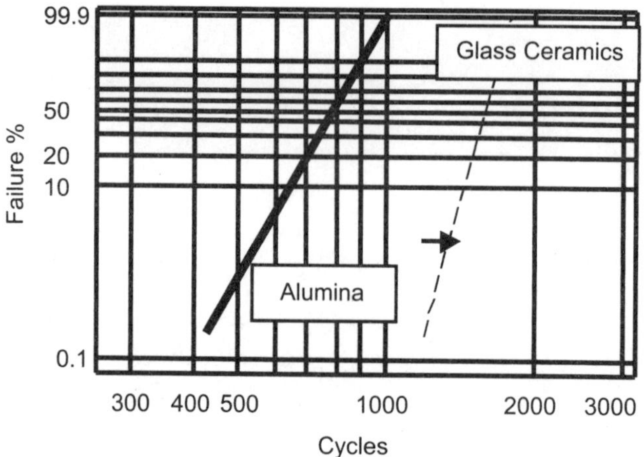

Figure 10.24 Thermal cycle test for glass–ceramic CSPs and alumina CSPs.

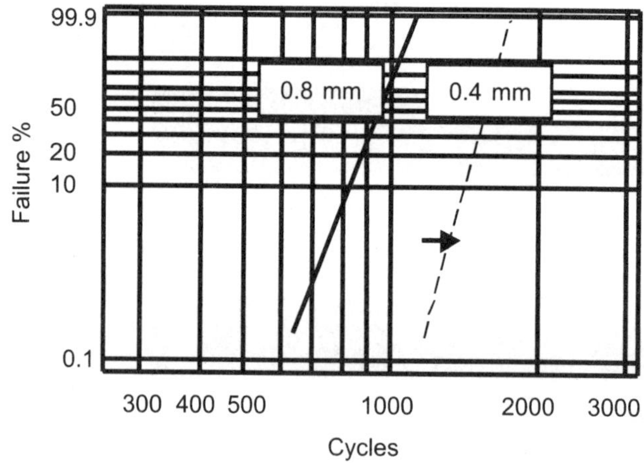

Figure 10.25 Thermal cycle test for thin and thick glass–ceramic CSPs.

Reliability test

The reliability of glass–ceramic CSPs was evaluated in a thermal cycle test in which the temperature ranged from −40 to +125°C, and from −65 to +150°C. Figure 10.24 compares the glass–ceramic package and the alumina package. There were no failures on the glass–ceramic packages before 1000 cycles, although failures were detected on the alumina packages between 500 and 1000 cycles. In this test, the size and the structure of the alumina package are identical to those of the glass–ceramic package, and the thickness of both substrates was about 0.4 mm.

Figure 10.25 compares the thickness of the glass–ceramic packages. The substrate thicknesses of the evaluated packages were 0.4 and 0.8 mm. Thin packages are more reliable than thick ones. These results show that thin glass–ceramic packages have high package reliability on the PWB base substrate.

Figure 10.26   Cross-sectional SEM image of solder bumps on alumina CSPs.

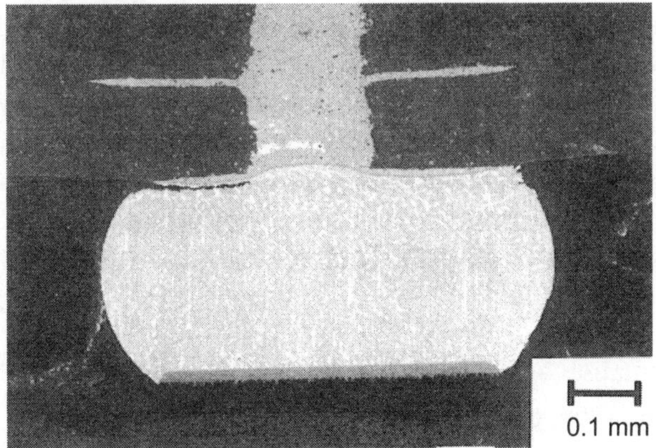

Figure 10.27   Cross-sectional SEM image of solder bumps on thick glass–ceramic CSPs.

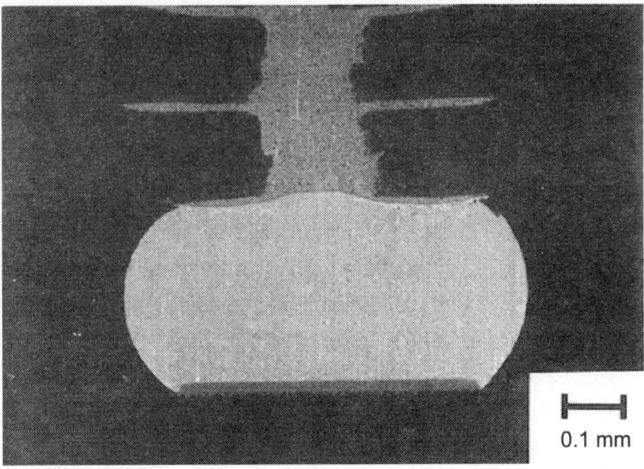

Figure 10.28   Cross-sectional SEM image of solder bumps on thin glass–ceramic CSPs.

Micro-structural analysis

The failures that occurred in the above-mentioned thermal cycle tests were analyzed. Figure 10.26 shows cross-sectional scanning electron microscope (SEM) images of the alumina CSP solder bumps that had undergone 500 cycles of a thermal cycle test. It was confirmed that fatigue fracture occurred in numerous solder ball bumps and several bumps were opened completely at 500 cycles. Cross-sectional SEM images of glass–ceramic CSP solder bumps are shown in Figures 10.27 and 10.28. In the thick package, several cracks were observed, though they caused no failure. On the other hand, no cracks were observed on the thin packages.

As a result, we can confirm that glass–ceramic CSP is the best interposer.

10.2.6.7   *Glass–ceramic package for high frequency*[18]

The radio frequency band from microwaves to millimeter waves allows high-speed, high-volume information transmission, but has previously been used for public communication and defense applications, and the devices for those bands are relatively expensive and large for the most part. Recently, however, a new consumer market has arisen for Bluetooth, wireless LAN, automobile radar systems, and other such applications in high-frequency bands; and there is now a strong demand for low-cost device technology for use in these high-frequency bands.

The issues associated with small, high-performance devices that integrate high-frequency analog circuits in large-scale integrated circuits include those listed below.

1.   Implementation of highly effective electrical shielding
2.   Protection of the GaAs chip
3.   Difficulty of large-scale integration of GaAs chips, integration on a module level

Because these issues are being dealt with one by one, the devices are large and expensive, as mentioned above. An approach that solves these problems together with good efficiency is needed. As such a method, we believe that low-temperature co-fired ceramic (LTCC) technology, which employs a glass–ceramic material, is optimum. First, concerning problem 1, LTCC technology allows vias to be formed at arbitrary positions within a multi-layer substrate; thus, it is possible to use these vias to implement shielding for the various function circuits of a multi-chip module or external shielding for the module. Then, for problem 2, LTCC technology is capable of forming complex cavity structures, thus that function can be implemented at the same time as the wiring board and GaAs chip packaging function. Furthermore, concerning problem 3, it is difficult to implement the kind of large-scale integration that is achieved on silicon by using GaAs alone, because of the nature of analog circuits and the nature of GaAs chips. In addition, the development period for such devices is long. To address this problem, too, it is possible to integrate the internal elements and the antenna; and by placing on the substrate side the functions that

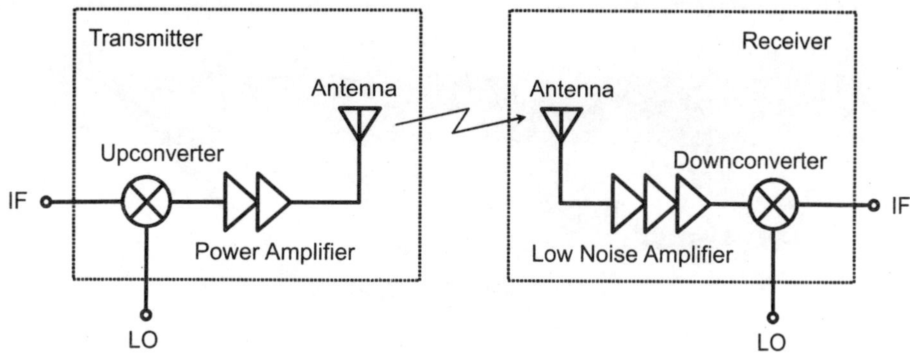

Figure 10.29 60 GHz transmitter and receiver block diagram.

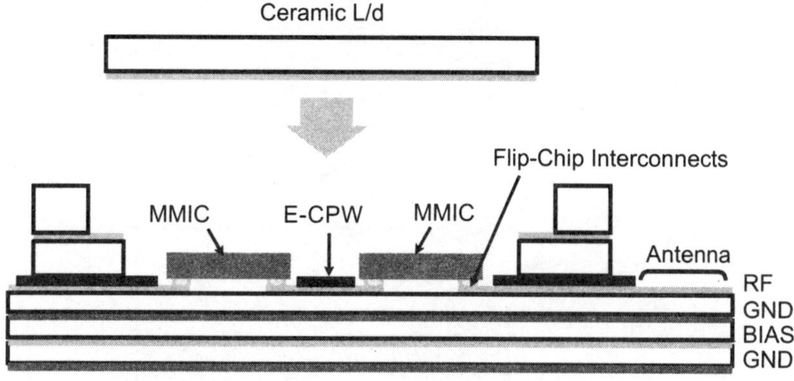

Figure 10.30 Schematic cross-sectional view of MCM.

cannot be integrated on the GaAs, integration on the module level is possible if LTCC technology is used.[19] Flip-chip bonding is necessary in order to avoid the parasitic inductance in the connection between the GaAs chip and the substrate in the 60 GHz band module that we have developed. This, of course, also contributes at the same time to making the module smaller. For the implementation of the flip-chip connection, precise positioning of the contact electrodes and flatness on the substrate side are important.

The development of LTCC technology for high-frequency bands is described in this section.

Module structure

A block diagram of the transmitter (TX) and receiver (RX) is shown in Figure 10.29. The MCM configuration includes an upconverter and power amplifier MMIC on the transmitter side and a downconverter and low-noise amplifier MMIC on the receiver side. The schematic cross-section view of the newly developed module is illustrated in Figure 10.30. The two MMICs are flip-chip bonded to the bottom of the cavity and

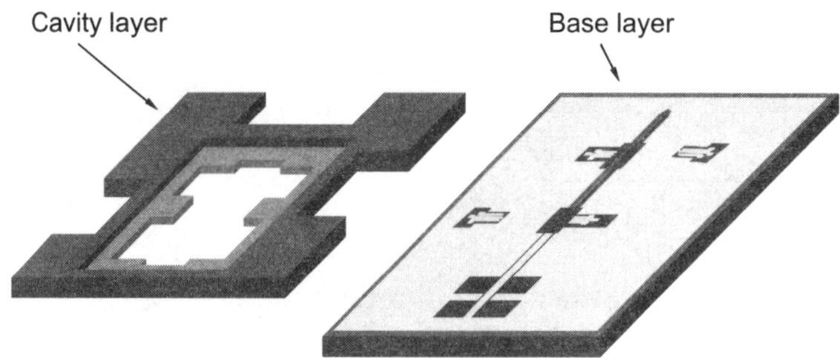

Figure 10.31   Cavity layer and base layer.

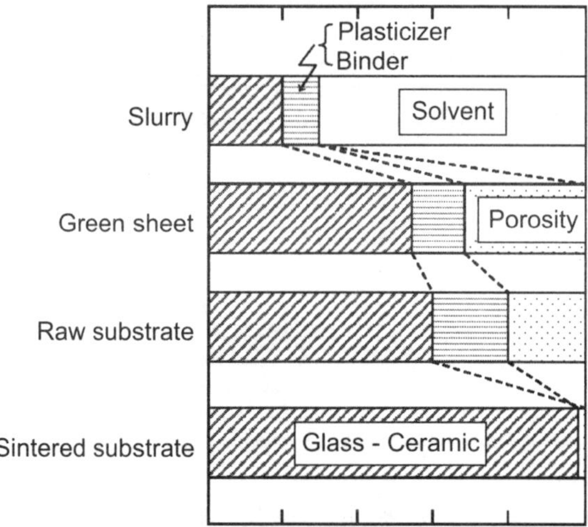

Figure 10.32   Constituent volume proportions in various stages.

connected by means of co-planar wave guide circuits. The between-chip circuits are embedded in the glass–ceramic material, suppressing the loss in this part to a small value. Furthermore, a double-slot antenna is formed by thick-film printing. To prevent the proximity effect between the chip and the substrate, the cavity has a further 80 μm hollow area directly below the chip. Also, to suppress resonances within the cavity, embedded resistors are formed. The configuration of the circuit layer within the MCM is shown in Figure 10.30. After making the chip connections, a metallized ceramic cap is attached to the cavity.

Multi-layer glass–ceramic substrate

What is most important in the flip-chip connection is the flatness and pattern position precision of the substrate. Also, because chips that operate at high frequencies are used, the chips are connected by gold–gold pressure welding, without the use of resins.

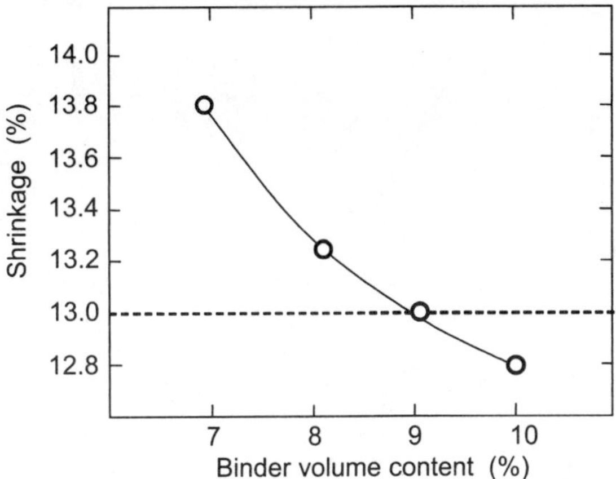

Figure 10.33 Binder volume content of slurry for cavity layer vs. shrinkage.

Board flatness. The bump height after flip-chip connection is about 15 to 20 μm; so when this method of making connections is used, the substrate warping within the chip area must be less than 20 μm. In the case of an MCM, as in the work described here, the warping must be less than 15 μm in multiple chip areas. However, because the shrinkage rate in the cavity area is generally different than that in the other glass–ceramic areas, it is difficult to sinter substrates that have a cavity structure. This substrate is separated into two parts, a cavity layer and a base layer, as shown in Figure 10.31. When these parts are pressed with a static hydraulic pressure of 300 kg/cm^2, an x–y shrinkage rate of the cavity layer is 14.5%, and that of the base layer is 13.0%, respectively. So the x–y shrinkage rate of the cavity layer is 1.5% larger. If the two layers are pressed together and then sintered, a warping of about 100 μm occurs in the bottom of the cavity, making it useless as a substrate for flip-chip connection.

The general methods of controlling the shrinkage rate of green sheet include those listed below:

- Particle diameter
- Pressing pressure
- Amount of binder in the slurry
- Powder composition
- Crystallization additives
- Sintering profile

Of those, selection of the slurry composition as the parameter is desirable for controlling the shrinkage of the base and cavity layer to obtain a smooth substrate.

The volumetric composition for each stage from slurry to substrate is shown in Figure 10.32. The change in shrinkage rate with the green sheet binder composition as the parameter is shown in Figure 10.33. Similarly, the warping of the cavity floor when the shrinkage of the cavity layer was controlled is shown in Figure 10.34.

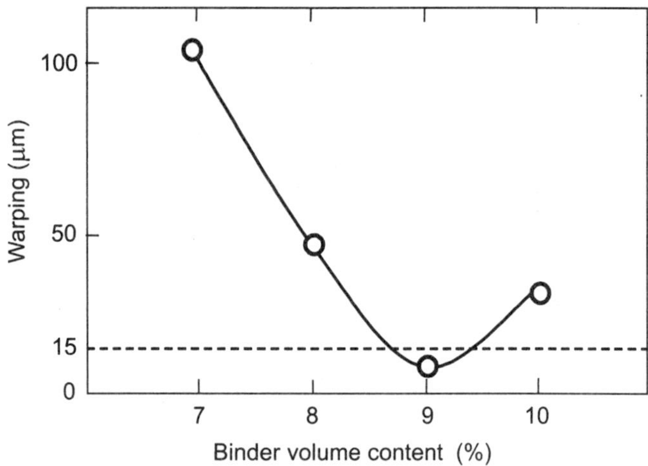

Figure 10.34 Binder volume content of slurry for cavity layer vs. warping of base layer.

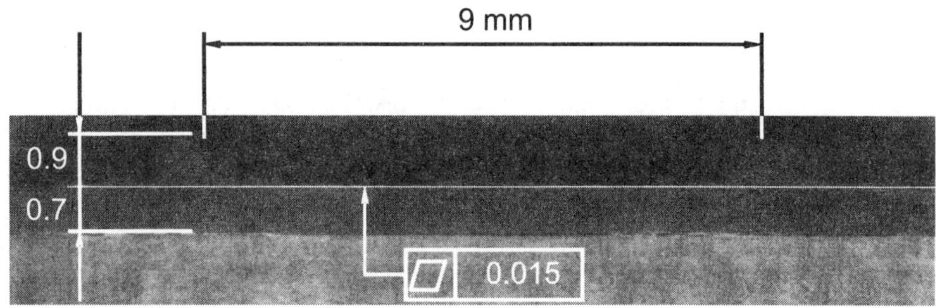

Figure 10.35 Substrate cross-section.

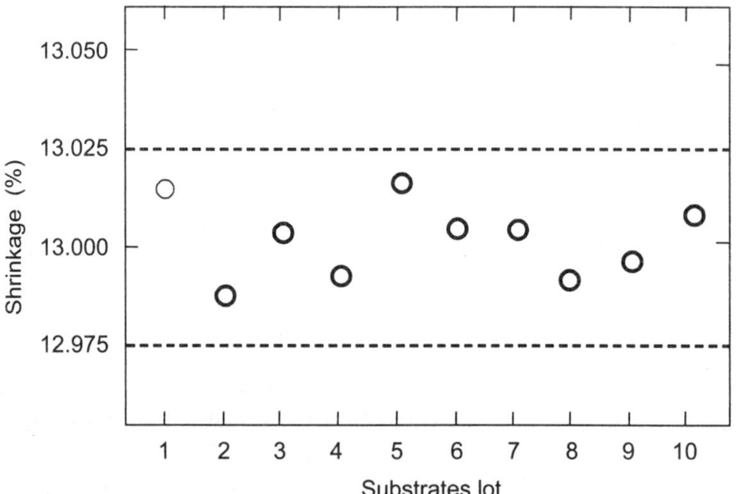

Figure 10.36 Shrinkage for substrates.

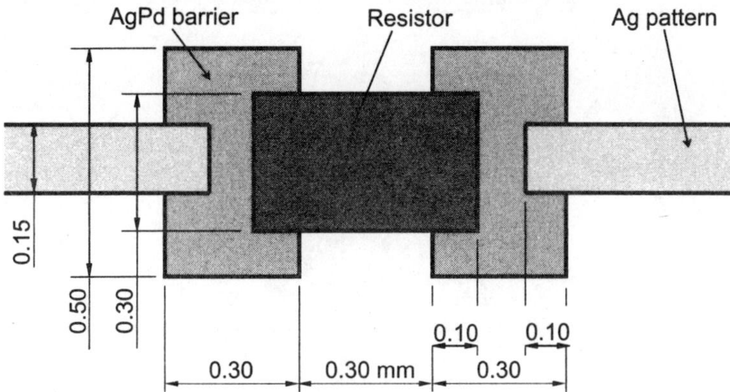

Figure 10.37   Schematic figure of embedded resistor.

These results show that a substrate that is suitable for flip-chip connection can be obtained by controlling the binder composition. The substrate cross section is shown in Figure 10.35.

Shrinkage rate. The shrinkage data for a substrate fabricated by the method described above is presented in Figure 10.36. Dispersion in the shrinkage rate data of 0.05%, or less than 50 μm at 100 mm was achieved.

Embedded resistors. This substrate has built-in resistors to prevent oscillations and resonances within the cavity. In the formation of the internal resistances, a Ag/Pd = 80/20 barrier layer for preventing diffusion is formed between the silver wiring and resistance elements with the structure shown in Figure 10.37. This measure suppresses the dispersion in the embedded resistances to ±20%.

Flip-chip connections

The flip-chip connection conditions are summarized in Table 10.8. The relation between the non-electrolytic (chemical) gold plating thickness and the shear strength of the chip is shown in Figure 10.38. If the gold plating thickness is at least 0.5 μm, the connection is sufficiently strong. A photograph of the connection formed on the chip is shown in Figure 10.39.

Table 10.8   Flip-chip bonding condition.

Temp. (chip side)	250°C
Temp. (substrate side)	300°C
Compression force	80 gf/bump
Time	30 sec

Module

Embedded co-planar circuitry. It is rather difficult to form 60 GHz band circuits with the mass production pattern accuracy of thick-film printing, but in this module, the

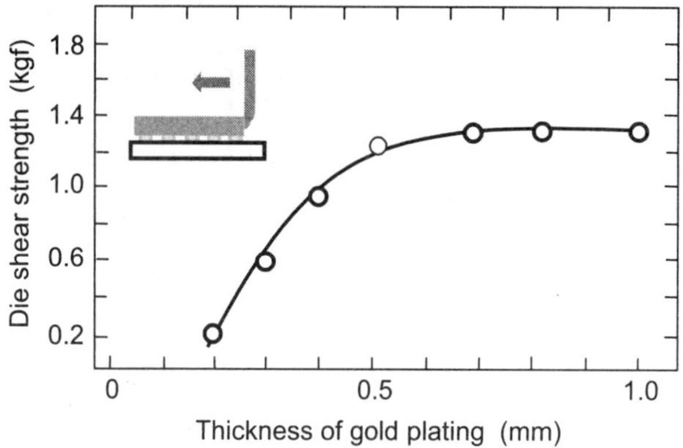

Figure 10.38 Thickness of gold plating vs. die shear strength.

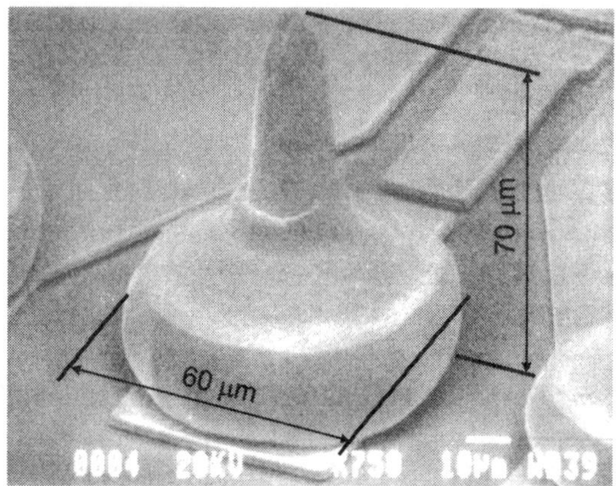

Figure 10.39 Figure of flip-chip bump.

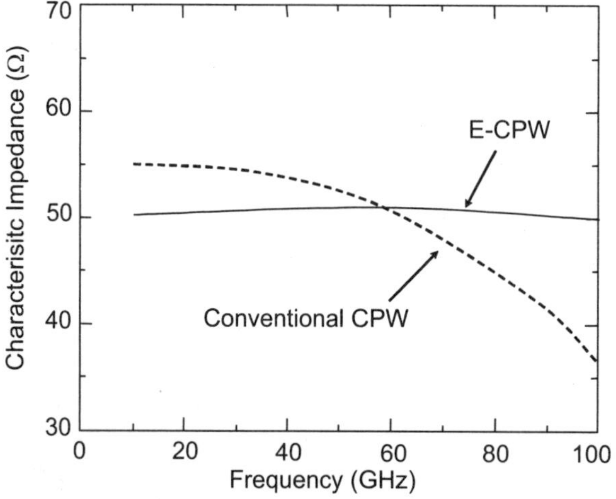

Figure 10.40 Characteristic impedance of embedded co-planar wave guide.

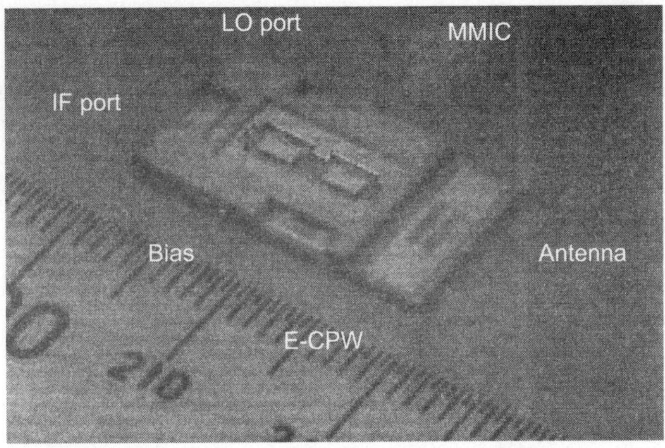

Figure 10.41   Photograph of antenna-integrated MCM substrate.

printing rules are relaxed by covering the co-planar circuits that make the connections between the chips with an insulation layer so that the circuits are internal to the layer. The result is that the calculated value of the characteristic impedance is nearly constant up to the 100 GHz band. The characteristic of the conventional co-planar circuits, on the other hand, depends greatly on the frequency. This result is shown in Figure 10.40. The insertion loss of these embedded circuits is 0.22 dB/mm at 60 GHz.

Module characteristics. After mounting on the substrate, the power amplifier, which consists of FETs, has an output of 12.2 dBm (16.6 mW) with a compression point of 1 dB. This corresponds to a transmission power of 10 mW or less, which has been confirmed to be sufficient for indoor wireless communication systems. The MCM figure is shown in Figure 10.41. The characteristics of the substrate and module are summarized in Table 10.9.[20]

The characteristics of the MMICs mounted on this substrate have been demonstrated to be sufficient for high-speed, high-volume indoor wireless communication. We believe that this technology is applicable to IEEE 1394 home net transceivers and other such equipment.

10.3   Advanced glass–ceramic package

10.3.1   *Materials and fabrication process*

MCM-D/C (a multi-chip module composed of deposited organic thin-film on ceramic substrate) technology[17,21] is an extremely promising approach to achieve high-wiring density, low dielectric constant, low dissipation factor, and high-flexural strength for high-speed and high-frequency multi-chip modules. A new MCM-D/C substrate has been developed for a high performance multi-chip module using the MGC substrate and Cu/photosensitive benzocyclobuten (BCB) thin-film technology. Table 10.10 shows typical properties of the photosensitive BCB used for the deposited insulator layer.

Table 10.9 Properties summary.

Substrate	Dielectric constant (at 60 GHz)	7.1
	Coefficient of thermal expansion ($\times 10^{-7}$/°C)	5.0
	Linear shrinkage (%)	13.0
	Inner resistor (Ω)	50±20%
	Minimum line width /pitch (μm)	80 μm / 80 μm
MCM	Frequency	50–60 GHz
	Output power (TX MCM)	10 mW (max)
	Conversion gain (RX MCM)	10 dB
PA	Output power (P 1 dB)	12 dBm
	Line gain	12 dB
LNA	Gain	18 dB
	Noise figure	5 dB
Upconverter	Conversion gain	−5 dB@IF frequency = 1 GHz
Downconverter	Conversion gain	−8 dB@IF frequency = 1 GHz
Antenna	Gain	5 dBi
	Port return loss	10 dB
E-CPW	Transmission loss	0.22 dB/mm@60 GHz
	Interconnect loss	0.9 dB (between MMICs)

Table 10.10 Typical properties of the photosensitive benzocyclobuten (BCB).

Photosensitive dielectric materials	*BCB*	*Polyimide*	*Epoxy*
Dielectric constant (1 MHz)	2.7	3.3	3.5
Dissipation factor (1 MHz)	0.08	0.3	–
Thermal expansion coefficient (ppm/°C)	60	40–50	69
Water absorption (%) (24 hours boiling water)	0.25	3.0	–
Glass transition temperature (°C)	>350	355	117
Curing temperature (°C) (in N_2 gas)	<250	>400	<150

This material shows low dielectric constant of 2.7 and low water absorption compared with polyimide and epoxy resin. Figure 10.42 shows the cross-section view of MCM-D/C substrate.

Cu/BCB layers are composed with thin-film technology, and so it is possible to achieve very high-wiring density on the glass–ceramic substrate. A fabrication process for a photosensitive-BCB thin-film multilayer is summarized in Figure 10.43.

After the MGC substrate surface is cleaned, BCB is spin-coated and baked. The exposed films are developed with solvent by the puddle method. The photosensitive BCB is cured at around 210°C. The fine-line fabrication process is summarized in Figure 10.44. After cleaning the substrate, a Cu thin-film is sputter-deposited. After

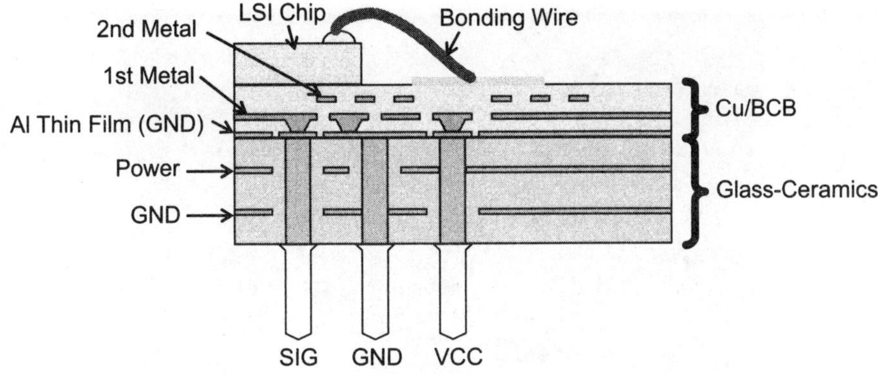

Figure 10.42 Cross-section view of MCM-D/C substrate.

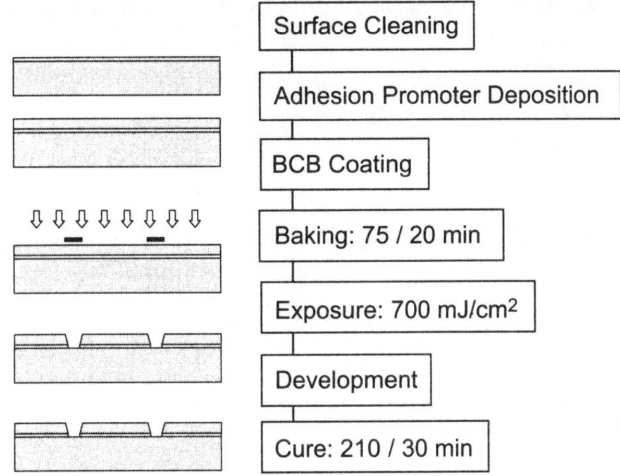

Figure 10.43 Fabrication process for a photosensitive BCB thin-film multilayer.

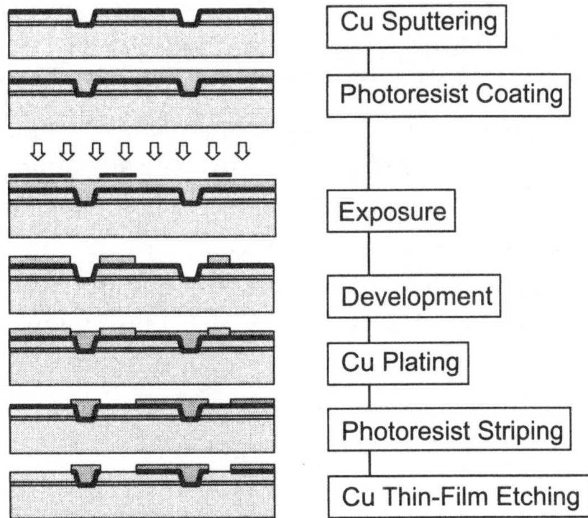

Figure 10.44 Fabrication process for fine line.

Figure 10.45 High-density R4400 RISC module fabricated with the MCM D/C based on the Cu/BCB thin-film technology.

an ordinary photoresist is spin-coated, the resist patterns are formed by standard photolithography process (exposure and development). A Cu conductor is formed by electroplating technology, and the photoresist is stripped off. Finally, the sputter-deposited layer is removed by Cu thin-film etching.

10.3.2 Application

10.3.2.1 High-speed RISC module[17]

Figure 10.45 shows a high-density R4400 RISC module fabricated with the MCM D/C based on the Cu/BCB thin-film technology. Table 10.11 summarizes the specification for an R4400 RISC module. Table 10.12 summarizes MCM-D/C substrate performance for the prototype. The prototype module is 65×65 mm and has 179 I/O pins for signal interconnection and power supply. The signal line is 40 μm wide and has a pitch of 108 μm. The two kinds of 40 μm via holes are formed in BCB layer; the one for the connecting upper and lower signal lines and the other for conducting the heat generated in an LSI chip to the thermal pin in the MGC substrate. A CPU and ten SRAMs are mounted to the MCM-D/C substrate and interconnected by wire bonding.

10.3.2.2 Chip carrier for chip size package (CSP)[17]

Figure 10.46 shows the cross-sectional structure of the newly developed glass–ceramic chip carrier for CSP. The fabrication process of the chip carrier is the same as for the MCM-D/C substrate.

Table 10.11   Specification for R4400 RISC module.

Module size		65×65 mm
Number of chips	CPU (bonding pad count)	1 (455)
	SRAM (bonding pad count)	10 (44)
	Chip capacitors	13
	Chip resistors	4
Conductive layers	BCB thin-film layers: top	1
	BCB thin-film layers: signal	2
	Glass–ceramic substrate: power	2
	Glass–ceramic substrate: GND	2
I/O pin	Number	179
	Pitch	2.54 mm
Bounding strength (WB)		7 g
Characteristic impedance		50 Ω
Delay time		55 ps/cm
Clock rate		75 MHz

Table 10.12   MCM-D/C substrate performance for the prototype.

Thin film layer (BCB)		
	Line width	40 μm
	Via hole	40 μm
	Line pitch	108 μm
	Number of layers	3(15/5/10 μm)
	Conductor	Copper
	Resistivity of conductor	1.8 μΩ cm
Base substrate (glass–ceramic substrate)		
	Via hole	0.15 mm
	Via hole pitch	0.423 mm
	Number of conductive layers	4
	Green sheet layers	14

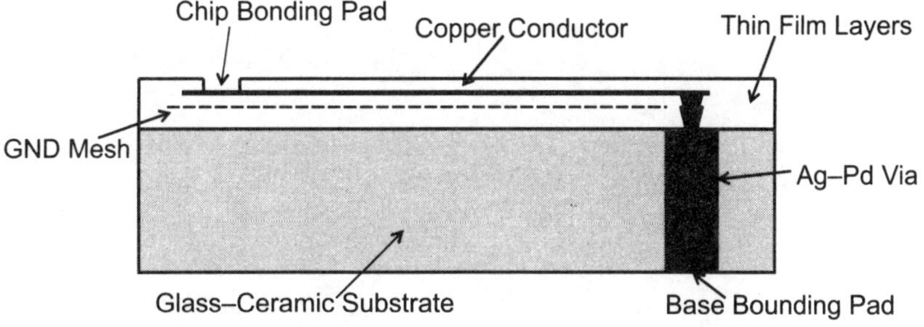

Figure 10.46 The cross-sectional structure of the newly developed glass–ceramic chip carrier for chip size package (CSP).

A mesh ground is used for the micro-strip line structure in deposited thin-film multi-layers of a chip carrier in order to obtain reliable adhesion between layers. Table 10.13 indicates the wiring rule and dimension of the high-density wiring chip

Table 10.13 Wiring rule and dimension of the high-density wiring chip carrier.

Deposit layer		
	Signal line width	25 μm
	Signal line width pitch	50 μm
	Via hole	60 μm
	Insulation separation	12 μm
	Chip bonding pad	80 μm φ
	Bonding pad pitch	108 μm
Base substrate (glass–ceramic substrate)		
	Base bonding pad (Via)	300 μm φ
	Bonding pad pitch	500 μm φ
	Number of I/O pads	525
	Size	14×11.3×0.64 mm
	Camber	<10 μm/10 mm

carrier. Figure 10.47 shows the photograph of a conventional QFP, and developed CSP using the newly developed glass–ceramic chip carrier for the same RISC chip. The developed CSP shows very small size compared with conventional QFP.

10.4 Summary

VLSI packaging modules for high-speed computer systems and optical communication systems require new technology with high-propagation speed and high-wiring density. The MGC substrate can meet these requirements, and is advantageous for bare-chip packaging because of its mechanical and thermal characteristics.

This substrate is applied to some modules for computer and communication systems, and it has become possible to achieve high system performance. The low

Figure 10.47 Photograph of a conventional QFP, and developed CSP using the newly developed glass–ceramic chip carrier for the same RISC chip.

temperature co-fired multilayer glass–ceramic packaging has the following outstanding characteristics:

1. Glass–ceramic materials with low dielectric constant have been developed for high propagation speed.

2. Low electrical resistivity can be realized using Ag–Pd or Au on the MGC package. It is advantageous for achieving a low voltage drop.

3. The MGC substrate has a high-flexural strength, because anorthite crystal plays a reinforcing role in the glass–ceramic body.

4. The MGC package can support high-wiring density in a large-scale area through use of highly accurate processes with shrinkage control technology.

5. An MGC multi-chip package with polyimide layers can support high-density wiring and high-speed signal transmission for high-speed computer systems.

6. The MGC substrate is advantageous in bare-chip MCM packaging and miniaturization for optical communication module.

7. The following new multilayer glass–ceramic substrates were developed for high-performance multi-chip modules packaging:

    *   Multilayer glass–ceramic substrate for the microwave multi-chip module
    *   MCM-D/C substrate for the high-speed RISC module
    *   Multilayer glass–ceramic chip carrier for the CSP
    *   3-DM module

References

1. T. Watari, "Computer packaging technology for system performance", *NEC Res. Devel.*, No. 98, pp. 49–59, 1990.
2. T. Watari, "The NEC SX supercomputer technology", Proc. 1988 IEEE Int. Conf. Comput. Design, pp. 54–57, Nov. 1986.
3. K. Kata, A. Sasaki, Y. Shimada and K. Utsumi, "New fabrication technology of dielectric permittivity multilayer ceramic substrate", Proc. ISHM'90, pp. 308–315, 1990.
4. Y. Shimada, Y. Kobayashi, K. Kata, M. Kurano and H. Takamizawa, "Large scale multilayer glass–ceramic substrate for supercomputer", IEEE Trans. Comp., Hybrids, Manuf. Technol., Vol. 13, No. 4, pp. 751–758, 1990.
5. A. Dohya, T. Watari, and H. Nishimori, "Packaging technology for the NEC SX-3/SX-supercomputer", Proc. 40th Electric Components & Technol. Conf., pp. 525–533, 1990.
6. Y. Shimada, A. Dohya, K. Kata, and J. Inasaka, "High speed computer packaging with multiplayer glass–ceramic substrate", Proc. ISHM'91, pp. 176–182, 1991.
7. Y. Uda, T. Shin'e, and K. Takahashi, "High speed optical fiber data-link transmitter/receiver using 1.3 LED and PIN-PD", Proc. EFOC-LAN '87, pp. 246–250, 1987.
8. Y. Shimada, T. Shin'e, Y. Uda, and R. Nagaoka, "Application of multilayer ceramic components for fiber-optic transmitter/receiver", *NEC Res. Devel.*, Vol. 32, No. 2, pp. 195–206, April 1991.
9. H. Tanaka, R. Nagaoka, T. Makabe, F. Yoshimura, Y. Kobayashi, T. Tamura, Y. Suzuki, and F. Suzuki, "Multi-chip module for 156MB/S optical interface", Proceedings of 1993 Japan IEMT Symposium, pp. 81–84, 1993.

10. H. Tanaka, R. Nagaoka, T. Makabe, F. Yoshimura, Y. Kobayashi, T. Tamura, Y. Suzuki, and F. Suzuki, "Multi-chip module for 156MB/S optical interface", Proceedings of 1993 ICEMM Symposium, pp. 247–251, 1993.
11. K. Ikuina, M. Kimura, and K. Utsumi, "Glass–ceramic multi-chip module for satellite microwave communications systems", Proceedings of 1995 International Conference on Multi-chip Modules, pp. 483–488, 1995.
12. N. Senba, Y. Shimada, K. Tokuno, and I. Morizaki, "Stack memory module", Proceedings of the Electronics Society Conference of IEICE, [2]17, 1995.
13. N. Takahashi, N. Senba, Y. Shimada, I. Morizaki, and K. Tokuno, "3-Dimensional memory module", Proceedings of 46th Electronic Components and Technology Conference, pp. 113–118, 1996.
14. N. Senba, N. Takahashi, Y. Shimada, I. Morizaki, K. Tokuno, M. Ohtsuka, and K. Hashimoto, "Application of 3-Dimensional Memory Module", ISHM'96 Proceedings, pp. 279–284, 1996.
15. N.Takahashi, N. Senba, Y. Shimada, I. Morisaki, and K. Tokuno, "Three-dimensional memory module", *IEEE Trans. CPMT, Part B*, 21, No. 1, pp. 15–19, 1998.
16. I. Hazeyama et al., "Flip chip bonding reliability of advanced glass ceramic chip size package", 1998 IEMT/IMC Proceedings, pp. 210, 1998.
17. A. Shibuya, I. Hazeyama, T. Shimoto, N. Takahashi, N. Senba, M. Kimura, Y. Shimada, H. Matsuzawa, and F. Mori, "New MCM composed of D/L base substrate, high-density-wiring CSP and 3D memory modules", Proceedings of 1997 Electronic Components and Technology Conference, pp. 491–496, 1997.
18. K. Ikuina et al., "Glass Ceramic Module for 60GHz-band Wireless Communication Systems", Proceedings of 2001 IMAPS, 2001.
19. K. Ikuina et al, "Glass–ceramic multi-chip module for satellite microwave communication systems", *Int. J. Microcircuits Electron. Packag.*, pp. 431 1995.
20. K. Maruhashi et al., "Low-cost 60 GHz-band antenna-integrated transmitter/receiver modules utilizing multi-layer low-temperature co-fired ceramic technology" 2000 International Solid State Circuits Conference, CA. Feb. 2000.
21. K. Matsui, T. Shimoto, Y. Shimada, and K. Utsumi, "MCM-D/L Technology based on Cu/benzocyclobuten thin film multilayer structure", ISHM'95 Proceedings, pp. 396–401, 1995.

11 Electrochemical processes in the fabrication of multichip modules

Sol Krongelb, Lubomyr T. Romankiw, Eric D. Perfecto, and Keith K.H. Wong

11.1 Introduction

Multichip modules (MCMs) are the building blocks of the computer which provide the transition between the printed circuit board and the silicon chips containing the semiconductor switching devices. These structures must meet a variety of stringent electrical, thermal, and reliability requirements dictated by the system design. Current computer designs call for progressively higher levels of semiconductor integration and increasing numbers of input/output (I/O) pins which in turn demand high wiring density on the MCM.[1] These high wiring densities are obtained by utilizing cofired multilayer ceramic substrates. The typical combinations are Mo conductors with Al_2O_3 ceramic or Cu conductors with cordierite ceramic. The metal conductor and ceramic combination is dictated by sintering considerations and is outside the scope of this chapter. The addition of thin-film layers on top of the multilayer ceramic generally reduces the total number of layers required in the ceramic substrates and also leads to improved pulse transmission between chips. Electrical modeling studies have shown that Cu/polyimide multilevel thin-film (MLTF) structures deposited on cofired ceramics provide the required characteristic impedance of 40 to 50 ohms and necessary short signal propagation delays.[2-4]

The performance requirements of recent, first level package designs demand a lower line resistance and lower delta-I noise than previous versions. Reduced line resistance implies an increase in Cu line cross-sectional area; noise reduction, on the other hand, requires wide spaces between the lines as well as a low dielectric constant. The wiring pitch shrinkage has required a switch from subtractive etch technology to pattern electroplating; the reduction in via dimensions was made possible by moving from photosensitive polyimide to laser ablation of the dielectric.[5,6] An example of the most demanding requirements which had to be addressed to date in fabricating multichip modules may be seen in the IBM eServer z900. This server was introduced in December 2000 and used the densest electronic package in the industry. Tables 11.1 and 11.2 summarize the physical attributes and thin-film ground rules of the MCM in the IBM eServer z900.

The advantage of pattern electroplating in meeting package fabrication requirements relates to the fact that the interconnections which carry the pulses between

Table 11.1   Parameters of the MCM modules in the IBM eServer z900.

	z900 Server
Ceramic	
Material	Cordierite/glass
Dielectric constant	5
Wiring metal	Copper
Size (x/y in mm)	127.5
Number of layers	101
Number of plane pairs	29
Thin films (BSM)	
Number of layers	6
Number of plane pairs	1
Thin films (BSM)	
I/O pins	4,224
TF pad size (mm)	1.35
Tightest pitch (mm)	1.64
Process	"Polycushion" structure

Note: Redistribution wiring only.

chips in a high-performance multichip module are essentially low-loss transmission lines with precisely controlled propagation characteristics. This means that the metal thickness and line widths, as well as the thickness of the dielectric, must be precisely controlled. Electrochemical technology combined with optical lithography, which was first used in the fabrication of magnetic thin-film heads,[7-10] provides an effective way to achieve the precise control needed for the conductor dimensions in the most advanced package structures. However, cost is always a factor in the choice of a manufacturing process, and the tooling for electrochemical processes tends to be somewhat more costly than that for subtractive chemical etching. Furthermore, a viable manufacturing process can only be achieved by integrating the electrochemical techniques with traditional wet and dry technologies, so that the final process selection must take into account interactions between the individual process steps.

This chapter provides the background for incorporating electrochemical processes into multichip module fabrication. Much of the material is drawn from work at IBM in producing multichip modules. Section 11.2 describes recent requirements and structural designs for multichip modules. Section 11.3 outlines the full range of technologies which have been used to fabricate conductors in packaging structures and then focuses on issues relating to the implementation of electrochemical processes for this application. The effective use of electrochemical technology in packaging applications is illustrated in Sections 11.4 and 11.5. Section 11.4 deals with issues which must be considered when integrating electrochemical technology with traditional wet and dry processes to form the thin film wiring layers on a sintered ceramic multichip module. Section 11.5 presents those aspects of the IBM eServer

Table 11.2   Thin-film wiring parameters within the MCM modules in the IBM eServer z900.

	z900 Server
Wiring pitch (μm)	33
Min. line width (μm)	16
Min. spacing (μm)	17
Metal thickness (μm)	6
Wiring process	Pattern electroplating
Wiring metal	Cu
Top surface metal	Cu/Ni/Au
Polyimide	BPDA-PDA
Thickness (μm)	9
Via diameter (μm)	20
Patterning	Laser
Dielectric const.	3.1
TCE (ppm/C)	5
Water uptake (%)	0.5
Modulus (kpsi)	1,280

z900 package which use plate-up techniques to form thin-film wiring conductors on the ceramic module.

11.2   MCM structures, process issues, and design trade-offs

Processes used for fabricating thin-film multilayers on ceramic or organic carriers entail a sequence of steps which are repeated to fabricate each of the desired number of levels of metal. The key to low-cost fabrication lies in

1. Minimizing the number of processing steps for wiring and terminal metal layers
2. Selecting robust processes to achieve high yields

Wiring layers can be fabricated in a variety of ways as outlined in Section 11.3. The fabrication process may or may not include steps to achieve a planarized surface upon completion of each level, and the structure may accordingly be classified as planar or non-planar.

In a multilevel package metal vias provide the connections between conductor levels. As shown in Figure 11.1, via-to-via connections may be stacked or staggered. Typically, stacked vias require planarization while staggered or staircase vias can be used in either planar or non-planar structures. Stacked via designs tend to have lower noise than staggered via structures and thus contribute to better overall system performance. Additionally, stacked vias provide paths of lower thermal and electrical

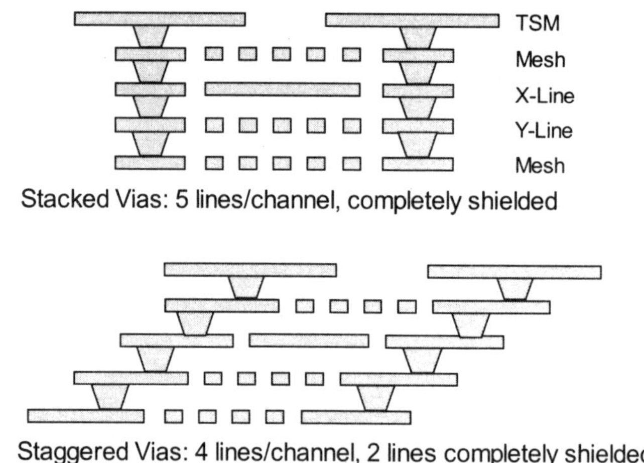

Figure 11.1 Triplate cross section of stacked vs. staggered vias. (From E.D. Perfecto et al., "Increased Thin Film Density by Stacked Vias," *IEEE 52nd Electronic Components and Technology Conference,* © 2002 IEEE. Reprinted with permission.)

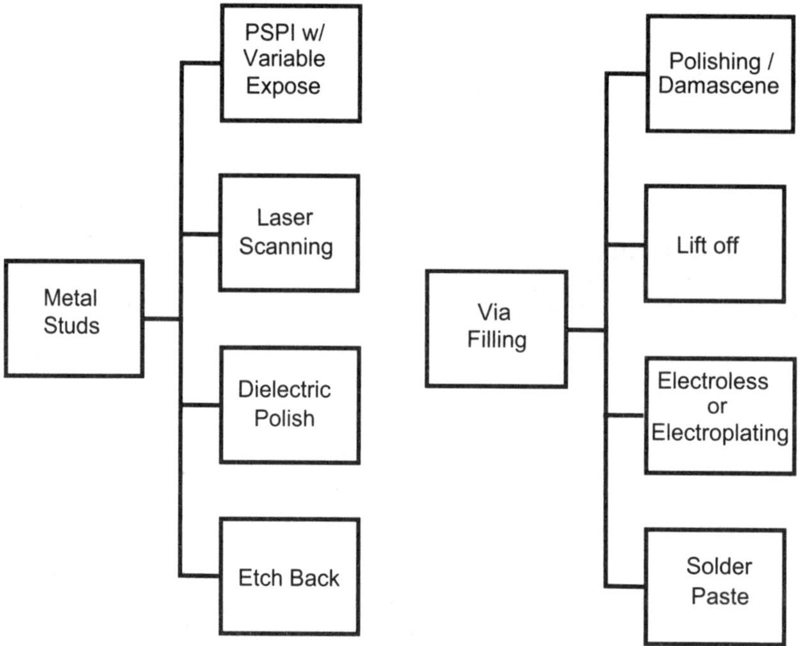

Figure 11.2 Processes for stacked via fabrication. (From E.D. Perfecto et al., "Increased Thin Film Density by Stacked Vias," *IEEE 52nd Electronic Components and Technology Conference,* © 2002 IEEE. Reprinted with permission.)

resistance between the substrate and the chip. The designer has to weigh the trade-offs between the higher cost and better electrical characteristics of planar structures against the lower fabrication cost but poorer performance parameters of the non-planar design.

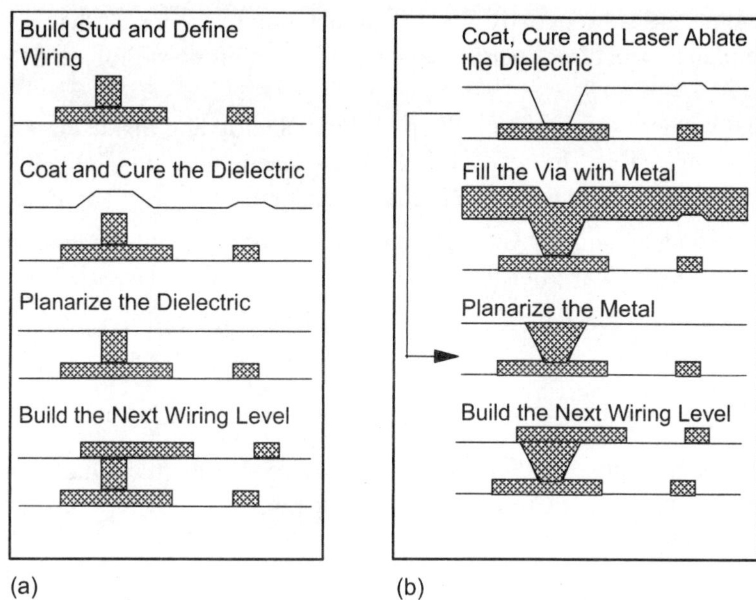

Figure 11.3 Process flow for (a) metal stud process and (b) via fill process. (From E.D. Perfecto et al., "Increased Thin Film Density by Stacked Vias," *IEEE 52nd Electronic Components and Technology Conference,* © 2002 IEEE. Reprinted with permission.)

11.2.1 *Via processes for planar structures*

Planar structures are being used in both MCM-D and MCM-L buildup. In a planar structure, stacked vias can be categorized into two process types: metal studs, and via fill. Figure 11.2 shows the various fabrication options that have been described in the literature. Metal stud processes, also known as metallic via post or via pillars, are described in Figure 11.3a. In this approach, a metal pillar is first fabricated on top of the bottom capture pad by pattern electroplating followed by dielectric polymer deposition and subsequent planarization of the dielectric to expose the top of the metal stud. In the metal stud process, planarization of the dielectric is achieved by one of the following methods:

1. Using photosensitive dielectric along the via area so that upon development, the dielectric hump on top of the metal stud is eliminated[11,12]
2. Laser scanning of the dielectric film to provide a planar structure[13]
3. Polishing of the dielectric to expose the metal studs[14–17]
4. Depositing multiple coats of dielectric (or lamination of the dielectric) to produce a planar dielectric surface and etching it back to expose the metal studs[18]

Of these methods, dielectric polishing is the most commonly used.

In the via fill approach the via openings in the dielectric film are defined first.

The openings are subsequently filled with metal up to the top of the dielectric, producing a planar structure. Four processes have been proposed for via filling:

1. Blanket deposition of Cu on top of the dielectric and inside the via, as shown in Figure 11.3b, followed by chemical-mechanical polishing (CMP) of the excess Cu back to the dielectric surface[19,20]

2. Blanket deposition of Cu using a lift-off mask on top of the dielectric to leave Cu inside the vias after lift off[21]

3. Formation of metal inside the via by electroless or electrolytic plating,[22,23] and

4. Copper/epoxy paste, solder or conductive paste fill such as of a laminate sheet[24,25]

Of these methods, blanket deposition followed by chemical/mechanical polishing to remove the excess Cu metal is most popular.

11.2.2 Non-planar structures

Non-planar thin-film structures are more cost-effective than planar structures since they avoid the planarizing steps and can be fabricated by either subtractive etching or pattern electroplating.[26,27] However, critical process steps must be optimized for the desired metal thickness to effectively control line widths and spacing between lines. These processes are shown schematically in Figure 11.4.

11.2.3 Stacked vias on non-planar structures

Non-planar DLM stacked vias have been demonstrated successfully at IBM.[28] Figure 11.5 shows the process flow used to achieve such stacked vias. Notice that pattern electroplating is used to fill the vias and at the same time to deposit the wiring metal on the subsequent level. The vias and the conductor lines are formed using only one plating step. The key elements for the successful fabrication of stacked vias are: (1) the aspect ratio of via diameter to via height and (2) optimization of the plating additives to achieve preferential fill of the via. Aspect ratios greater than 2:1 will favor good via fill, thus reducing the via dimple at the Cu surface. Plating optimization is required in order to avoid pinching off the plated metal on top of the via. Reverse pulse plating can promote selective fill of the via and relax the constraint on metal height.[29]

Evaluation of test vehicles consisting of four levels of stacked via chains with bottom via diameters ranging from 5 to 25 µm has been reported in the literature.[30] The vias were fabricated by laser ablation of 10 µm poly(bisphenyl dianhydride-para-phenyl diamine) (BPDA-PDA) polyimide. The four inner levels had a nominal Cu thickness of 5.0 µm. The top surface metal (TSM) level, used for controlled collapse chip connection (C4) joining, consisted of 4 µm Cu and 2.0 µm Ni. Figure 11.6 shows the cross section for various stacked via diameters. This work indicated that good via fill for vias of all dimensions could be achieved using a Cu bath with organic

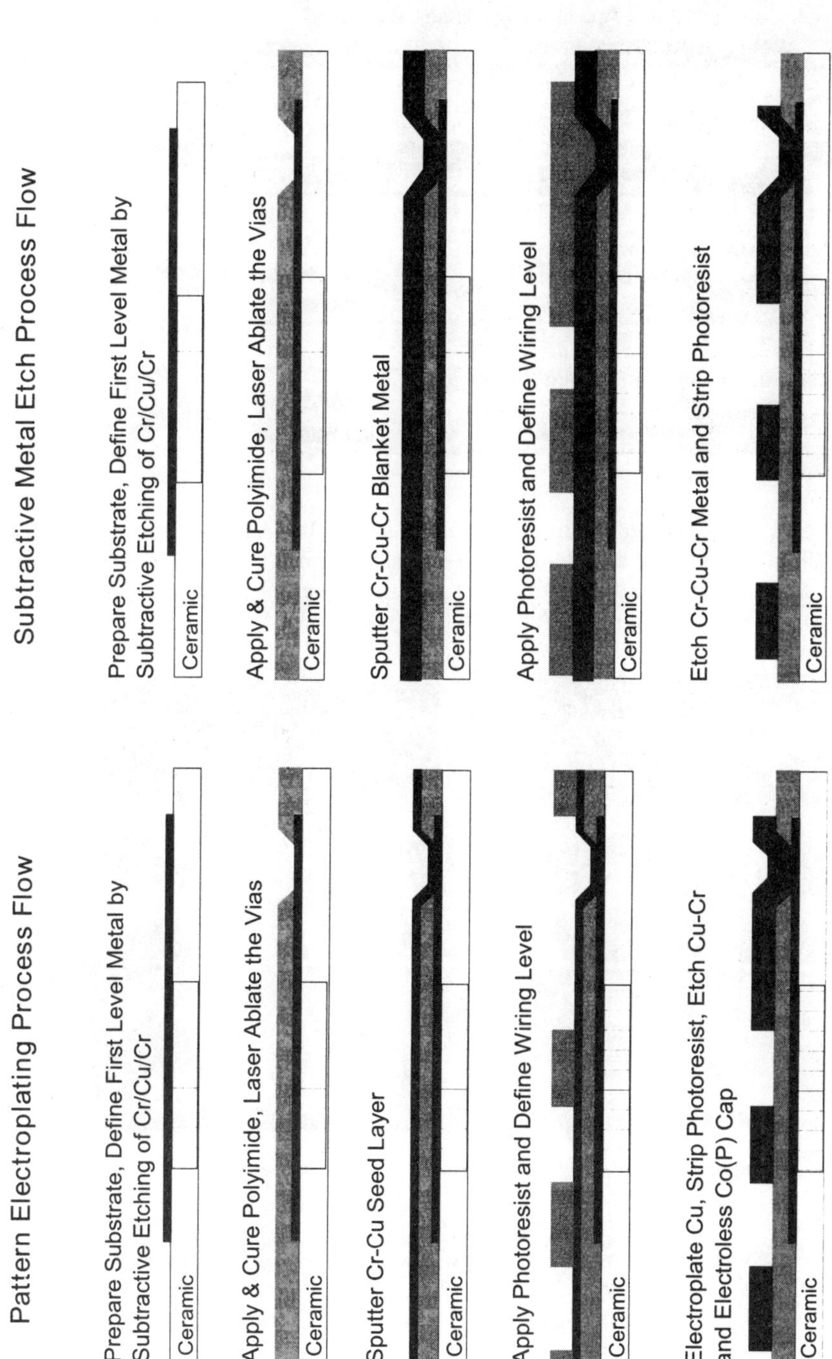

Figure 11.4 Process flow for pattern electroplating and subtractive etch.

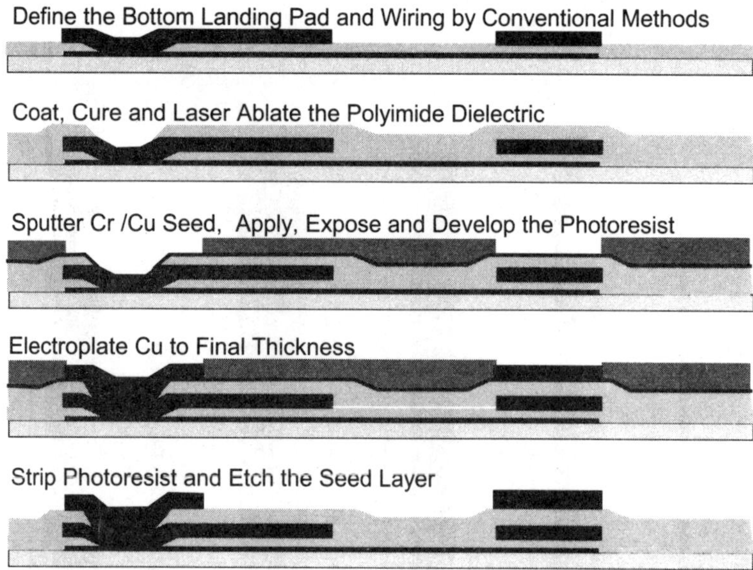

Figure 11.5 Non-planar stacked via process flow. (From E.D. Perfecto et al., "Increased Thin Film Density by Stacked Vias," *IEEE 52nd Electronic Components and Technology Conference,* © 2002 IEEE. Reprinted with permission.)

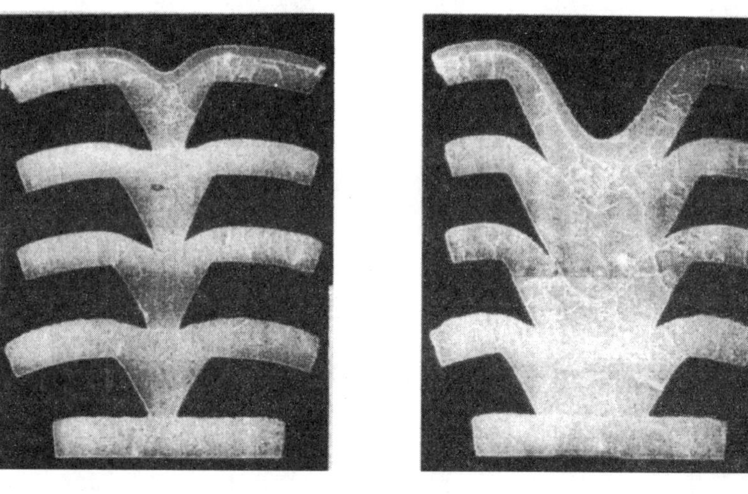

Figure 11.6 Cross section of four-level stacked vias with 5 and 20 μm diameters. (From E.D. Perfecto et al., "Increased Thin Film Density by Stacked Vias," *IEEE 52nd Electronic Components and Technology Conference,* © 2002 IEEE. Reprinted with permission.)

additives. However, the Ni plating on the top metal level needed improvement, perhaps by optimization of additives to give a better via fill.

11.2.4 *Copper–polyimide interface*

Good adhesion is required between the various interfaces to maintain structural integrity. When Cu is deposited on cured polyimide, an adhesion layer (typically chrome) is used prior to Cu deposition. The chrome (Cr) layer thickness is usually between 10 and 50 nm, and is deposited by sputtering or evaporation. The polyimide surface is also pre-treated by either an ion beam or light radio frequency (RF) plasma process prior to the sputtering of the Cr/Cu metals. This treatment modifies the surface bonds in the polyimide to further improve adhesion.

The process of creating a polyimide dielectric on top of the copper can lead to interface reactions which adversely affect the performance of the structure. In forming the polyimide layer, a polyamic acid (PAA) precursor is spin applied to the Cu surface and is subsequently cured at 400°C. During the cure cycle, Cu reacts with the PAA to form Cu_2O, which diffuses into the polyimide (PI) layer to form Cu_2O precipitates.[31] Kowalczyk[32] demonstrated that the polyimide precursor solvent, *N*-methyl-pyrrolidinone (NMP), causes mobility and enhances aggregation of Cu precipitates. When photosensitive polyimide is used, the reacted Cu leaves a residue in the developed pattern openings which is very difficult to remove. Cu diffusion has also been reported on preimidized polyimides,[33] but none has been observed when poly(amic ethyl ester), m-Paete, or BCB are used.

The diffused Cu in the polyimide can increase the dielectric constant of the polyimide and also degrade its mechanical and thermal properties. To prevent Cu diffusion into the polyimide, various metal capping layers such as Cr, Pt, Pd, Ti, and Co(P), have been proposed over the years.[34–37] Additionally, these copper diffusion barrier layers enhance the polyimide to Cu adhesion. They also facilitate auto-inspection by providing good optical contrast with the polyimide. It should be noted that when the metal wiring is defined by the subtractive etching or by lift off of a Cr/Cu/Cr sputtered film, the Cr protection layer is only present on the top and bottom of the conductor. There is no diffusion barrier or adhesion promoting layer on the edges of the Cu features. For pattern electroplated films, electroless Co or chromate treatment have been shown to successfully cap both the top and sides of the Cu conductors.[38–40] The steps in an electroless process for the formation of Ni(P), Co(P), or CoW(P) are described in Section 11.3. Interactions with other process steps that can arise when electroless barrier deposition becomes part of a complete MCM fabrication process are discussed in Section 11.4.

11.2.5 *Effect of wet processes on the integrity of BSM pads*

Back side metallurgy (BSM) features on ceramic substrates are deposited by selective activation followed by electroless Ni and Au plating on sintered Mo pads. Multilevel thin-film processing on alumina substrates involve more than ten exposures of the metal features on the top and bottom surfaces to various acids and alkaline solutions. Typically, Cr etching takes place in an alkaline $KMnO_4$ solution. This step is followed by an oxalic acid rinse. This process was found to cause corrosion of the completed BSM pads and affected the reliability of brazed pins.

Table 11.3 IBM and military standard stress conditions used to qualify MLTF modules.

1. Thin-film opens and latent opens	
Thermal cycling	0/100°C, 3–5 kc (IBM)
	–25/125°C, 3–5 kc (IBM)
	–55/125°C, 1 kc (Mil-Std 883)
Thermal shock	–65/150°C, 1–3 kc (Mil-Std 883)
Temperature age	150–200°C/1–2 khr (Mil-Std 883)
Electromigration	150°C/300 mA, 1–3 khr (IBM)
2. Thin-film shorts and latent shorts	
Temperature/humidity/bias	85°C/85%RH/5-15 V, 1–4 khr (IBM)
Temperature bias	130–150°C/35-75 V, 1–2 khr (IBM)
Highly accelerated stress test	140°C/85%RH, 100–500 hr (IBM)
3. Bottom-surface metal and pin integrity	
Reflows	13× reflows
Thermal shock	–65/150°C, 1 kc (Mil Std 883)
Temperature/humidity/bias	85°C/85%RH/5 V, 2 khr (IBM)
Highly accelerated stress test	140°C/85%RH, 100–500 hr (IBM)
Mechanical stressing	Pin flexing/bending (1–5 lb forces)

However, the extent of the BSM pad corrosion could not be readily detected at the final stage of MCM processing.

Since different metals on the top and on the bottom surfaces are exposed to both high- and low-pH solutions during the process, localized galvanic cells may be expected to form. The electromotive force varies from region to region because of varying metal interconnection lengths between top- and bottom-surface features. If there are pinholes in the Ni and Au films on the BSM pads, Mo, Ni, and Au will come into contact with wet etchants. In these regions one would expect different galvanic reactions to take place. Predicting the steady-state electromotive potentials and currents of such cells and their effect on BSM corrosion is not possible since the cathodic and anodic areas will vary according to how much Au, Ni, and Mo is exposed. One can use Pourbaix diagrams[41] for various metals to qualitatively assess the corrosion potentials which exist due to processing conditions. For example, Pourbaix diagrams show that Mo exposed to pH 7 solutions and Ni exposed to solutions with pH below 8 and above 12 can lead to corrosion reactions. Since the alkaline $KMnO_4$ solutions used to etch Cr have a pH greater than 10, and photo-lithography processing solutions have a pH range of 10 to 14, these process steps can induce considerable BSM corrosion. To prevent such in process corrosion, the BSM metal features are protected by a temporary resist overcoat during processing.

11.2.6 *Reliability assessment and product qualification*

Non-planar structures can create localized variations in polyimide thicknesses. Tight ground rules and high wiring density allow only very limited tolerance for contamination and process variability since both can affect the yield and the long-term reliability. The complex processing steps and multiple thermal excursions associated with MLTF fabrication can cause cracking or delamination of the ceramic metallurgy. If the BSM features are not well protected during Cr etch processing, the underlying metals, Ni and Mo, can be corroded, leading to voids and delamination as discussed above. Contamination in the form of either fluid trapped within the pores or salts formed on the surface can lead to corrosion and migration-induced shorts. These considerations dictate that reliability assessment and product qualification of a multichip module with thin-film wiring evaluate BSM features and pin integrity as well as opens and shorts in the thin-film and ceramic wiring. Table 11.3 gives a summary of the IBM and military standard stress conditions used to qualify MLTF modules from a reliability point of view.

11.3 Conductor formation and related process steps

This section, after summarizing the merits and limitations of a variety of methods for depositing metal features for packaging, will focus on issues relevant to the use of electrochemical processes in MCM fabrication. Further discussion on each of these processes can be found in the literature.[42]

11.3.1 Process options

11.3.1.1 Subtractive etch

For features with relatively relaxed ground rules (e.g., line width >25 μm, line spacing >50 μm, and width to thickness aspect ratio >3), a sub-etch process can be used (Figure 11.7A). In this process, a blanket film is deposited onto a substrate by sputtering or evaporation. The film is then covered with a lithographically patterned photoresist and a wet chemical etch process is used to remove the unwanted copper. The chemical etching process is frequently used in printed circuit board fabrication and in the fabrication of some thin-film ceramic packages. Wet etching tools are relatively inexpensive. However, the wet etching process is isotropic in nature. The resulting undercut limits its applications to relatively thin films and wide features. Replacing wet chemical etch by ion milling alleviates the undercut problem, but the technique is expensive, and minimizing back-sputtering of copper over other areas of the substrate is difficult. Limited success of copper patterning by RIE has been reported, but this process is not useful in manufacturing because of its high cost.

In a typical packaging application, the blanket metal is a sputter-deposited Cr/Cu/Cr layer which is chemically etched through a lithographically defined photoresist mask. The Cr layer, typically 25 nm thick, is used to enhance the adhesion between the metal and the photoresist and polyimide. The Cr also provides good optical contrast with the polyimide, facilitating autoinspection. The Cu is etched with ammonium persulfate solution in a spin/spray tool with an endpoint detector used to minimize undercut and to ensure proper line width control. Subtractive processes for

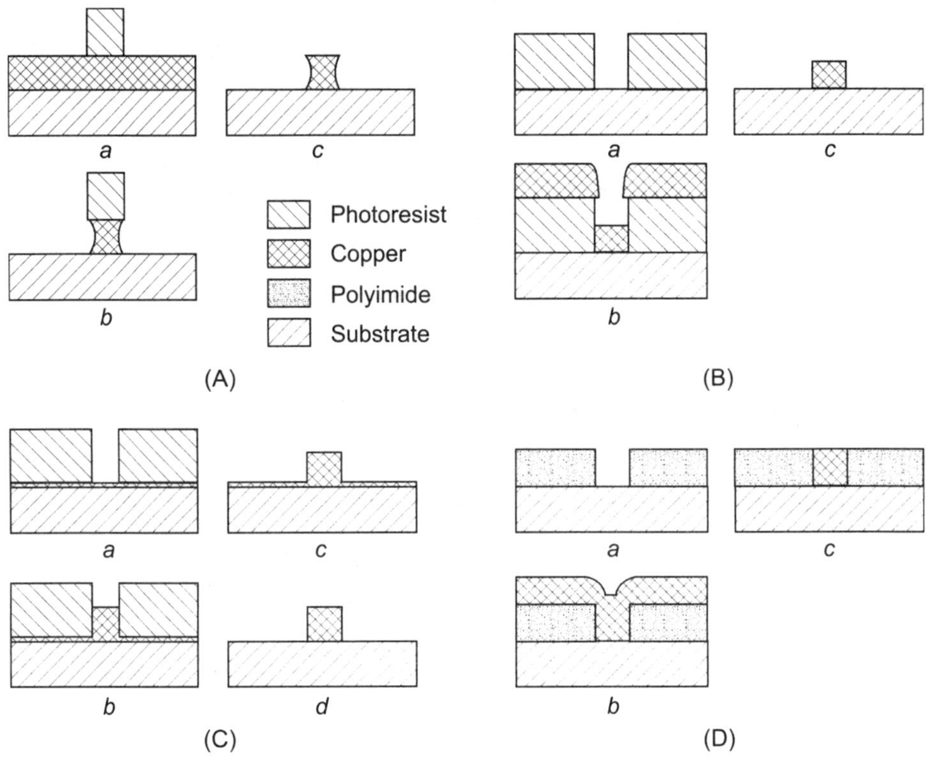

Figure 11.7 Schematic drawings of metallization process options for thin-film wiring: (A) subtractive etch, (B) lift off, (C) electrolytic through-mask plating, (D) Damascene. (Republished with permission from the *IBM Journal of Research and Development*.)

3 to 6 μm thick Cu layers can be optimized to yield an undercut of 1.2 times the thickness of the Cu; 10 μm features (minimum dimension) can be successfully etched for Cu thicknesses of 2 μm or less. The combined tolerances associated with the sputtered metal thickness, resist lithography, and etching result in a ±5-μm variation in the line width.

11.3.1.2   *Electrolytic through-mask plating*

The electroplating of metal lines over areas not covered by a photoresist is referred to as through-mask plating. As noted in the introduction, this process was first invented for use in magnetic thin-film head fabrication. The technology has since been adopted for packaging fabrication and also for micro-electro-mechanical system (MEMS) applications. Figure 11.7C depicts this process, which begins with the blanket deposition of a thin adhesion/barrier metal (typically 25 nm of Cr) followed by another blanket deposition of a copper plating seed layer. A patterned photoresist layer which serves as a plate-through mask is then formed on the copper. A 250 nm thickness of Cu is sufficient to ensure adequate electrical connection for

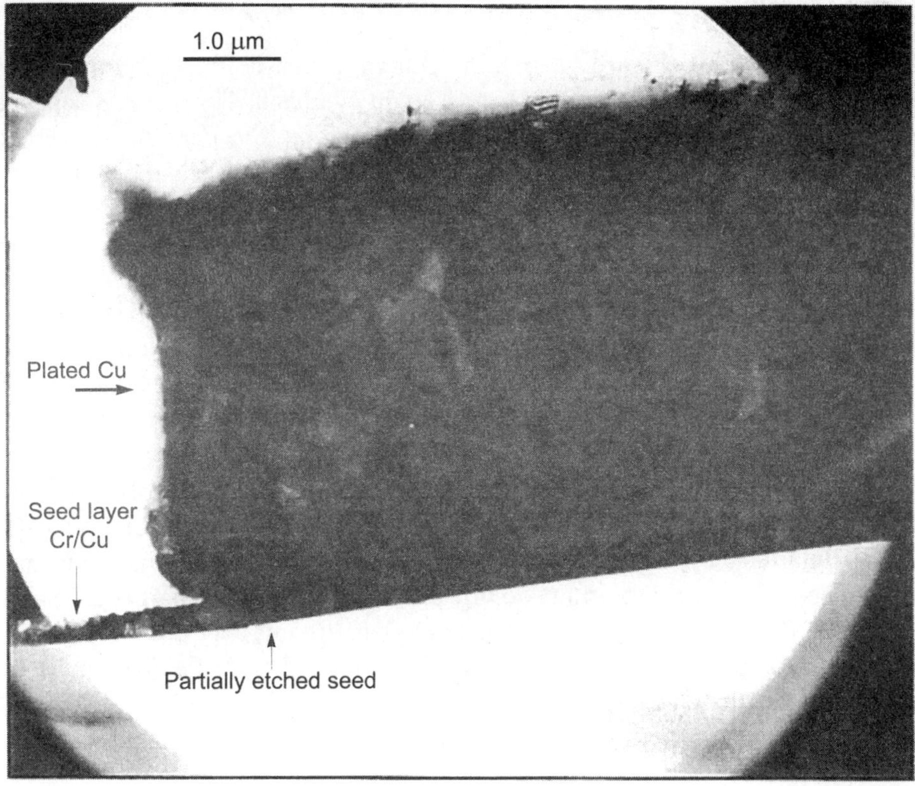

Figure 11.8 Cross-sectional transmission electron micrograph of 5 μm thick electroplated Cu on Cr/Cu sputtered seed. (Republished with permission from the *IBM Journal of Research and Development.*)

plating to all parts of the pattern on a 127×127 mm substrate, even though the seed layer is usually partially etched in the plating solution before the current is switched on to start electrolytic deposition. (This etch step assures uniform initiation of plating by removing any oxide layer which may have formed on the copper.) After the desired metal thickness is achieved by plating, the photoresist is removed, and the seed layer and adhesion/barrier film are etched away.

Electrolytic through-mask plating has particular appeal for packaging since this technique can achieve greater conductor packing density, a larger height-to-width aspect ratio, and more faithful pattern reproduction than any other patterning technique. Lines can have near vertical profiles, and line widths can be controlled to better than ± 3 μm. With proper optimization of the seed layer etching, the width control and reproducibility are determined primarily by the photolithographic tolerances. Copper is the material of choice for wiring because of its electrical properties, plating chemistry, and cost. The transmission electron micrograph in Figure 11.8 shows the plated wiring metal and a partially etched Cu seed. After the seed metals are etched, the plated Cu wiring is often capped using electroless Co(P) chemistry[43] for reasons already discussed in Section 11.2.

11.3.1.3   Lift off

In the lift-off process (Figure 11.7B), the pattern is first defined by a photoresist pattern which has been processed to have a slight overhang. The metal is deposited by a line-of-sight deposition process such as evaporation. Since the resist overhang ensures that there is practically no deposition of metal on the sidewalls of the resist, an organic solvent can be used to undercut the resist and float away the metal deposited on it. The metal deposited directly on the substrate remains after the resist is removed and forms the conductor pattern as defined by the openings in the resist. While lift-off has had some use for relatively thin Al conductors on silicon devices, the low resolution, low throughput, and high cost of the process limit its use for thin-film packaging. The prime advantage of the lift-off technique is that it can be easily adapted to create a stack of multiple metals including those which may not be easy to electroplate.

A variant of lift off is deposition through a metal stencil. This process was used for the terminal metal layers and lead-tin flip-chip bonding with C4 in early IBM servers. Unfortunately, depositing enough material into patterns of high aspect ratio and with good thickness uniformity control is difficult to achieve, particularly as the wafer size increases. Even in C4 manufacturing, lift off has been replaced by through-mask electroplating, which proved to be a much more flexible, lower cost process.

11.3.1.4   Electroless plating

Electroless plating offers an alternative to electroplating. Copper, nickel–phosphorus, nickel–boron, and gold are among the materials of interest to packaging which can be formed electrolessly. Here the substrate is seeded with a solution containing a catalyst, usually a salt of palladium. The palladium seed initiates the reduction of metallic ions by the reducing agents in the electroless plating solution. Freshly formed metal such as copper is reactive and acts as a catalyst for further reduction of copper ions. A major advantage of electroless plating is that a conducting plating seed layer is not needed. However, the rate for electroless plating is typically lower than that for electrolytic plating, and plating solution process control is more complicated. Still, electroless copper plating is the key process for fabricating printed circuit boards (second level packaging.) In the first level package, electroless nickel and gold plating are widely used to provide the bridge between the sintered Mo in the ceramic and the thin-film metallurgy on top. Electroless Co(P) is also used as a capping layer on Cu as discussed later in this section. Due to environmental health concerns, the once widely used formaldehyde based electroless copper bath is rarely used today. Newer formulations, such as the lower pH baths based on hypophosphite and dimethylaminoborane, have been reported recently.[44] These formulations, which present fewer environmental concerns and which are compatible with polyimide because of their lower pH, may lead to wider use of electroless plating of copper in packaging again.

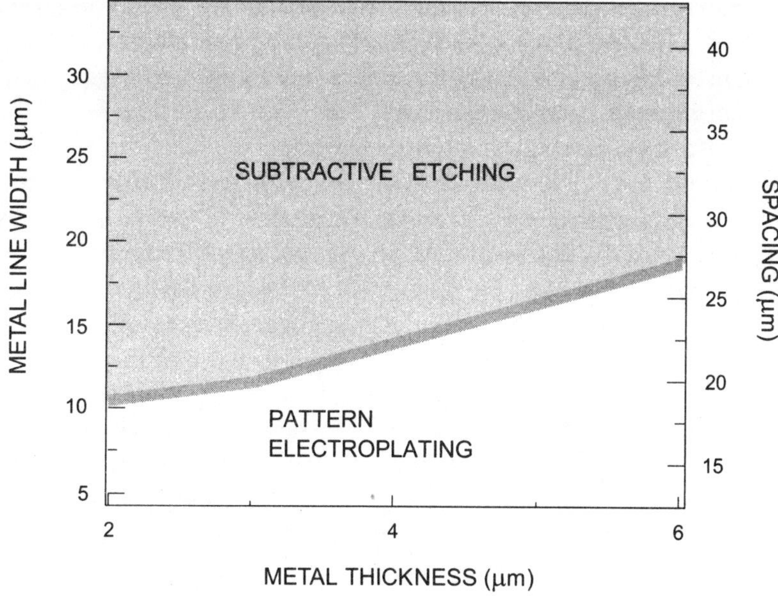

Etch factor = 1.5 for metal thickness < 3 µm
1.2 for metal thickness > 3 µm
13 µm represents 10 µm of minimum mask feature
spacing plus 3 µm develop process bias

Figure 11.9 Line parameters governing the choice of subtractive etch vs. electroplating. (Republished with permission from the *IBM Journal of Research and Development*.)

11.3.1.5   Damascene

The Damascene process is widely used for copper wiring in semiconductor back-end-of-the-line interconnect. This process has the advantage that it can be used with a broad range of metal deposition techniques. As shown in Figure 11.7D, the conductor features are formed by defining openings in the dielectric, backfilling the openings with metal, and finally removing the excess metal by a planarization step. Metal deposition can be done by chemical vapor deposition or by electroplating as in the plate-through-mask process. Ray et al. have described the use of the Damascene process in multi-chip module fabrication.[45]

11.3.2   *Process choice considerations*

Subtractive etching and through-mask electroplating are the processes most commonly used to form conductors in packaging structures. Cost, thickness and width of the conductors, spacing between them, and tolerances set by the product performance requirements are among the factors that determine which process is used. Subtractive etching is a lower cost process and is therefore the preferred method for less demanding packaging structures. Pattern electroplating has proven to be the most reliable and cost-effective process when closely spaced conductors with high aspect ratios are required. Figure 11.9 shows the typical line widths and spacing that can be achieved under optimal etching conditions[46] and can be of help in choosing the conductor formation process for a particular structure.

For advanced packages, particularly those with thick conductors and with precise shape and dimension requirements, the process of choice is electroplating through a lithographically defined mask. In contrast to subtractive etch or lift off, through-mask plating is an atom-by-atom mold replication process which precisely duplicates the cross-section of the lithographic stencil down to atomic dimensions.[47-49]

The viability of through-mask plating as a precise manufacturing process has been proven by nearly 25 years of use in thin-film head manufacture and by the fact that virtually all the inductive thin-film heads today are produced by this process. In recent years, through-mask electroplating has also been used extensively in fabricating structures for MEMS applications.[50-54] The fabrication requirements for demanding packaging structures are comparable to those for thin-film heads and MEMS devices[55-57] so that the experience that was gained and the extensive scientific understanding of electrochemical processes that was developed in the thin-film head work and is now also coming out of the MEMS community has a direct bearing on the application of electrochemical technology to MCM fabrication.

11.3.3   Electrolytic through-mask plating

In a typical packaging product, copper, nickel, and gold are the most commonly plated metallurgies. Copper is used for internal wiring because of its low electrical resistivity. Nickel and gold are used at the terminal level for wire bonding and tin−lead solder joints. The nickel acts as a diffusion barrier between the copper and the Sn in the solder. The gold prevents oxidation of the Ni and also allows the solder to wet the terminal without the use of an acid organic flux. The major requirement for plated metal conductors in packaging are summarized in Table 11.4. These properties are primarily controlled by the chemistry of the plating bath, the plating parameters (effective current density, temperature, mass-transport conditions), the nature of the seed layer (material, thickness), and the postplating processing conditions.

Table 11.4   Requirements for plated metal conductors in packaging.

Uniform plating thickness (global and local scale)	$\pm 10\%$
Low electrical resistivity	$<1.86\ \mu\text{ohm cm}$
Smooth surface	$<0.05\ \mu\text{m}$ (peak to peak)
High ductility	$>10\%$
Low residual stress	<200 MPa (copper)
Bath compatible with photoresist	

Choosing the appropriate plating bath is a critical first step to achieving the desired properties in the electroplated deposit. Table 11.5 shows the essential components of the copper, nickel, and gold plating baths used in packaging. However, additives are usually required to improve the metallurgical properties and the thickness uniformity of the deposit. Plating baths used in industry are generally supplied by vendors who formulate and market proprietary solutions designed to meet the requirements of specific applications. Vendors are not anxious to talk about the makeup

Table 11.5　　Essential components of typical copper, nickel, and gold plating baths used in packaging.

Acid copper plating bath	
Copper sulfate	50–75 g/l
Sulfuric acid	150–200 g/l
Chloride	50–80 mg/l
Brightener	Vendor specific
Leveler	Vendor specific
Nickel sulfamate bath	
Nickel	60–80 g/l
Boric acid	20–50 g/l
Anode activator	Vendor specific
Wetting agent	20–50 dyn/cm
pH	3–3.4
Gold sulfite bath	
Gold	7–18 g/l
Sulfite	20–45 g/l
Arsenic	40–60 ppm
pH	8.5–9.0

of their proprietary solutions, the details of which strongly affect the deposit's metallurgical structure, crystal size, and orientation and, therefore, its functional properties. However, the vendor will often work with the customer to optimize a proprietary bath to achieve a specific set of parameters in the customer's tools.

11.3.3.1　Plating tools

Precise control of the thickness of the conductors must be maintained across the entire substrate at each level, since the height of the plated copper influences the characteristic impedance of the transmission lines and thus, the electrical behavior of the structure. This requirement dictates that the plating system be designed to achieve uniform mass-transfer and uniform secondary current distribution across the entire substrate. Fortunately, the tool design concepts and the in-depth understanding of electrochemical processes that have been developed to address the uniformity

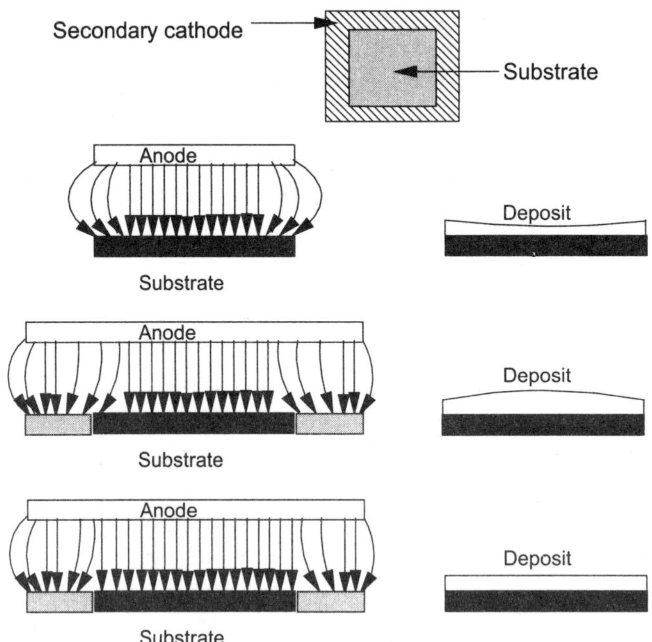

Figure 11.10 Schematic illustration showing how adjusting the current to a secondary cathode influences the lines of current flow to the substrate to affect the uniformity of the deposit.

requirements of the thin-film head used in magnetic storage are applicable to the plating operations for the multilevel package.

Efficient and uniform mass-transport of plating solution to the substrate can be achieved by enforcing a jet of plating solution towards the substrate,[58] by bubbling air or N_2 through the solution, by mechanically agitating the substrate or by moving a paddle. The paddle cell[59-61] which has proven itself as the workhorse of the thin-film head industry, is particularly appropriate when the most demanding copper plating requirements must be met. The parallel plate electrode configuration of the paddle cell and the insulating sidewalls, which are perpendicular to the electrodes, promote straight field lines and uniform current density. The paddle, which moves in close proximity to the workpiece, provides the controlled, uniform agitation necessary for uniform mass-transfer right at the surface where it is necessary to keep a very uniform diffusion layer thickness. The extensive modeling which has already been done on processes in this type of cell provide the understanding necessary to achieve optimum uniformity in packaging applications.[62-64]

11.3.3.2 *Current distribution*

It is usually desirable to add an independently powered secondary cathode around the substrate to provide additional control over the lines of current flow and thus optimize plating uniformity. Figure 11.10 shows how the edge to center global plating

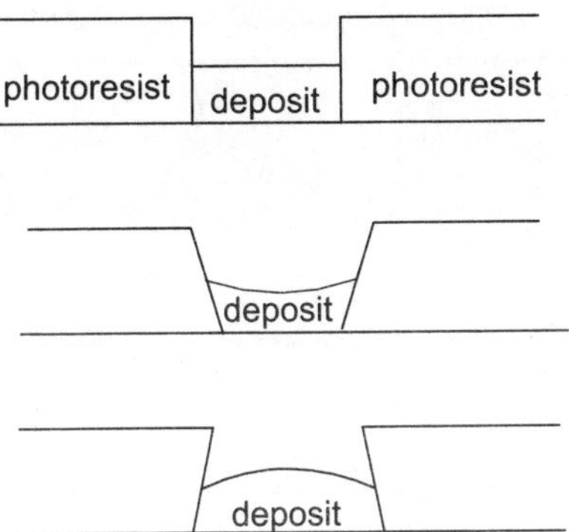

Figure 11.11	The surface profile of the deposit is affected by the wall angle of the resist.

thickness distribution can be modulated with the current passing through this auxiliary cathode. The work of Mehdizadeh et al.,[65,66] which shows how the current distribution across the substrate can be predicted quantitatively in the presence of an auxiliary electrode, provides valuable guidance in optimizing uniformity, but plating and measurements on actual substrates are still needed to establish the final condition.

The density of the resist pattern (i.e., the fraction of the surface covered by resist) and variations in this density from point to point across the substrate also affect the uniformity that can be achieved in the plating process.[67,68] Mehdizadeh et al.[69] have considered these effects by describing the lithographic pattern as a continuous distribution of "active-area density" and developing a model to predict the current to individual features as a function of the active-area density of the surrounding region. Factors which minimize pattern-driven nonuniformities include a low plating rate, high bath conductivity, and mild variations in the active-area density across the substrate. It is advisable to examine the lithographic pattern in the context of the work of Mehdizadeh et al. for any through-mask plating application where plating uniformity is important and to add dummy features if necessary. Fortunately, the conductor patterns in the typical wiring level of advanced packaging structure show little variation in active-area density across the wafer. This characteristic, in combination with the high conductivity of the acid copper bath and a low plating current density, facilitates excellent thickness control across the substrate in packaging applications.

The wall angles of the pattern resist can also affect the top surface profile of the deposit. This effect is shown in Figure 11.11.

11.3.3.3   *Bath chemistry control and plated metallurgy*

The chemistry of a plating bath must be routinely monitored since experimental evidence shows that some of the additives are chemically unstable, and their breakdown products play a role in the plating process. Commonly used techniques for additive analysis include high-performance liquid chromatography, microcapillary electrophoresis, cyclic voltammetric plating, stripping analysis, and Hull Cell analysis.[70-73]

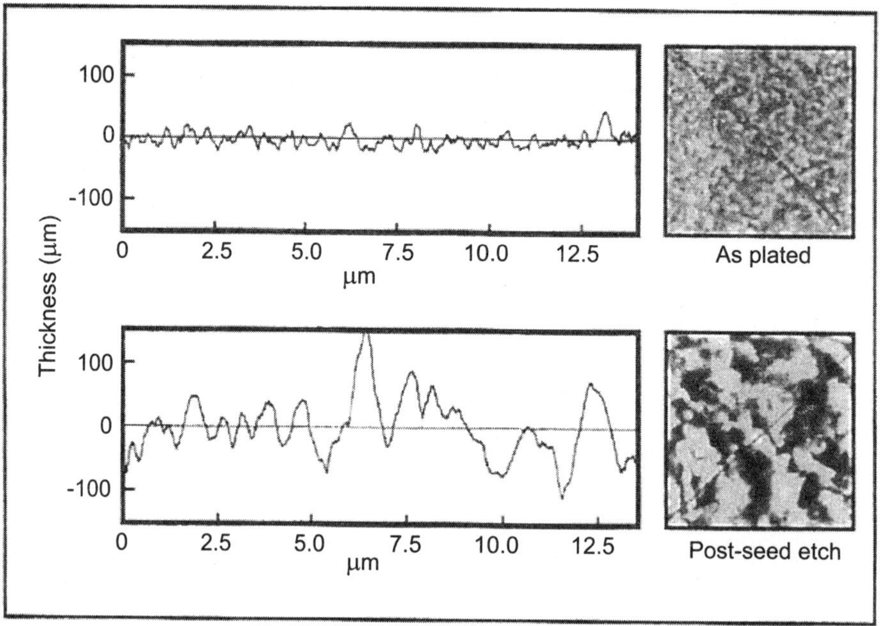

Figure 11.12 Comparison of atomic force microscope measurements of surface roughness of Cu as plated and after seed etching. (Republished with permission from the *IBM Journal of Research and Development*.)

Because of their sensitivity to changes in the chemistry of the plating bath, the metallurgical properties of deposited films such as electrical resistivity, grain structure, crystallographic orientation, surface roughness, and residual stress of the deposit should be periodically measured and evaluated. In the case of the Cu–Ni–Au stack used for terminal wiring bonding, a functional test such as pull strength is also required.

The subsequent temperature cycle in the fabrication process can significantly affect the metallurgy of the plated deposit. For example, significant grain growth, as well as a reduction of twinning, are observed in plated copper.

11.3.3.4 *Seed etching process optimization*

Through-mask electroplating entails the use of a seed layer which must ultimately be removed. The etching of the seed layer presents a challenge since it is not practical to protect the plated features from the etchant during this operation. Fortunately, the sputtered Cu seed etches at a slightly faster rate than the plated Cu features, which reduces thickness loss of the Cu plated features during the seed etching. The etch-rate differences seen for sputtered films vs. plated films can be explained in terms of structural differences (Figure 11.8). For example, sputtered films prepared under low Cu adatom mobility conditions produce a columnar structure,[74] while plated films produce equiaxed large-grain structures.

Another effect, this one unfavorable, arises from the oxide films formed on the Cu seed which delay the onset of seed etching. The consequent increase in total etching time results in increased plated metal roughness which can cause problems in the layer-to-layer via interconnections and also adversely affects autoinspection. Other factors that influence the seed etching time are Cu seed thickness, solution pH, and percentage of over-etching. Figure 11.12 shows atomic force microscope (AFM) measurements of Cu after plating and after seed etching. The AFM measurements after seed etching can vary from ± 20 to ± 150 nm.

During the etching of the Cu in the seed layer, the endpoint detector is triggered as soon as featureless areas are clear. To ensure that the Cu seed is etched in the high-density feature areas, the "standard etch process" calls for a 20% over-etch. An accelerated etching of the Cu plated features occurs during the over-etch portion of the cycle. Galvanic coupling between the Cr and the Cu etch solution has been suggested as the mechanism for this effect.[75,76] The effects of accelerated etching are more pronounced with increased numbers of wiring levels and interconnections. For example, accelerated etching is observed at the second wiring level, but is non-existent on the mesh levels where the pattern is uniformly dense. IBM has developed a proprietary etching process to eliminate accelerated etching.[77]

11.3.3.5 Capping

To prevent copper reacting with polyimide causing degradation of the dielectric, and to provide a good adhesion between copper with the polyimide, a thin capping layer is put in between the plated copper and subsequent polyimide layer. Electroless NiP, CoP, or CoWP are the most commonly used capping material. The steps in a typical electroless capping process are shown below.

- Deionized (DI) water rinse
- Pre-etch surface copper oxide
- DI water rinse
- Palladium seeding
- DI water rinse
- Complexing agent rinse to remove excess palladium
- DI water rinse
- NiP/CoP/CoWP plating
- DI water rinse
- Dry

Another example of capping is the use of nickel between copper and soft gold in the wire-bonding pads. Since copper and gold readily interdiffuse into each other, particularly at elevated temperatures, a nickel layer diffusion barrier is needed.

Details on the metallurgy of electrolytic and electroless plated capping material have been reported in the literature.[78,79] The choice of one material over others depends on the product reliability requirements such as the integrity of the structure after a given number of temperature reflows, and on manufacturing costs. An extensive discussion of capping is given in Chapter 3 of this volume.[80]

11.4   Integrating electrochemical technology with traditional microelectronic fabrication processes

A viable manufacturing process can only be achieved by integrating the electrochemical techniques with traditional wet and dry technologies, so that the final process selection must take into account interactions among the individual process steps. The dual layer polyimide (DLP) process[81] and the dual level metal (DLM)[82] process are examples of such integrated combinations of technologies. (The name DLP highlights the fact that the insulation layers for a wiring level and its associated via level are formed concurrently in a single operation after the conductors and vias have been created by electroplating through photoresist masks. In the DLM process, via openings and wiring channels are first defined in the respective insulation layers. These openings are then filled by electroplating so that all metallization is done concurrently in a single plating operation.) In this section, the steps of the DLP process will be described with emphasis on some typical issues which must be addressed when electrochemical technology is integrated with other microelectronic fabrication processes.

11.4.1   MCM structure

A schematic cross section of the multilevel interconnection wiring whose fabrication is described in this section is shown in Figure 11.13. The wiring structure consists of a ground plane, two planes with orthogonal conductors referred to as x and y planes, respectively, a second ground plane, a level of wiring for engineering changes (EC), and a top layer which provides test pads and chip attachment metallurgy. The thin-film wiring structure is built on top of a ceramic substrate which contains the power distribution wiring and an associated array of pads. Since slight distortion of the pad array inevitably occurs during firing of the ceramic, an additional level of oversized capture pads is included in the multilevel structure to ensure registration to the studs in the cofired ceramic. Vias interconnect the various levels according to the needs of the circuit design. Many of these interlevel connections, such as those in Figure 11.13 which connect the top surface metallurgy to the y plane or to the substrate wiring, entail stacks of vias.

Typical interconnection wiring is designed to have a characteristic impedance of 40 to 50 ohms. (Since there is no ground plane separating the x and y levels, the actual characteristic impedance depends on the degree of fill in the adjacent wiring level.) The diameter of the vias and the width and pitch of the conductors depend on the particular product design. Some of the more aggressive structures call for conductors which are 6 μm thick and 12 μm wide on a 25 μm pitch. The typical dielectric is polyimide which, with its dielectric constant of 3.5, needs to be about 6 μm thick to achieve the desired characteristic impedance.

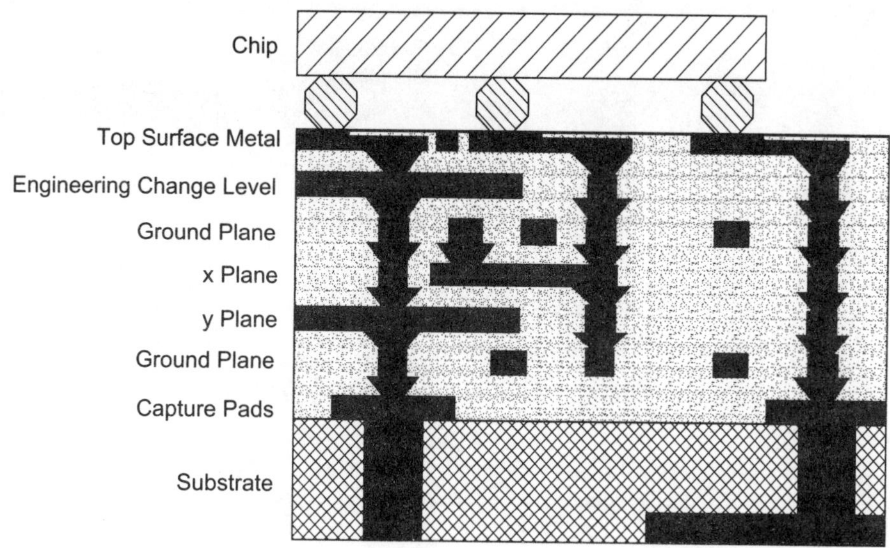

Figure 11.13 Schematic cross section of the multilevel thin-film wiring on a high-performance multichip module. (Republished with permission from the *IBM Journal of Research and Development*.)

From a fabrication point of view, meeting the performance requirements for the interconnect structure translates into (1) maintaining precise spacing between conductor planes, (2) fabricating conductors with precisely defined cross-sectional shape and dimensions, and (3) incorporating a low-loss dielectric insulator with relatively low dielectric constant which can withstand a 400°C temperature during C4 solder joining of the chips. The polyimide that was used to meet the dielectric requirements of this structure is a version of Du Pont PI-2611 BPDA-PDA, which has a polyamic acid precursor. As noted in Section 11.2, this choice of polyimide leads to the potential of an adverse interaction between the copper and H_2O, a by-product of the polyimide cure, which forms Cu_2O at the Cu–polyimide interface. With time, the Cu_2O migrates into the polyimide resulting in Cu_2O particles in the dielectric near the Cu interface which degrade the dielectric properties of the polyimide.[83] To prevent this effect it is necessary to create a Co(P) barrier layer on the copper prior to forming the polyimide insulation.[84]

11.4.1.1 Fabrication steps and related issues

Further issues in the use of electrochemical processes to build a packaging structure can be brought to light by following through the sequence of steps in the DLP process. In this process, the fabrication of any given level entails forming a conductor pattern for that level on a planar surface, building via studs on the conductors as required, filling the structure with a dielectric, and planarizing the surface. (The ground plane is actually a patterned mesh, so from the fabrication point of view the operations to build a ground plane or a wiring plane are identical.) Building the

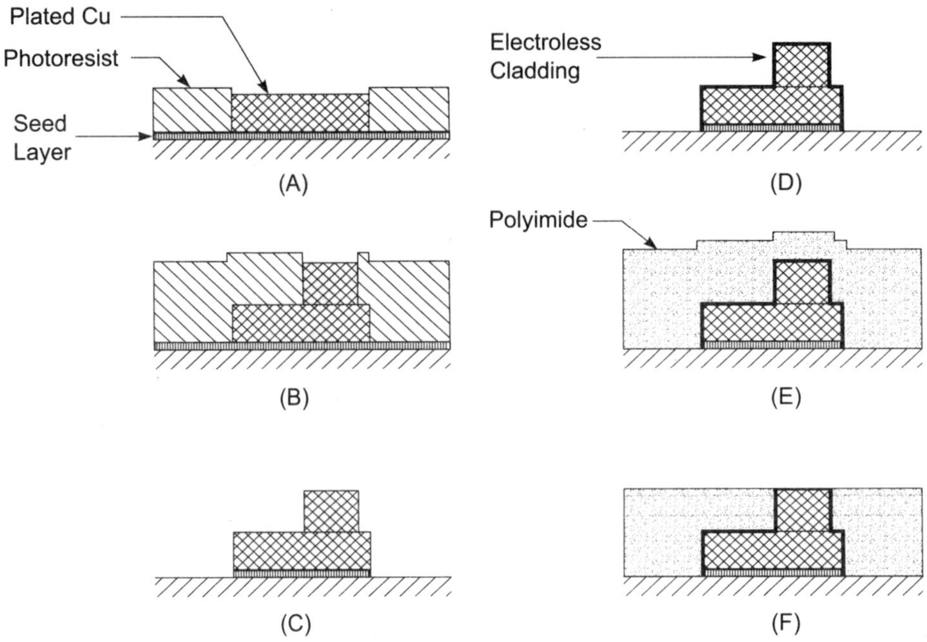

Figure 11.14 Drawings (A) through (F) represent the process sequence used to fabricate one conductor and via level by the DLP process. This sequence of steps is repeated on top of each previously completed level until the multilevel structure is completed. (Republished with permission from the *IBM Journal of Research and Development.*)

structure of Figure 11.13 can then be regarded as a repetition of these process steps for each desired level.

Figure 11.14 shows the process sequence used to build one level of conductors and vias. The starting surface for any level is the planarized polyimide surface of the preceding level with flush, exposed vias. Formation of the wiring pattern begins by sputter depositing a Cr–Cu seed layer on the planarized surface of the preceding level (Figure 11.14A). A 20 nm Cr layer is used to promote adhesion of the Cu to the polyimide. A Cu thickness of 300 nm is more than enough to ensure negligible ohmic drop across the typical 127×127 mm packaging substrate during plating. A layer of photoresist is then spun on, exposed, and developed to form openings for the wiring pattern. It is advisable to briefly etch the Cu surface in dilute sulfuric acid prior to applying the resist in order to ensure a clean, standardized surface. The thickness of the photoresist has to be slightly greater than the thickness of the plated wiring to prevent the Cu from mushrooming above the top of the resist. After development, an ash step in an argon–5% oxygen plasma is required to ensure that there is no residual photoresist in the openings which would interfere with uniform initiation of plating.

The conductors are now plated in a paddle cell which, as noted in Section 11.3, is well able to achieve uniform plating across large substrates. With the inclusion of an auxiliary secondary cathode and attention to other design details to ensure uniform current density across the substrate, the thickness of the copper wiring can be controlled by the deposition time and current. A commercial acid copper sulfate bath such as Sel-Rex CuBath-M[85] which produces deposits with near bulk resistivity is an appropriate bath for this operation.

Vias to the next level are built after plating the wiring pattern by removing the wiring conductor lithography, applying a new layer of photoresist of sufficient thickness to provide at least 6 μm of resist (the height of the stud) on top of the copper, and exposing and developing the stud pattern. The ashing and plating steps used to form the wiring are then repeated to create the studs (Figure 11.14B).

The next operation, removal of the seed layer, uses wet spray etching rather than the ion milling or sputter etching. This process choice merits special discussion because it illustrates how the side effects of a particular operation can adversely affect subsequent processing steps. When performance requirements override cost considerations, sputter etching or ion milling is preferred for seed layer removal because of the absence of electrochemically driven etch rate modification and because of the better resolution and control they can achieve compared to wet etching.[86,87] However, physical sputtering processes, including ion milling, suffer from redeposition of the etched material onto other parts of the structure, particularly the vertical walls of the conductors. This redeposition adversely affects the initiation of the electroless Co(P) plating and the adhesion of the Co(P) to the sidewalls of the conductors, which must have the barrier layer to prevent the interaction of Cu with polyamic acid. Furthermore, copper is also redeposited on the insulator surface and leads to spurious Co(P) deposition on the polyimide wherever there are high levels of redeposited Cu. Thus, in spite of the usefulness of ion milling and sputter etching in other applications of through-mask plating, the wet spray etch process for seed layer removal is needed in this application to achieve compatibility with the electroless Co(P) cladding operation.

The Cu and Cr layers are each removed in separate etch steps. After stripping the resist, the copper layer is first spray etched with a solution containing ferric ammonium sulfate and sulfuric acid. The Cr adhesion layer is then spray etched with a hot solution of potassium permanganate and tribasic sodium phosphate. This etch is followed by a spray rinse in oxalic acid to remove solid residue from the $KMnO_4$ process. The wiring pattern does not need to be masked for this etch step since even with a factor of two over-etch to ensure complete removal of the copper seed, only about 0.5 μm of copper is lost from the lines. Adequate control of the final line thickness can be maintained simply by depositing sufficient additional Cu during the plating operation to compensate for the Cu lost during seed layer removal. Figure 11.14C represents the structure after seed layer removal.

The Co(P) barrier layer can now be formed on the Cu metallization to prevent the reaction of the Cu with the H_2O produced during the cure of the polyimide which will be applied in the next step. A properly oxidized Co(P) deposit, in addition to forming a barrier layer, will also enhance the adhesion between the copper and the polyimide. Details of the electroless deposition process for the Co(P) layer and the oxidation step to improve adhesion to the polyimide have been described by O'Sullivan et al.[88] As depicted in Figure 11.14D, the cladding process provides a Co(P) barrier on the tops and sides of all conductors and studs.

The next operation is to concurrently form the insulation for both the via and stud levels by a series of spinning and curing steps designed to ensure complete filling of all the crevices in the wiring structure and to achieve partial planarization of the cured polyimide surface. Before applying the polyimide, the part is subjected to a water plasma treatment to enhance the adhesion of the applied polyimide to the cured polyimide of the previous layer.[89] The first step in the polyimide application is to spin on a highly diluted polyimide solution (about 5%) to provide initial wetting of all surfaces of the structure. Following a low temperature bake, a second application of polyimide at 50% dilution is spun on using a slow ramp-up of the spin speed to achieve complete filling of the narrow spaces of the pattern. After a second low temperature bake, polyimide at full strength is applied and cured according to the manufacturer's recommended cure cycle. The final result is the partially planarized insulation depicted in Figure 11.14E. The thickness of the layer is determined by the spin speeds, which should be adjusted so that the thinnest portions of the cured polyimide are 1 to 2 μm above the final polished thickness.

The final planarization and polishing are done mechanically using a 0.05 μm alumina slurry. Table rotation speed and quill rotation speed and pressure are optimized for uniform polishing across the part. The topography in the polyimide above the metal features is removed in the initial stages of polishing. Continued polishing then brings the dielectric to its final thickness. Since the polyimide application is controlled so that the prepolish thickness is within 1 to 2 μm of the final value, very little polyimide has to be removed. This fact, combined with the excellent thickness control achievable in the electrodeposition process, results in tight control of the critical vertical dimensions of the structure. The completed level with planarized surface is shown in Figure 11.14F. From this point, fabrication of the next level can continue by depositing a new seed layer on the planarized surface and repeating the sequence of steps in Figure 11.14A through Figure 11.14F.

11.4.1.2 *Functional evaluation*

Scanning electron micrographs (SEMs) of the structure built by Krongelb et al.[90] using the DLP process are shown in Figures 11.15 and 11.16. The SEM work was carried out on a sample in which the polyimide in a region of the structure was completely removed by extended ashing in an oxygen plasma. The SEM in Figure 11.15 shows contact pads, which are part of the top surface wiring, and the upper mesh ground plane. Two rows of via chains can also be seen in the foreground. Figure 11.16 shows the stacked vias with 11 levels of metallurgy in one of these chains. The well-defined edges of the conductors and their highly vertical profiles are characteristic of plated through-mask patterns.

The package structure also incorporated test patterns to measure the resistivity of the deposited copper, the resistance of the vias which interconnect conductors on different levels, the isolation resistance between adjacent conductors, and the pulse propagation characteristics of the transmission lines. These test patterns included a range of aggressive pattern dimensions down to 9 μm line width. Copper resistivity on all samples fell between 1.6 and 1.7 μohm cm, the resistivity of bulk copper.

Figure 11.15   SEM photograph of completed structure showing top-surface metallurgy and top side of upper ground plane. Polyimide has been ashed away to allow the copper structure to be seen in the SEM. (Republished with permission from the *IBM Journal of Research and Development.*)

Chains of 18.4 μm vias showed a typical resistance of 20.6 ohms, which was at the low end of the expected range of 20 to 24 ohms. Isolation resistance as measured on patterns with an effective length of 6 cm and a separation of 29 μm was in the giga-ohm range for all samples. These measurements indicate that such processing deficiencies as incomplete removal of the seed layer, spurious electroless deposition of the Co(P) capping material on the polyimide, underplating due to lifting of the resist during electrodeposition, gross lithography problems which create unwanted openings in the plating mask, and process-induced changes in the polyimide which alter its insulating qualities can all be controlled. Pulse propagation measurements showed the rise time of the transmitted pulse to be within the expected range and provided further confirmation that the fabrication process faithfully replicated the structure design.

The matter of via resistance merits additional discussion since there is always concern in building multilayer structures that residues and oxidation or other tarnishing layers from previous process steps would contribute to excess resistance between levels. The photolithography steps used to define the pattern openings for both the via and the conductor levels are particularly likely sources of residue, and care has to be taken to ensure complete removal of all traces of resist at the bottom of the pattern openings. This requirement can be met by optimizing the exposure and development parameters to favor complete development (sometimes at a slight sacrifice in replicating the lateral dimensions of the mask) and by introducing an ashing step before plating. The Cu plating electrolyte in combination with the

Figure 11.16 SEM photograph showing a via stack with 11 levels of metallurgy. (Polyimide has been ashed away as in Figure 11.15.) (Republished with permission from the *IBM Journal of Research and Development*.)

pre-plating etch step, is sufficiently aggressive to remove any slight oxidation on the underlying copper including that caused by the ashing process. The fact that the measured values of via chain resistance were in the low end of the expected range demonstrates that the DLP process deals effectively with issues contributing to interface resistance between levels.

The through-mask plating process offers the advantage that the effects of any residue are visibly apparent as soon as the part is removed from the plating bath. If any residue is present, plating may not take place in the affected regions of the pattern or, if plating does occur, the deposit will be rough with a cauliflower-like structure. In either case, the defect can be seen under visual inspection. In a lift-off process, on the other hand, the metallization would appear to be normal, although the interface could have a very high contact resistance which would not be detected until the completed structure was tested electrically.

11.5   z900 Server: IBM's latest MCM-D/C package

11.5.1   Server overview

IBM's latest high-end server, the z900 and its predecessor, the S/390 G6 Enterprise Server, utilize non-planar thin-film structures on a 127.5×127.5 mm glass–ceramic substrate. The thin films consist of a six-level triplate structure (Tables 11.1 and 11.2).

The sintered copper wiring embedded in the cordierite glass–ceramic provides a signal speed which is comparable to that in the thin-film wiring, allowing design flexibility. Additionally, the z900 also includes a thin-film structure on the bottom side (BSM) of the substrate. This thin-film structure consists of a polyimide cushion layer with evaporated I/O pads. Pins which are brazed to the thin-film structure in turn provide the connection to the system board.

The z900 is configured as a "16 + 4-way" system and operates at 2,694 million instructions per minute (MIPS).[91] This represents an extraordinary leap in performance over its predecessor, the G6 system, which operates at 1,644 MIPS. The z900 interconnects 35 chips and 367 decoupling capacitors on the top surface metal (TSM) with the processor chips. The latter are based on CMOS technology operating at 459 MHz. The hermetically sealed modules utilize the C-ring technology, in which the seal is formed directly on the surface of the outside edge of the ceramic substrate. This seal, in combination with the module base plate, cap, and heat sink assembly, provides encapsulation for the chips. The hermetic nature of the package, enhances both chip to thin-film interconnection reliability and ensures consistent thermal performance. The maximum module power runs in the range of 1300 W, which requires module cooling.

11.5.1.1 Top surface thin-film structure

The non-planar thin-film structure not only provides signal wiring capability, but also serves several other critical functions, namely

1. Ceramic and thin-film net repairability
2. Engineering change capability
3. Microchip connectivity to the MCM
4. Shielding for noise reduction[92]
5. Pin braze capability for high density I/O

This package has the largest number of layers (101 ceramic layers and 6 thin-film layers) and the largest wiring length (997 m) of any MCM package announced by IBM to date.

The first metallized thin-film layer of the z900, the M0 level, is the capture pad level. The capture pads comprise a 2 μm thick Cr/Cu/Cr metal film which has been patterned into 330 μm diameter pads and serve as the electrical interface between sintered ceramic metallurgy and the thin-film structure. The next metal level, M1, is a mesh layer. The first wiring level, M2, which is the X wiring level of the X/Y plane pair, follows. It consists of 5 μm thick Cu lines which are 18 μm wide on a 33 μm pitch. The Y level, M3, runs perpendicular to the X layer to reduce capacitance.

The fine pitch of the thin-film wiring allows the z900 MCM to support chips with as many as 1667 signal I/Os and chip size as larger as 17.6×18.3 mm. A very tight placement of chips minimizes the critical path lengths which is essential for achieving the 2.18 ns system speed.

Following the *X/Y* wiring is the M4 mesh level, which is a ground plane. The uppermost thin-film layer, M5, has the usual C4 pads which connect the chip to the MCM and additional lines which permit net repairability for both the thin-film and ceramic levels. Net repairability is accomplished by laser line deletion and line reconnection to form alternate paths. The Ni/Au metallurgy in the final layer, M5, is used to provide a good intermetallic connection after the C4 reflow process. There are a total of 101,000 connections, which are made with 97/3 lead/tin C4 bonds.

All levels of thin-film wiring, with the exception of the capture pad level, are formed by through-mask plating on a sputter deposited seed layer. The capture pads are formed by a subtractive etch process since their 2 μm thickness and 330 μm diameter put them well within the capabilities of that process.

Each thin-film metal layer is insulated by a layer of BPDA-PDA. This polyimide was chosen for its low cost, 3.1 dielectric constant, low moisture absorption, and low thermal expansion coefficient, which closely matches that of silicon and glass–ceramic. During the fabrication process, the polyimide precursor formulation is spun on to the desired thickness, soft baked to remove solvent and create a tack-free film, and then thermally cured to convert the precursor to polyimide. Each dielectric layer is patterned by projection laser ablation. Approximately 95,000 vias are formed which are subsequently metallized to create interconnections between successive metal layers. The vias are 20 μm in diameter with a height equal to the polyimide thickness of 6 to 10 μm.

It should be noted that although the thin-film fabrication process described above does result in an interconnect structure with some topography, there have not been any issues with electrical parametric data or other performance factors arising from non-planarity of the wiring layers. In-line measurements of the metal and polyimide thicknesses and related process controls are used to ensure tight tolerances on critical dimensions which influence electrical performance.

11.5.1.2   *Bottom side thin-film structure*

The bottom side of the MCM has a pin grid array (PGA) structure for connection to the system board. A total of 4224 pins are distributed on an interstitial pattern with the minimum pitch of 1.64 mm between pins.[93] The 4224 pins are symmetrically arranged into four quadrants, each with an equal number of pins.

The BSM of the z900 system features a structure referred to as a "poly-cushion"[94] which is capable of absorbing the stresses created during pin braze as well as during I/O socket insertion of the 4224 pins on this module. Figure 11.17 shows a schematic of the polycushion structure. The first thin-film layer of the z900 BSM consists of 1.35 mm capture pads formed by a one-step metal evaporation through a patterned metal mask. These pads function as an interface between the glass–ceramic and the terminal pad structure. Each of the capture pad overlays two vias which are 88 μm in diameter and are distributed at a nominal pitch of 400 μm in the glass–ceramic substrate. The next level consists of a BPDA-PDA polyimide layer with nominally 300 μm diameter vias fabricated in a manner similar to that used for the TSM dielectric layers previously described. The thin-film structure is then

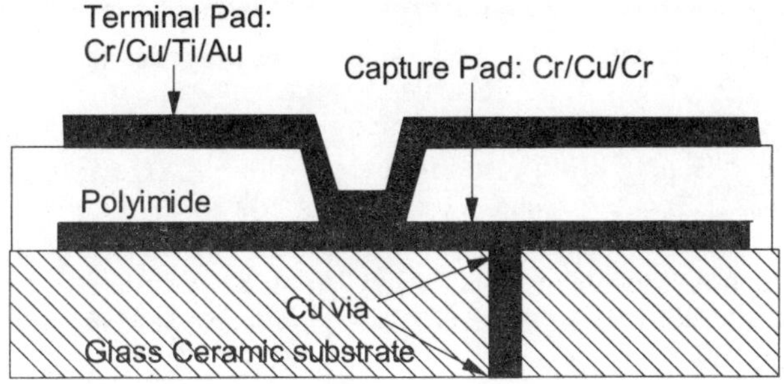

Figure 11.17   Schematic of the "polycushion" structure formed by the thin-film layers on the backside of the multichip module in the z900 server. (From E.D. Perfecto et al., "MCM D/C packaging solution for IVBM latest S/390 servers," *IEEE Transactions on Advanced Packaging*, © 2000 IEEE. Reprinted with permission.)

completed by depositing the Cr/Cu/Ti/Au terminal metal pads using a one-step evaporation through a molybdenum mask. Tapered gold coated Kovar pins are brazed on the terminal metal pads using preattached gold–tin braze. The terminal metal pads, like the capture pads, have a diameter of 1.35 mm. The bottom side processing is completed by applying a layer of protective coating over the entire BSM. This final layer serves to protect the exposed Cu on the sidewalls of the terminal metallurgy from potential corrosion damage in the field.

References

1. K. Prasad and E.D. Perfecto, "Multilevel Thin Film Applications and Processes for High End Systems," *IEEE Trans. Components, Packag. Manuf. Technol. B*, **17**, No. 1, pp. 38–49 (1994).
2. Y. Tsukada et al., "Surface Laminar Circuit and Flip Chip Attach Packaging," *IMC Proc.*, pp. 325–358 (1992).
3. E.D. Perfecto, H. Longworth et al., "Improving Wireability on Non-Planar Structures for the Next MCM-D Generation," *Int. J. Microcircuits Electron. Packag.*, **18**, No. 4, pp. 375–381 (1995).
4. N. Iwasaki and S. Yamaguchi, "A Pillar-Shaped Via Structure in a Cu–Polyimide Multilayer Substrate," *IEEE Trans. Components, Hybrids Manuf. Technol.*, **13**, No. 2, pp. 440–443 (1990).
5. K. Prasad and E.D. Perfecto, "Multilevel Thin Film Applications and Processes for High End Systems," *IEEE Trans. Components, Packag. Manuf. Technol. B*, **17**, No. 1, pp. 38–49 (1994).
6. Y. Tsukada et al., "Surface Laminar Circuit and Flip Chip Attach Packaging," *IMC Proc.*, pp. 325–358 (1992).
7. L.T. Romankiw, "Elimination of Undercut in Anodically Active Metal During Chemical Etching," U.S. Patent 3,853,715 (December 10, 1974).
8. I.M. Croll, A.T. Pfeiffer, and L.T. Romankiw, "Variable Prespin Drying Time Control of Photoresist Thickness for Plating," U.S. Patent 4,281,057 (July 28, 1981).
9. L.T. Romankiw, "Thirty Years of Thin Film Magnetic Heads for Hard Disk Drives," *J. Mag. Soc. Jpn.*, pp. 1–4, January 1 (2000).
10. L.T. Romankiw, "Integrated Magnetoresistive Read-Inductive Write, Batch Fabricated Magnetic Head," U.S. Patent 3,908,194 (September 23, 1975).

11. A. Iqbal, M. Swaminathan, M. Nealon, and A. Omer, "Design Trade-offs Among MCM-C, MCM-D and MCM-D/C Technologies," *IEEE Trans. Components Packag. Manuf. Technol. B*, **17**, No. 1, pp. 22–29 (1994).

12. T.F. Redmond, C. Prasad, and G. Walker, "Polyimide–Copper Thin Film Redistribution on Glass Ceramic/Copper Multilevel Substrates," *Proc. IEEE 41st Electronic Components and Technology Conference*, Atlanta, May 13–15, pp. 689–693 (1991).

13. E.D. Perfecto, K. Kelly, K. Lidestri, H. Longworth, T. Wassick, J. Pennacchia, M. Ellsworth, and A. Merryman, "MCM-D/C Application for a High Performance Module," *Proc. ISHM/IEPS 5th International Conference on Multichip Modules*, Denver, April 17–19, pp. 69–74 (1996).

14. M. Ellsworth, H. Hammel, E. Perfecto, and T. Wassick, "A High Density, High Performance MCM-D/C Package: A Design, Electrical and Process Perspective," *Proc. IEEE 46th Electronic Components and Technology Conference*, Orlando, May 28–31, pp. 821–828 (1996).

15. W. Shutler, H. Longworth, J. Pennacchia, E. Perfecto, and R. Shields, "A Family of High Performance MCM-C/D Packages Utilizing Cofired Alumina Multilayer Ceramic and Shielded Thin Film Redistribution Structure," *Int. J. Microcirc. Electron. Packag.*, **20**, No. 3, pp. 289–296 (1997).

16. G. Katopis, D. Becker, H. Smith, and H. Stoller, "MCM-C/D Design for the CMOS Implementation of the S/390 System," *Proc. IEEE 47th Electronic Components and Technology Conference*, San Jose, May 18–21, pp. 479–485 (1997).

17. E.D. Perfecto, K. Prasad, C. Osborn, M. Swaminathan, G. White, and C. Prasad, "Comparison of Planar and Non-Planar Multi-Level Cu–Polyimide Interconnect," *Proc. IEEE 43rd Electronic Components and Technology Conference*, Orlando, June 1–4 (1993).

18. K. Prasad and E.D. Perfecto, "Multilevel Thin Film Applications and Processes for High End Systems," *IEEE Trans. Components Packag. Manuf. Technol. B*, **17**, No. 1, pp. 38–49, February (1994).

19. T.E. Dinan and M. Datta, "Kinetics of Copper Etching in Acid Ammonium Persulfate Solutions," *Proc. Symposium on High Rate Metal Dissolution Processes*, PV 95–19, The Electrochemical Society, Pennington, NJ, pp. 189–201 (1995).

20. K.K.H. Wong, S. Kaja, and P.W. DeHaven, "Metallization by Plating for High-Performance Multichip Modules," *IBM J. Res. Develop.*, **42**, No. 5, pp. 587–596 (1998).

21. Y.-H. Kim, G.F. Walker, J. Kim, and J. Park, "Adhesion and Interface Studies Between Copper and Polyimide," *J. Adhes. Sci. Technol.* **1**, No. 4, pp. 331–339 (1987).

22. Y.-H. Kim, J. Kim, G.F. Walker, C. Feger, and S.P. Kowalczyk, "Adhesion and Interface Investigation of Polyimide on Metals," *J. Adhes. Sci. Technol.* **2**, No. 2, pp. 95–105 (1987).

23. S.K. Ray, K. Beckham, and R. Master, "Flip Chip Interconnection Technology for Advance Thermal Conduction Modules," *Proc. IEEE 41st Electronic Components and Technology Conference*, Atlanta, May 11–16, pp. 772–778 (1991).

24. E.D. Perfecto, S. Ray, T.A. Wassick, and H. Stroller, "Evolution of EC and Repair Technology in High Performance Multi-Chip Modules at IBM," presented at the IEEE 48th Electronic Components and Technology Conference, Seattle, May 26–28 (1998).

25. R. Messier, A.P. Giri, and R.A. Roy, "Revised Structure–Zone Model for Thin Film Physical Structure," *J. Vac. Sci. Technol. A*, **2**, No. 2, p. 500 (1984).

26. E.D. Perfecto, K. Prasad, C. Osborn, M. Swaminathan, G. White, and C. Prasad, "Comparison of Planar and Non-Planar Multi-Level Cu–Polyimide Interconnect," *Proc. IEEE 43rd Electronic Components and Technology Conference*, Orlando, June 1–4 (1993).

27. K. Prasad and E.D. Perfecto, "Multilevel Thin Film Applications and Processes for High End Systems," *IEEE Trans. Components Packag. Manuf. Technol. B*, **17**, No. 1, pp. 38–49, February (1994).

28. M. Swaminathan, E. Perfecto, and K. Prasad, "Thin Film Multichip Module Technology at IBM — Design and Electrical," *Proc. IEPS International Electronics Packaging Conference and Exhibition*, San Diego, September 12–15, pp. 10–17 (1993).

29. S. Wakabayashi, S. Koyama et al., "A Build-Up Substrate Utilizing a New Via Fill Technology by Electroplating," *Proc. 4th International Conference on Adhesive Joining and Coating Technology in Electronics Manufacturing*, pp. 280–288 (2000).

30. E.D. Perfecto and L. Goldmann, "Increased Thin Film Wiring Density by Stacked Vias," *IEEE 52nd ECTC Proceedings*, San Diego, CA, pp. 651–656, May (2002).

31. M. Swaminathan, E. Perfecto, and K. Prasad, "Thin Film Multichip Module Technology at IBM — Design and Electrical," *Proc. IEPS International Electronics Packaging Conference and Exhibition*, San Diego, September 12–15, pp. 10–17 (1993).

32. A. Iqbal, M. Swaminathan, M. Nealon, and A. Omer, "Design Trade-offs Among MCM-C, MCM-D and MCM-D/C Technologies," *IEEE Trans. Components Packag. Manuf. Technol. B*, **17**, No. 1, pp. 22–29 (1994).

33. T.F. Redmond, C. Prasad, and G. Walker, "Polyimide–Copper Thin Film Redistribution on Glass Ceramic/Copper Multilevel Substrates," *Proc. IEEE 41st Electronic Components and Technology Conference*, Atlanta, May 13–15, pp. 689–693 (1991).

34. E.D. Perfecto, K. Kelly, K. Lidestri, H. Longworth, T. Wassick, J. Pennacchia, M. Ellsworth, and A. Merryman, "MCM-D/C Application for a High Performance Module," *Proc. ISHM/IEPS 5th International Conference on Multichip Modules*, Denver, April 17–19, pp. 69–74 (1996).

35. M. Ellsworth, H. Hammel, E. Perfecto, and T. Wassick, "A High Density, High Performance MCM-D/C Package: A Design, Electrical and Process Perspective," *Proc. IEEE 46th Electronic Components and Technology Conference*, Orlando, May 28–31, pp. 821–828 (1996).

36. W. Shutler, H. Longworth, J. Pennacchia, E. Perfecto, and R. Shields, "A Family of High Performance MCM-C/D Packages Utilizing Cofired Alumina Multilayer Ceramic and Shielded Thin Film Redistribution Structure," *Int. J. Microcirc. Electron. Packag.*, **20**, No. 3, pp. 289–296 (1997).

37. G. Katopis, D. Becker, H. Smith, and H. Stoller, "MCM-C/D Design for the CMOS Implementation of the S/390 System," *Proc. IEEE 47th Electronic Components and Technology Conference*, San Jose, May 18–21, pp. 479–485 (1997).

38. G. Katopis, D. Becker, H. Smith, and H. Stoller, "MCM-C/D Design for the CMOS Implementation of the S/390 System," *Proc. IEEE 47th Electronic Components and Technology Conference*, San Jose, May 18–21, pp. 479–485 (1997).

39. E.D. Perfecto, K. Prasad, C. Osborn, M. Swaminathan, G. White, and C. Prasad, "Comparison of Planar and Non-Planar Multi-Level Cu–Polyimide Interconnect," *Proc. IEEE 43rd Electronic Components and Technology Conference*, Orlando, June 1–4 (1993).

40. V.M. Dubin, S. Lopatin, A. Kohn, N. Petrov, M. Eizenberg, and Y. Shacham-Diamand, "Electroless Barrier and Seed Layers for On-Chip Metallization," Chapter 3, this book.

41. M. Pourbaix, *The Thermodynamics of Aqueous Solutions* (Translation), Edward Arnold Publishers, London (1949).

42. L.T. Romankiw, "Pattern Generation in Metal Films Using Wet Chemical Techniques: A Review," *ECS Proceedings of Symposium on Etching for Pattern Definition*, H.G. Hughes and M.J. Rand, eds., Electrochemical Society, Princeton, NJ, pp. 161–193 (1976).

43. E.D. Perfecto, K.-W. Lee, H. Hamel, T. Wassick, C. Cline, M. Oonk, C. Feger, and D. Mcherron, "Evaluation of Cu Capping Alternatives for Polyimide–Cu MCM-D," *IEEE 51st ECTC Proceedings*, Orlando, FL, May (2001).

44. R. Jagannathan and M. Krishnan, "Electroless Plating of Copper at a Low pH Level," *IBM J. Res. Develop.*, **37**, pp. 117–123 (1993).

45. S. Ray, D. Berger et al., "Dual-Level Metal (DLM) Method for Fabricating Thin Film Wiring Structures," *Proc. 43rd Electronic Component and Technology Conference*,

pp. 538–543 (1993).
46. K. Prasad and E.D. Perfecto, "Multilevel Thin Film Applications and Processes for High End Systems," *IEEE Trans. Components Packag. Manuf. Technol. B*, **17**, No. 1, pp. 38–49, February (1994).
47. L.T. Romankiw, "Pattern Generation in Metal Films Using Wet Chemical Techniques: A Review," *ECS Proceedings of Symposium on Etching for Pattern Definition*, H.G. Hughes and M.J. Rand, eds., Electrochemical Society, Princeton, NJ, pp. 161–193 (1976).
48. L.T. Romankiw, "Electrochemical Technology in Electronics Today and its Future: A Review," *Oberflache/Surfaces*, **25**, pp. 238–247 (1984).
49. L.T. Romankiw, "A Path from Electroplating Through Lithographic Masks in Electronics to LIGA in MEMS," *Electrochem. Acta*, **42**, No. 20/22, pp. 2985–3005 (1997).
50. A.B. Frazier, R.O. Warrington, and C. Friedrich, "The Miniaturization Technologies: Past, Present, and Future," *IEEE Trans. Indust. Electron.*, **42**, pp. 423–430 (1995).
51. E.W. Becker, W. Eherfeld, P. Hagmann, A. Maner, and D. Munchmeyer, "Fabrication of Microstructures with High Aspect Ratios and Great Structural Heights by Synchrotron Radiation Lithography, Galvanoforming, and Plastic Molding (LIGA process)," *Microelectron. Eng.*, **4**, pp. 35–56 (1986).
52. L.T. Romankiw, "Electroforming of Electronic Devices," *J. Plating Surface Finishing*, **84**, pp. 10–16 (1997).
53. L.T. Romankiw, "A Path from Electroplating Through Lithographic Masks in Electronics to LIGA in MEMS," *Electrochem. Acta*, **42**, No. 20/22, pp. 2985–3005 (1997).
54. B. Lochel, A. Maciossek, H.J. Quenzer, and B. Wagner, "Ultraviolet Depth Lithography and Galvanoforming for Micromachining," *J. Electrochem. Soc.* **143**, pp. 237–244 (1996).
55. L.T. Romankiw, S. Krongelb, D.A. Thompson, R. Anderson, E.E. Castellani, P.M. McCaffrey, A.T. Pfeiffer, and B.J. Stoeber, "Electrodeposition in Thin-Film Recording Head Fabrication," *Extended Abstracts Electrochemical Society*, **79**, No. 2, pp. 1170–1172, October (1979).
56. S. Krongelb, J.O. Dukovic, M.L. Komsa, S. Mehdizadeh, L.T. Romankiw, P.C. Andricacos, A.T. Pfeiffer, and K. Wong, "The Application of Electrodeposition Processes to Advanced Package Fabrication," International Symposium on Advances in Interconnection and Packaging, *SPIE Proceedings*, **1389**, pp. 249–256 (1990).
57. P.C. Andricacos and L.T. Romankiw, "Magnetically Soft Materials in Data Storage: Their Properties and Electrochemistry," *Advances in Electrochemical Science and Engineering*, H. Gerischer and C.W. Tobias, eds., VCH Publishers, New York, pp. 230–321 (1994).
58. C. Karakus and D.T. Chin, "Metal Distribution in Jet Plating," *J. Electrochem. Soc.*, **141**, pp. 691–697 (1994).
59. J.V. Powers and L.T. Romankiw, U.S. Patent 3,652,442.
60. P.C. Andricacos, K.G. Berridge, J.O. Dukovic, M. Flotta, J. Ordonez, H.R. Poweleit, J.S. Richter, L.T. Romankiw, O.P. Schick, F. Spera, and K.-H. Wong, "Vertical Paddle Cell," U.S. Patent 5,516,412 (May 14, 1966).
61. P.C. Andricacos, K.G. Berridge, J.O. Dukovic, and L.T. Romankiw, "Electroplating Workpiece Fixture," U.S. Patent 5,522,975 (June 4, 1966).
62. S. Mehdizadeh, J. Dukovic, P.C. Andricacos, L.T. Romankiw, and H.Y. Cheh, "Optimization of Electrodeposit Uniformity by the Use of Auxiliary Electrodes," *J. Electrochem. Soc.*, **137**, pp. 110–117 (1990).
63. S. Mehdizadeh, J. Dukovic, P.C. Andricacos, L.T. Romankiw, and H.Y. Cheh, "The Influence of Lithographic Patterning on Current Distribution: A Model for Microfabrication by Electrodeposition," *J. Electrochem. Soc.*, **139**, pp. 78–91 (1992).
64. J.O. Dukovic, "Computation of Current Distribution in Electrodeposition, a Review," *IBM J. Res. Develop.*, **34**, pp. 693–704 (1990).
65. S. Mehdizadeh, J. Dukovic, P.C. Andricacos, L.T. Romankiw, and H.Y. Cheh, "The Influence of Lithographic Patterning on Current Distribution: A Model for

Microfabrication by Electrodeposition," *J. Electrochem. Soc.*, **139**, pp. 78–91 (1992).

66. S. Mehdizadeh, J. Dukovic, P.C. Andricacos, L.T. Romankiw, and H.Y. Cheh, "Optimization of Electrodeposit Uniformity by the Use of Auxiliary Electrodes," *J. Electrochem. Soc.*, **137**, pp. 110–117 (1990).

67. J.O. Dukovic, "Feature–Scale Simulation of Resist Patterned Electrodeposition," *IBM J. Res. Develop.*, **37**, pp. 125–141 (1993).

68. S. Mehdizadeh, J.O. Dukovic, P.C. Andricacos, LT. Romankiw, and H.Y. Cheh, "Influence of Lithographic Patterning Current Distribution in Electrodeposition: Patterning Current Distribution in Electrodeposition: Experimental Study of Mass-Transfer Effects," *J. Electrochem. Soc.*, **140**, pp. 3497–3505 (1993).

69. S. Mehdizadeh, J. Dukovic, P.C. Andricacos, L.T. Romankiw, and H.Y. Cheh, "The Influence of Lithographic Patterning on Current Distribution: A Model for Microfabrication by Electrodeposition," *J. Electrochem. Soc.*, **139**, pp. 78–91 (1992).

70. C. Madore, D. Landolt, C. Hassenpflug, and J.A. Hermann, "Application of the Rotating Cylinder Hull Cell to the Measurement of Throwing Power and the Monitoring of Copper Plating Baths," *Plating Surf. Finish.*, **82**, pp. 36–41 (1995).

71. R. Hac, C. Ogden, and D. Trench, "Cyclic Voltammetric Stripping Analysis of Acid Copper Sulphate Plating Baths, I. Polyether-Sulphide Based Additives, II. Sulfonium-alkanesulfonate Based Additives," *Plating*, **63**, pp. 62–66 (1982).

72. D.R. Gave and G.D. Wilcox, "Hull Cell," *Trans. Inst. Metal Finish*, **71**, pp. 71–73 (1993).

73. L.T. Koh, G.Z. You, C.Y. Li, and P.D. Foo, "Investigation of the Effects of Byproduct Components in Cu Plating for Advanced Interconnect Metallization," *Microelectron. J.*, **33**, pp. 229–234 (2002).

74. R. Messier, A.P. Giri, and R.A. Roy, "Revised Structure–Zone Model for Thin Film Physical Structure," *J. Vac. Sci. Technol. A*, **2**, No. 2, p. 500 (1984).

75. L.T. Romankiw, "Pattern Generation in Metal Films Using Wet Chemical Techniques: A Review," *Proceedings of Symposium on Etching for Pattern Definition*, H.G. Hughes and M. Rand, eds., The Electrochemical Society, Pennington, NJ, pp. 161–193 (1976).

76. L.T. Romankiw, "Pattern Generation in Metal Films Using Wet Chemical Techniques: A Review," *ECS Proceedings of Symposium on Etching for Pattern Definition*, H.G. Hughes and M.J. Rand, eds., Electrochemical Society, Princeton, NJ, pp. 161–193 (1976).

77. E.D. Perfecto, A.P. Giri, R.R. Sheilds, H.P. Longworth, J.R. Pennacchia, and M.P. Jeanneret, "Thin-Film Multichip Module Packages for High-End IBM Servers," *IBM J. Res. Dev.*, **42**, pp. 597–606 (1998).

78. E.J. O'Sullivan, A.G. Schrott, M. Paunovic, C.J. Sambucetti, J.R. Marino, P. Bailey, S. Kaja, and K. Semkow, "Electrolessly Deposited Diffusion Barriers for Microelectronics," *IBM J. Res. Dev.*, **42**, pp. 607–620 (1998).

79. E.D. Perfecto, K.-W. Lee, H. Hamel, T. Wassick, C. Cline, M. Oonk, C. Feger, and D. Mcherron, "Evaluation of Cu Capping Alternatives for Polyimide–Cu MCM-D," *IEEE 51st ECTC Proceedings*, Orlando, FL, May (2001).

80. V.M. Dubin, S. Lopatin, A. Kohn, N. Petrov, M. Eizenberg, Y. Shacham-Diamand, "Electroless Barrier and Seed Layers for On-Chip Metallization," Chapter 3, this book.

81. S. Krongelb, L.T. Romankiw, and J.A. Tornello, "Electrochemical Process for Advanced Package Fabrication," *IBM J. Res. Dev.*, **42**, pp. 575–585 (1998).

82. S. Ray, D. Berger et al., "Dual-Level Metal (DLM) Method for Fabricating Thin Film Wiring Structures," *Proc. 43rd. Electronic Component and Technology Conference*, pp. 538–543 (1993).

83. Y.H. Kim, G.F. Walker, J. Kim, and J. Park, "Adhesion and Interface Studies between Copper and Polyimide," *J. Adhes. Sci. Technol.*, **1**, pp. 331–339 (1987).

84. E.J. O'Sullivan, A.G. Schrott, M. Paunovic, C.J. Sambucetti, J.R. Marino, P. Bailey, S.Kaja, and K. Semkow, "Electrolessly Deposited Diffusion Barriers for Microelectronics," *IBM J. Res. Dev.*, **42**, pp. 607–620 (1998).

85. Trade name of Sel-Rex, a division of Enthone-Omni, New Haven, CT.
86. L.T. Romankiw, S. Krongelb, D.A. Thompson, R. Anderson, E.E. Castellani, P.M. McCaffrey, A.T. Pfeiffer, and B.J. Stoeber, "Electrodeposition in Thin-Film Recording Head Fabrication," *Extended Abstracts Electrochemical Society*, **79**, No. 2, pp. 1170–1172, October (1979).
87. S. Krongelb, J.O. Dukovic, M.L. Komsa, S. Mehdizadeh, L.T. Romankiw, P.C. Andricacos, A.T. Pfeiffer, and K. Wong, "The Application of Electrodeposition Processes to Advanced Package Fabrication," International Symposium on Advances in Interconnection and Packaging, *SPIE Proceedings*, **1389**, pp. 249–256 (1990).
88. E.J. O'Sullivan, A.G. Schrott, M. Paunovic, C.J. Sambucetti, J.R. Marino, P. Bailey, S.Kaja, and K. Semkow, "Electrolessly Deposited Diffusion Barriers for Microelectronics," *IBM J. Res. Dev.*, **42**, pp. 607–620 (1998).
89. R.D. Goldblatt, L.M. Ferreiro, S.L. Nunes, R.R. Thomas, N.J. Chou, L.P. Buckwalter, J.E. Heidenreich, and T.H. Chao, "Characterization of Water Vapor Plasma Modified Polyimide," IBM Internal Publication, November 19 (1991).
90. S. Krongelb, L.T. Romankiw, and J.A. Tornello, "Electrochemical Process for Advanced Package Fabrication," *IBM J. Res. Dev.*, **42**, pp. 575–585 (1998).
91. Special issue on the IBM eServer z900, *IBM J. Res. Dev.*, **46**, pp. 361–644 (2002).
92. A. Iqbal, M. Swaminathan, M. Nealon, and A. Omer, "Design Trade-offs Among MCM-C, MCM-D and MCM-D/C Technologies," *IEEE Trans. Components Packag. Manuf. Technol. B*, **17**, No. 1, pp. 22–29 (1994).
93. R. Shields, T. Redmond, H. Stoller, E. Perfecto, and J. Pennacchia, "Thin Film Technology Enhances IBM's Newest Mainframe — the S/390 G5 Server," *IBM Microelectronics*, Fourth Quarter (1998).
94. D.Y. Shih, P. Palmateer, Y. Fu, S. Kapur, B. Ghosal, P. Brofman, P. Lauro, and M. Norcott, "Polyimide Stress Cushion for Multichip Glass–Ceramic Module Packaging," *IEEE, 45th ECTC Proceedings*, Las Vegas, NV, pp. 728–38, May 21–24 (1995).

Acronyms used in the text

BCB	benzocyclobutene (an organic material which may be used as a dielectric)
BPDA-PDA	biphenyl tetracarboxylic acid dianhydride - *p*-phenylene diamine form of polyimide
BSM	backside metallurgy
C4	controlled collapse chip connection
DLM	dual level metal fabrication sequence
DLP	dual level polyimide fabrication sequence
MEMS	micro-electro-mechanical system
MLTF	multilevel thin film
MCM	multichip module
MCM-D	multichip module which includes thin-film dielectric and metal layers
MCM-L	multichip module formed with laminated, organic board technology
TSM	top surface metal (the metal level used for C4 joining)
PAA	polyamic acid
PGA	pin grid array
PI	polyimide
PSPI	photosensitive polyimide

12 Bumping technology for advanced packages

Shinichi Wakabayashi

12.1 Introduction

The semiconductor industry has continued to develop higher performance devices with smaller size and lower cost. [1,2] For the reduction of chip and package sizes, peripherally designed bonding pads on chips and leads on packages are inappropriate. [3] It is considered that the area array configuration of bonding pads on the entire surface of the chip and the flip-chip connection in ball grid array (BGA) package are expected to produce good results of size reduction. Therefore, in recent years, this technology has been very widely accepted in the industry, and has been intensively studied in research and development laboratories. [1]

Bumping technology for packages is mainly required for flip-chip attach to supply a sufficient amount of solder on bonding pads as pre-solder. Several technologies such as solder plating, screen printing, and micro-ball attach are available in the market. In this chapter, the present technical status of flip-chip package manufacturing technology and the bumping technologies on the packages are discussed in detail.

12.2 Process for flip-chip assembly

The flip-chip assembly process is very simple, as shown in Figure 12.1. A chip, which has bumps on it, is flipped and the position of the bumps is aligned to the pads on a package. The chip and package are then put into a furnace and the solder bond is made by reflow. After solder bonding, the interstice of the chip and the package is filled with underfill resin. This underfill resin fastens the chip to the package bonding and distributes the bonding stress throughout the bonding area. Usually, the resin is an epoxy-based material and contains silica fillers to obtain proper thermal coefficient of expansion. By this means, underfill resin improves the bonding reliability drastically. If the flux cleaning before underfill is not satisfactory, de-wetting and weak bonding of the underfill will occur. A flip-chip assembled device is shown in Figure 12.2a. The package is composed of a core layer and buildup layers on both sides of the core. The package and the chip are bonded by the solder bumps and the interstice is filled with resin-containing fillers, as shown in Figure 12.2b. [4]

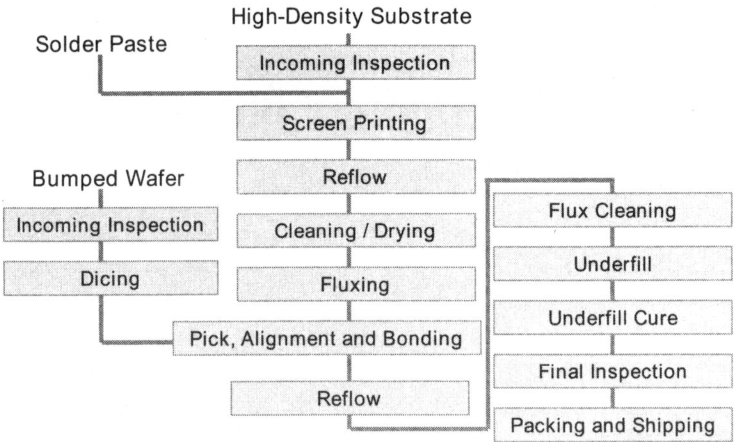

Figure 12.1 A flip-chip assembly process.

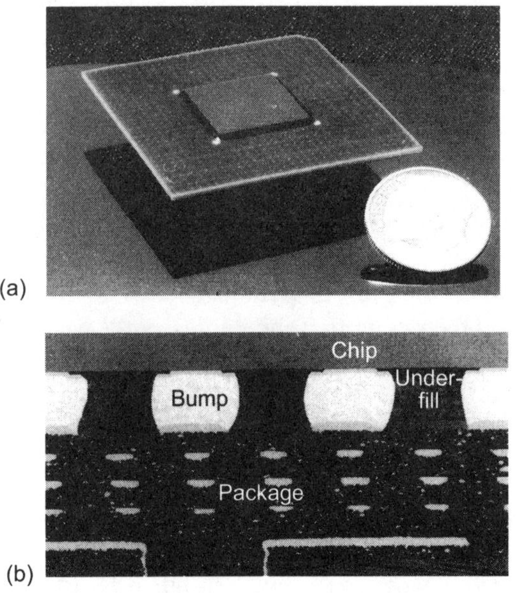

Figure 12.2 A flip-chip assembled device (a) and its cross section (b).

12.3   Manufacturing process for flip-chip packages

Recent flip-chip packages are led by micro-processor unit (MPU) package technology, due to the big demand in the market. In this field, Alumina ceramics have long been used for the package material. However, for the Pentium processor generation, the package material was changed from ceramics to plastic, and the printed circuit board (PCB) manufacturing technology was introduced.[5] This change has brought big success to the technology. In the course of this change, the PCB changed from

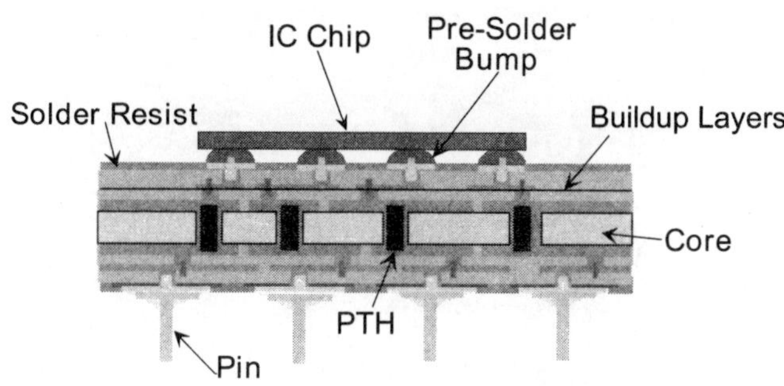

Figure 12.3 A structure of flip-chip pin grid array (FC-BGA) package.

the existing technology to the buildup technology. In accordance with this technology change, the patterning for the package was changed from the subtractive to the semi-additive method. The semi-additive method consists of a photolithographic method for defining patterns and electroless and electrolytic copper plating for making patterns. This method is superior for making fine line patterns for high-density buildup packages. The schematic structure of the flip-chip package is shown in Figure 12.3.[6] The package is composed of a core layer and two buildup layers on each side of the core. This composition is called 2–2–2 structure. Typical design rules for buildup layers adapted to MPU package are line/space 25 µm, via size 65 µm, and flip-chip pad pitch 200 µm, respectively.

The manufacturing process of this package is shown in Figure 12.4. By this process, at first the glass-epoxy core is drilled mechanically, and the through holes are plated with copper and filled with epoxy-plugging resin. The pattern is then defined by a photoresist and etched chemically. After this process, film type buildup layers made of epoxy are vacuum laminated on both sides of the core, and micro-vias are perforated with a CO_2 laser. The vias are then filled with copper by electroplating, and a pattern is formed by the semi-additive method. By repeating this process, a two-layer buildup structure is obtained. The features of the technology used for this package are vacuum lamination of the dielectric layers, resin plugging into the plated through holes, and stacking via on plugged through holes (via on via structure).

According to the roadmap published by International Technology Roadmap for Semiconductor (ITRS),[1] the design rules for line/space and flip-chip pad pitch will reach 20 µm and 150 µm, respectively, in a few years. In recent developments, already established material, processes, and equipment were used to realize the basic design rules mentioned above. The 12 µm line and space patterns fabricated in a laboratory are shown in Figure 12.5.[7] In this case, two buildup layers were fabricated on the core layer, which had copper patterns made by the subtractive method. The lines on the buildup layers were designed as 12 µm line and space and were made by the semi-additive plating method. The feature of these buildup layers is a very smooth surface, which is suitable for making fine patterns. At this time, there are some adhesion problems of copper on the smooth layer. In this photo, the seed layers for semi-additive patterns were made by sputtering technology to achieve good adhesion of copper.

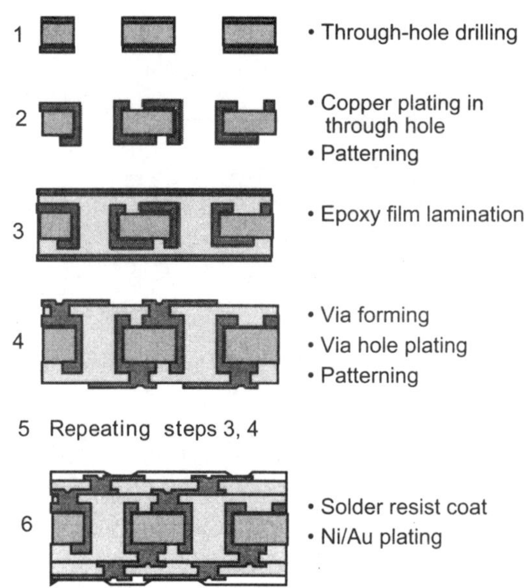

Figure 12.4 The manufacturing process of the flip-chip package.

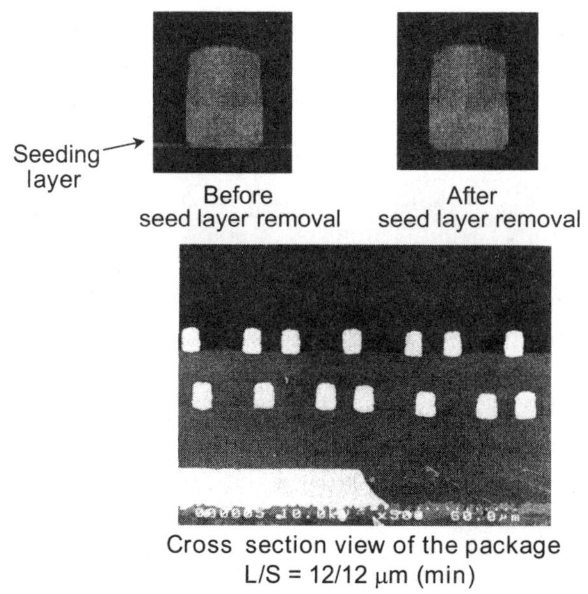

Figure 12.5 A cross-sectional view of the package fabricated with fine lines.

Via fill plating for micro-vias connecting the buildup layers is a new technology introduced recently in actual production lines.[7] In the case of via fill plating, optimized plating solutions and conditions enabled the copper to be preferentially deposited inside the vias during plating. In Figure 12.6, the step-by-step via fill propagation process is shown. To control the deposition rate inside and outside the vias,

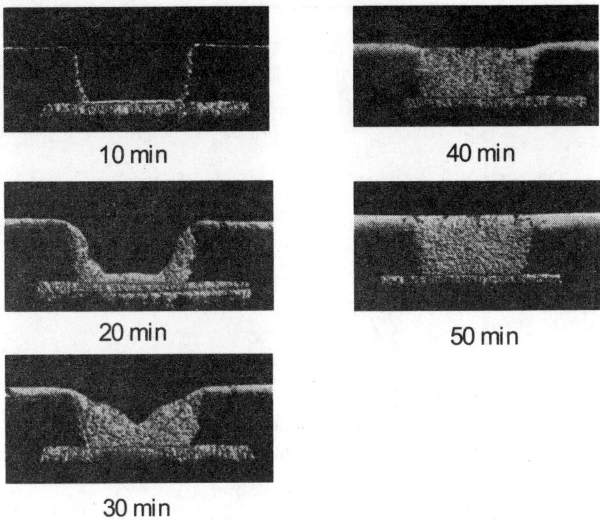

Figure 12.6 The step-by-step via fill process by copper electroplating.

several additives were introduced into the plating solution. In general, the additives for the copper sulfate solution are chloride ion, brightener, polymer, and dye.[8] The brightener is an organic compound containing sulfur, and the polymer is a polyoxy-ethylene-type polymer. These compounds work as an enhancer and suppressor of the copper deposition. The dye is a grain refiner, which causes the grain structure to become more regular and finer than the deposit obtained from the only brightener containing solution.[9] The current–potential curves of the solution, which is commercially available and specially designed for copper via fill, are shown in Figure 12.7. The additives are named brightener, polymer, and leveler, respectively. The compositions of each additive are not clear, but they are mainly composed of *bis*(3-sulfopropyl) disulfide (SPS), polyethylene glycol (PEG), and Janus Green B (JGB), respectively. The current–potential curves were obtained with a rotating disk electrode at 0 and 1000 revolutions per minute (rpm). The disk was preplated with copper at 20 mA/cm^2 for 1 min before each measurement. At this time, a saturated calomel electrode was used as the reference electrode. The standard solution for the polarization measurements is 120 g/l H_2SO_4 and 120 g/l $CuSO_4$. The addition amount of the additives is 10 ml/l of each formulation. By comparison of curve (2) to curve (1) in Figure 12.7, it is clear that the brightener shifts curve (1) to the positive side by its catalytic effect (enhancement) on the copper deposition. On the other hand, the polymer suppresses the copper deposition strongly and shifts curve (1) to the very negative side, as shown in curve (3). Curve (4), which is obtained from the solution containing both the brightener and the polymer, is located at the mean position of curve (2) and curve (3); but, the curve obtained at 0 rpm shows a positive shift from curve (4) that is obtained at 1000 rpm (under strong agitation). This is a unique phenomenon and reverse shift from that which occurs in general current–potential curves. This behavior can be explained as the competitive adsorption of the

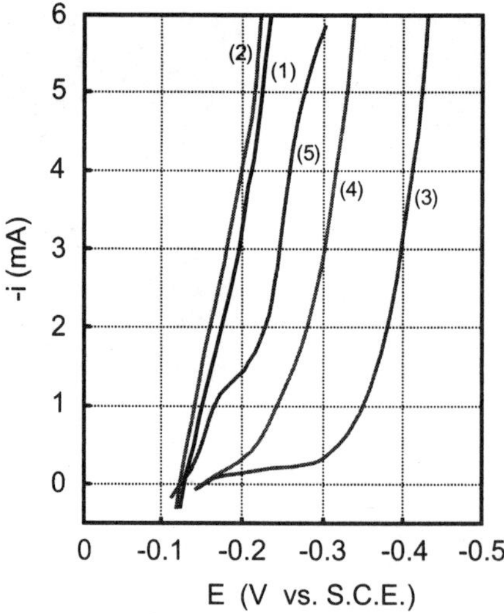

1) Additive free, 1000 rpm   2) Brightener, 1000 rpm

3) Polymer, 1000 rpm   4) Brightener + Polymer, 1000 rpm

5) Brightener + Polymer, 0 rpm

Figure 12.7   The current–potential curves of the copper plating solution containing additives.

additives. Some portion of the brightener (enhancer) adsorbed at the cathode in a still solution is substituted by the polymer (suppressor) caused by strong agitation, and as a result, adsorption of the polymer becomes predominant and suppresses copper deposition more strongly. This is the evidence of diffusion control on the inhibiting action of the polymer, and this action is predominant compared with copper diffusion to the electrode. The data using SPS and PEG as additives for standard sulfate copper solution showed almost the same phenomenon as Figure 12.7, but the phenomenon was not typical compared with Figure 12.7. In this case, the curve containing leveler (JGB) is not shown in Figure 12.7, because the addition of leveler (JGB) is confirmed as being not essential to cause this phenomenon and copper via fill.

In this case, the mechanism of via fill can be explained as follows. At first, SPS and PEG are adsorbed at the cathode with some proportions in still solution, but once the solution is agitated, PEG is supplied to the cathode more than the former condition and replaces the SPS adsorbed at the cathode. As a result, behavior of PEG (diffusion control on inhibiting action) becomes predominant and suppresses copper deposition to a greater extent. The SPS replacement with the PEG was also confirmed by electron spectroscopy for chemical analysis (ESCA). From curves (4) and (5), it is clear that this diffusion control on the inhibiting action is predominant at the strong agitation area rather than the weak agitation area. This result leads to the conclusion that the weak agitation area (inside the via) keeps a higher population of adsorbed SPS (enhancer of copper deposition) than the strong agitation area (outside the via) and induces higher current density onto the inside of the via.

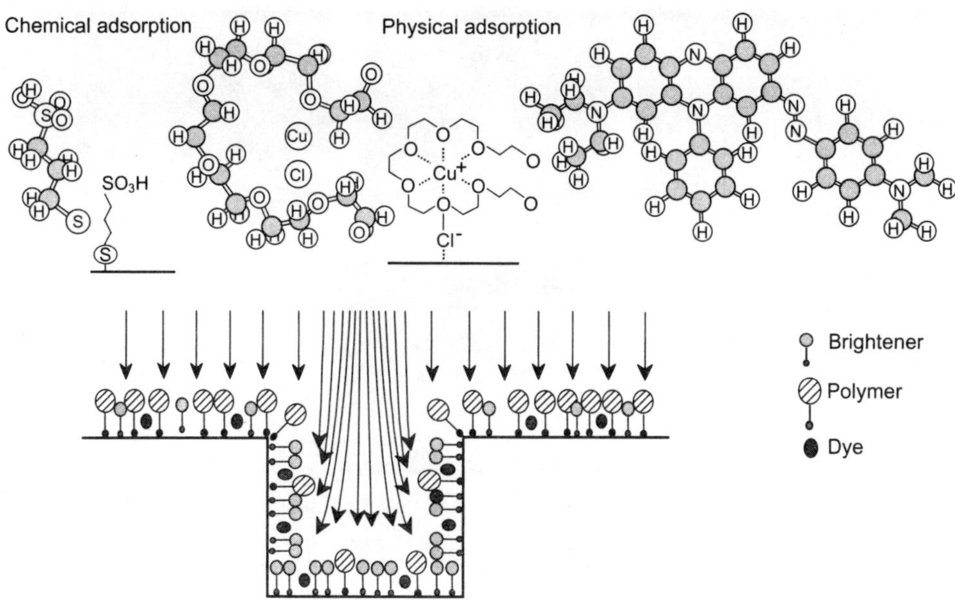

Figure 12.8   The schematic explanation of via fill plating with the adsorption model of the additives.

In the actual plating condition, this mechanism works preferentially under suitably controlled conditions at the inside and the outside of the via, respectively; i.e., a weak agitation area (inside the via) introduces higher current density than a strong agitation area (outside the via), and the via proceeds to fill properly. The schematic explanation of via fill plating with the adsorption model of the additives is shown in Figure 12.8. By previous studies, sulfur compounds like SPS were confirmed to be adsorbed at the cathode by chemical adsorption.[10,11] PEG was also studied very widely and was accepted to make a copper metal complex which makes a further complex structure with Cl⁻ ion that is specifically adsorbed at the electrode.[12] This is a physical adsorption like the PEG adsorption case. The adsorption structure model of the additives (i.e., SPS, PEG, Cl⁻ ion, and JGB) on the via shows a higher population of adsorbed SPS on the inside of the via than on the outside. This structure induces higher current flux onto the inside of the via than onto the outside and fill the via. The JGB is adsorbed with uniform intervals to make regular grains. Once via fill starts, this phenomenon is enhanced automatically because of the increase of the relative SPS concentration by the decrease of the area size in the via, and the same number of SPS still remains at or near the electrode during plating. The detailed electrochemical mechanism of the copper via fill was reported in our previous paper.[6] The evidence of the strong enhancement of the copper deposition by the brightener is shown in Figure 12.9. The copper test piece was cleaned and dipped into the SPS solution for few seconds to adsorb the SPS on half of the copper surface, and then the sample was plated with the via fill plating solution containing no SPS. The plating thickness of the copper on the SPS adsorbed area is almost 17 times higher than that of the SPS-free area.

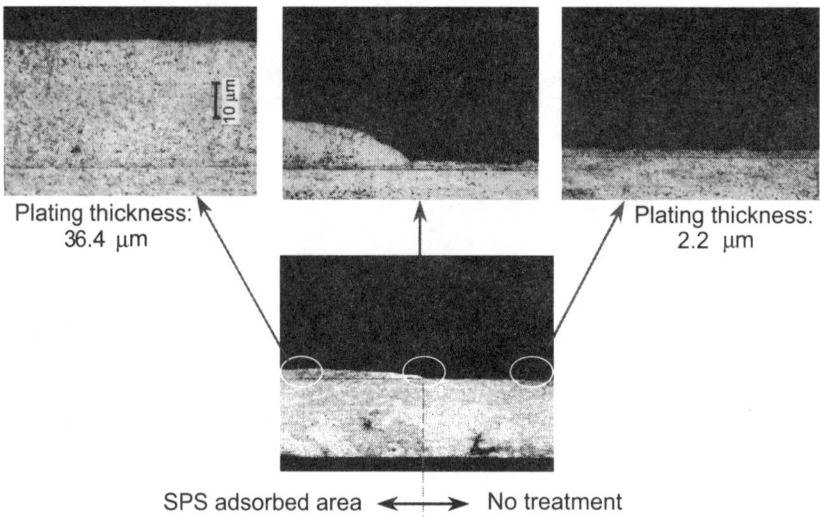

Figure 12.9   The enhancement of copper deposition by brightener.

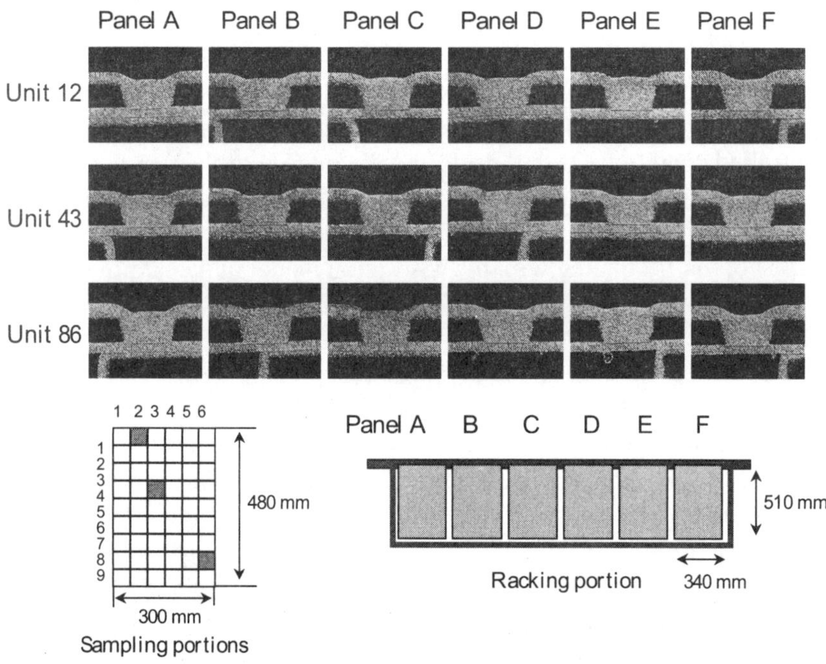

Figure 12.10   The uniformity of vias filled with copper plating.

Via fill technology has already been transferred to actual production lines. An example of the vias satisfactorily filled by copper plating in a production line is shown in Figure 12.10. In this case, six large panels (510 × 340 mm) are placed on a rack and three points of each panel are selected to evaluate the via filling. The results are satisfactory for actual application.

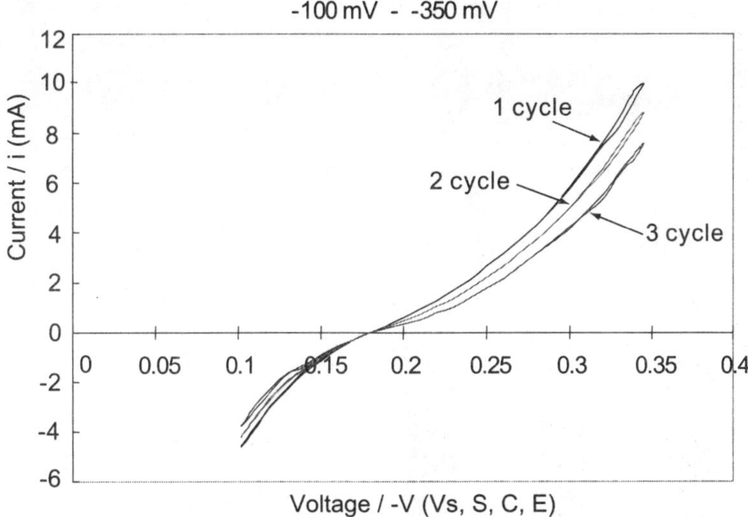

Figure 12.11   The cyclic voltammogram of the copper plating solution contains SPS and PEG with SPS adsorbed copper electrode (scan speed 50 mV/sec).

Recently several new technologies have been developed. These are multi-stacked vias, simultaneous plating of via and line, and uniform plating of the inner walls of very high aspect ratio through vias [13] by a control system based on the understanding of the via fill mechanism. For micro-through via plating technology, pulse reverse plating technology was applied to control the adsorption structure of the brightener and the polymer at the inside and outside of the vias, which has a very high aspect ratio.

This is based on the fact that the adsorption and desorption behavior can be controlled by the anodic scan of the electrode. The cyclic voltammogram of the copper solution contains SPS and PEG with SPS adsorbed copper electrode and is shown in Figure 12.11. The cyclic voltammogram obtained by the negative scan from the equilibrium potential shows almost the same current–potential curve as curve (2) in Figure 12.7, but, by the following positive scan to 0.12V vs. SCE and the second negative scan, the voltammogram is different from the first scan. The current–potential curve shifted negative like curve (3). This means desorption (stripping) of the SPS at the positive potentials and the replacement of SPS with PEG during the following negative scan. At this replacement, the chloride ion that is specifically adsorbed at the electrode is considered to contribute to the PEG adsorption to a certain extent.

In general, via filling is optimized by the adjustment of agitation and pulse reverse conditions. By the agitation, the outside and the inside of the via will have a bigger difference of flow rate of the solution, and a relatively SPS-rich adsorption structure is produced in the inside of the via. The use of periodic pulse reverse (PPR) current will make almost the same adsorption structure in micro-through via by the stripping of SPS and replacement with PEG. The gradient of the SPS amount adsorbed at the inside and outside of the via can be properly controlled by the geometrical effect

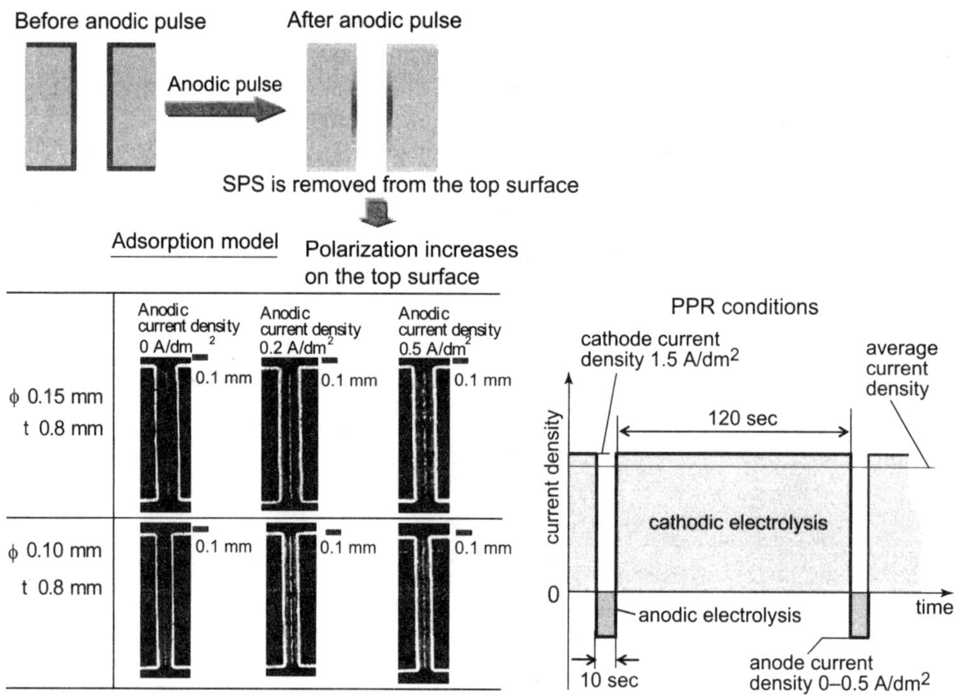

Figure 12.12   Cross-sectional views of the copper plating in very high aspect ratio vias.

of the potential gradient and applied current. In the case of micro through via plating (Figure 12.12), the inside of the via is first plated with electroless plating to form a seed layer for electroplating. Then the micro through via is treated with the SPS containing solution to adsorb SPS at the inside of the via. After this, the panel is treated properly by the anodic pulse and makes the SPS-rich adsorption structure in the inside of the via. This adsorption layer induces higher current density in the inside of the via for the plating and cancels the geometrical current distribution effect (higher current concentration at the outside) to form uniform thickness plating. The experimental results shown in Figure 12.12 are very uniform for vias of high aspect ratio.[13]

The importance of multi-stacked vias is generally understood as an effective way of space saving for a high-density package, as shown in Figure 12.13. In this case, four or three filled vias are stacked and the drilled through hole is filled with plugging resin. The most important point of this design is that it enables the reduction of loop inductance of the power lines by the shortening of the lines. The simulation data of the loop inductance reduction of stacked vias are shown in Figure 12.13. The inductances of the stacked and the off set vias models are 47 and 191 pH, respectively. Therefore, the stacked vias structure is very advantageous for the thin package because of the availability of the short path.

After lamination and patterning of the top layer and bottom layer of the packages, flip-chip pads were pre-soldered. Finally chip capacitors and pins were assembled. In Figure 12.14, one of the most advanced flip-chip packages which has via-on-via structure and stacked vias with pre-solder bumps is shown.

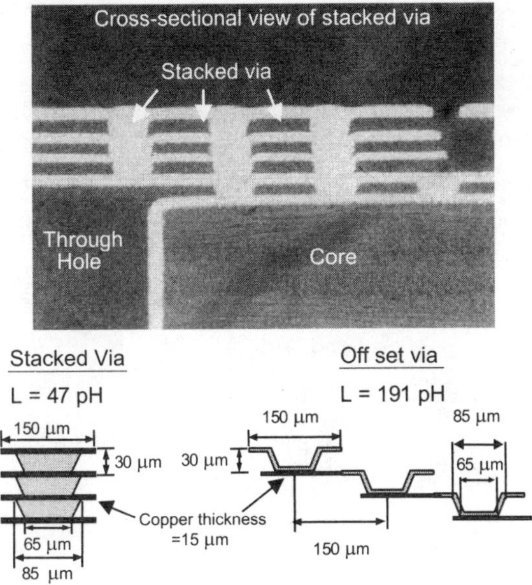

Figure 12.13 The stacked vias and the simulation results of the loop inductance.

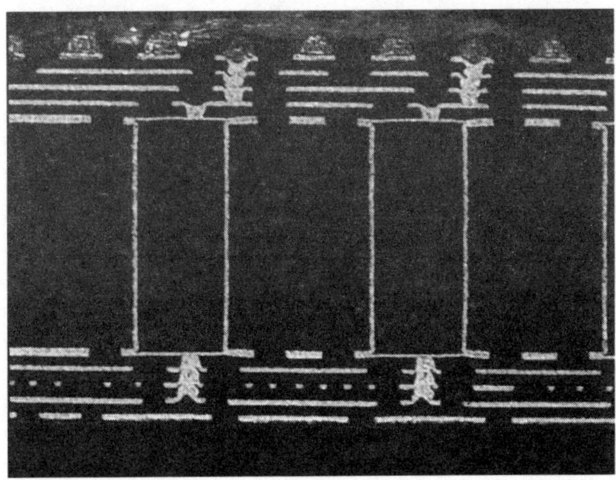

Figure 12.14 The cross-sectional view of an advanced flip-chip package.

12.4 Pre-soldering technology for flip-chip pads on packages

12.4.1 Screen printing

Pre-soldering on flip-chip pads uses screen-printing technology very widely. This technology is a cost competitive and has high productivity for present design rules.[14] The process flow and design rules are shown in Figures 12.15 and 2.16. The printing process is composed of the printing of the solder paste with a screen and a squeeze,

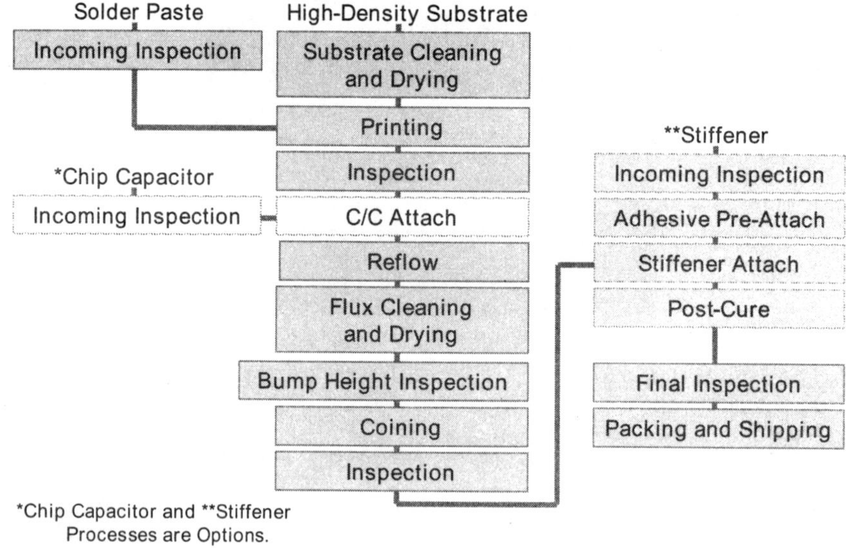

Figure 12.15 A pre-soldering process flow by screen printing.

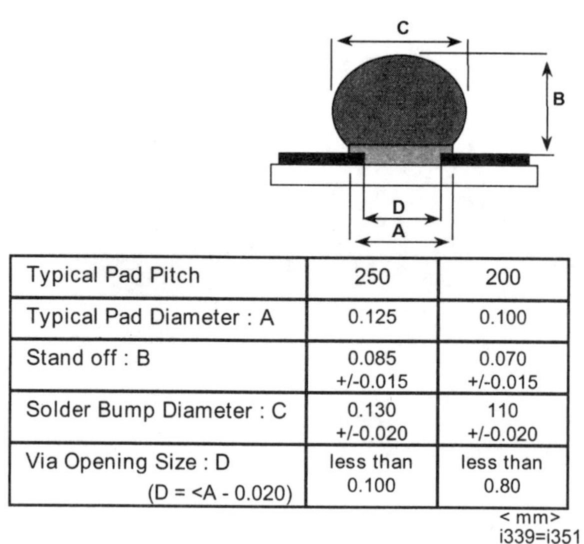

Typical Pad Pitch	250	200
Typical Pad Diameter : A	0.125	0.100
Stand off : B	0.085 +/-0.015	0.070 +/-0.015
Solder Bump Diameter : C	0.130 +/-0.020	110 +/-0.020
Via Opening Size : D (D = <A - 0.020)	less than 0.100	less than 0.80

Figure 12.16 The present design rules for bumps.

reflow of the printed paste, cleaning of the flux residues, and coining of the solder bumps to obtain uniform bump height. Solder paste, masks, and printers needed by this technology are easily available in the market. If the machines and the processes are properly installed, bump fabrication by this technology is not as difficult at present. The typical pad pitch at this time is around 200 μm and the height is around 100 μm. The solder pads, printing pastes, and reflowed and coined bumps by this technology are shown in Figure 12.17. The data of bump height and distribution are

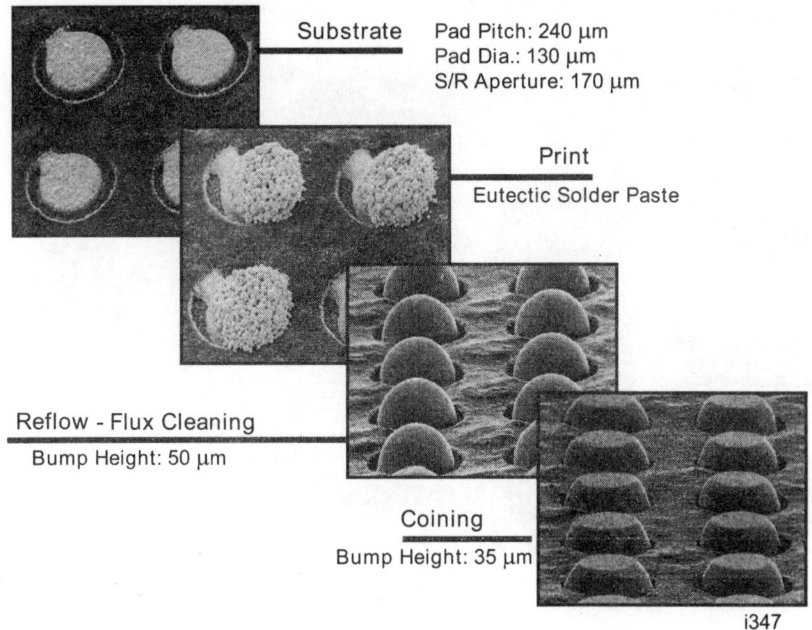

Figure 12.17 The pre-soldered bumps.

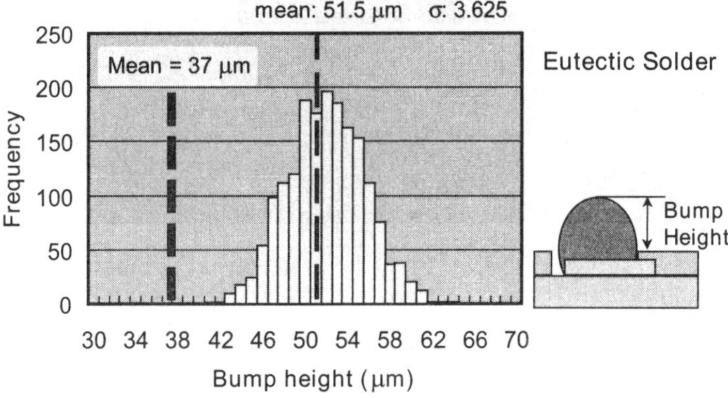

Figure 12.18 The bump height distribution data.

shown in Figure 12.18. The values of σ – 3.6 μm of printed 50 μm high bumps and 2.1 μm after coining (bump height 35 μm) – are considered to be satisfactory for an actual flip-chip attach.

The limitation of this technology seems to be around 150 μm, and several new technologies for finer pitch printing have been investigated. In this technology, a photoresist is used for defining the printing position and areas. This means the primary technology constraint to achieve fine printing is determined by the resolution of the photoresist. Therefore, this is a very hopeful technology for future buildup packages. This technology needs specially designed paste to fill the openings of the

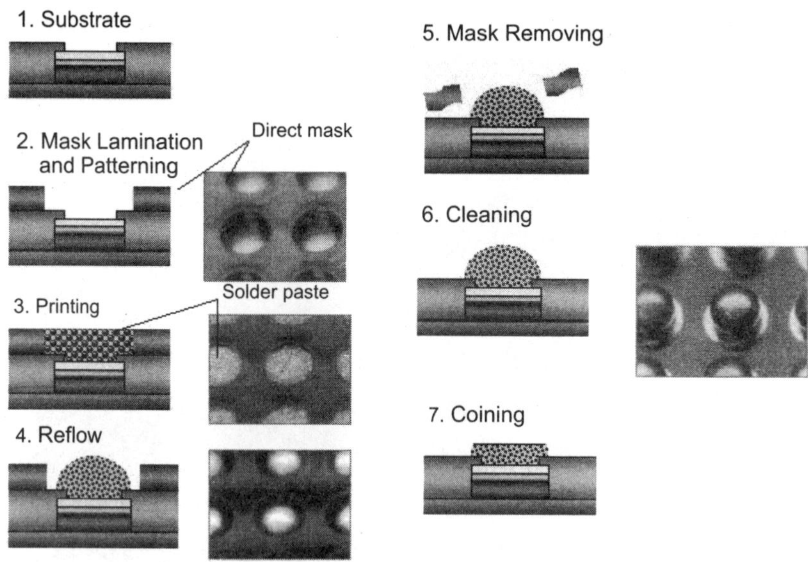

Figure 12.19   A new screen-printing technology for fine pitch fabrication.

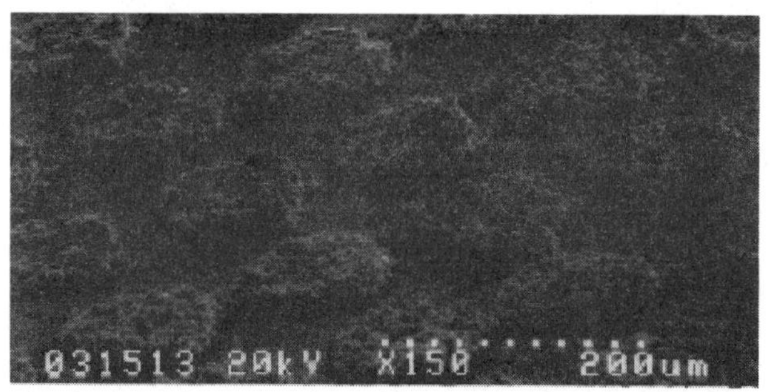

Solder	Tm, °C	Reflow conditions, °C
Sn–Ag(3.5)	221	255 peak, 220 over 30 sec
Sn–Ag(3.0)–Cu(0.5)	217–220	255 peak, 220 over 30 sec
Sn–Ag(2.5)–Cu(0.5)–Bi(1.0)	214–221	250 peak, 230 over 30 sec
Sn–Zn(7.8)–Bi(3.0)	187–197	
Sn(63)–Pb(37)	183	230 peak, 183 over 40 sec

Figure 12.20   The candidate materials for lead-free bumps and printed Sn–Ag paste.

photo mask and mask removal. This process is shown in Figure 12.19, and the bumps fabricated by this technology are also shown in Figure 12.19. The technologies classified in this category are now in development status at various laboratories.

A recent requirement in the market is lead-free bumps for packaging. The candidate materials for lead-free bumps are shown in Figure 12.20. Mainly, Sn–Ag and Sn–Ag–Cu materials are considered to be solders used for a wide range of applications.

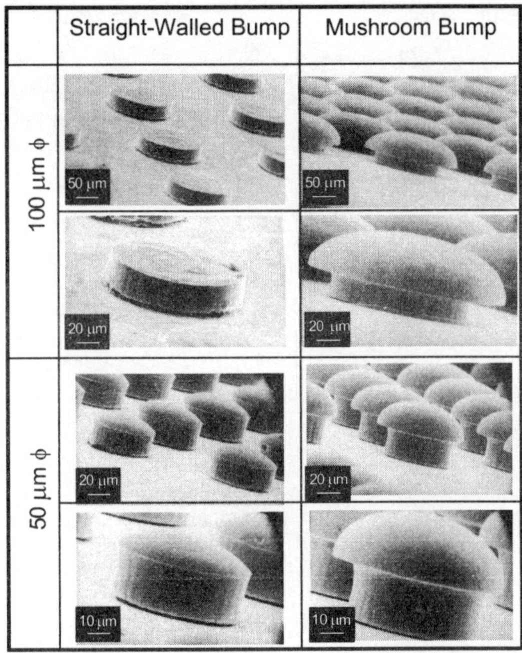

Figure 12.21 The Ag–Sn mushroom bumps made by plating.

12.4.2 *Plating*

Solder plating for bumping is a popular technology for chip bumping because the technology easily makes very fine bumps.[15–17] Depending on photolithographic technology, the mechanical damage of the wafer is very limited. The bump plating process for the package side is almost the same as the process shown in Figure 12.19 except for printing. The defining solder pad areas by the photoresist is plated with solder plating. The plating solution should consist of low attack materials for the photoresist and should have a high deposition rate.

Plating uniformity all over the substrate is very important. To achieve good uniformity of the bump height, specially designed machines are used for wafer bumping.[18] In ordinary cases of package pre-soldering, rack plating with specially designed jigs to make uniform current distribution is used. As a special case needing a relatively large amount of solder on the pad, mushroom bumps are required. Mushroom bumps made by plating are shown in Figure 12.21.

In this plating bump case, it is also possible to make lead-free bumps like printing. Lead-free solutions such as Sn–Ag and Sn–Ag–Cu are available in the market.

12.4.3 *Micro-ball attach*

The ball mounting technology for existing BGAs was attempted to attach micro-balls as a pre-solder on a package. The process flow is shown in Figure 12.22. At first the

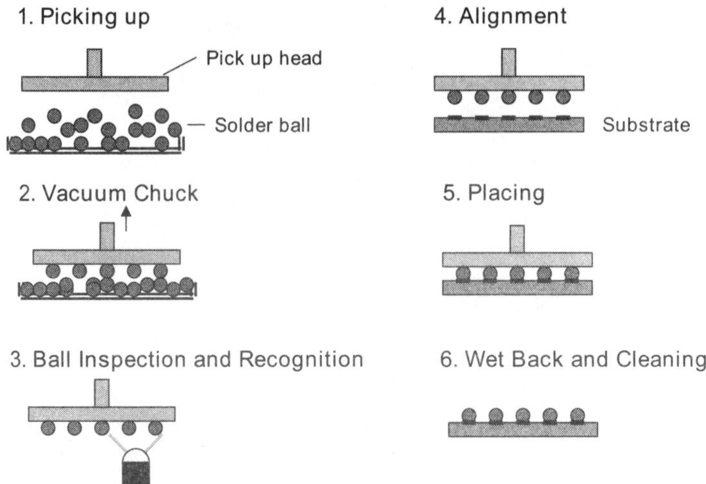

Figure 12.22   The process flow of the micro-ball attach.

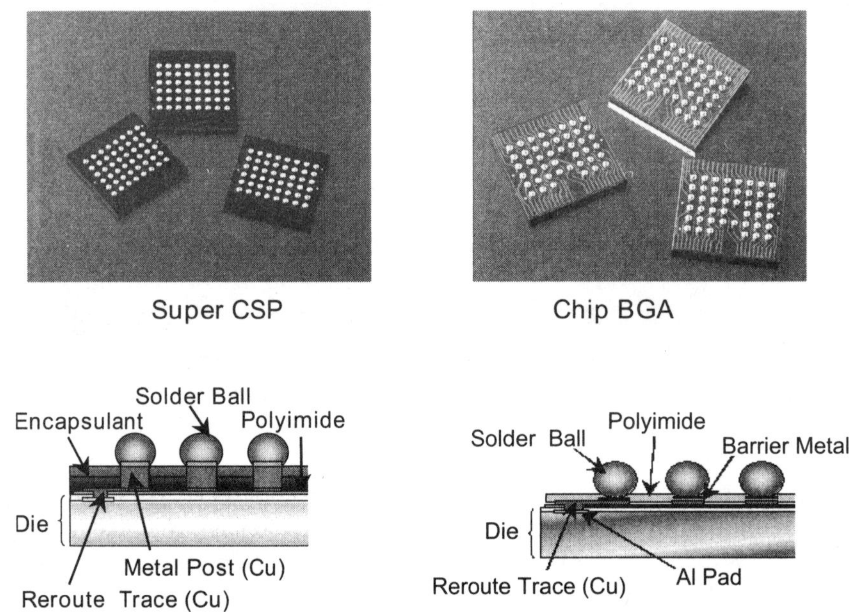

Figure 12.23   Super-CSP and chip BGA.

micro-balls are aligned on a specially designed jig, which has small holes to pick the micro-balls up by vacuum. The ball inspection and the recognition of the lack of balls on the jig are very important in this process. The balls are then transferred to the package and reflowed. The technology has been recently refined to handle finer balls. The available ball size for this technology ranges from 50 to 200 µm in diameter, and the distribution of the size is ±2 µm. Lead-free solder balls are also available for use in this case.

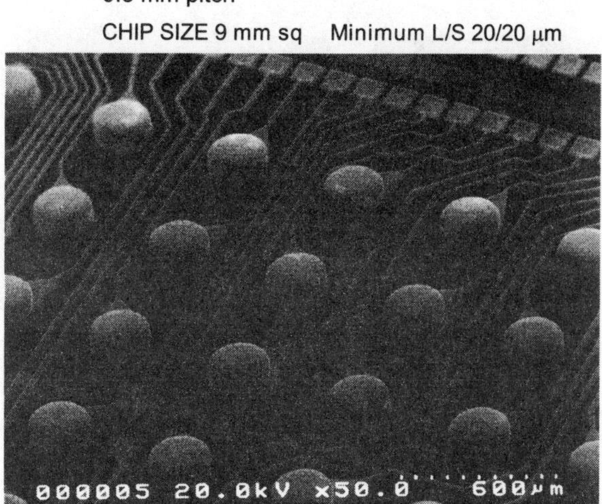

Figure 12.24 The copper bumps with rerouting traces of super-CSP.

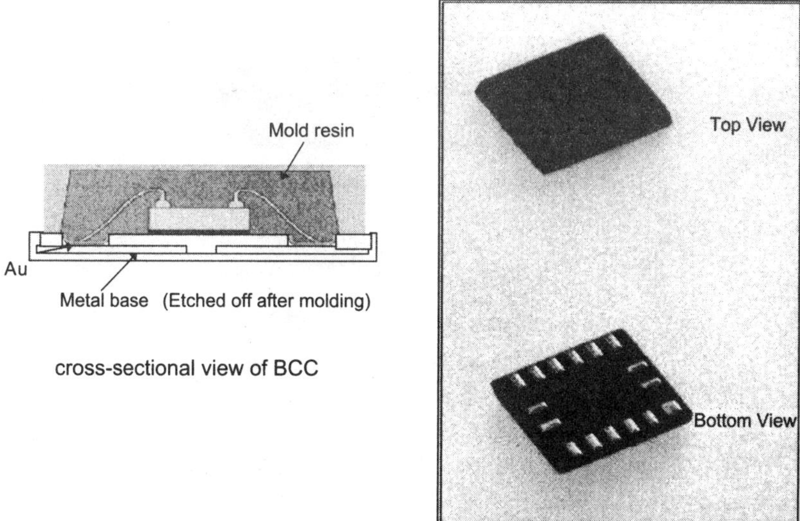

Figure 12.25 The bumped chip carrier.

12.5 Evolving package bumping technologies

Recently, a new packaging idea called wafer-level package (WLP) was evolved in
the manufacturing processes of the chip size packages (CSP). [19,20] The WLP covers
all packaging processes with a wafer form. Super-CSP (SCSP) and chip BGA are
typical WLP CSPs. The photos of them are shown in Figure 12.23, and the scanning
electron microscopic (SEM) image of the wafer with copper bumps and rerouting
traces is shown in Figure 12.24. To manufacture the SCSP, the copper traces and
bumps are fabricated with photolithography and semi-additive plating technology on

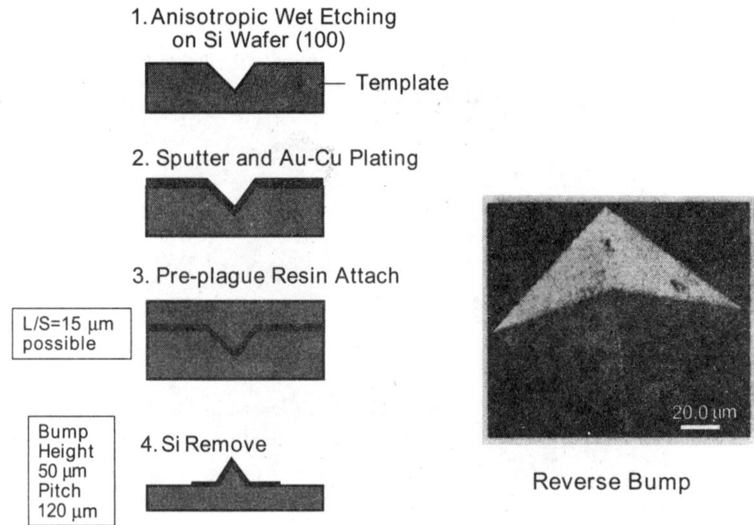

Figure 12.26 The reverse bump manufacturing process.

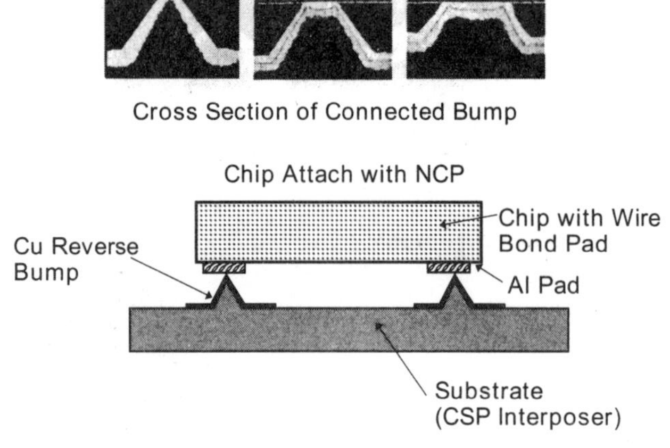

Figure 12.27 The reverse bump interconnection.

a wafer. Then, the wafer is molded with encapsulation resin, and the balls are mounted on the top of the copper bumps. The singulation process by dicing is the final process for this type of WLP CSP. The bump height of the copper in this case is controlled within 15 µm for 100 µm bumps. The solder balls mounting on the bumps are 350 µm in diameter and the mounting method is almost the same as micro-ball attach described in the former section. The chip BGA has no copper bumps and molding resin and therefore it can be classified as a simplified type of the SCSP.

The bumped chip carrier (BCC) is a CSP manufactured by leadframe and assembly technology.[21] The chip is mounted on a base metal frame, and wire is bonded on half etched bonding pads. After molding the chip, the metal frame was etched off completely. As shown in Figure 12.25, the result is small resin bumps which have metal layers on the top of them.

Recently, a new technology called reverse bump technology was developed.[22] This technology is a new idea, which has bumps on the package or substrate. The process is a type of bump transfer technology. At first, holes are made on a silicon wafer by anisotropic chemical etching. After deposition of a thin nickel seed layer, gold and copper are electroplated on the wafer to make patterns and bumps. Then, the resin board is laminated and peeled from the wafer. The manufacturing process of this bump and the interconnection of the reverse bump are shown in Figures 12.26 and 12.27. The pyramid-like bumps and patterns are transferred on the resin board by this process. Reverse bumps are also made by screen-printing technology used for B^2IT layer to layer interconnect. These two bumps were confirmed to show good reliability results.

References

1. International Technology Roadmap for Semiconductor, ITRS '99.
2. R. Tummala, K. Shintotani, V. Sundaram, S. Bhattacharya, G. White, F. Liu, P. Raji, S. Hee and D. Ravi, What is the future? Is it SOC or SOP? Proc. ISEPT 2001 pp. 5–11, Beijing, China, August 8–11, 2001.
3. P. Elenius and L. Levine, Comparing flip-chip and wire-bond interconnection technology, *Chip Scale Rev.*, July/August, 2000, 85–91 (2000).
4. S. Wakabayashi and M. Shimizu, Technical trends of LSI packaging, Proc. 1999 Symposium on VLSI Circuits, pp. 69–72, Kyoto, June 17–19 (1999).
5. B. Siu, The blurring boundaries of VLSI and System packaging, Proc. 1997 IEMT/IMC SYMPOSIUM, pp. 1–4, Omiya, April 16–18 (1997).
6. S. Wakabayashi, S. Koyama, T. Iijima, M. Nakazawa, and N. Kaneko, A build-up substrate utilizes a new via fill technology by electroplating, Proc. 4th International Conference on Adhesive Joining & Coating Technology in Electronics Manufacturing, pp. 280–288, Helsinki, June 18–21 (2000).
7. S. Wakabayashi, T. Iijima, and S. Koyama, New technologies for advanced build-up packages and substrates, Proc. of the electroniikan valmistus 2002, pp. 179–183, Pori, May 23–24 (2002).
8. Y.M. Loshkarev and E.M. Govorova, The electrodeposition of copper in the presence of brightening and smoothing agents, *Prot. Metals*, 34(5), 399–415 (1995).
9. J.J. Kelly, C. Tian, and A.C. West, Leveling and microstructural effects on additives for copper electrodeposition; *J. Electrochem. Soc.*, 146(7), 2540–2545 (1999).
10. T. Osaka, T. Sawaguchi, F. Mizutani, T. Yokoshima, M. Takai, and Y. Okinaka, Effects of saccharin and thiourea on sulfur inclusion and coercivity of electroplated soft magnetic CoNiFe film, *J. Electrochem. Soc.*, 146(9), 3295–3299 (1999).
11. K. Uosaki, Formation and novel functions of self-assembled monolayers of thiol derivatives, *Denki Kagaku*, 67(12), 1105–1113 (1999).
12. M. Yokoi, S. Konishi, and T. Hayashi, Adsorption behavior of polyoxyethyleneglycol on the copper surface in an acid copper sulfate bath, *Denki Kagaku*, 52(4), 218–223 (1984).
13. K. Nakamura, K. Ban, M. Nakazawa, S. Wakabayashi, and N. Kaneko, Application of pulse reverse plating to realize uniform plating films in very high aspect ratio via, Proc. 11th Micro Electronics Symposium, pp. 99–102, Osaka, October 18–19 (2001).
14. K. Toriyama, S. Fujiuchi, and Y. Orii, Development of Flip Chip Joint Technology with Paste Printing Method, Proc. 5th Symposium on Microjoining and Assembly Technology in Electronics, pp. 177–182, Yokohama, February 4–5 (1999).

15. S. Watanabe, Y. Ihara, Y. Kitahara, T. Kobayashi, and S. Wakabayashi, Solder bump fabrication on wafers by electroplating process, Proc. 1997 IEMT/IMC Symposium, pp. 110–115, Omiya, April 16–18 (1997).
16. K. Keven Yu and Francisca Tung, Solder bump fabrication by electroplating for flip-chip application, Proc. IEEE/CHMT Int'l Electronics Manufacturing Technology Symposium, vol. 15, pp. 277–281 (1993).
17. A. Gemmler, W. Leonhard, H. Richter, and K. Rauess, Preventive process maintenance in tin/lead micro-plating for flip-chip mounting technology, *Plating Surface Finishing*, 84(10), 64–72 (1997).
18. R. Schetty, Emerging plating processes for advanced electronic applications, *Metal Finishing*, 100, 18–21 (2002).
19. M. Hou, Wafer level packaging for CSPs, *Semiconductor Int.*, 305–308, July (1998).
20. D.S. Patterson, Some fundamentals of wafer-level package, *Solid State Technol.*, June (2002).
21. Y. Yoneda, Development of BCC(TM) (Bump Chip Carrier), IEEE CPMT, Third VLSI Packaging Workshop of Japan Technical Digest, pp. 71–72, Kyoto, Dec. 2–4 (1996).
22. S. Denda, New concepts in flip chip bump technology, Proc. 2002 ICEP, pp. 364–369, Tokyo, Apr. 17–19 (2002).

13 Plated through-hole technology for boards

Haruo Akahoshi

13.1 Introduction

Through-hole interconnection by copper plating is one of the essential technologies of printed wiring board (PWB) fabrication. Plating technology provides reliable low resistance layer to layer interconnections in the various designs of multilayer structures. The plated material employed here is usually copper because of its low electrical resistivity accompanied by good processability both for mechanical and chemical (plating/etching) manufacturing processes. Also, the ductile nature of copper provides acceptable interconnect reliability, absorbing thermal stress and plastic/elastic deformation through the manufacturing process and under user operation of printed circuits boards in electronic equipment. The wiring density and interconnect reliability of the PWBs are both closely influenced by the performance of the plating technology. Since plated through-hole technology already has a long history, many books and reviews have appeared discussing this technology.[1–9]

The common insulating material of the PWB is a thermoset resin impregnated with a reinforcing glass cloth. The thermal expansion coefficient of the insulating layer in the depth direction, which is usually in the range of 50 to 100 ppm/K, is several times larger than that of copper. The large difference of the thermal expansion coefficient causes significant thermal stress from environmental temperature change during the manufacturing process, such as resin curing and soldering operations, and under user operation of the printed circuits boards in electronic equipment. The mechanical properties of the plated copper deposits and the plated copper thickness at the inside wall of the plated through holes both greatly influence the through-hole interconnect reliability. Therefore, improvement of the mechanical properties, ductility and tensile strength, and improvement of the plated thickness inside the through holes, are important points of interest in through-hole plating technology. Additionally, the uniformity of the plated thickness on the surface plane, and thickness ratio of the surface to the inside of the holes are also important issues in terms of manufacturing performance. The surface plane copper thickness and the thickness distribution have a large influence on the circuit line width deviation and the resolution of fine line circuits as thinner, uniform copper is desired. On the other hand, sufficient thickness is required to achieve reliable interconnection.

 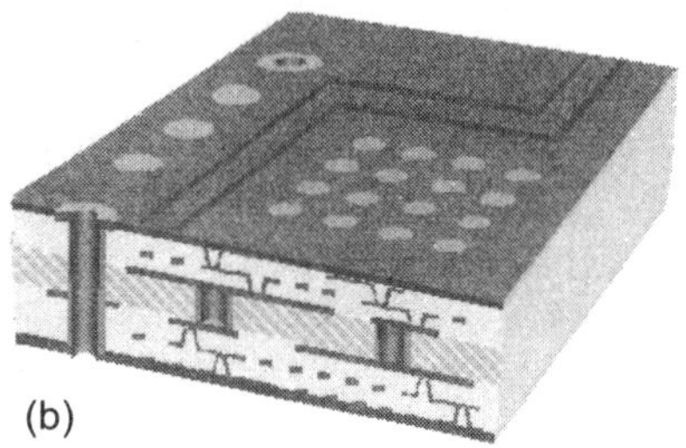

Figure 13.1   (a) Conventional through-hole printed wiring board, (b) sequential buildup printed wiring board with microvia interconnections.

Through recent advancement in PWB manufacturing technology, a new structure of the layer to layer interconnection has been developed. Laser drilled or photo imaged dead ended microvia holes are employed in high-density interconnect (HDI) technology, through the sequential buildup manufacturing process.[1,2,10,11] Figure 13.1 shows a cross-sectional view of a typical structure of the sequential buildup PWB compared with conventional through-hole PWBs. The hole size of these microvias is much smaller than the conventional mechanically drilled through holes, ranging from ϕ 50 to 200 µm in diameter. However, the absolute value of the thermal stress on these microvia structures is relatively smaller than in the case of high aspect ratio through holes on high layer count thick boards. The depth of the microvia is not as large, because they usually connect only to the next layer of the plated circuit, and also because the aspect ratio of the diameter to the depth of the via is usually less than 0.5 from the limitation of the plated copper uniformity in the via. The dead ended structure causes a thickness distribution problem near the bottom part of the via. Improvement of plating uniformity in the microvias, and filling up the microvia with plated copper, are both emerging technologies in plating interconnection.

13.2   Electroplating of copper through hole

The electrolytic plating method is most widely employed in copper plated through-hole copper plating, in the printed circuit board industry. Four types of electrolytic copper plating bath are well known and used in the plating industry: cyanide, fluoroborate, pyrophosphate, and acid sulfate baths.[7,12] Historically, the pyrophosphate bath was once widely employed for through-hole plating, because of its good throwing power and the excellent mechanical properties of the deposit compared to the acid sulfate bath in those days. Today, acid sulfate copper is the most common plating bath for through-hole plating on PWB. The performance of the acid sulfate bath was remarkably improved by the development of a high throwing power acid sulfate bath. Typical composition and the properties for these baths are summarized in Refs. 1, 2, 4, 6. Table 13.1 shows an example of the composition and the plating condition of an acid sulfate bath used for through-hole plating. Several types of additives are usually used together in combination: brightener, leveler, polymeric surfactant, and chloride ion. These additives appear in patents and are listed in Ref. 4. The plating

Table 13.1   Composition of acid sulfate electroplating copper bath.

	Conventional solutions	*High throw solutions for through-hole plating*
$CuSO_4$, mol/L	0.7–1.0	0.22–0.4
H_2SO_4, mol/L	0.5–1.0	1.5–2.5
Cl^-, m mol/L	0.15–2.5	1.0–2.5
Additives	Trace	Trace
Temperature, °C	20–30	20–30

performance, both of the mechanical properties of the deposits, and thickness distribution are strongly influenced by the additive concentration. Proper control of the additives is essential. Unfortunately, the direct analytical determination of the concentration of the additives is difficult in many cases. Several kinds of indirect evaluation methods such as cyclic stripping voltammetry (CSV) have been developed.

13.2.1   *Pulse electroplating for high aspect ratio through hole*

According to the continuous demand for increase of wiring density, supported by the improvement of small hole drilling technology, the aspect ratio of the through holes to be plated has become very high for high layer count multilayer PWBs. On these applications, throwing power is insufficient even for high throwing power acid copper baths, under conventional DC plating conditions. The pulse plating technique has been recognized as one of the effective methods which improves the uniformity of the plated copper thickness. Many studies have been conducted on the influence of the pulse wave form and the bath formulation on uniformity and on the morphology of the plated copper.[13–26] However, it has not been applied widely for through-hole plating in actual PWB production until recently. The expected advantages of pulse plating method are considered to be as follows.[21–23]

1. Increase of the average current density (reduces plating period)
2. Improvement in thickness uniformity of the deposit
3. Improvement in physical properties of the deposit: fine grained structure, reduced porosity, improved ductility and hardness, and smooth surface
4. Better adhesion to the substrate with reduction of oxide films at the interface

An example of the wave form for pulse plating is shown in Figure 13.1. Simple pulse plating (unipolar pulsed deposition) with on/off cycle of cathodic current and periodical pulsed reverse (PR) current plating (bipolar pulsed deposition) are both investigated. Typical pulse conditions are several hundreds of hertz to several kilohertz in pulse frequency with a duty cycle of 25 to 75% for pulse plating. In the case of PR plating, 10 to 20 ms forward (cathodic) period followed by 0.5 to 2 ms reverse (anodic) period, which usually has a higher current density than cathodic current, has been examined.[21,23] The bath performance, distribution of plated thickness, surface morphology, and quality of the deposits are closely related to both the pulse wave form and electrolyte composition. The optimum pulse condition was influenced by the type of electrolyte. Therefore, careful optimization and process control will be necessary. Application of pulse plating (pulse reversal) to high aspect ratios through-hole plating in practical production has been reported recently.[26] High throwing power value (82%) was reported on high aspect ratio through holes (AR: 25, ϕ 0.25 mm holes on 5.0 mm thick boards of 400×500 mm size) with proper control of plating conditions and optimized uniform jet-flow agitation.

13.2.2 Simulation and modeling

The plated thickness distribution problem based on the current distribution is intrinsic nature for the electroplating method. Therefore, mathematical modeling and numerical simulation of through-hole electroplating have gained much interest. Thickness distribution focused inside a through hole has been analyzed based on the Butler–Volmer equation combined with fluid dynamics and mass-transportation covering: mass transport and ohmic limitation,[28] local kinetic parameter variations,[29] additive effects,[30,31,38] convection and periodic reversal flow,[31–33] periodic electrolysis,[34] axisymmetric two-dimensional flow,[35] and electrical migration.[39,40] Simulated results were compared with experimental ones using an actual through hole, cylindrical electrode, and gap cell.[9] Good agreements were generally obtained under controlled experimental conditions on these studies.

Recently, supported by the remarkable improvement of computing power on small size computers, practical simulation of plating distribution has become very easy even on desktop computers.[41–48] They can be used for predicting the thickness distribution not only inside through holes but also on the surface plane, in blanket plating, and pattern plating. Design optimization is achieved for plating conditions (current density, bath composition, etc.), plating pattern layout and plating tank structure; and the layout of the anode, auxiliary electrode, and baffle plate. CAD software was also applied for wafer plating of Damascene Cu wiring LSI.[45,46]

13.2.3 Plating microvia /via filling

A most interesting aspect of plating interconnection in high-density PWB is microvia plating for sequential buildup PWB.[1–3,10,11] A cross-sectional view of a typical structure of the sequential buildup PWB compared with conventional through-hole PWBs is shown in Figure 13.1. Introducing laser drilled or photo-imaged microvia on layer to layer interconnection makes it possible to obtain a remarkable increase in wiring density. The smallest diameter size of mechanically drilled through holes is ϕ 0.15 mm in current production levels, while the diameter size of laser drilled or photo-imaged microvias reached ϕ 100 to 50 μm. Interstitial via-hole structure enables another circuit wiring (or soldering pad) to be located just above (or beneath) the via on another circuit layer. These features of the microvia structure enable a very high density to be achieved on the wiring boards. Copper plating small diameter microvias having a high aspect ratio are one of the challenging technologies, similar

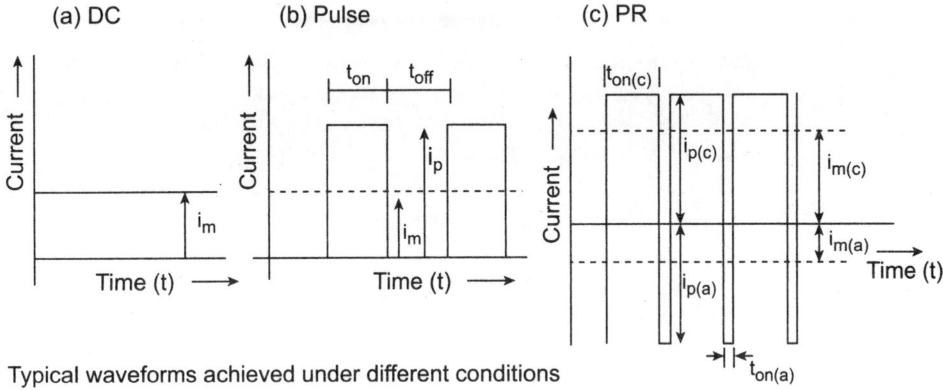

Figure 13.2   Typical current wave forms for dc and pulse plating (from Kalantary and Gabe[23]): (a) dc plating, (b) unipolar pulse plating, (c) bipolar pulse plating (periodical reversal pulse: PR).

to high aspect ratio through-hole plating. Furthermore, filled via structure described in Figure 13.2 provides a much higher wiring density than the "staggered" design microvia of buildup PWBs. It enables perpendicular layer to layer interconnection at the same geometry via layout, and chip interconnection just on the vias at the surface layer.

The early attempt to obtain stackable filled via structures was the use of the plate-up process, a kind of through-mask plating.[49–52] Via "post" was plated up from the bottom of the vias 1 dimensionally from the bottom to the top portion. This structure can be realized with a small modification of through-mask pattern plating. When a full build electroless copper is applied to this process, the process sequence is very much simplified, and the interconnect structure completes just after finishing the electroless plating step. However, because of the plating rate limitation of practical full build electroless copper plating bath (5 to 6 μm/h), it requires a long plating period to fill up whole vias especially in the case of a PWB with a thick insulating layer.

Periodical reverse current plating was applied to filling microvias by using commercial acid sulfate copper bath developed for through-hole plating.[52] Microvias of ϕ 60 μm with 60 μm depth were successfully filled in a properly selected condition (Figure 13.3). During the forward current period, the growth rate of the surface or top edge portion of the via is much larger than that of the via bottom as a result of Cu ion diffusion control and current distribution. On the other hand, during the reverse current period, anodic dissolution is suppressed at the bottom of the via, while excess Cu deposit at the top surface is effectively dissolved out. Proper selection of F/R period and each current density enables the via to be filled up by preferred Cu growth from via bottom. This method of plate up is not only from the bottom portion but also from the sidewall of vias. Plating time required to fill up the via was remarkably reduced compared with a "post" structure via formed by full build electroless copper plating, even on PWBs having a thick insulating layer. However, the reverse (anodic) current period still reduces the average plating rate

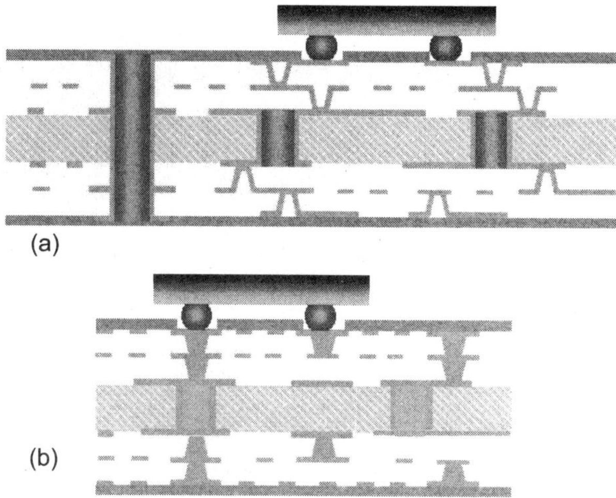

Figure 13.3   (a) "Staggered" layout microvia in sequential buildup PWBs; (b) "stacked" layout microvia in sequential buildup PWBs.

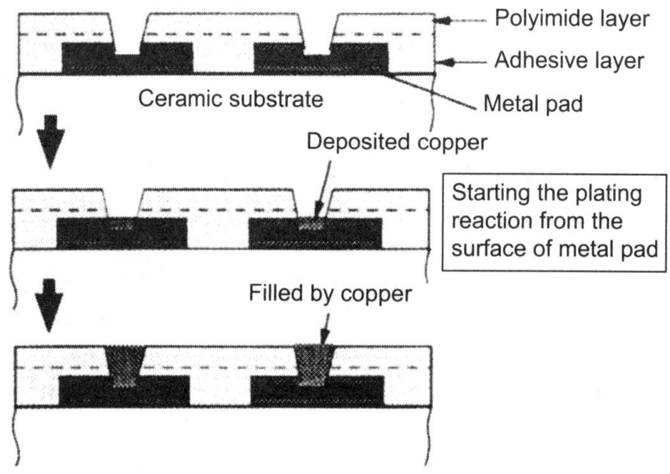

Figure 13.4   Cross-sectional image of via "post" plating process. (From Takahashi et al.[50])

compared to the case of continuous DC plating. The overall plating period required to fill up the vias is still longer than bottom-fill DC plating described later. The leveling effect of a gap structure like a microvia was studied on Ni plating earlier than these attempts on copper plating.[54–57] When a leveling agent which has an inhibiting effect on the deposition reaction was in a transport limited flux condition, non-uniform concentration distribution occurs on the plated surface having a gap structure. Consumption of the leveling agent at the plated surface in the plating condition is an essential feature to give the transport limited flux of the leveling agent. The concentration of the leveling agent at the bottom of the gap should be lower than the top edge or the planar top surface portion. This results in preferred growth at the gap

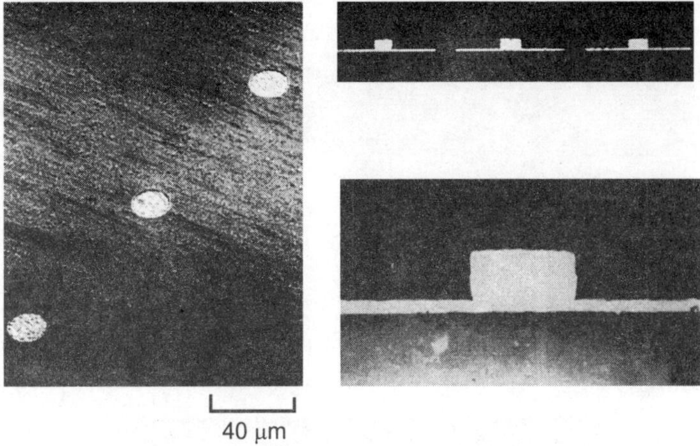

Figure 13.5   Top view and cross sectional view of the plated-up via "post" using full build electroless copper plating (From Takahashi et al.[50])

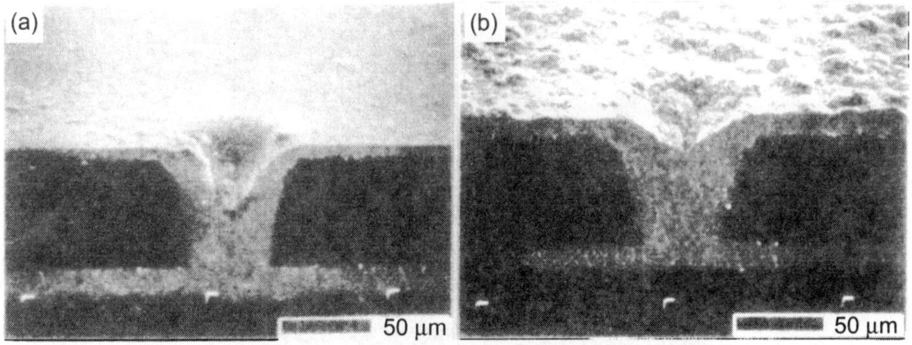

Figure 13.6   Cross-sectional views of plated copper using PR plating followed by 10 μm DC plating in 1.2 A/dm^2. (From Abe et al.[52])
PR current: Cathodic current: 60 s, 1.2 A/dm^2; anodic current: 30 s, 1.2 A/dm^2.

bottom. According to this model, the leveling behavior was numerically simulated using boundary element theory.[56] Simulation using electrochemically measured surface reaction rate constants and diffusion coefficients of leveling agents was also reported.[59] Coumarin for the Ni plating Watts bath, and thiourea for the acid copper were examples of the leveling agents which showed gap-filling behavior. This shape evolution technique was first applied for the "Damascene process" on-chip Cu interconnection of high-speed large-scale integrated circuits having sub-millimeter features to be filled, in order to reduce the wiring resistance and to improve electromigration resistance from conventional Al (alloy) wiring.[58,59,67–73] An acid sulfate bath was employed for this application. This technology was also extended to ϕ 100 μm scale filled microvia fabrication for high density sequential buildup PWBs.[60–66] Microvias of ϕ 50 μm to ϕ 100 μm with an aspect ratio of 0.5 to 1.0 are successfully filled in properly selected plating conditions. Many studies have been done using a

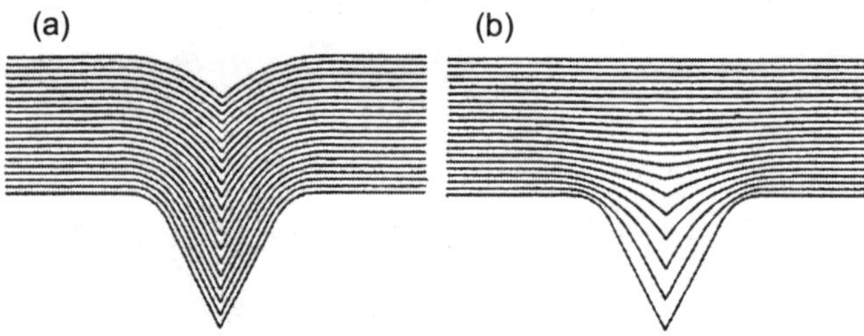

Figure 13.7   Simulated leveling (gap fill) in a Watts-nickle bath without (a) and with (b) coumarin. (From Dukovic and Tobias.[56])

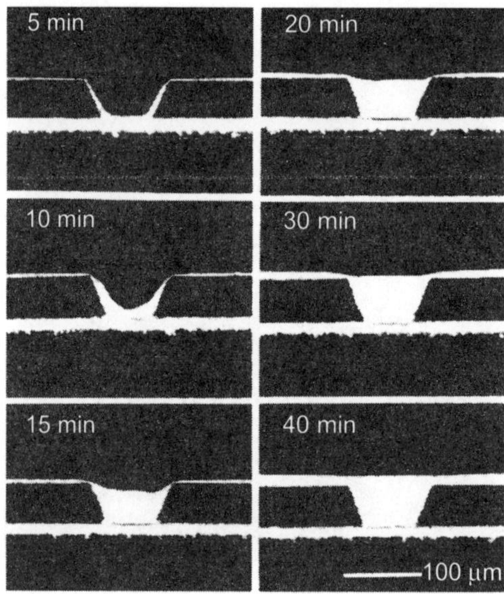

Figure 13.8   Filling process of microvia. (From Matsunami et al.[65])

combination of additives: Janus Green (JG) ("leveler"), *bis*-(3-sulfopropyl)-disulfide (SPS) or 3-mercapto-1-propanosulfonic acid (MPSA) ("brightener"), polyethylene glycol ("polymer" or "suppressor"), and a trace amount of Cl^- ion in an acid sulfate bath.[60–63,66,70,73] The effects of plating current wave form, superimposed pulsed or stepped function current, were also studied.[60,64] The "superfilling" behavior of this via-filling technology was explained by the combination effects of suppression at the top surface and acceleration at the via bottom, but precise details of the shape evolution mechanism and quantitative interpretations are still under investigation. In addition to the "consumed leveler (inhibitor)" mechanism, the accumulation effect of accelerator (brightener) is suggested.[67–73] Surface coverage of the accelerating additive (usually brightener) for small features is considered to rapidly increase due to the decrease of the local surface area during the deposition reaction. According to this model, experiments and simulations have good agreement on micro-scale features, filled without using any "consumed leveler" components.

One of the further issues of via-filling plating technology is the compatibility with the filling property and the uniformity of the plated thickness on the surface plane and inside through holes. Micro-scale leveling effects sometimes fall into a trade-off with macro-scale uniformity. Careful optimization and maintenance of the plating conditions including the bath formulation and proper control of liquid flow (circulation/agitation) will be necessary. The analysis and control of the concentration of the additives are therefore other important issues in the application of via-fill plating on practical production.

13.3　Electroless copper for through-hole plating

The most attractive feature of full build electroless copper plating method in through holes is the plated copper thickness uniformity inside the holes and also on the surface plane. Excellent throwing power in excess of 95% can be achieved while the value of optimized electroplated through holes stay between 80 and 85%. However, the quality of the electroless copper was not considered to meet the requirements of mechanical properties during early development. Through much investigation, the quality of the full build electroless copper deposits has remarkably improved.[74–82] It reached the same level as acid electrolytic copper used for through-hole plating. Table 13.2 shows an example of the properties in comparison with electrolytic copper. Optimized electroless copper through holes sometimes showed a higher reliability level than electroplated through holes because of the superior throwing power

Table 13.2　Performance of electroless copper plating solutions developed for full build through-hole plating.

	Electroless copper		Electroplating
	High-speed bath	*High reliability bath*	*Acid sulfate bath*
Ductility (%)	10–20	15–25	15–20
Tensile strength (MPa)	250–300	300–350	300–350
Plating rate (μm/h)	5–6	2–3	20–40

of the full build electroless copper through holes, and they were applied on high layer count, high-density multilayer PWBs (Figure 13.9).[78] The plating rate of the electroless copper plating is rather lower than electrolytic copper. It is limited to under 5 to 6 μm/h in the practical full build bath. However, the loading factors of the electroless copper plating are much higher than electroplating. The tank structure for the electroless copper bath is free from geometrical limitation of anode/cathode layout which is necessary in electroplating, and a large number of boards can be put in the electroless bath simultaneously in baskets. Therefore, in the case of high speed baths (~5 to 6 μm/h), the productivity during certain periods is almost the same as common acid copper plating.

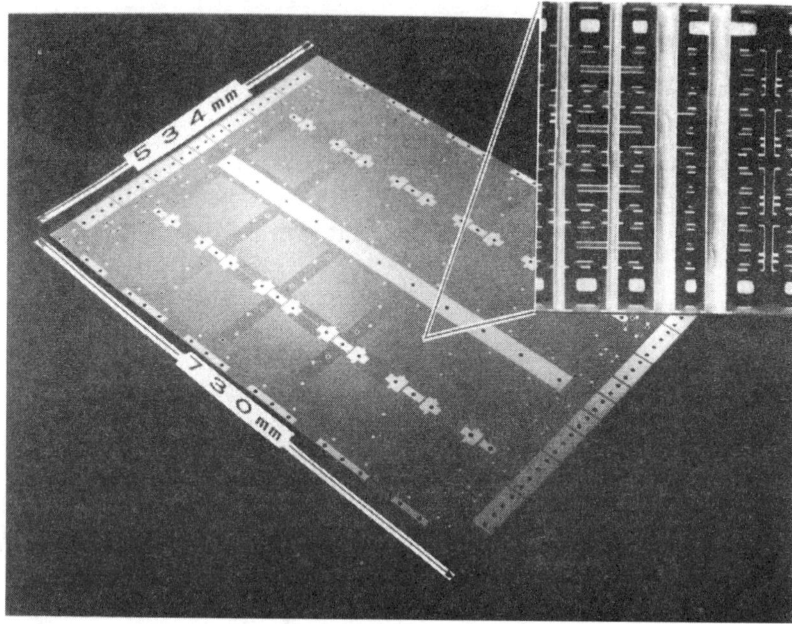

Figure 13.9 The high aspect ratio plated through holes plated by full build electroless copper plating solution on the high layer count high density multilayer printed wiring boards. Board thickness is 7.1 mm with 46 wiring layers. Diameter of the smallest drilled through hole is 0.3 mm. (From Takahashi et al.[78])

The major components of the electroless copper plating solution are copper salts, reducing agents, and complexing agents. Small amounts of several kinds of additives are usually added simultaneously, in order to improve the bath performance. In electroless copper plating, Cu ion is consumed by the plating reaction, but it is not supplied from the anode as in electroplating. Replenishment of Cu ions with some chemical compounds is necessary to maintain the Cu ion concentration in the plating bath. Copper sulfate is commonly used. Formaldehyde is almost solely employed as the reducing agent in practical electroless plating solution. Recently, alternative reducing agents have been studied because of environmental considerations. They are described separately. The actual reacting species in this system is the methylene glycol anion delivered from formaldehyde in a highly alkaline solution ($<$pH 11.5).[4,5,86] Therefore, the pH of the electroless copper plating solution using formaldehyde is very high. The plating temperature for the full build electroless plating is 60 to 80°C to achieve a sufficient plating rate and an improved deposit quality. The highly alkaline and high temperature plating bath sometimes causes material damage on the epoxy laminates and the photoresists, used in the circuit wiring fabrication. Because of the high pH of the plating solution, a complexing agent is necessary to maintain the solubility of Cu ions in the plating solution and to avoid generating precipitates of copper hydroxide. The complexing agent does not simply form a complex with Cu ion, but also has a large influence on the performance of the bath, plating rate, stability of the bath, and quality of the deposit. Careful selections of the complexing agent are required especially for the full build bath. Examples of evaluated complexing agents are ethylenediaminetetraacetic acid (EDTA), potassium sodium tartarate (Rochelle salt), tertakis-2-hydroxypropyl-ethylene-diamine (Quadrol), and cyclohexane-diaminetetraacetic acid (CDTA). Full build high temperature baths usually employ EDTA. Trace amounts of additives are used to improve the bath stability and the deposit quality. Cyanide was one of the most commonly used additives in the early baths. The plating rate of the cyanide additive bath is relatively slow: 1 to 2 μm/h, because of the insufficient stabilizing effect of cyanide. The sulfur containing compounds such as 2-mercaptobenzothiazole (MBT) or thiourea have good stabilizing effects; however, the mechanical properties of the deposits obtained from the bath using sulfur-containing additives are generally poor, and they cannot be employed for the full build bath. Abundant work has gone into the survey of the additives to improve bath performance.[4,76–83] The modern high-speed bath employs 2,2′-dipyridyl instead of cyanide, in conjunction with poly-ethyleneglycol or its derivatives.[77] The plating rate was improved to 2 to 3 μm/h by using a combination of these. A series of inorganic oxoacid salts (vanadate, silicate, germanate) have also been developed as the additives. The plating bath employs silicate or germanate as the additive which provides very ductile deposits with high plating rates (4 to 6 μm/h).[79,81,83] The deposit obtained from these improved bath shows very high elongation values 10 to 20% (on 30 μm), and 20 to 30% (on 40 μm), comparable to the current acid sulfate electroplated copper. They can be applied even for a high aspect ratio through-hole plating on high layer counts thick multilayer PWB that requires high interconnect reliability.[78]

The electroless copper plating reaction proceeds with two coupled reactions: reduction of Cu ion to metallic copper film and oxidation of formaldehyde, releasing electrons and forming formate ions.

The overall reaction for the electroless copper plating using formaldehyde is:

$$[Cu^{2+}-L] + 2HCHO + 4OH^- \rightarrow Cu^0 + 2HCOO^- + 2H_2O + H_2 + [L] \qquad (1)$$

The oxidation of formaldehyde proceeds in aqueous alkaline solution catalyzed by noble metals such as Au, Pt, Pd, and Cu. This reaction only holds true for these catalytic metal surfaces, and so the direct reaction between the copper ion (complex) and formaldehyde in the solutions usually does not occur even if these two components are present in the same plating solution simultaneously. At the plated surface, the plated copper itself acts as the catalyst, and the plating reaction proceeds continuously. Figure 13.10 shows these reactions schematically. Formaldehyde is oxidized at the Cu surface generating H_2, formate ions, and electrons. The electrons move to the depositing site and are transferred to the Cu ions (complex) in the plating solution. The Cu ions are reduced to metallic copper and then deposited on the plated copper surface. These two reactions are supposed to be independent, and to be only coupled electrochemically, in the mixed potential theory.[87] A recent investigation suggested they are not actually independent, and certain interactions exist on both reactions mutually.[88] There are some other side reactions that proceed sponta-neously. Cuprous oxide is considered to be formed from the following reaction:

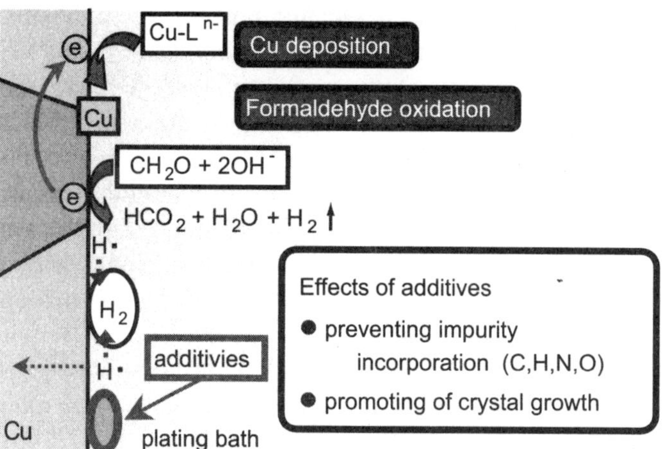

Figure 13.10 A schematic model of electroless copper plating reaction.

$$2[Cu^{2+}-L] + HCHO + 5OH^- \rightarrow Cu_2O + HCOO^- + 3H_2O + [L] \qquad (2)$$

The generated cuprous oxide results in copper particles by the disproportionation reaction and further reduced by formaldehyde.

$$Cu_2O + H_2O + [L] \rightarrow Cu^0 + [Cu^{2+}-L] + 2OH^- \qquad (3)$$

$$Cu_2O + 2HCHO + 2OH^- \rightarrow 2Cu^0 + 2HCOO^- + H_2O + H_2 \qquad (4)$$

When enough oxygen was dissolved in the plating solution, these cuprous oxide and copper particles readily dissolve into the plating solution.

$$Cu_2O + 1/2O_2 + 2H_2O + [L] \rightarrow 2[Cu^{2+}-L] + 4OH^- \qquad (5)$$

$$Cu^0 + 1/2O_2 + H_2O + [L] \rightarrow [Cu^{2+}-L] + 2OH^- \qquad (6)$$

The generation of cuprous oxide particles in the plating solution has a significant influence on the bath stability and deposit quality.

Another side reaction is the disproportionation reaction of formaldehyde (Cannizzaro reaction):

$$2HCHO + OH^- \rightarrow CH_3OH + HCOO^- \qquad (7)$$

The Cannizzaro reaction rate increases according to the increase of the plating temperature and the bath pH. The consumption of formaldehyde (and so the generation of formate) exceeds a negligible level in the full build copper bath operated at high temperatures around 70°C.

According to the principle of the electroless copper plating reaction, plating is expected to proceed only on the plated copper surface. However, nodule formation and extraneous deposition, as a result of copper particle formation in the plating solution, sometimes occur under improper operating conditions. Cuprous oxide generation as described above, and the interfused dust particles are possibly the cause of these phenomena. Continuous filtration is essential to maintain the bath stability. Excess accumulation of Fe ions in the plating bath also causes extraneous deposition through Cu ion reduction by $Fe(OH)_2$ precipitation.[95]

Dissolved oxygen plays a very important role in stabilizing the electroless copper bath. Stabilization by air bubbling was known at an early stage.[89–90] Oxygen in the plating solution dissolves copper or cuprous oxide particles and prevents the growth of these particles. Oxygen in the plating solution also dissolves copper from the normally plated copper surfaces. However, the solubility of the oxygen gas is relatively low. It is 10 to 20 ppm around room temperature and approximately 4 ppm at 70°C. Therefore, the dissolving reaction rate at the plane surface is limited by the linear diffusion of oxygen, and is much smaller than the plating rate. On the other hand, at the small particle surface, the oxygen flux became a non-linear spherical diffusion, and the dissolving reaction rate exceeded the plating rate under sufficient oxygen supply. Oxygen in the plating solution is consumed by the above copper particle dissolving reaction.[91,92] It is also exhausted by the hydrogen gas continuously generated on the plating reaction. Once the air agitation stops, the oxygen gas concentration immediately runs down, and the bath becomes unstable. Under the high temperature plating condition of the full build bath, the saturated concentration of the oxygen gas is low as described above. Continuous and uniform air agitation is therefore very important to maintain the bath stability. The complexing agents and additives also influence the dissolved oxygen concentration. In the case of Rochelle salts and monoethanol amine, rapid decrease of the dissolved oxygen concentration was reported, while less decrease was observed with EDTA (Figure 13.11).[93] The former complexing agents were estimated to generate greater amounts of cuprous oxide. Cyanide and 2,2′-dipyridyl are well-known additives having good stabilizing effects on the electroless copper plating solutions. They are considered to form a Cu(I) complex and prevent the generation of cuprous oxide particles. Suppression of the decrease in the dissolved oxygen concentration was also observed with the addition of these additives (Figure 13.12). The accumulation of the plating reaction by-products, sulfate and formate ions, also reduces the dissolved oxygen levels. This is the dominant parameter that determines the plating bath life time.

Factors affecting the mechanical properties of electroless copper deposits are summarized in Ref. 76. The incorporated impurities, voids found in the crystal grain or on the grain boundary, are considered to be the major causes of reducing the ductility of the electroless copper deposits. The grain size also has a large influence on the ductility and the tensile strength of the deposits. The impurity level of the electroless copper deposits is generally higher than those of the electroplated copper, even in a deposit having excellent ductility. The hydrogen gas bubbles incorporated

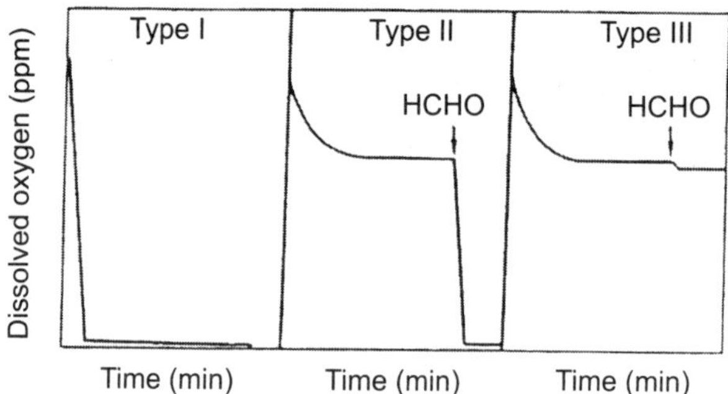

Figure 13.11　Time-dependent change of dissolved oxygen concentration in alkaline solutions containing various complexing agents. (From Matsuoka and Hayashi.[93]) Arrows indicate timing of formaldehyde addition.
Type I: Monoethanolamine, diethanolamine, diethylenetriamine, triethylenetetramine, tetraethylenepentamine, gluconic acid salt, Rochelle salt;
Type II: Triethanoleamine, nitrilotriacetic acid;
Type III: Ethylenediamineteraacetic acid, diethylenetriaminepentaacetic acid.

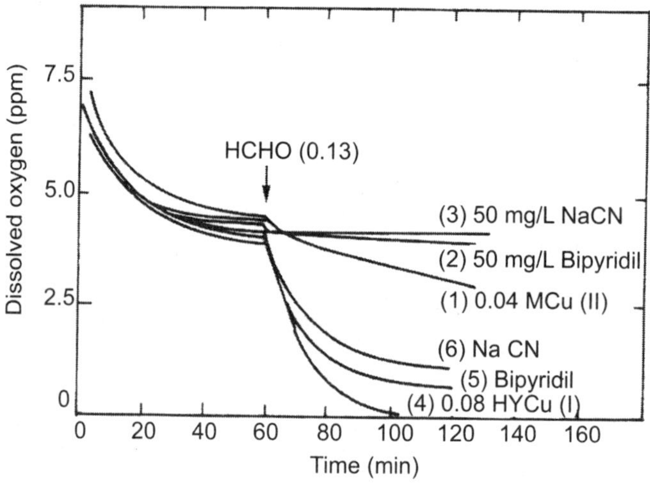

Figure 13.12　Effects of stabilizer on the time dependent change of dissolved oxygen concentration. (From Matsuoka and Hayashi.[93])
(1) to (3): [Cu(II)] = 0.04 mol/L without stabilizer (1), with 50 mg/L of 2,2′-dipyridyl (2), and with 50 mg/L of NaCN (3);
(4) to (6): [Cu(II)] = 0.08 mol/L without stabilizer (4), with 50 mg/L of 2,2′-dipyridyl (5), and with 50 mg/L of NaCN (6).

in the copper deposits have a large influence on the ductility of the deposits. Some of the atomic hydrogen generated by the formaldehyde oxidation in the plating reaction diffuses from the surface to the inside of the copper deposits, are trapped in voids (size of several nanometers to submillimeters) and charged up to high pressures. These bubbles were first observed by a transmission electron microscope (TEM) in Ref. 94. The calculated hydrogen gas pressure was up to ~700 atm, which

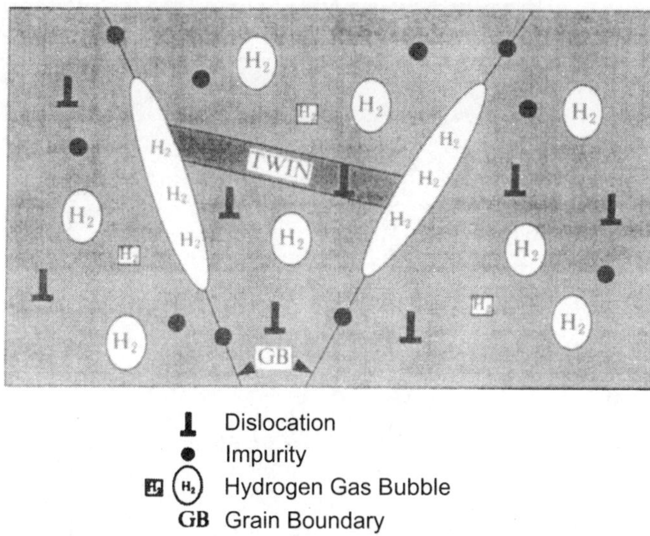

⊥ Dislocation
● Impurity
(H₂) Hydrogen Gas Bubble
GB Grain Boundary

Figure 13.13 Microstructure of electroless copper deposits showing various interfacial defects. (From Nakahara and Okinaka.[76])

is the same level of yield stress of polycrystalline copper.[76,94] By low temperature annealing ($<200°C$), the molecular hydrogen incorporated in these voids (diffusible hydrogen) diffused out from the deposit, and the embrittlement effect by the hydrogen gas pressure became smaller. Measured ductility of the copper deposits was improved correspondingly. Large voids (~0.2 mm) can still also be the cause of reduced ductility even after low temperature annealing. Figure 13.13 shows a schematic illustration of the voids and impurities in the electroless copper deposits cited from Ref. 76. The incorporation of Fe also reduced the ductility. Since the solubility of the Fe ion in the highly alkaline plating solution is very low, trace amounts of Fe ion ($\sim$ppm) form colloidal iron hydroxide, as mentioned previously. Co-deposition of the iron hydroxide particles are considered to be the cause of the defects in the crystal grain or the grain boundary. The crystal grain size also has a deep influence on the ductility of the copper deposit. Figure 13.15 shows the relationship of the crystal grain size and ductility as a function of additive concentration in the plating solution.[96]

13.3.1 Formaldehyde free bath

A part of the recent investigation on electroless copper plating solutions is focused on the replacement of the reducing agent formaldehyde by other chemicals. The reducing agent of the electroless copper plating solution was limited to formaldehyde in practical applications for a long period. The major motivations for replacement of formaldehyde are considered to be: (1) health and environmental considerations, (2) lowering the plating solution pH or the plating temperature to avoid materials damage, and (3) eliminating hydrogen evolution. The major candidates of alternative

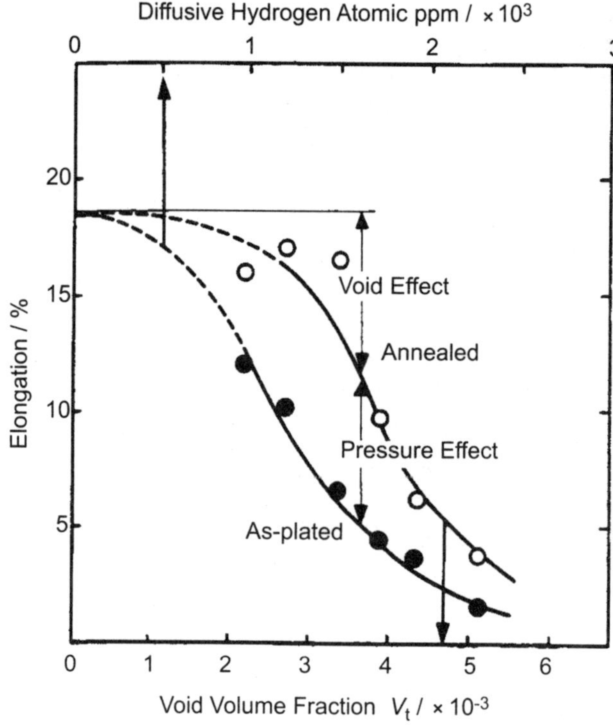

Figure 13.14 Ductility (% elongation) of various as-deposited (solid circles) and annealed (open circle) electroless copper deposits plotted against diffusible hydrogen content. The annealing treatment was performed at 150°C for 24 h. As this annealing drives out all the diffusible hydrogen, the horizontal axis (upper scale) can be expressed only in terms of void volume fraction for the annealed deposits. (From Nakahara and Okinaka.[76])

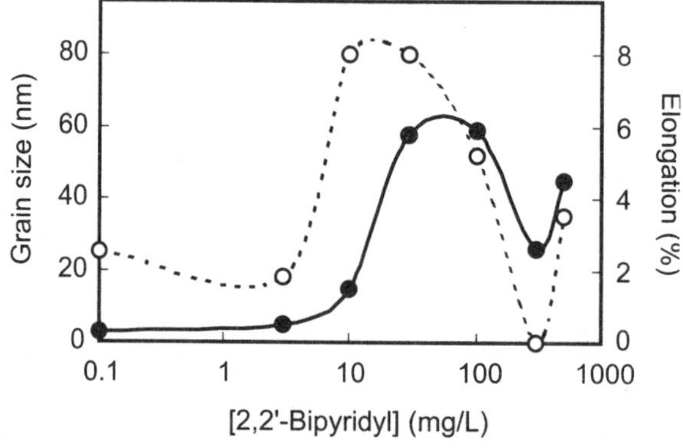

Figure 13.15 Ductility (% elongation) (solid circles) and crystal grain size (open circles) of the electroless copper deposits plotted against additive concentration. The crystal grain size were calculated from the half height width of diffraction line from ⟨1,1,1⟩ lattice. (From Akahoshi.[96])

reducing agents to formaldehyde are listed in Table 13.3 with their typical formulations and plating conditions.

Table 13.3 Typical formulations of formaldehyde-free electroless copper plating solutions. (Source: Refs. 97, 112, 120, 128.)

Reducing agent	Dimethylamine borane	Sodium hypophosphite	Glyoxylic acid	Co(II) salts
Complexing agent	Tetraazadodecane	Sodium citrate	Ethylenediamine-tetraacetic acid	Ethylenediamine
Cu salt	$CuSO_4$	$CuSO_4$	$CuSO_4$	$CuCl_2$
pH buffer	Triethanolamine	HBO_3	KOH	HNO_3
Additives	2,2′-Dipyridyl	$NiSO_4$	2,2′-Dipyridyl polyethylene glycol	2,2′-Dipyridyl
pH	9	9	12.5	6–7
Temperature	65°C	60°C	60–70°C	50°C
Plating rate	2–2.5 μm/h	2–5 μm/h	2–4 μm/h	3 μm/2hr

13.3.1.1 Hypophosphite bath

An electroless copper plating solution using Hypophosphite has been reported.[97–109] Hypophosphite is a reducing agent widely used in electroless Ni plating. In the case of electroless Ni plating, a hypophosphite bath can be operated at a pH range of 4 to 9. It could be a significant advantage over the conventional formaldehyde-type electroless plating solution, which is operated in a strong alkaline solution, with a pH range of 11.5 to 13. The benefit of avoiding high pH operation is minimizing the material damage by an alkaline solution. However, Cu metal itself does not act as an electro-catalyst for the hypophosphite oxidation reaction. Therefore, this system employs certain modifications. A small amount of Ni ion is added to the plating solution as a "catalyzer" or a "mediator." Ni is deposited on the plated surface simultaneously with the copper deposition, and this Ni portion acts as a catalyst for the hypophosphite oxidation reaction. Schematic plating reaction mechanism is illustrated in Figure 13.16, and the obtained plating rate as a function of Ni ion concentration is shown in Figure 13.17.

Because of this mechanism, the deposited Cu from this type of plating solution is expected to include small amounts of Ni. The relationship between Ni ion concentration in the plating solution and Ni content in the copper deposit was reported in Refs. 103 and 108. Several percent of Ni and P were included in the copper deposit.[108] It was also reported in Ref. 103 that Pd acts as a mediator similar to Ni, and hypophosphite reducing electroless plating is possible using Pd ion instead of Ni. As a result of the inclusion of Ni or Pd, an increase of electrical resistance and embrittlement of the copper deposit are both expected. The resistivity of the Ni/Cu deposit was reported in Ref. 99. The resistivity increased ten times larger or more to the value of pure copper. The mechanical properties of the deposit of this system have not been reported. Because of the high resistivity, these systems are not suited for

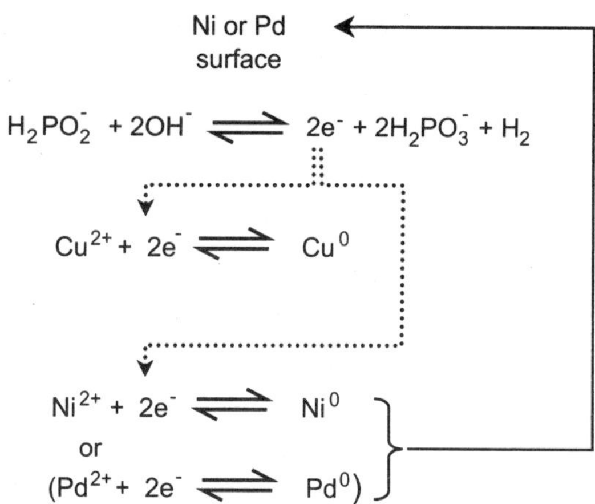

Figure 13.16 Schematic plating mechanism of an electroless copper plating solution using hypophosphite as a reducing agent.

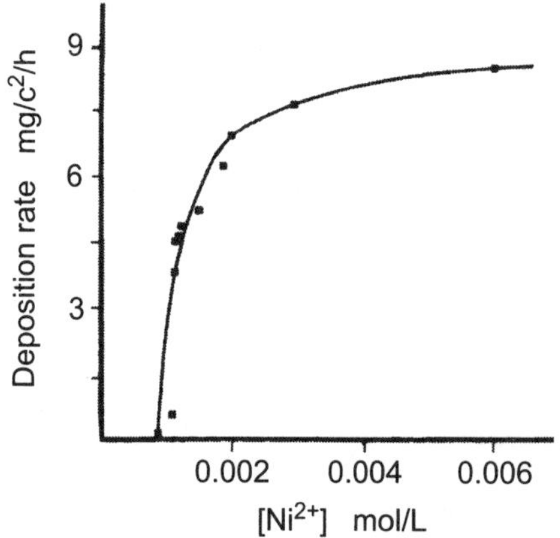

Figure 13.17 Deposition rate of a hypophosphite reducing electroless copper plating bath as a function of Ni^{2+} "mediator" ion concentration. (From Hung.[97])

bulk wiring materials. The application to a surface treatment for adhesion improvement[108] has been investigated. The morphology of the deposit from hypophosphite copper bath was nodulose or dendritic.[108] These results indicate that the application of this system should be limited to seed layer deposition, prior to electrolytic copper plating, and the application to the full build plating would be difficult.

The plating method reported in Refs. 104 to 106 uses a hypophosphite bath without the Ni ion. The Ni ion free hypophosphite plating bath is not autocatalytic, and the plating reaction only proceeds on the Pd catalyst surface. No deposition occurs on the copper foil, and the plating solution is very stable. The plating reaction is terminated after the whole Pd catalyst surface has been covered by the copper deposit. After a very thin coating (~0.1 μm) of electroless plating, low density cathodic current was applied and additional thickness was plated electrolytically. It was reported that by following these application sequences, adhesion of the plated copper foil to the resin surface was superior to that obtained from the formaldehyde

Table 13.4　　Composition of a hypophosphite reducing electroless copper plating bath. (From Hung.[97])

Copper sulfate	0.024 mol/L
Sodium citrate	0.052 mol/L
Sodium hypophosphite	0.27 mol/L
H_3BO_3	0.5 mol/L
Nickel sulfate	0.002 mol/L
pH	9.2
Temperature	65°C

bath. It is estimated that the substrate catalyzed nature of this bath is advantageous to plating copper throughout minute rough pits. It results in good mechanical inter-locking effects and good adhesion of the plated copper.[109]

13.3.1.2　DMAB bath

Dimethylamine borane (DMAB) is already used as a reducing agent in electroless Ni(Ni–B) plating,[110] and electroless Au plating. DMAB is a strong reducing agent and is expected to work in a lower pH region than formaldehyde. It is a preferred feature to avoid material damage with highly alkaline plating baths such as the form-aldehyde electroless copper plating bath. However, few studies have been done using DMAB as a reducing agent for electroless copper plating. Electroless copper plating at a low pH using DMAB as a reducing agent was reported in Refs. 111 to 114. It is important to use a strong pH buffer in order to maintain the bath stability in low pH levels (≤ 9). EDTA, a chelating agent commonly used in the formaldehyde electroless copper plating bath, is a strong chelating agent but a weak buffer around pH 9. Tri-ethanolamine was employed as an excellent pH buffer with EDTA as a strong complexant, and the bath stability was markedly improved. It also indicates that multidentate nitrogen donor compounds as a series of new complexants, such as tetraazadodecane are suitable ligands for the DMAB system (Figure 13.18),

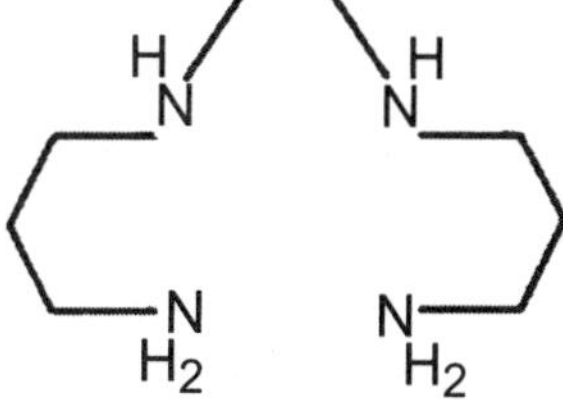

1,5,8,12-Tetraazadodecane

Figure 13.18　　Examples of strong nitrogen donor complexants for Cu^{2+} in electroless copper plating solutions using DMAB as a reducing agent. (Source: Ref. 112.)

1,10-phenanthroline and 2,2′-dipyridyl, both commonly examined as additives in the formaldehyde bath, were also used for this system. The deposition rate of the stabilized bath was 2 to 3 μm/h in pH 9.0. The resistivity of the deposit from this bath was 1.83–1.95 μΩ cm. The author's interest was to apply this bath to microelectronic pattern fabrication. Mechanical properties (ductility and ultimate tensile strength) were not reported. Therefore, the applicability of this bath to through-hole plating on PWBs still remains unclear.

13.3.1.3   Glyoxylic acid

Electroless copper plating solutions employing glyoxylic acid were investigated in works.[115–120] Glyoxylic acid is a kind of aldehyde, the same as formaldehyde. However, it has almost no vapor pressure from room temperature to plating temperature (70 to 80°C). It has no irritant odor at all such as formaldehyde. The oxidation reaction of glyoxylic acid is quite similar to that of formaldehyde. An amount of 1 mol of glyoxylic acid releases one electron, and 1 mol of oxalate ion. The overall plating reaction is written as follows:

$$CHOCOO^- + 2OH^- \rightarrow C_2O_4^{2-} + e^- + H_2O + 1/2H_2 \qquad (8)$$

The behavior of the plating reaction is quite similar to the plating solution using formaldehyde. The plating solution, almost having the same composition as the formaldehyde bath except for the reducing agent, results in having a good quality of copper deposits with the same level plating rate.[118,120] Therefore, the plating solution pH is in the range of 11.5 to 13.0 as high as the formaldehyde bath. The advantage of a lowering bath pH could not be obtained in this system. Another disadvantage of this system is the large reaction rate constant of the Cannizzaro reaction. The Cannizzaro reaction is a disproportionation reaction of aldehydes. In the case of formaldehyde, it disproportionates to formate and methanol by the Cannizzaro reaction. Glyoxylic acid produces glycolate and oxalate, as expressed in the following equation:

$$2CHOCOO^- + OH^- \rightarrow C_2O_4^{2-} + CH_2OHCOO^- \qquad (9)$$

The Cannizzaro reaction rate is reduced by using KOH as the pH adjusting agent instead of NaOH.[118] The rate constant of the Cannizzaro reaction of glyoxylic acid was reported through quantitative measurements of the reaction product, oxalate and glycolate (Table 13.5).[120] The plating solution formulation was optimized to minimize the side reaction. The consumption rate of glyoxylic acid was reduced by using KOH and lowering the solution pH, while maintaining the same level of plating rate by increasing the plating temperature and Cu^{2+} ion concentration (Table 13.6).[120] The solubility of the potassium salt of oxalate is 51.4 g/100 mL at 60°C, that is much higher than that of sodium salt (4.8 g/100 mL). This is an additional advantage of using KOH instead of NaOH, to avoid precipitate formation during continuous operation of the plating bath in practical use, enabling the bath lifetime to be prolonged. Mechanical properties of the plated copper and the interconnect reliability of the

Table 13.5   Rate constant of Cannizzaro reaction of glyoxylic acid. (Source: Ref. 120.)

Alkali	*Rate constant* $\times 10^3$ $(L^{-1}\ mol\ sec)$	
	pH = 12.0	*pH = 12.5*
NaOH	7.0	1.8
KOH	3.5	1.2

plated through holes on test PWBs by the optimized plating solution are examined in Ref. 6. Excellent performance has been proved for practical application. Glyoxylic acid is the most promising candidate to replace formaldehyde in electroless copper plating bath for full build through-hole plating.

The low temperature (~25°C) deposition process of the electroless copper plating solution using glyoxylic acid was examined in Ref. 119. The induction period prior to plating initiation on some different chelating agents was measured. Rochelle

Table 13.6   Optimized formulation of glyoxylic acid bath. (Source: Ref. 120.)

Component	*Concentration or condition*
$CuSO_4$	0.06 mol/L
EDTA	0.1 mol/L
Glyoxylic acid	0.03 mol/L
pH	12.1 adjusted by KOH
Additives	2,2′-Dipyridyl polyethylene glycol
Plating temperature	74°C

salts (potassium–sodium tartrate) and β-alaninediacetic acid showed a very long induction period at the glyoxylic acid bath while other chelating agents showed a short induction period, the same as those of the formaldehyde bath. These behaviors are quite different from the high temperature EDTA bath, where the glyoxylic bath shows similar plating performance to the formaldehyde bath. It is suggested that careful optimization of the plating solution formulation is required when the bath employs tartarate as a chelating agent.

13.3.1.4   Co(II) complex system

Many organic reducing agents applied for electroless plating including formaldehyde, hypophosphite, and dimethylamine borane, have a common nature to generate

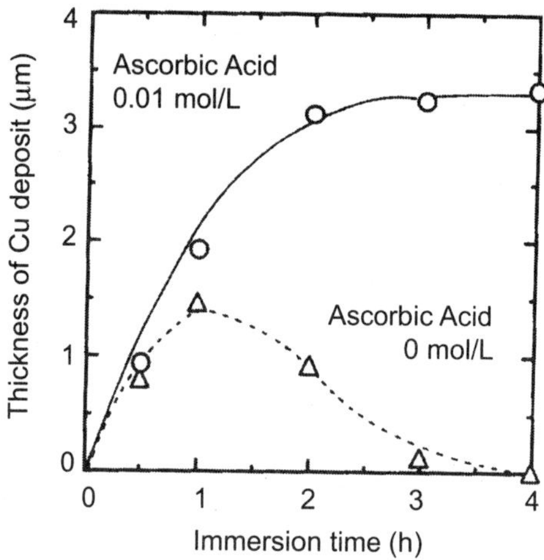

Figure 13.19   Relation between plating time and thickness of deposit obtained from Co(II) complex reducing electroplating solution. (From Nawafune et al. [128])

hydrogen gas as the reaction product during its oxidation reaction. Because of the low solubility of hydrogen gas in aqueous electrolyte, the generated hydrogen forms gas bubbles at the plated surface. These gas bubbles have the possibility of causing gas plugging at small features on the plated surface. The hydrogen embrittlement effect is also reported on formaldehyde-reduced electroless deposits. [76] A new type of auto-catalytic electroless copper plating system which is free from hydrogen generation

Table 13.7   Bath composition and plating condition of Co(II) complex reducing electroless copper bath. (From Nawafune et al. [128])

$CuCl_2$	0.05 mol/L
Ethylenediamine (en)	0.60 mol/L
$Co(NO_3)_2$	0.15 mol/L
Ascorbic acid	0.01 mol/L
2,2′-Dipyridyl	20 mg/L
pH	6.75
Temperature	50°C

has been proposed using Co(II) complex as the reducing agent. [122–128] The plating reaction is based on a simple electron exchange between $Cu(en)_2^{2+}$ and $Co(en)_2^{2+}$ and is expressed in the following equation:

$$Cu(en)_2^{2+} + 2Co(en)_2^{2+} \rightarrow Cu + Co(en)_3^{3+} + en \qquad (10)$$

(en: ethylenediamine)

The additional feature of this system is that the plating reaction proceeds near a neutral pH of 6 to 7, and so the material damage of highly alkaline plating solution could be avoided. The addition of ascorbic acid improved the plating performance as shown in Figure 13.19 and Table 13.7.[128] The application of this bath for Cu wiring fabrication ultra-large-scale integrated circuit was studied in Ref. 128. In order to apply this system to practical through-hole (via-hole) plating on high density PWBs, much more improvement seems to be required on bath lifetime, bath stability, quality of the deposit, and material cost issues.

References

1. "Printed Circuits Handbook", 5th edition, edited C.F. Coombs, Jr., McGraw-Hill, New York (2001).
2. "Printed Circuit Board Materials Handbook", edited M.W. Jawitz, McGraw-Hill, NewYork (1997).
3. "High Performance Printed Circuit Boards", edited C.A. Harper, McGraw-Hill, NewYork (2000).
4. "Modern Electroplating", 4th edition, edited M. Schlesinger and M. Paunovic, John Wiley & Sons, NewYork (2000).
5. "Electroless Plating: Fundamentals and Applications", edited G.O. Mallory and J.B. Hajdu, American Electroplating and Surface Finish Society (1990).
6. "New Trends and Approaches in Electrochemical Technology", edited N. Masuko, T. Osaka, and Y. Fukunaka, Kodansha, Tokyo (1993).
7. "The Properties of Electrodeposited Metals and Alloys A Handbook", edited W.H. Safranek, American Elsevier Publishing, New York (1974).
8. E.K. Yung, L.T. Romankiw, and R.C. Alkire, *J. Electrochem. Soc*, **136**, 206 (1989).
9. E.K. Yung and L.T. Romankiw, *J. Electrochem. Soc*, **136**, 756 (1989).
10. Y. Tsukada, S. Shohei, and Y. Mashimoto, Proc. 41th Electron. Comp. Technol. Conf. 22 (1992).
11. H. Holden, *Circuit World*, **23**, No. 2, 14 (1997).
12. J.W. Dini, "Modern Electroplating", 4th edition, edited M. Schlesinger and M. Paunovic, p. 68, John Wiley & Sons, NewYork (2000).
13. J. Mann, *Trans. Inst. Metal. Finish.*, **56**, 70 (1978).
14. H. Hager, *J. Appl. Electrochem.*, **16**, 189 (1986).
15. J.R. White and R.T. Gaiasco, *Pat. Surf. Finish.*, **71**, 122 (1988).
16. G. Holmbom and B.E. Jacobsson, *Surf. Coat. Technol.*, **35**, 333 (1988).
17. A. Montgomery, *Circuit World*, **15**, No. 2 (1989).
18. O. Chene and D. Landolt, *J. Appl. Electrochem.*, **19**, 188 (1989).
19. T. Pearson and J.K. Dennis, *J. Appl. Electrochem.*, **20**, 196 (1990).
20. M.R. Kalantary, D.R. Gabe, and M.R. Goodnough, *Metal Finish.*, **89**, No. 4, 21 (1991).
21. M.R. Kalantary, D.R. Gabe, and M.R. Goodnough, *J. Appl. Electrochem.*, **23**, 231 (1993).
22. S. Coughlan, *Trans. I.M.F.*, **73**, 54 (1995).
23. M.R. Kalantary and D.R. Gabe, *Surf. Eng.*, **11**, 246 (1995).
24. S. Roy and D. Landolt, *J. Appl. Electrochem.*, **27**, 299 (1997).
25. E.J. Taylor, J.J. Sun, and M.E. Inman, *Plat. Surf. Finish.*, **83**, No. 12, 68 (2000).
26. J. Kamiya, S. Fujimaki, K. Shishido, and H. Takakusaki, The 16th National Convention Record Jpn. Inst. Electron. Packag., 1 (2002).
27. M. Matloz, C. Creton, C. Clerc, and D. Landolt, *J. Electrochem. Soc.*, **134**, 30115 (1987).
28. O. Lanzi and U. Landau, *J. Electrochem. Soc.*, **135**, 1922 (1988).
29. O. Lanzi, U. Landau, J.D. Reid, and R.T. Galasco, *J. Electrochem. Soc.*, **136**, 1923 (1989).
30. D.A. Hazlebeck and J.B. Talbot, *J. Am. Inst. Chem. Eng.*, **36**, 1145 (1990).
31. D.A. Hazlebeck and J.B. Talbot, *J. Electrochem. Soc.*, **138**, 1985 (1991).
32. D.A. Hazlebeck and J.B. Talbot, *J. Electrochem. Soc.*, **138**, 1998 (1991).
33. M.Pesco and H.Y. Cheh, *J. Electrochem. Soc.*, **136**, 399 (1989).

34. M. Pesco and H.Y. Cheh, *J. Electrochem. Soc.*, **136**, 408 (1989).
35. J.-W. E. Chern and H.Y. Cheh, *Chem. Eng. Comm.* **114**, 175, (1992).
36. J.-W. E. Chern and H.Y. Cheh, *Chem. Eng. Comm.* **114**, 191, (1992).
37. J.-W. E. Chern and H.Y. Cheh, *Chem. J. Electrochem. Soc.*, **143**, 3139 (1996).
38. J.-W. E. Chern and H.Y. Cheh, *Chem. J. Electrochem. Soc.*, **143**, 3144 (1996).
39. S.H. Chan and H.Y. Cheh, *J. Appl. Electrochem.*, **31**, 605 (2001).
40. S.H. Chan and H.Y. Cheh, *J. Appl. Electrochem.*, **31**, 617 (2001).
41. P. Paumelle and C. Senlis, *Galv. Org. Trait. Surf.* **68**, 692 (1999).
42. G. Nelissen, A. Van Theemsche, S. Lecho, and J. Deconinck, Proc. AESF Surf. Fin, 363, (1999).
43. G. Nelissen, A. Van Theemsche, and J. Deconinck, Proc. AESF Sur. Fin. (2000).
44. C. Dan, B. Van den Bossche, L. Bortels, G. Nelissen, and J. Deconinck, *J. Electroanal. Chem.*, **505**, No. ER1-2, 12 (2001).
45. U. Landau, Abstract #505, Electrochem. Soc. Meeting, Phoenix, AZ, October 22–27 (2000).
46. U. Landau, J. D'Urso, A. Lipin, Y. Dordi, A. Malik, M. Chen, and P. Hey, Extended Abstracts, vol. 99-1, No. 263, Electrochem. Soc. Meeting, Seattle, WA, May 2–6 (1999).
47. CAD Software for Electrochemical Cells; "Cell-Design" L-Chem, Inc., U.S.A.; "Elsy+", ElsyCa N.V., Belgium, "Castor Elec 3D", CETIM, France.
48. M. Trade, *Metal Finish.*, No. 4, 24 (2001).
49. A. Takahashi and T. Itabashi, *J. Jpn. Inst. Interconnect Packag. Electron. Circuit*, **11**, 356 (1996).
50. A. Takahashi, T. Itabashi, R. Watanabe, T. Miwa, and H. Akahoshi, Proc. of the 7th Printed Circuit World Convention (PCWC VII), P13-1 (1996).
51. S. Abe, T. Fujinami, T. Aono, and H. Honma, *J. Surf. Finish. Soc. Jpn.*, **48**, 433 (1997).
52. S. Abe, M. Ohkubo, T. Fujinami, and H. Honma, *Trans. IMF*, **76**, 12 (1998).
53. T. Fujinami, K. Kobayashi, A. Maniwa, and H. Honma, *J. Surf. Finish. Soc. Jpn.*, **48**, 660 (1997).
54. S.S. Kuglikov, N.T. Kudryavstev, G.I. Medvedev, and T.M. Izmailova, *Prot. Met., Engl. Trans.*, **8**, 668 (1972).
55 S.S. Kuglikov, N.T. Kudriavtsev, G.F. Vorobiova, and A.Ya. Antonov, *Electrochim. Acta*, **10**, 253 (1965).
56. J.O. Dukovic and C.W. Tobias, *J. Electrochem. Soc.*, **137**, 3748 (1990).
57. K.G. Jordan and C.W. Tobias, *J. Electrochem. Soc.*, **138**, 1251 (1991).
58. P.C. Andricacos, C. Uzoh, J.O. Dukovic, J. Horkans, and H. Deligianni, *IBM J. Res. Develop.*, **42**, 567 (1997).
59. K. Kobayashi, A. Sano, H. Akahoshi, T. Itabashi, T. Haba, S. Fukada, and H. Miyazaki, Proc. 2000 Int. Interconnect. Technol. Conf., 34 (2000).
60. T. Kobayashi, J. Kawasaki, J. Ishibashi, K. Tanaka, and H. Honma, *J. Surf. Finish. Soc. Jpn.*, **49**, 1332 (1998).
61. T. Kobayashi, J. Kawasaki, K. Mihara, T. Yamashita, and H. Honma, *J. Jpn. Inst. Electron. Packag.*, **3**, 324 (2000).
62. K. Kondo, K. Hayashi, Z. Tanaka, and N. Yamakawa, *J. Jpn. Inst. Electron. Packag.*, **3**, 607 (2000).
63. K. Kondo, N. Yamakawa, T. Tanaka, and K. Mano, *J. Jpn. Inst. Electron. Packag.*, **4**, 37 (2001).
64. T. Kobayashi, J. Kawasaki, K. Mihara, and H. Honma, *Electrochimica Acta*, **47**, 85 (2001).
65. T. Matsunami, T. Ito, Y. Iwamoto, and S. Yamato, *J. Jpn. Inst. Electron. Packag.*, **4**, 629 (2001).
66. S. Miura, K. Mihara, T. Fukushi, and H. Honma, *J. Jpn. Inst. Electron. Packag.*, **5**, 246 2002).
67. J.J. Kelly, C. Tian, and A.C. West, *J. Electrochem. Soc.*, **146**, 2540 (1999).
68. A.C. West, *J. Electrochem. Soc.* **147**, 227 (2000).
69. A.C. West, S. Mayer, and J. Reid, *Electrochem. Solid-State Lett.*, **4**, C50 (2001).

70. Y. Cao, P. Taephaisitphongse, R. Chalupa, and A.C. West, *J. Electrochem. Soc.*, **148**, C466 (2001).

71. T.P. Moffat, D. Wheeler, W.H. Huber, and D. Josell, *Electrochem. Solid-State Lett.*, **4**, C26 (2001).

72. D. Josell, D. Wheeler, W.H. Huber, J.E. Bonevich, and T.P. Moffat, *J. Electrochem. Soc.*, **148**, C767 (2001).

73. D. Josell, D. Wheeler, and T.P. Moffat, *Electrochem. Solid-State Lett.*, **5**, C49 (2002).

74. H. Nakahara, Printed Circuits Handbook, 5th edition, edited C.F. Coombs, Jr., McGraw-Hill, New York (2001).

75. K. Minten, "Printed Circuit Board Materials Handbook", 26.1, edited M.W. Jawitz, McGraw-Hill, N.Y. (1997).

76. S. Nakahara, and Y. Okinaka, "New Trends and Approaches in Electrochemical Technology", 39, edited N. Masuko, T. Osaka, and Y. Fukunaka, Kodansha, Tokyo (1993).

77. S. Imabayashi, I. Tanaka, H. Kikuchi, M. Watanabe, H. Oka, S. Izumi, Y. Taniguchi, and S. Fujita, Proc. 41st Electron. Comp. Technol. Conf., 1053 (1992).

78. A. Takahashi, N. Ooki, A. Nagai, H. Akahoshi, A. Mukoh, and M. Wajima, *IEEE Trans. Comp. Hybrids Manuf. Technol.*, **15**, 418 (1992).

79. H. Akahoshi, M. Kawamoto, T. Itabashi, O. Miura, A. Takahashi, S. Kobayashi, M. Miyazaki, T. Mutoh, M. Wajima, and T. Ishimaru, *IEEE Trans. Comp. Packag. Manuf. Technol.*, Part A, **18**, 127 (1995).

80. H. Morishita, M. Kawamoto, M. Wajima, and K. Murakami, U.S. Patent 4,099,974 (1978).

81. H. Akahoshi, K. Murakami, M. Kawamoto, M. Wajima, R. Toba, S. Kawakubo, and A.Tadokoro, U.S. Patent 4,632,852 (1986).

82. H. Akahoshi, K. Murakami, M. Wajima, and S. Kawakubo, *IEEE Trans. Comp. Hybrids and Manuf. Technol.*, CHMT-9, p. 181 (1986).

83. H. Akahoshi, *Denki Kagaku*, **61** 843, (1993).

84. K. Minten and J. Toth, Proc. of the 5th Printed Circuit World Convention (PCWC V), B 5/2 (1990).

85. J. Cisson, J. Seigo, and K. Minten, *Circuit World,* **18**, No. 4, 5 (1992).

86. K. Muller, *Metallobeflache*, **14**, 65 (1960).

87 M. Paunovic, *Plating*, **55**, 1161 (1968).

88. H. Wise and K.G. Weil, *Ber. Bunsenges. Phys. Chem.*, **91**, 619 (1987).

89. M.C. Agens, U.S. Patent 2,938,805 (1960) (GE).

90. W.A. Alphaugh and T.D. Zucconi, U.S. Patent 4,152,467 (1979).

91. J.W.M. Jacobs and J.M.G. Rikken, *J. Electrochem. Soc.*, **135**, 2822 (1988).

92. T. Itabashi, H.Akahoshi and A. Takahashi, *Circ.Tecnol.*, **9**, 112 (1994).

93. M. Matsuoka and H. Hayashi, *Metal Finish.*, **83**, No.6, 87 (1985).

94. Y. Okinaka and S. Nakahara *J. Electrochem. Soc.*, **123**, 475 (1976).

95. F.W. Schneble, Jr., J.F. McCormack, and R.J. Zeblisky, U.S. Patent 3,959,531 (1976).

96. H. Akahoshi, K. Murakami, and M. Wajima, Ext. Abs. 166th ECS Meeting, p. 640 (1984).

97. A. Hung, *Plat. Surf. Finish*, **75**, 62 (1988).

98. A. Hung, *Plat. Surf. Finish.*, **75**, 74 (1988).

99. A. Hung, *Plat. Surf. Finish.*, **76**, 60 (1989).

100. A. Hung and K-M Chen, *J. Electrochem. Soc.*, **136**, 72 (1989).

101. A. Hung and I. Ohno, *Plat. Surf. Finish.*, **77**, 54 (1990).

102. A. Hung and I. Ohno, *J. Electrochem. Soc.*, **137**, 918 (1990).

103. J.G. Gaudiello and G.L. Ballard, *IBM J. Res. Dev.*, **37** 107 (1993).

104. P.E. Kukanskis, J.J. Grunwald, D.R. Ferrier, and A. Sawoska, U.S. Patent 4,209,331 (1981).

105. R. Goldstein, P.E. Kukanskis, and J.J. Grunwald, U.S. Patent 4,265,943 (1981).

106. J.G. Grunwald, P.E. Kukanskis, R.A. Leitze, D.A. Sawoska, and H. Rhodenizer, Proc. AES 2nd Electroless Plating Symp. F1 (1984).

107. D.R. Ferrier, U.S. Patent 4,325,990 (1982).

108. H. Honma and T. Fujinami, *Circuit Technol.*, 6, 209 (1991).
109. T. Mizuguchi, T. Kobayashi, T. Fujinami, and H. Honma. *J. Surf. Finish. Soc. Jpn.*, **49**, 1327 (1998).
110. G.O. Mallory, "Electroless Plating: Fundamentals and Applications", edited G.O. Mallory and J.B. Hajdu, *Amer. Electroplat. Surf. Finish. Soc.*, p. 11 (1990).
111. F. Pearistein and R.F. Weightman, *Plating*, **60**, 474 (1973) (DMAB).
112. R. Jagannathan and M. Krishnan, *IBM J. Res. Develop.*, **37**, 117 (1993).
113. R. Jagannathan, M. Krishnan and G.P. Wandy, U.S. Patent 4,818,286 (1989).
114. R. Jagannathan, R.F. Knarr, M. Krishnan, and G.P. Wandy, U.S. Patent 5,059,243 (1991).
115. J. Darken, *Trans. Inst. Metal Finish.*, **69**, 66 (1991).
116. J. Darken, U.S. Patent 4,617,205 (1986).
117. H. Honma, M. Komatsu, and T. Fujinami, *J. Surf. Finish. Soc. Jpn.*, **42**, 913 (1991).
118. H. Honma and T. Kobayashi, *J. Electrochem. Soc.*, **141**, 730 (1994).
119. H. Watanabe, K. Chiba, Y. Hukuda, and Y. Kurokawa, *Circuit Technol.*, **10**, 118 (1995).
120. T. Itabashi, H. Akahoshi, T. Iida, E. Takai, and N. Nishimura, *J. Jpn. Inst. Electron. Packag.*, **5**, 252 (2002).
121. A. Vaskelis, E. Norkus, G. Rozovskis, and H.J. Vinkevicius, Proc. Interfinish 96 World Congress, 2, 229 (1996).
122. A. Vaskelis, E. Norkus, and J. Vinkevicius, *Trans. Inst. Metal Finish.*, **75**, 1, (1997).
123. A. Vaskelis, G. Stalnionis, and Z. Jusys, *J. Electroanal. Chem.*, **465**, No. 2, 142 (1999).
124. A. Vaskelis and E. Norkus, *Electrochim. Acta*, **44**, 3667 (1999).
125. E. Norkus, P. Norkus, J. Vaiciuniene, and J. Jaciauskinene, *Trans. Inst. Metal Finish.*, **79**, 77 (2001).
126. A. Vaskelis, A. Jagminiene, and L. Tamasauskaite-Tamasiunaite, *J. Electroanal. Chem.*, **521**, No.1-2, 137 (2002).
127. H. Nawafune, *J. Surf. Finish. Soc. Jpn.*, **49**, 1180 (1998).
128. H. Nawafune, S. Nakao, S. Mizumoto, Y. Murakami, and S. Hashimoto, *J. Surf. Finish. Soc. Jpn.*, **50**, 374 (1999).

14 Electroplated contact materials for connectors and relays

Yutaka Okinaka

14.1 Introduction

Connectors and electromechanical relays are essential components of modern electronic instruments. Contact materials used on those components include hard gold, palladium, palladium–nickel, and palladium–cobalt alloys; and all of these contact finishes are prepared by electroplating as part of a continuous high speed processing line. Material properties of the contacts play a critical role in determining performance reliability of the products, and thus electroplating constitutes an important part of the overall manufacturing process of connectors and relays.

This chapter reviews selected topics on electroplated contact materials for connectors and relays with emphasis on results of recent studies performed to clarify factors affecting critical physical properties of those materials.

14.2 Connectors

14.2.1 Plating baths and techniques

Electroplated hard gold is most widely used as a contact finish for connectors. Commonly employed hard gold plating baths contain $KAu(CN)_2$ as the source of gold, a citrate buffer of pH 3.5 to 4.0, and a small amount of a Co or Ni salt as the hardening agent. Those baths, abbreviated in this chapter as CoHG and NiHG baths, respectively, were developed in the late 1950s,[1] and they are still being used extensively by the industry. Hard gold can also be plated from a bath containing no hardening additive (designated as AFHG bath), if the bath is operated under a specific set of conditions.[2-4] The AFHG bath consists of $KAu(CN)_2$ and a phosphate buffer of pH 7. A suitable gold concentration and an optimum operating temperature are chosen for each bath depending on the current density or the plating speed desired. Examples of the compositions and operating conditions of various hard gold plating baths are shown in Table 14.1.[1-4]

Palladium and palladium alloys, especially Pd–Ni and Pd–Co, capped with a thin layer of hard gold are also employed as contact finishes.[5-8] Examples of bath compositions and operating conditions of these baths are given in Table 14.2.[9] Pd

421

and Pd alloy plating baths were originally developed for the sole purpose of substituting for gold when the price of gold rose excessively in the 1970's. At the time of writing this chapter, however, palladium is more expensive than gold, and therefore, there is no economic reason why palladium should be substituted for gold. From the technological standpoint, however, it should be realized that palladium does have certain advantages over gold depending on applications, and this aspect will be discussed in a subsequent section.

Table 14.1   Examples of bath compositions and operating conditions for plating hard gold.[1–4]

	CoHG or NiHG bath		AFHG bath	
	Low speed	*High speed*	*Low speed*	*High speed*
$KAu(CN)_2$ (as Au)	8 g/L	30 g/L	25 g/L	30–40 g/L
Citric acid	100	74		
K-citrate		111		
KOH	40			
KH_2PO_4			100	100
$CoSO_4$ (as Co) or $NiSO_4$ (as Ni)	0.1–0.5	0.25		
pH	3.5–4.2	4.2	4.3–4.5	6.5–7.5
Temperature	30°C	65°C	25°C	40°C
Agitation	Mild	Vigorous	Mild	Vigorous
Current density (A/dm^2)	1–2	10–40	1–2	20–30

It is assumed that the reader is familiar with the general construction and configuration of sliding connectors used widely in various electronic devices. A typical connector assembly consists of two separable parts which are mated by insertion when in use. There is a variety of configurations and dimensions for such connectors, and it is beyond the scope of this chapter to describe geometrical and physical details of these connectors. However, a brief general description is in order concerning the technique of manufacturing connector terminals. The terminals, generally made of a copper-based alloy, are electroplated first with nickel, followed by thin strike gold and then hard gold (or an alternative) after suitable pre- and post-treatment processes such as electropolishing, acid dip, water rinse, and drying. These operations are performed by using a continuous reel-to-reel strip plating machine, in which a strip of substrate on a pay-off reel is transported at a constant speed (as fast as 30 to 80 cm sec^{-1}) from one end of the machine to a take-up reel at the other end, while the strip travels through cells in which it is subjected to the chemical, electrochemical, and cleaning operations mentioned above. The strip plating is done on either a pre-punched or unpunched strip of the terminal material. Plating of most connector terminals is done on punched strips with terminals attached to tie bars, which are removed during connector assembly. For further details of the strip plating technology, the reader is referred to a review article written by Turner.[10] Photographs of

Table 14.2   Examples of bath compositions and operating conditions for plating Pd, Pd–Ni, and Pd–Co alloys.[8,9]

	Pd	$Pd_{80}Ni_{20}$	$Pd_{80}Co_{20}$	
			Bath A	Bath B
$PdCl_2$ (as Pd)	1–40 g/L			
1,3-Diaminopropane	‡			
K_2HPO_4	50–100			
$Pd(NH_3)_2Cl_2$ (as Pd)		25 g/L	11 g/L	40 g/L
$Ni(NH_3)_4SO_4$		‡		
$CoSO_4$ (as Co)			2	5
$(NH_4)_2SO_4$		50	53	‡
NH_4OH		100		
3-Pyridine sulfonic acid		5		
pH	10–13	7.7	7–10	7.5
Temperature	40–70°C	60°C	‡	‡
Current density (A/dm^2)	0.1–5.0	0.3–25	‡	30

‡ Values not given in Ref. 9.

sections of terminal strips and assembled connectors of different dimensions can also be found in a recently published review article.[11]

14.2.2   Required properties

Contact materials on reliable connectors must meet the following requirements: (1) high hardness and resistance to mechanical wear, (2) low electrical contact resistance, and (3) high material integrity and stability in the specific environment to which the connector is to be exposed.

14.2.2.1   Hardness and wear resistance

The CoHG and NiHG deposits contain less than 0.4% by weight of Co and Ni, respectively, and yet the deposit hardness (170 to 200 kg mm^{-2} in Knoop hardness) is more than twice as high as that of electroplated, pure soft gold deposits (70 to 80 kg mm^{-2}). It is known that the high hardness of CoHG cannot be reproduced by simple metallurgical dissolution or alloying of such a small amount of metallic Co in pure gold metal. A detailed study of the hardening mechanism performed by Lo et al.[12] shows that the major factor determining the hardness of CoHG is the grain size. The hardening mechanism of NiHG is presumed to be the same as that of CoHG. Both CoHG and NiHG deposits consist of extremely small grains measuring 20–30 nm, compared to 1 to 2 µm for soft gold.[13] The hardness and the grain size of the AFHG deposit are comparable to those of CoHG or NiHG.[14]

The small grain size of the hard gold is attributed to the inhibition of crystal growth caused by incorporated impurities or inclusions. Examples of impurity contents of CoHG and AFHG are listed in Table 14.3. Considerable efforts were

Table 14.3   Typical contents of elements included in electroplated hard gold.

	CoHG		AFHG	
	wt. %	*at. %*	*wt. %*	*at. %*
Co	0.20	0.57	—	—
C	0.28	3.93	0.055	0.88
H	0.041	6.90	0.001	0.19
O	0.092	0.97	0	0
N	0.16	1.92	0.03	0.41
K	0.28	1.20	0.22	1.08
Total	1.05	15.5	0.31	2.56
Purity as Au	98.9	84.5	99.7	97.4

expended in the past to establish the identity of included compounds, but the identification is still far from complete. The inclusion of AuCN has been well established in both CoHG and AFHG by transmission electron microscopy.[14–16] CoHG has been shown to contain $K_3Co(CN)_6$ and/or $K_4Co(CN)_6$ in addition to metallic Co. However, the presence of these cyanide compounds does not account for the large excess of C over N in CoHG, and the source of the excess carbon has not been identified. In spite of the incomplete chemical identification, it has been shown conclusively that the inclusions play a highly significant role in determining critical physical properties of hard gold (i.e., hardness, resistance to mechanical wear, and electrical contact resistance). The CoHG, which contains much greater amounts of impurity elements than AFHG, has a significantly higher wear resistance than the latter,[2,4] in spite of the fact that both gold deposits have a similar hardness. Obviously, hardness is not the only factor controlling wear resistance.

Antler[17] compared the two types of hard gold for the variation of dimensionless wear index defined by Holm[18] with a number of sliding motion of the wear tester, in which a solid gold rider was made to slide at a load of 100 g against a sheet of copper plated with AFHG or CoHG. His results are reproduced in Figure 14.1: after only 15 passes the wear index of AFHG was as high as 95, whereas that of CoHG increased very slowly to reach this value only after approximately 100 passes.

Attempts were made to correlate the wear resistance of CoHG with other physical properties to find the reason why this type of hard gold exhibits the exceptionally high resistance to wear. Antler[19] noted that the wear of AFHG is due to the adhesion between the sliding surfaces (called "adhesive wear" as opposed to "abrasive wear" encountered with much harder materials), and that the relatively high ductility is responsible for the adhesion. On the other hand, Celis et al.[20] reported that wear resistance is related to internal stress. For CoHG they showed that

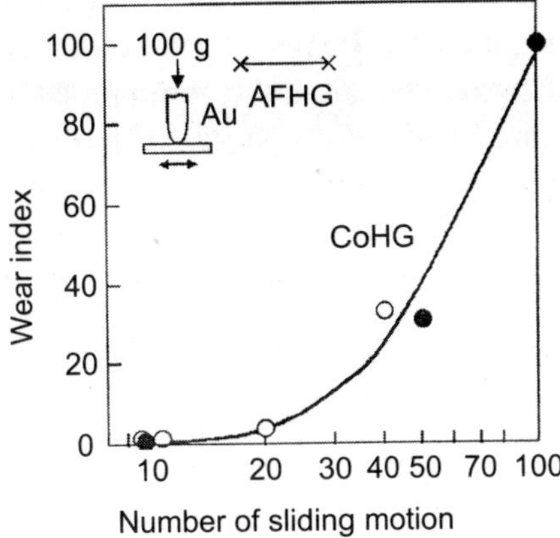

Figure 14.1   Wear resistance of CoHG and AFHG.[17]
(Substrate, Cu sheet; load, 100 g; unlubricated.)

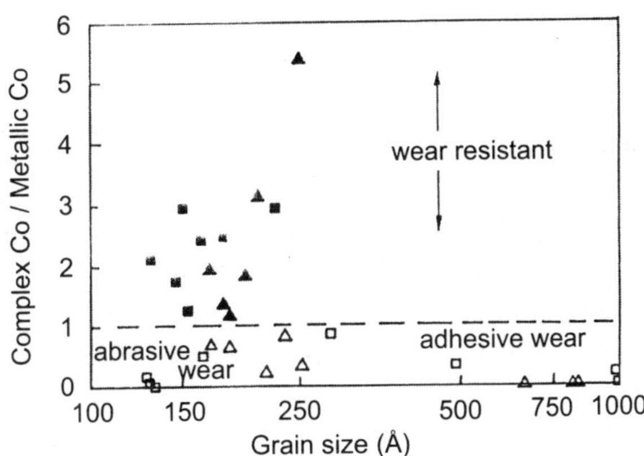

Figure 14.2   Effect of the ratio of complexed Co to metallic Co in CoHG on wear resistance
as a function of grain size.[20]

the internal stress varied with metallic Co content of the deposit, and that wear resistant deposits were obtained only at a metallic Co content in the range of 0.05 to 0.1% in weight, where the internal stress was in the range of 100 to 250 N mm^{-2}. When the internal stress was too low, adhesive wear resulted; when it was too high, abrasive wear was observed. These empirical relationships are of interest, but no theoretical interpretation has been offered to the author's knowledge.

From the practical standpoint, it is important to note the finding reported by De Doncker and Vanhumbeeck[21] that the wear resistance of CoHG is closely related to the ratio of the amount of complexed Co (e.g., Co in $K_3Co(CN)_6$) to that of metallic Co in the gold deposit. As illustrated in Figure 14.2,[20] when the ratio was less than unity, either abrasive wear or adhesive wear took place depending on the grain size of the deposit; and wear resistant deposits were obtained only when the ratio was greater than unity. The above authors developed a method for the separate determination of complexed Co and metallic Co, which should be useful to perform the analysis to ascertain that the CoHG plating process in use is being operated under conditions to produce the gold deposit in the optimum composition range.

Abys et al.[8] compared various properties of hard gold (NiHG), Pd–Ni, and Pd–Co deposits; and showed that both hardness and wear resistance increase in the order NiHG < Pd–Ni < Pd–Co, clearly indicating the superiority of the palladium alloys compared to NiHG. Table 14.4 quantitatively compares various properties of

Table 14.4 Material properties of NiHG, Pd, Pd–Ni, and Pd–Co alloys.[8]

	NiHG	*Pd*	*Pd–Ni*	*Pd–Co*
Wear resistance (number of cycles to failure at 100 g load)	20 K	>80 K	35 K	>80 K
Knoop hardness, kg mm^{-2}	140–200	150–500	450–550	590–640
Ductility, % elongation (2.5 µm thick film)	<3	3 to >10	3–10	3–7
Friction coefficient (at 10 K cycles)	0.60	0.31	0.55	0.43
Porosity index (on 1.5–2.5 µm thick Ni on connector pins)	0.5	—	0.11	0.02

the three different materials and pure Pd. It should be noted that both Pd–Ni and Pd–Co possess much higher ductility than NiHG, and yet the wear resistance of both alloys of Pd is significantly greater than that of NiHG. Apparently, the relevance of ductility to wear resistance proposed by Antler[19] for hard gold does not hold for the Pd alloys. The much higher hardness of the Pd alloys indicates that there is interplay among hardness, ductility, and wear resistance. Abys et al.[22] measured friction coefficients of the four different contact materials. The values obtained at 10 K sliding cycles are included in Table 14.4. The results appear to indicate that friction coefficient may serve as a measure of wear resistance.

14.2.2.2 Contact resistance

For obvious reasons low and stable electrical contact resistance is an important property required of the contact material used on connectors. It has been known for many years that the contact resistance of CoHG can increase to an unacceptable level especially at elevated temperatures. The cause of this phenomenon has been studied by many investigators. Originally, it was thought that the surface segregation of an included organic substance (often called "polymer"), which was assumed to be derived from cyanide species present in the plating bath, was responsible for the increased contact resistance.[23] Later studies, however, showed that it is cobaltous oxide (CoO) formed on the gold surface that is responsible for the increase in contact resistance.[24] The CoHG film is known to contain cobalt in two different forms: metallic Co and Co-cyanide complex.[25] De Doncker and Vanhumbeeck[21] conclusively showed that the contact resistance is determined by the content of metallic Co rather than that of complexed Co, indicating that the CoO forms as a result of the

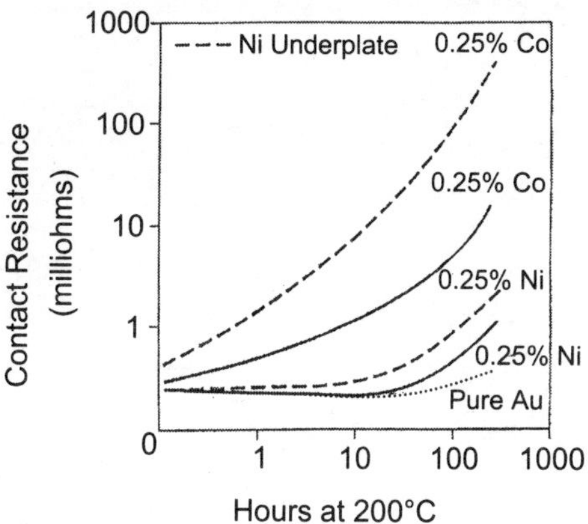

Figure 14.3 Comparison between CoHG, NiHG, and pure soft gold for thermal stability of contact resistance at 200°C.[26]
(Substrate, Cu sheet; solid lines, 2.5 µm thick gold without Ni underplate; dashed lines, 2.5 µm gold with 2.5 µm Ni underplate; dotted line, 50 µm thick pure soft gold without Ni underplate.)

oxidation of metallic Co, not the complexed Co. The nickel present on the surface of aged NiHG has been identified as NiO by an X-ray photoelectron spectroscopy (XPS) analysis,[26] which also is believed to have formed from metallic Ni rather than from complexed Ni.

For applications involving exposure to elevated temperatures, thermal stability of contact resistance is important. Connectors measuring only 15 mm in length with pin width and pitch as small as 0.5 to 0.8 mm are now in use. These small connectors, which are often used on surface-mount circuit boards, are often required to withstand temperature excursions up to 240°C, which is a solder reflow temperature. Exposure to elevated temperatures up to 350°C is encountered in connectors used in automobiles.

The superior contact resistance stability of NiHG compared to CoHG has been known for a long time in the industry, but actual data demonstrating it were published only relatively recently.[26–28] Some data reported by Antler[26] are reproduced in Figure 14.3, in which contact resistance values measured on 2.5 µm thick CoHG and NiHG films plated directly on copper coupons with and without a Ni underplate are plotted against time at 200°C. The gold films contained 0.25% in weight of Co or Ni. The data for a 50 µm thick pure gold film plated directly on copper are also included in the figure for comparison. These results clearly demonstrate that (1) the contact resistance of NiHG is significantly more stable than that of CoHG; and (2) the presence of an underplated layer of Ni increases the contact resistance of both CoHG and NiHG, the extent of the increase being more significant for CoHG than for NiHG. Huck[27] and Kumakura and Sekiguchi[28] reported similar sets of data showing that the stability of contact resistance at elevated temperatures is significantly better for NiHG than CoHG. It should be noted that the stability is highly dependent on temperature. For example, Huck[27] found that at 300°C a significant increase in contact resistance began to occur within a few minutes for both CoHG and NiHG,

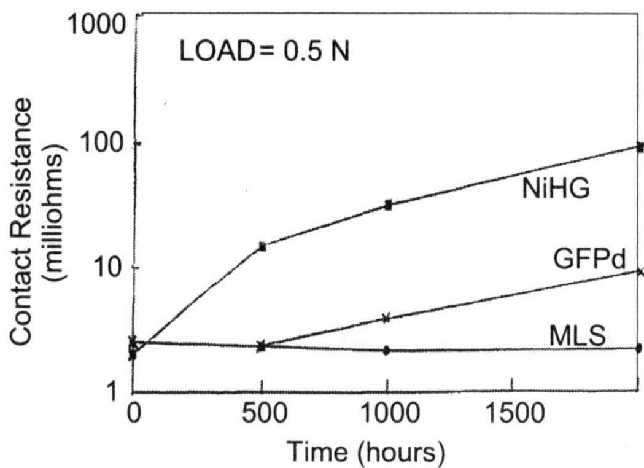

Figure 14.4 Comparison between NiHG, Au-flashed Pd (GFPd), and Pd-based multilayered structure (MLS) for thermal stability of contact resistance at 200°C.[31] (Average of 50 readings for each data point.)
Film compositions:
NiHG: NiHG(2.5 μm) / Ni(2.5–3.8 μm)
GFPd: CoHG(0.1 μm) / Pd(2.5 μm) / Ni(2.5–3.8 μm)
MLS: CoHG(0.1 μm) / Pd(0.025–2.5 μm) / $Pd_{80}Ni_{20}$(0.1–2.5 μm) / Pd
or Au(0.1 μm) / Ni(2.5–3.8 μm)

while at 125°C the rapid increase did not begin until after about 1000 h. At both temperatures, the rate of increase in contact resistance was much slower for NiHG than for CoHG.

Kumakura and Sekiguchi[28] carried out XPS analyses of the surfaces of CoHG and NiHG, and discovered the following differences between the two types of hard gold: (1) the surface film on aged CoHG consisted mainly of Co hydroxide, whereas that on NiHG was composed primarily of Ni oxide (the finding on CoHG is in agreement with that of Schubert[29]), (2) the surface film on CoHG was approximately four times thicker than that on NiHG, (3) the film thickness on NiHG grew to a constant thickness during thermal aging, whereas on CoHG the surface film continued to grow with the time of aging. These three differences between the surface films on CoHG and NiHG appear to be closely related to the difference in thermal stability of contact resistance between the two kinds of hard gold.

As was mentioned already, some connectors in automobiles may be exposed to a temperature higher than 300°C, which makes even NiHG unsuitable for such applications. It has been known that palladium-based materials exhibit a significantly better contact resistance stability than hard golds.[30,31] Kudrak et al.[31] compared the contact resistance stability of Pd-based finishes with that of NiHG. Their results at 200°C are reproduced in Figure 14.4, where pure Pd and a multilayered structure consisting of $Pd/Pd_{80}Ni_{20}/Pd$ exhibit considerably better stabilities than NiHG. It should be noted here that both Pd finishes were capped with a thin flash (0.1 μm) of CoHG. This gold flash is necessary for all practical Pd-based contact finishes to protect the surface of Pd from oxidation and from contamination with polymers catalytically produced from atmospheric organic contaminants. The gold flash appears to serve also as a lubricant. In this context, the use of the CoHG flash would seem problematic because, when used alone, the contact resistance of CoHG is not thermally stable, as mentioned repeatedly above. In practice, however, the thin flash of CoHG plated over Pd has not been a cause of any significant increase in contact resistance, apparently because the thickness of the CoHG flash is so small that even the total amount of metallic Co present in the entire gold layer is insufficient to bring about any measurable increase in contact resistance.

14.3 Relays

Several different materials including Cu and Ag alloys, Pd metal, and Au metal have been used as the contact material on electromechanical relays (hereafter called simply "relays"). The service life of such relays is known to be limited by two different causes: (1) increased contact resistance and (2) inability to open closed contacts. Cause 1 is similar to the phenomenon encountered with connectors described in the preceding section. Cause 2 is known to be partly due to adhesion, or "sticking" between two contact surfaces, and partly due to other problems such as material fatigue and decreased or lost power of the electromagnet used to drive the relay. The phenomenon of sticking of the various relay contacts has been investigated extensively in the past, primarily from the viewpoint of physics rather than chemistry.[32–37] Prompted by the interest in this problem occasionally encountered with miniature relays with electroplated gold contacts, which are mounted directly on circuit boards to switch high frequency signals,[38] the present author and collaborators at Waseda University conducted a systematic study of the sticking of various electroplated gold surfaces from the electrochemical and materials viewpoint.[39]

14.3.1 Sticking of relay contacts in general

It is well known that when two clean surfaces of pure, soft gold are made to slide against each other, the contact surfaces become worn very easily because of the adhesion of the two surfaces and the resultant material transfer from one surface to the other and vice versa. This phenomenon is known as adhesive wear, as described in the preceding section on connectors. The sticking of relay contacts is phenomenologically identical to adhesive wear, although the sliding distance of the relay contacts is only a small fraction of that of connectors. To understand factors causing the sticking, it is useful to measure the force of adhesion between a pair of contact surfaces as a function of variables such as the load applied to contact surfaces, the sliding distance, and the number of open–closure cycles. Kobayashi et al.[35–37] carried out a series of intensive investigations on this subject using 99.99% pure gold as the contact material. They showed that initially the adhesive force increases with the number of sliding cycles, reaching a plateau after a certain number of sliding cycles. As expected, the adhesive force was found to increase with increasing contact force. One of the most significant findings they reported was that the adhesive force decreased with increasing contact resistance, and that plotting the former against the latter yielded a linear logarithmic relationship, which is reproduced in Figure 14.5. This result was interpreted[35] by assuming that the real area of contact between the

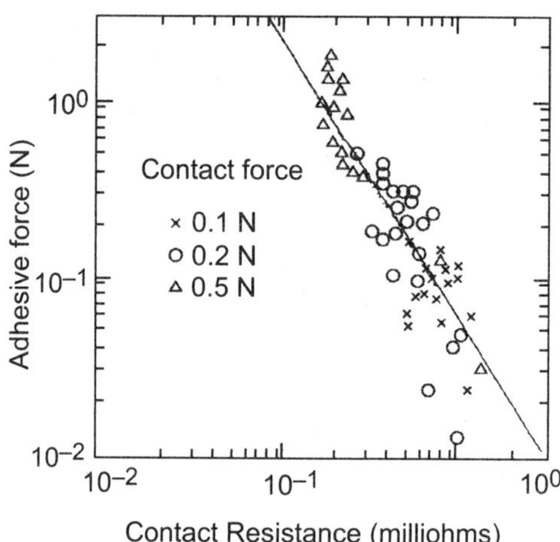

Figure 14.5 Relationship between adhesive force and contact resistance for pure gold contacts.[35]

two gold surfaces is the factor determining both adhesive force and contact resistance; namely, the adhesive force increases and the contact resistance decreases with increasing real contact area. Qualitatively, this type of relationship can also be anticipated under the condition that the presence of non-conducting or highly resistive contaminants is the cause of increase in contact resistance, because the presence of such contaminants is expected to decrease the adhesive force. This aspect will be discussed further in the subsequent section.

14.3.2 Sticking of electroplated gold contacts

The work conducted at the author's laboratory[39] was focused on investigating the sticking behavior of different kinds of gold (NiHG, CoHG, AFHG, and Au–Ni alloy) plated on the contact pair in various combinations. Compositions of the baths used to plate NiHG, CoHG, and AFHG were similar to those listed in Table 14.1; and the Au–Ni alloy was plated in a bath in which the Ni content of the NiHG bath was increased by a factor of 20. Experiments involved measurements of contact resistance and adhesive force by using a commercial electric contact simulator, in which the contact pair consisted of a copper rivet and a NiP-coated aluminum plate. Both members of the contact pair were plated with a 0.2 μm thick gold and allowed to slide against each other for a distance of 50 μm at a contact force of 0.5 N. After 3000 sliding cycles at the speed of 10 mm min^{-1}, the strength of adhesion was determined by measuring the force required to separate the contact pair.

The result of adhesive force measurements carried out with various combinations of gold films plated on the contact pair is summarized in Table 14.5. It is seen that with the three different kinds of hard gold, sticking was found to occur only when both members of the contact pair (both rivet and plate) are plated with the same kind of gold. Among the three kinds of gold, the highest adhesive force was observed with NiHG, followed by AFHG and then by CoHG. No sticking occurred when members of the contact pair were plated with different kinds of hard gold. With Au–Ni alloy no sticking was ever observed even when the same alloy was plated on both members of the contact pair.

For practical purposes these results are considered to be highly significant in selecting a suitable gold to be plated on each member of the contact pair. Intuitively, it seems natural that a contact pair with identical surfaces tends to suffer from sticking. However, this intuition does not apply to the case where both members of a contact pair are plated with the Au–Ni alloy, which did not exhibit any tendency of sticking. Furthermore, the result that the magnitude of adhesive force found was different for the different kinds of hard gold (Table 14.5) should be explained. To

Table 14.5 Results of measurement of adhesive force between various electroplated contact finishes[39] (measurements performed after 3000 sliding cycles).

Rivet → Plate ↓	NiHG	AFHG	CoHG	Au–Ni alloy
NiHG	0.115 N	0	0	0
AFHG	0	0.083 N	0	0
CoHG	0	0	0.012 N	0
Au–Ni alloy	0	0	0	0

Notes:
1. Each value of adhesive force is the average of five measurements.
2. Ni and Co contents of NiHG and CoHG were nominally equal to 0.25 wt. %.
3. Au–Ni alloy contained 3.8 wt. % of Ni.
4. "0" indicates no adhesion.

understand these results, effects of various factors such as hardness, surface roughness, and friction coefficient were investigated. Briefly, it was found that the tendency of sticking was less when the surface roughness was low and the friction coefficient was small. Details will be published elsewhere.

Finally, the relationship between adhesive force and contact resistance will be discussed for the case of electroplated hard gold contacts. Figure 14.6 shows a logarithmic plot of the data obtained with NiHG. In this experiment the data points for various values of adhesive force and contact resistance were obtained at different numbers of sliding cycles. It is remarkable that the relationship obtained from this series of experiment is qualitatively similar to that of Figure 14.5 reported by Kobayashi et al.[35] If it is assumed that the presence of an insulating or highly resistive material existing as an impurity on the contact surfaces was the cause of the high contact resistance and the low adhesive force at the initial period of sliding cycles, it is likely that as the sliding motion was repeated with the applied load, the surface impurity tended to be removed, leading to decreased contact resistances and increased adhesive force; the latter in turn resulted in an increased possibility of sticking.

It should be noted that an ideal relay contact should have a low contact resistance combined with a low adhesive force. The presence of surface contaminants should be beneficial for lowering the adhesive force and hence for minimizing the tendency to stick; however, if the contaminants are highly resistive or insulating, the

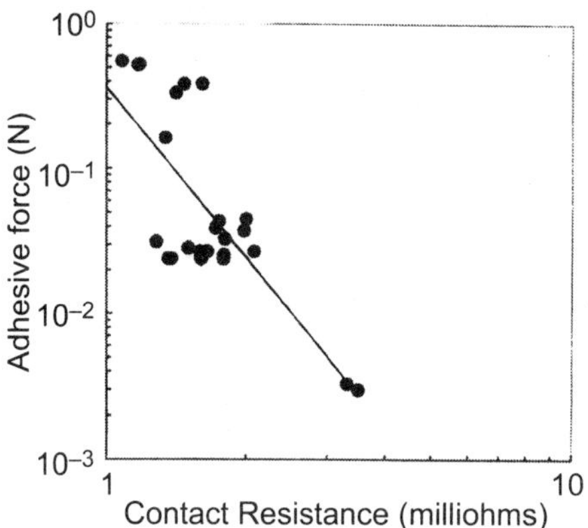

Figure 14.6 Relationship between adhesive force and contact resistance for NiHG contacts.[36]

presence of such contaminants is undesirable because it increases contact resistance. Therefore, for optimum performance of relay contacts with a given set of contact materials, compromise must be made between the tendency of sticking and the magnitude of contact resistance.

14.4   Summary

Electrochemical technology in the form of electroplating plays an essential role in the manufacture of connectors and relays, which are important components of modern electronic instruments and devices. Electroplated hard gold is the contact material most widely used in those components. Electroplated palladium and palladium alloys are advantageous over hard gold, particularly in applications at high temperatures.

It is emphasized that in manufacturing connectors and relays with high reliability, it is important to select a contact material with an optimized composition suited to the specific application. For the hard gold to be used on connectors, the detailed composition including not only the element of hardening additive (Co, Ni, or none) but also the form and quantity of the inclusion (complexed and metallic Co or Ni) determines critical properties such as the resistance to wear and electrical contact resistance. For miniature relays, selection of a proper combination of different kinds of hard gold on the two members of a contact pair is important in minimizing the occurrence of sticking. Results of some recent studies exemplifying the importance of the material selection in the above applications have been introduced and reviewed briefly in this chapter.

References

1. Gold Plating Technology, edited F.H. Reid and W. Goldie, Electrochemical Publications, Ltd., Ayr, Scotland (1974).

2. F.B. Koch, Y. Okinaka, C. Wolowodiuk, and D.R. Blessington, *Plating*, **67(6)**, 50 (1980).

3. F.B. Koch, Y. Okinaka, C. Wolowodiuk, and D.R. Blessington, *Plating*, **67(7)**, 43 (1980).

4. Y. Okinaka and F.B. Koch, in "Proceedings of the 10th World Congress on Metal Finishing (Interfinish 80)", pp.48–52 (1980).

5. J.A. Abys, Y. Okinaka, H.S. Trop, G.J. Russ, and B.T. Kerns, 35th Meeting of International Society of Electrochemistry, Berkeley, CA (1984).

6. J.A. Abys, Connectors '87, Institute of Metal Finishing, Coventry, England (1987).

7. J.A. Abys, J.J. Maisano, C. Wolowodiuk, and H.K. Straschil, Connectors '89, Institute of Metal Finishing, Coventry, England (1989).

8. J.A. Abys, G.F. Breck, H.K. Straschil, I. Boguslavsky, and G. Holmbom, *Plating Surf. Finish.*, **86(1)**, 108 (1999).

9. J.A. Abys and C.A. Dullaghan, in "Modern Electroplating," Fourth Edition, ed. M. Schlesinger and M. Paunovic, John Wiley & Sons, New York, Chapter 12 (2000).

10. D.R. Turner, in "Modern Electroplating," Fourth Edition, edited M. Schlesinger and M. Paunovic, John Wiley & Sons, New York, Chapter 24 (2000).

11. Y. Okinaka and M. Hoshino, *Gold Bull.*, **31(1)**, 3 (1998).

12. C.C. Lo, J.A. Augis, and M.R. Pinnel, *J. Appl. Phys.*, **50**, 6887 (1979).

13. Y. Okinaka and S. Nakahara, *J. Electrochem. Soc.*, **123**, 1284 (1976).

14. S. Nakahara and Y. Okinaka, *J. Electrochem. Soc.*, **128**, 284 (1981).

15. S. Nakahara, *J. Electrochem. Soc.*, **136**, 451 (1989).

16. G. Holmbom and B.E. Jacobson, *J. Electrochem. Soc.*,**135**, 787 (1988).

17. M. Antler, Private communication (1989).

18. R. Holm, "Electric Contacts", Springer-Verlag, Berlin (1946).

19. M. Antler, IEEE Transactions on Components, Hybrids, and Manufacturing Technology, CHMT-4, No. 1, 15 (1981).

20. J.P. Celis, J.R. Roos, W. Van Vooren, and J. Vanhumbeeck, *Trans. Inst. Met. Finish.*, **67**, 70 (1989).

21. R. De Doncker and J. Vanhumbeeck, *Trans. Inst. Met. Finish.*, **62(2)**, 59 (1985).

22. J.A. Abys, E.J. Kudrak, C. Fan, I. Boguslavsky, and G. Holmbom, *Connector Specifier*, pp. 12–14 (February 1999).

23. G.B. Munier, *Plating*, **56**, 1159 (1969).

24. J.H. Thomas and S.P. Sharma, *J. Electrochem. Soc.*, **126**, 445 (1979).

25. Y. Okinaka, F.B. Koch, C. Wolowodiuk, and D.R. Blessington, *J. Electrochem. Soc.*, **125**, 1745 (1978).

26. M. Antler, *Plating Surf. Finish.*, **85(12)**, 85 (1998).

27. M. Huck, *Metallwiss. Technik.*, **46**, 132 (1992).

28. H. Kumakura and M. Sekiguchi, *IEICE Trans. Electron.*, **E82-C,** No.1, 13 (1999).

29. R. Schubert, *J. Electrochem. Soc.*, **128**, 126 (1981).

30. K. Horibe and T.Hirano, Proceedings of AESF Annual Conference SUR/FIN '91, Toronto, Ontario, Canada (1991).

31. E.J. Kudrak, J.A. Abys, and V.A. Chinchankar, Application of Multiplexing Technology, SP-1137, SAE Technical Paper Series 960396, Society of Automotive Engineers (1996).

32. A. Kobayashi and T. Kubono, Technical Report of the Institute of Electronics, Information, and Communication Engineers (IEICE), EMD 93–70 (1993).

33. H. Iwata, H. Maruyama, and T. Otani, Technical Report of the Institute of Electronics, Information, and Communication Engineers (IEICE), R93-59, EMD 93–89 (1994).

34. H. Iwata, H. Maruyama, and T. Otanai, Technical Report of the Institute of Electronics, Information, and Communication Engineers (IEICE), EMD 94–44 (1994).

35. A. Kobayashi, S. Takano, and T. Kubono, Technical Report of the Institute of Electronics, Information, and Communication Engineers (IEICE), EMD 95–57 (1995).

36. A. Kobayashi and T. Kubono, Technical Report of the Institute of Electronics, Information, and Communication Engineers (IEICE), EMD 96–84 (1996).

37. A. Kobayashi and T. Kubono, Technical Report of the Institute of Electronics, Information, and Communication Engineers (IEICE), EMD 97–60 (1997).

38. See, for example, Teledyne Relays Data Book, Teledyne Relays, Hawthorne, CA (1997).
39. K. Yoshizawa, Y. Okinaka, T. Homma, T. Osaka, and T. Tsukamoto, Proceedings of the 16th Annual Meeting of Electronics Packaging (JIEP), Yokohama, Japan, March 18–20, 2002, pp. 83–84.

Part V

Processing tools

15 Chemical–mechanical planarization: From scratch to planar

David K. Watts, Norio Kimura, and Manabu Tsujimura

> The smaller the particles of the substance (abrasives), the
> smaller will be the scratches by which they continually fret
> and wear away the glass until it is polished …
> Sir Isaac Newton, *Optiks,* **1652**[1]

15.1 Introduction

15.1.1 Historical perspective on polishing

Polishing of surfaces with fine abrasives has been around for centuries. Many scientists have studied the mechanisms for material removal, with references dating as far back as 1652[1] and models continue to be refined today.[2–6] The continued interest in polishing over the years stems from the fact that it has been used for several applications, from fine finishing of stone artwork and architecture to polishing of optics and semiconductor processing, and has been used in a variety of forms that incorporate free abrasives or bonded abrasives,[7] spanning the entire spectrum of mechanical (e.g., lapping), chemomechanical (e.g., chemical–mechanical polishing — CMP), or chemical (e.g., electropolishing).[8] Depending on the application, varied requirements might be imposed on surface flatness, roughness, and defects (surface and subsurface damage).

Polishing entered into semiconductor device fabrication when IBM introduced CMP in the early 1980s to meet the tighter tolerances on surface topographies.[9] The superior capability for flatter surfaces of CMP over the existing glass reflow or spin-on glass (SOG) techniques rapidly propelled CMP into the mainstream of semiconductor processing. Since their introduction into semiconductor device processing, CMP processes have been developed for a wide range of applications and materials. In some more recent applications, multiple materials are simultaneously made coplanar. In such cases, we use the term "chemical–mechanical planarization." So the words "From Scratch to Planar" in the title can be viewed as spanning the beginnings of CMP as a means of surface finishing, or scratch removal (see the opening quote from 1652) to current applications in which planarization is required.

Additionally, "From Scratch to Planar" can be an indication of the application of CMP to microelectronics, where both surface defects or scratches and planarity are critical areas of current research and development.

Applications of CMP in microelectronics can be found throughout the semiconductor device manufacturing process from the front end (transistors) to the back end (interconnect and packaging). Examples of CMP processes applied to the front end of devices include shallow trench isolation (STI)[10] or more recently metal gate formation,[11] and to the back end (e.g., interlayer dielectric — ILD)[4,9] contacts and vias,[5,12] or Damascene lines and bond pads.[13,14] The materials that require CMP are also highly varied, including silicon, silicon dioxide, silicon nitride, tungsten, tungsten nitride, titanium, titanium nitride, copper, tantalum, tantalum nitride, platinum, iridium, organic and inorganic polymers, and others. Continuous development of CMP is ongoing for existing applications in order to meet the stricter requirements for within die planarity due to the shrinking dimensions of semiconductor devices. Further, advanced applications and new materials requiring CMP continue to be introduced, driving process research and development efforts. Even for applications in which CMP has become more mature and reliable, increasing pressure has been placed on equipment and consumable suppliers to reduce the cost of ownership for the CMP processes. This chapter will present an overview of the CMP process and its application to microelectronics. While an exhaustive description of all technical issues in every area is beyond the scope of this chapter, we attempt here to give an understanding of the range of CMP issues and applications, both in current and in future microelectronic devices.

15.1.2   *Principles and motivation for CMP*

The development, cost, and performance of microelectronic devices have been transformed from mostly focusing on the front-end transistors to being dominated by the back-end interconnects and packaging. This can be understood when we consider the sources of time delay in signal propagation. As dimensions in semiconductor devices shrink, there is a corresponding decrease in device delay, which would correspond to improved performance. In contrast, the time delay in the interconnect (RC delay) increases with shrinking dimensions. As the dimensions decrease beyond 0.5 µm the interconnect delay becomes the dominant contributor to the total delay which is a measure of device performance.[15,16] Further, as devices become more complex, new materials and multilevel metallization schemes are required to minimize the interconnect RC delay.

Ultra-large-scale integration (ULSI) multilevel metallization introduces the inherent issue that non-planarity in each level is cumulative. The success of this integration scheme requires strict specifications on surface planarity. Further, shrinking dimensions have lead to higher aspect ratios that cause limitations in depth of field for lithography and create challenges of consistent etching across a wafer. These issues required a process to flatten the ILD which insulates and supports the interconnect vias and lines. CMP was proposed as a possible process to achieve these specifications.[9] The introduction of CMP into semiconductor device processing brought a significant amount of wet chemistry research and development into a traditionally dry, high-vacuum industry. Because of the nature of the process, it was initially considered a "dirty" process when compared to the other fabrication processes. Significant development effort was required to convince the industry of the advantages of CMP over the alternative glass reflow or SOG techniques, and that the process would have no negative impact on yield. It was soon determined that CMP of the ILD was the only process capable of polishing the necessary range of structures and pattern densities to meet the tighter tolerances on both local and global planarity.[4,9] Further, the introduction of the CMP process enabled smaller dimensions and multilevel metallization design schemes required for future generation devices.

Although new ILD materials are currently being developed, the material of choice in the semiconductor industry for the last couple of decades has been SiO_2. As mentioned in the introduction, polishing has been around for centuries. The application of CMP to ILD polishing was able to take advantage of the existing understanding of glass polishing fundamentals. Compared to glass polishing, however, the ILD CMP application required tighter specifications on global flatness (uniformity) with less material removed. Further, semiconductor devices have varying pattern densities across the wafer, which introduce greater challenges in achieving the same high level of local flatness. While many in the semiconductor industry viewed CMP as a "black box" process, a strong understanding of the fundamentals is important to developing a CMP process that meets these progressively tighter tolerances for each specific application.

We briefly discuss some polishing fundamentals here. A more comprehensive discussion can be found in the literature.[17,18] There is much debate in the literature as to the details of the CMP mechanism, but there is clear agreement that polishing occurs based on chemical modification of the surface coupled with mechanical removal of the chemically modified surface.[2–6] Wafer-level polishing issues can be separated into two categories, *local,* within a die (a few microns) and *global,* across a wafer (typically 200 and 300 mm). A critical global process parameter in ILD CMP is the *removal rate,* which is a measure of the material removal per unit time. By measuring the removal rate as change in the thickness (dT) of the material with time (dt), Preston empirically obtained an equation showing that removal rate (R) was proportional to the applied pressure (P) and the linear velocity (v) at the interface:[19]

$$R = dT/dt = K_p P v \qquad (1)$$

where K_p is a proportionality coefficient. Equation 1 has been shown to be widely applicable to CMP of ILD,[20] but variation from the Preston equation has been observed depending on CMP slurry chemistry, or pad hardness.[6,21] Brown et al.[2] have shown K_p to be inversely proportional to Young's modulus, or the hardness of the material. Reagents in solution must react with the surface to create a soft surface film that is removed by shear abrasion. Removal rate is related to the hardness of the surface film that is chemically produced during polishing, and therefore the removal rate will also be chemically dependent.

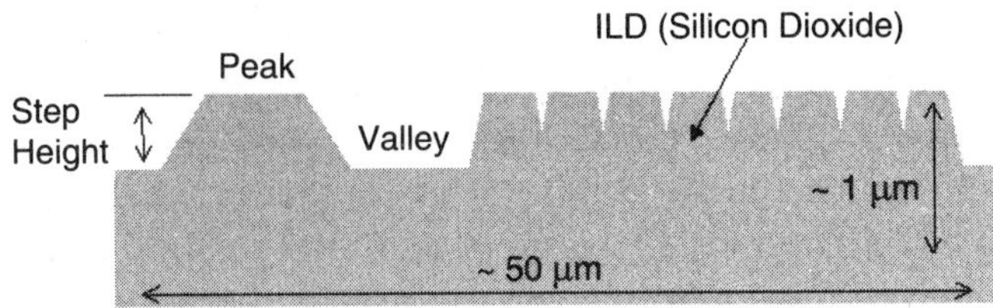

Figure 15.1 Simple schematic cross section of local area of ILD film.

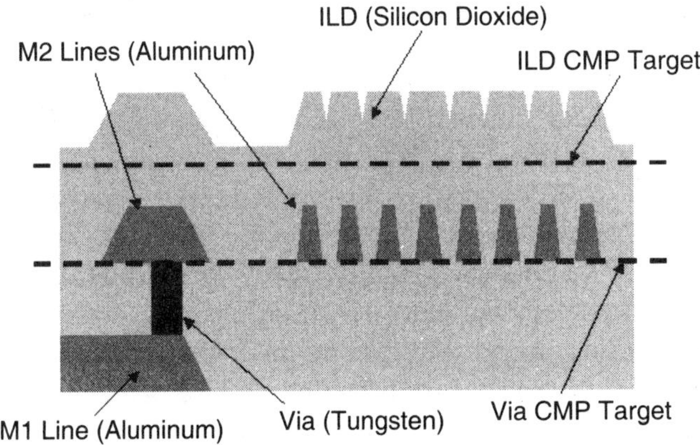

Figure 15.2 Schematic of a local cross section through multilevel interconnect.

On a local scale, it is also important to recognize that in CMP, it is not material removal alone, but selective material removal that is critical to achieving polishing. This concept can be easily understood by referring to Figure 15.1, which represents a cross section of an ILD film. In order to achieve polishing at the wafer surface, protruding areas ("Peak" in Figure 15.1) must be removed at a significantly more rapid rate than recessed areas ("Valley" in Figure 15.1). Higher pressure in the peaks compared to the valleys of the device pattern as well as chemical passivation of valleys allows for selective removal of the protruding areas until a flat film is obtained. To characterize polishing capability, we can define a polishing ratio:

$$P = R_v/R_p \qquad (2)$$

where R_v is the removal rate in the valley areas, and R_p is the removal rate in the peak areas. The smaller the value of P, the better the polishing process is. When the value of P is one, no polishing is achieved as in the case, for example, of a wet chemical etch. Various test structures of different sizes and pattern densities may be patterned on a semiconductor wafer and the value of P can be obtained for each structure and used to characterize the polishing capability of a particular CMP process.

With the success of CMP for ILD polishing, application of the CMP process was broadened and CMP processes have been developed for additional applications such as metal gates and shallow trench isolation (STI) in the front end; or the formation of contacts, vias, and metal lines in the interconnect.[4,5,9–14,22,23] While the fundamentals of CMP remain similar with all these processes, the new materials and multiple surface materials introduced by these applications require additional factors to be considered that are critical to process performance. Significant research and development effort has been and continues to be required. We will review some of these CMP processes in the next section.

15.2 Applications of CMP in semiconductor device processing

15.2.1 *CMP of interlevel dielectric layers*

The application of ILD (and metal via) CMP to a semiconductor device interconnect can be discussed in reference to Figure 15.2. After the metal lines (aluminum) are formed (M1 or M2), the ILD material (SiO_2) is deposited globally across the entire semiconductor wafer. Topography is transferred into the ILD film due to the underlying metal. ILD CMP is required to remove the topography leaving a flat film both locally and globally at the target thickness. Via holes are then patterned, etched, lined with an adhesion/barrier film, typically Ti/TiN (not shown in Figure 15.1), and filled with metal (W) deposited across the entire semiconductor wafer. CMP is required to selectively remove the W and barrier film leaving the via holes filled with metal and coplanar with the dielectric at the target thickness. The next level of metal (e.g., M2) can then be deposited and patterned, followed by the ILD deposition, ILD CMP, and so on as the processes are repeated for multiple levels in the interconnect.

The CMP process must be optimized to obtain a high global removal rate to allow for maximum process throughput, while maintaining a selective (peak/valley) local removal allowing for optimum polishing. These two goals are typically a trade-off in process development and require optimization of polishing pad, abrasive, chemistry, and process parameters (pressure, velocity, etc.). Further, the layout of the device pattern can present significant challenges to the local polishing capability. Variations in pattern density within each die, in both small isolated and large dense structures, have different peak/valley selectivity (Equation 2). This will cause different local rates of polishing leading to a large within-die thickness range.

Conventionally, a resilient pad is used for ILD CMP; and slurry, containing suspended fine abrasives and chemical, is introduced onto the pad. Recently an alternative to this loose abrasive process, utilizing fine abrasives "fixed" into the polishing pad, has received a significant amount of attention.[7] The emergence of this fixed abrasive CMP process into semiconductor processing can be seen most readily in the patent literature with over 40 patents issued on the use of fixed abrasive pads for semiconductor polishing since 1998, to companies such as Advanced Micro Devices, Applied Materials, Ebara, IBM, Micron Technologies, 3M, Obsidian,* Rodel, Sony, and others.

---

* Obsidian was purchased by, and is now part of, Applied Materials.

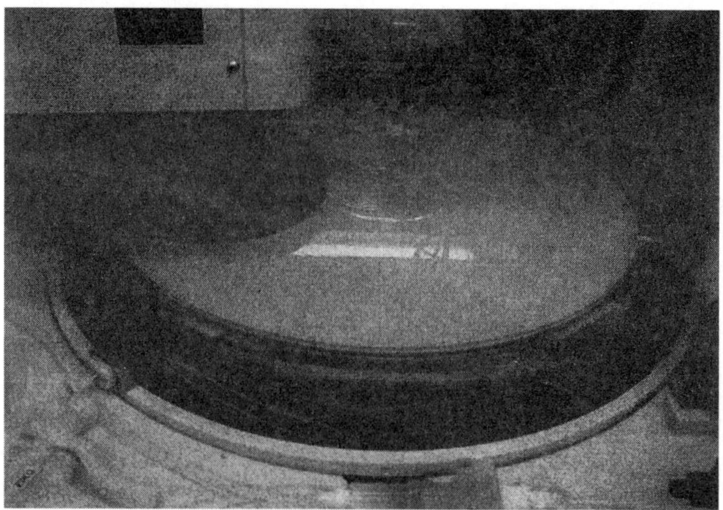

Figure 15.3 Picture of Ebara's rotary CMP platen with bonded abrasive pad.

Due to the potential advantages, fixed-abrasive CMP has been in development for several years; yet it has not been able to gain general acceptance as a production worthy process for semiconductor manufacturing. One of the biggest issues has been the low lifetime of the fixed abrasive material on the industry standard rotary platform in which a rotating carrier holding the wafer is used to press the wafer surface against a polishing pad on a rotating platen. Development efforts turned to the so-called "web-based" CMP tool platform, which requires proving out a new process technology on a new CMP tool platform retarding industry acceptance. Another serious issue with developing a production-worthy fixed abrasive CMP process has been the observance of greater surface damage compared to the conventional CMP process.

Some of our recent development efforts examined fixed abrasive processing using the industry-proven rotary CMP platform.[7a] We have developed a "bonded abrasive" CMP pad for the rotary platform (Figure 15.3). The bonded abrasive pad consists of a polymer material made to a specific hardness and porosity with embedded fine particles evenly distributed throughout the polishing pad during a single pad bonding process (Figure 15.4). This allows for consistent polishing and significantly longer lifetime compared to the conventional slurry/pad processes. A significant advantage of the bonded abrasive CMP approach over a slurry process is the potential to significantly reduce the cost of ownership (CoO) of the CMP process. The consumable costs in semiconductor device manufacturing can contribute to over 50% of the annual CoO for the conventional CMP process. Evidence for pad lifetimes that are more than an order of magnitude greater than those observed in production today has been observed using a low cost chemical solution. Additionally, fixed abrasive CMP can eliminate the need for careful slurry handling and distribution techniques to avoid detrimental effects such as coagulation that can cause serious defects at the wafer surface.

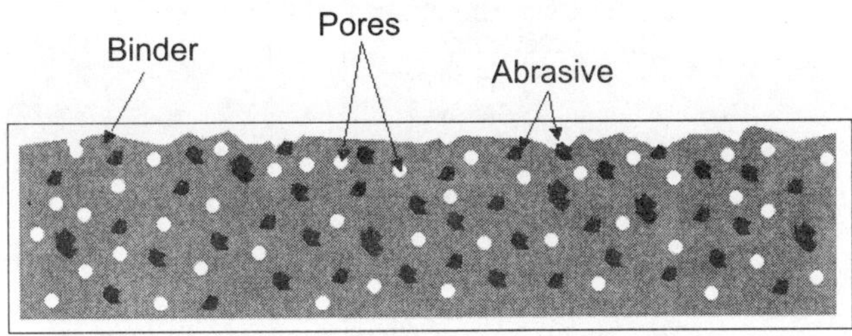

Figure 15.4　Schematic cross section of bonded abrasive pad.

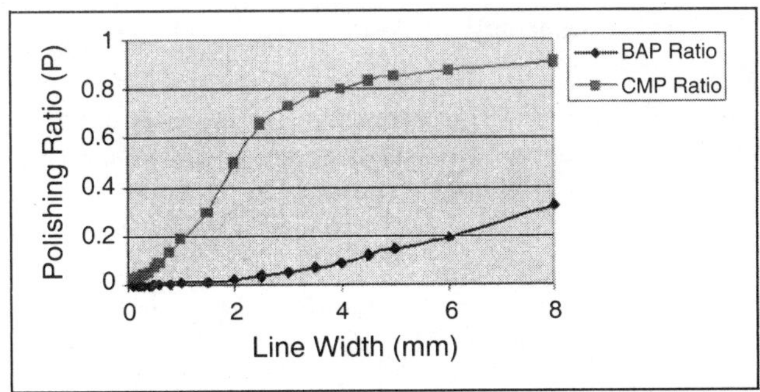

Figure 15.5　Variation in polishing ratio with feature size for bonded abrasive (BAP) versus conventional (CMP) polishing.

The local polishing performance of the bonded abrasive pad has proven to be superior to that of conventional slurry/pad processes. The bonded abrasive pad process was directly compared with a conventional silica slurry process on an identical platform. Evaluation of the polishing ratio (Equation 2) for different feature sizes was obtained using test TEOS wafers patterned with various line widths. The data are shown in Figure 15.5. The superior polishing capability of the bonded abrasive pad is reflected in the lower polishing ratio for all feature sizes measured.

Comparing the time required for different features to be made planar can also exhibit this improved planarity. Typical time-to-planarity (TTP) results for the same bonded abrasive versus conventional processes are shown in Figure 15.6. In this particular experiment, eight wafers were processed at the process times shown, and the step height was measured by profilometry on three different topography test structures for each wafer. The normalized areas of the test structures were 1, 100, and 400 squares. The data show the trends in decreasing step height with time. The bonded abrasive process shows a similar trend to a conventional CMP process where an increase in feature sizes requires a longer time to reach planarity, as shown in Figure 15.6a. Figure 15.6b shows the results comparing the planarity capability of the same feature size on wafer processes by the conventional process versus the bonded abrasive process and shows that the bonded abrasive process gives a significantly faster time to planarity and lower step height (as expected from the data

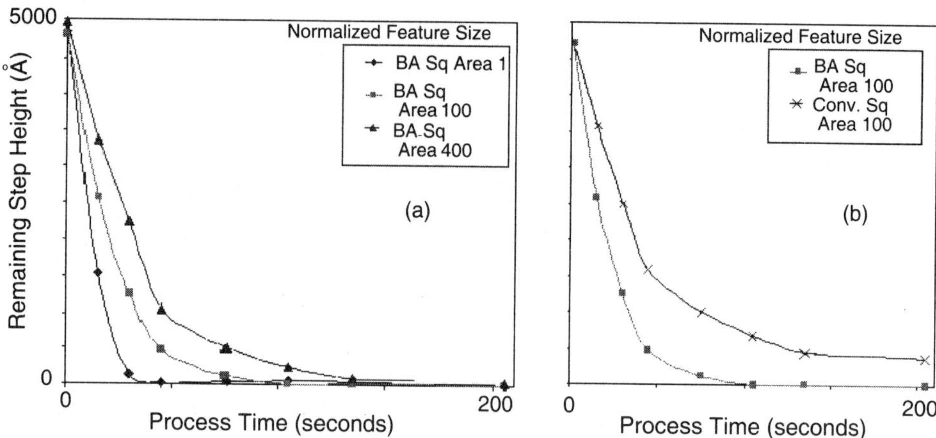

Figure 15.6 Time-to-planarity (TTP) experiments: (a) Dependence of TTP on normalized feature sizes for the bonded abrasive process. (b) Comparison of TTP on bonded abrasive versus conventional processes.

in Figure 15.5). The improvements in planarity over conventional CMP make the bonded abrasive process ideal for ILD CMP.

15.2.2   Shallow trench isolation

An example of CMP applied to the front end of semiconductor device processing can be found in device isolation techniques. As devices have continued to shrink, planarity of the isolation material (SiO_2) has become required in order to achieve acceptable line width control. The local oxidation of silicon (LOCOS) technique in which the silicon substrate was selectively oxidized, created an unacceptable level of topography for devices below the 0.5 µm technology node. Further, the "bird's beak" effect of the LOCOS integration scheme, in which the oxidation encroached into the active regions, interfered with subsequent gate processing. The STI integration scheme was developed to eliminate these issues and was potentially scalable to future generations of semiconductor devices.[10] Figure 15.7a shows a cross section of a trench prior to CMP. In the STI integration scheme, a CMP stop layer (Si_3N_4) is deposited (over a sacrificial SiO_2 film not shown) on the substrate wafer (Si) and then patterned and etched to create shallow isolation trenches. The isolation material (SiO_2) is then deposited as is shown in Figure 15.7a. CMP then planarizes the SiO_2 film, clearing it from the active Si_3N_4 surface while leaving the trenches filled with SiO_2, as shown in Figure 15.7b. After CMP, the Si_3N_4 layer can be selectively etched away prior to gate formation in the active regions between the trenches.

With STI CMP, all the challenges of an ILD CMP process remain (global uniformity, local planarity, pattern density effects, etc.), with the additional requirement of leaving two materials coplanar. This requirement introduces a challenge in STI CMP process development. Early research balanced the trade-off between high selectivity required to stop on the Si_3N_4 layer and low selectivity to prevent "dishing"

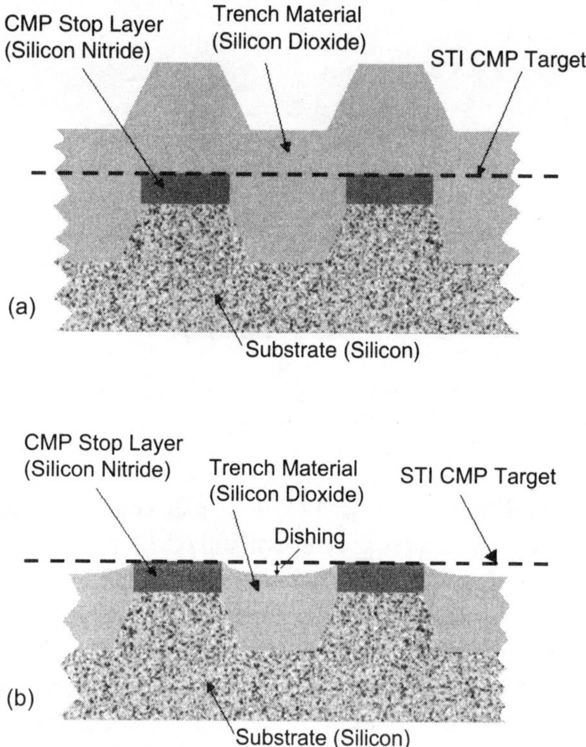

Figure 15.7 Schematic of a local cross section of a shallow trench (a) prior to STI CMP, and (b) after STI CMP.

into the trenches. Dishing, shown in Figure 15.7b, occurs because of the difficulty in preventing removal of SiO_2 in the trenches while clearing it from the Si_3N_4 active areas. Initial STI CMP processes were developed that used silica abrasive slurries in which the SiO_2/Si_3N_4 removal rate selectivity was in the neighborhood of about three to five. The moderate selectivity helped to minimize dishing, while still allowing the nitride film to act as a stop layer. Unfortunately significant challenges were observed with the ability to uniformly planarize the SiO_2 film so that the range of the remaining nitride was acceptable. To ease the burden on the CMP process, an additional RIE step was introduced in which a reverse mask was used to etch away most of the SiO_2 film above the active area. The additional process steps, especially the need for photolithography, added considerable processing cost to the device manufacturing. To eliminate this cost, considerable research effort was devoted to developing a CMP process that was capable of direct STI CMP. Highly selective slurries were developed using ceria abrasive with SiO_2/Si_3N_4 removal rate selectivity >20. CMP processes with selective slurries were capable of effectively stopping on the Si_3N_4 layer. The challenge with uniformity for a selective process does not significantly affect the remaining nitride range, but does impact variation in dishing across the wafer. With additional process development, dishing could be minimized and these processes could be put into production to eliminate the need for a reverse mask RIE etch.

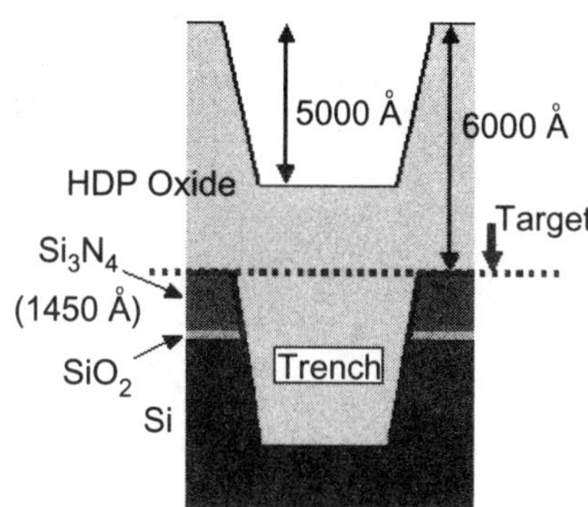

Figure 15.8 Schematic of a local cross section of film stack on MIT 864 prior to direct STI CMP using a bonded abrasive process.

While the capability for direct STI CMP of such selective processes is good, the high cost of the slurry still leaves pressure on the CMP process to reduce the cost of ownership. A bonded abrasive process was developed that responded to the demand for a direct STI CMP process at a reduced cost of ownership. A direct STI process using a bonded abrasive pad was demonstrated on STI test wafers with the MIT 864 mask that was patterned with various line widths and densities. Figure 15.8 shows a schematic cross section of the film stack including the dimensions for the 864 wafers used (the sacrificial SiO_2 film is shown). Post-bonded abrasive CMP data are shown in Figures 15.9 to 15.11. The data look at two different pattern densities (50 and 90%) as well as two different line widths at 50% density. Measurements were taken at nine different die across the diameter of the 200 mm wafer. Excellent uniformity of <1.5% at 1σ was observed in both the isolation trench (Figure 15.9) and the active nitride (Figure 15.10) regions. The bonded abrasive process was able to planarize to the target trench SiO_2 thickness across the wafer while stopping on the active Si_3N_4 layer with minimal nitride removal. The process achieved this with outstanding planarity. Profilometer measurements of the step height (Figure 15.11) showed <60 Å of dishing into the trenches even for the large (250 μm) line widths. Such superior planarity performance of the bonded abrasive process provides an ideal, low cost of ownership process for direct STI CMP and has been demonstrated in 0.18 μm production.[7b]

15.2.3 Metal CMP

Metal CMP is similar to STI CMP in that the process requires different materials be made coplanar. In metal CMP, however, there are three different materials to consider. One application of metal CMP is in via formation as was shown in Figure 15.2. The vias are typically lined with an adhesion/barrier layer, a thin Ti/TiN film, and filled with tungsten metal. Formation of the vias requires a CMP process that planarizes the tungsten film, and then removes the tungsten and the barrier material above the dielectric, leaving the vias filled and leaving coplanar metal, barrier, and dielectric surfaces. Such an approach is termed "Damascene" metal pattern

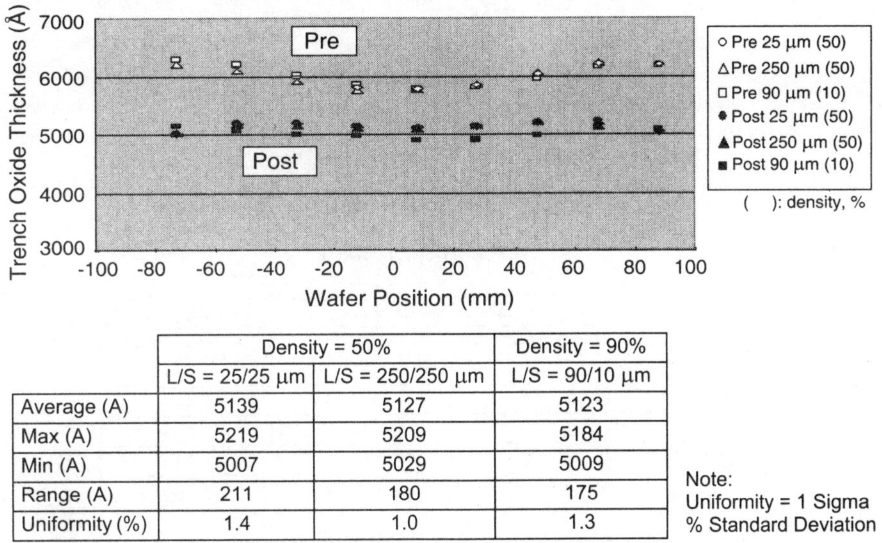

	Density = 50%		Density = 90%
	L/S = 25/25 μm	L/S = 250/250 μm	L/S = 90/10 μm
Average (Å)	5139	5127	5123
Max (Å)	5219	5209	5184
Min (Å)	5007	5029	5009
Range (Å)	211	180	175
Uniformity (%)	1.4	1.0	1.3

Note:
Uniformity = 1 Sigma
% Standard Deviation

Figure 15.9 Trench oxide uniformity after direct CMP with bonded abrasive process on MIT 864 patterned STI wafers.

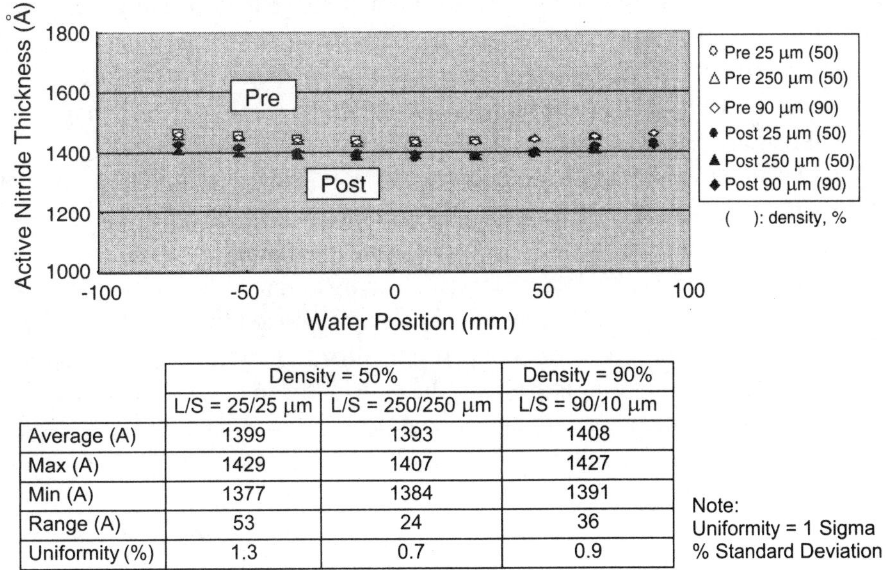

	Density = 50%		Density = 90%
	L/S = 25/25 μm	L/S = 250/250 μm	L/S = 90/10 μm
Average (Å)	1399	1393	1408
Max (Å)	1429	1407	1427
Min (Å)	1377	1384	1391
Range (Å)	53	24	36
Uniformity (%)	1.3	0.7	0.9

Note:
Uniformity = 1 Sigma
% Standard Deviation

Figure 15.10 Active nitride uniformity after direct CMP with bonded abrasive process on MIT 864 patterned STI wafers.

formation.* The tungsten via CMP process requires the development of chemistry, abrasive, and pad combination that can planarize tungsten and selectively remove the

---

* "Damascene" comes from the Damascus method of decorating swords with inlaid precious metals.

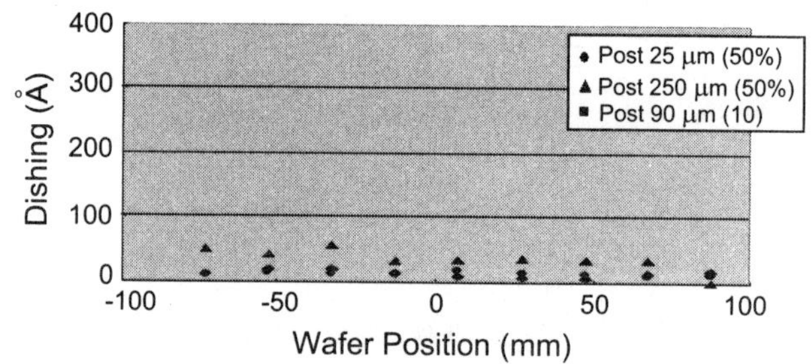

	Density = 50%		Density = 90%
	L/S = 25/25 µm	L/S = 250/250 µm	L/S = 90/10 µm
Average (Å)	13	41	13
Max (Å)	18	57	20
Min (Å)	10	33	6
Range (Å)	8	24	14

Figure 15.11 Trench dishing after direct STI CMP with bonded abrasive process on MIT 864 patterned wafers.

tungsten and Ti/TiN films over SiO_2. The Damascene integration scheme requires a CMP process that has a high W/SiO_2 removal rate ratio so that the dielectric material can act as a CMP stop layer. This presents a significant challenge since the process must not cause recess in the tungsten vias (see Figure 15.12). Further, the effect of pattern density is critical since the high selectivity of the CMP process can also cause localized erosion of the dielectric in dense via arrays.

Another more recent application of metal CMP is in metal line formation. The CMP process is required in a dual-Damascene integration scheme. This approach was taken with the introduction of copper to replace aluminum lines.[13,14] Copper was introduced because of its lower resistance compared to conventional aluminum. Dual-Damascene integration (Figure 15.13) was introduced to simplify the number of process steps in multilevel interconnect metallization. The "via-first" approach shown in Figure 15.13 patterns and etches the vias followed by the trenches, and then fills them both simultaneously with the copper deposition.* This approach has fewer process steps than the integration scheme for the aluminum line/tungsten via interconnect in Figure 15.2.

The Damascene integration scheme introduces two new processes. Copper electroplating is required as the deposition method in order to achieve void-free filling of the high aspect ratio vias and lines, and copper CMP is required for the formation of lines in the dual-Damascene approach.[8a] The impact of pattern density on the remaining topography is even more severe in dual-Damascene copper CMP than in tungsten via CMP. Line widths may vary from nanometers to microns and pattern densities from 10 to 90% depending on the device. Planarization challenges

---

* Alternatively, "trench-first" dual-Damascene approaches have been explored but are generally not the preferred approach due to issues with misalignment.

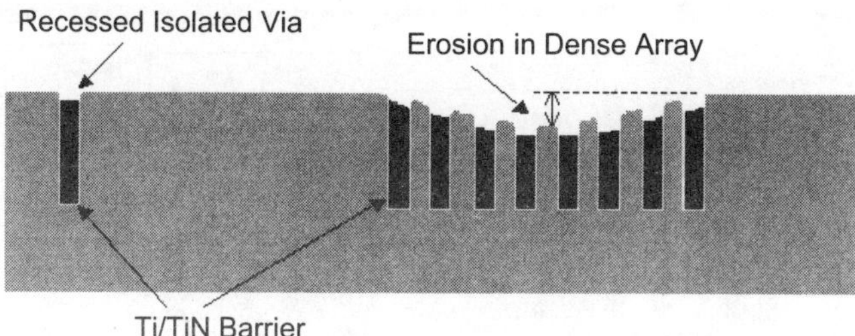

Figure 15.12 Schematic cross section of isolated and dense tungsten vias illustrating via recess and array erosion after CMP process.

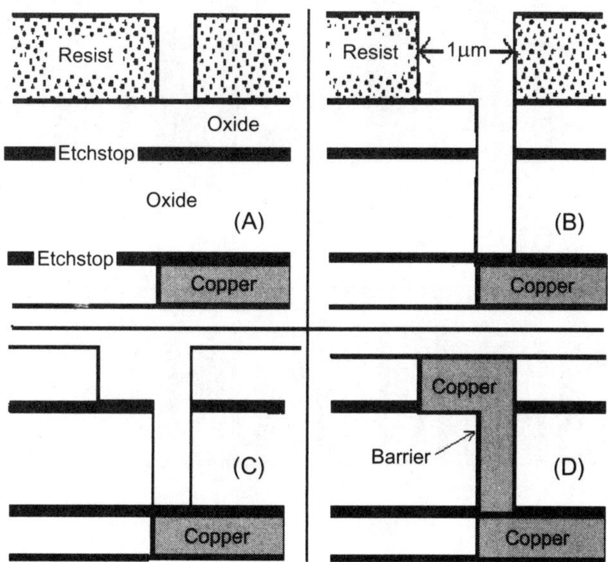

Figure 15.13 Schematic cross sections of four stages of a dual-Damascene integration scheme with vias formed first and trenches second (from Venkatesan et al.[13]). (Reproduced with permission from Ref. 13.)

after CMP include recess in narrow lines (<1 μm), dishing in wide lines (10 to 100 μm), and erosion in dense arrays (50–90%). Dishing and erosion are both critical measures of the copper CMP process capability and must both be minimized to prevent the cumulative effect of topography in multilevel interconnect metallization.

Figure 15.14 gives some typical data we have used to characterize a copper CMP process. The post-CMP dishing and erosion data in Figures 15.14a and 15.14b were measured using a surface profilometer on five different dies across the diameter of a 200 mm test wafer.* Dishing was measured across 100 μm lines, and erosion

---

* Test wafers were SKW copper Damascene with a modified MIT test pattern of various isolated lines and arrays.

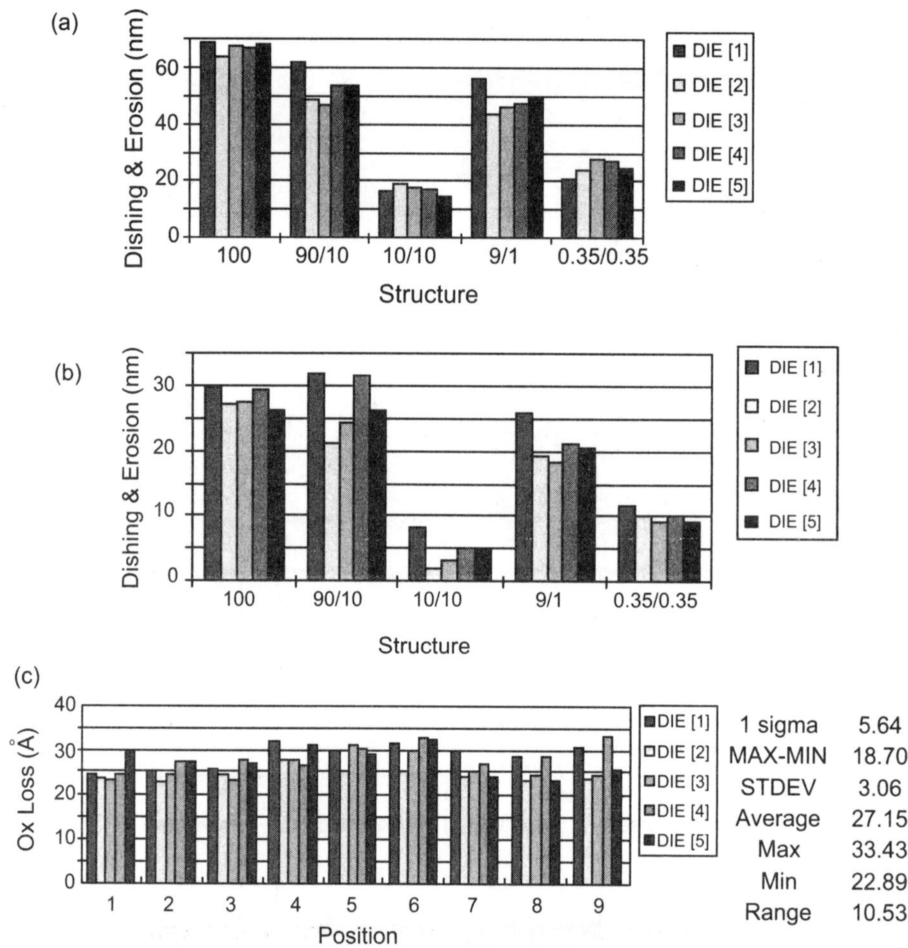

Figure 15.14 Copper CMP process characterization data. (a) Dishing and erosion for process A. (b) Dishing and erosion for process B. (c) Dielectric (SiO_2) loss at nine field sites per die after 30% overpolish with process B.

was measured on several test structures of different densities and line widths in order to evaluate the pattern dependence of the CMP process. Figures 15.14a and 15.14b show two line widths at 50% (10 μm lines and spaces and 0.35 μm lines and spaces) and two at 90% (9 μm lines — 1 μm spaces and 90 μm lines — 10 μm spaces). There is a clear dependence of erosion on pattern density, but not on line width.* Process A was developed for the first generation of semiconductor devices using copper interconnects. The dishing and erosion for process A in Figure 15.14a were acceptable for these devices; however, as devices and topography tolerances shrink, it was necessary to develop a copper CMP process with less dishing and erosion. The optimization of the consumables (pad and slurry) used in CMP has been shown to

---

* Dishing was shown to depend on line width: 10 μm lines showed 30% less dishing than 100 μm lines.

have the most impact on minimizing dishing and erosion. Process B was developed by optimizing the slurry chemistry and particles. This process exhibited a significant improvement in dishing and erosion, as shown in Figure 15.14b.

Another reason that dishing and erosion are critical in dual-Damascene CMP is because they both have a direct impact on the line resistance. It should be noted that any measurement of dishing and erosion is only meaningful if coupled with a measure of the dielectric ("oxide") loss in the widest, most challenging features ("field") on each die to confirm that the loss is minimized while ensuring that the oxide surface is completely cleared of all copper and barrier material. It is the sum of dishing, erosion, and field oxide loss that gives a true measure of total copper thinning, which correlates to the line resistance.* The oxide loss for process B was measured at nine different field areas in each of the five dies across the diameter of the wafer.** The data of Figure 15.14c show that all copper and barrier material was cleared with very minimal oxide loss (<30 Å).

The introduction into semiconductor processing of two sequential processes, copper electroplating and CMP, generated a great deal of electrochemical research and process development towards production worthy processes capable of meeting the stringent specifications of dual-Damascene integration. While this chapter focuses on CMP, in semiconductor process development it is necessary to consider interactions between the two processes. Our studies have shown that the electroplating and CMP processes have significant interaction in semiconductor wafer processing, at both the global and the local level, creating challenges in process integration.

An example of interaction between CMP and electroplating was observed on a local scale as the electroplating process was optimized to maximize gap filling in the high aspect ratios of dual-Damascene integration. The initial process exhibited good gap fill in dense lines (highest aspect ratio), but produced an overfill effect in the dense lines, as shown in Figure 15.15a. Wafers with this process can result in issues of residual copper in dense lines after copper CMP resulting in electrical failures due to shorting. Overpolishing of the wafers to clear the copper residual leads to excessive dishing in wide copper lines. Feedback of this process issue from the CMP development team to the electroplating development team led to optimization of the deposition process to eliminate overfill without compromising gap fill capability in dense lines, as shown in Figure 15.15b.

Global interaction between electroplating and CMP was observed in a study of the impact of annealing after copper electroplating on CMP copper removal. The need for annealing has been demonstrated previously. Some of our results are shown here for a series of 200 mm wafers with a 1 μm film of electroplated copper. Figure 15.16 shows the impact of anneal temperature and anneal time on copper sheet

---

* Direct measurements of line sheet resistance in test structures not only correlate to the total copper loss, but also can exhibit line shorting indicative of remaining copper or barrier residue.

** Dielectric loss was obtained from the difference between optical measurements of dielectric thickness before barrier and copper deposition and the thickness after CMP at the same field locations.

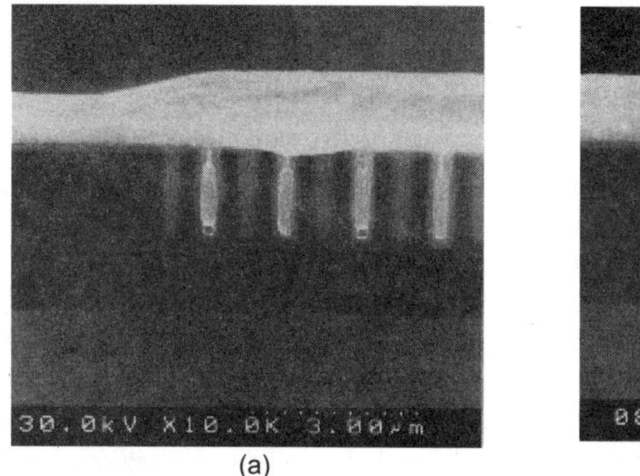
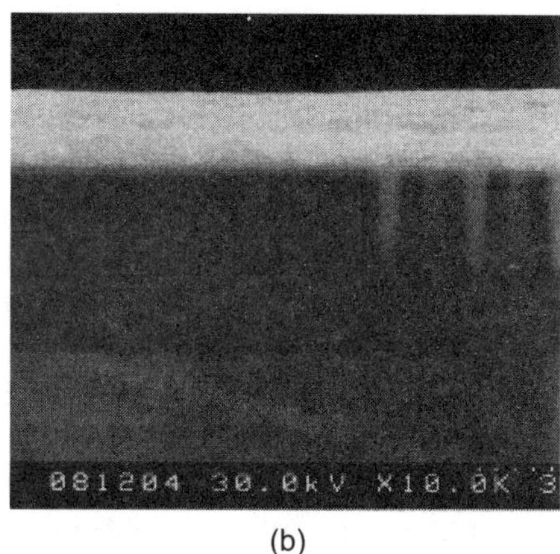

(a)   (b)

Figure 15.15 Scanning electron microscope (SEM) cross sections comparing electroplating processes optimized for (a) gap filling and (b) further optimized for CMP. (Reproduced with permission from Ref. 8(a).)

(a)

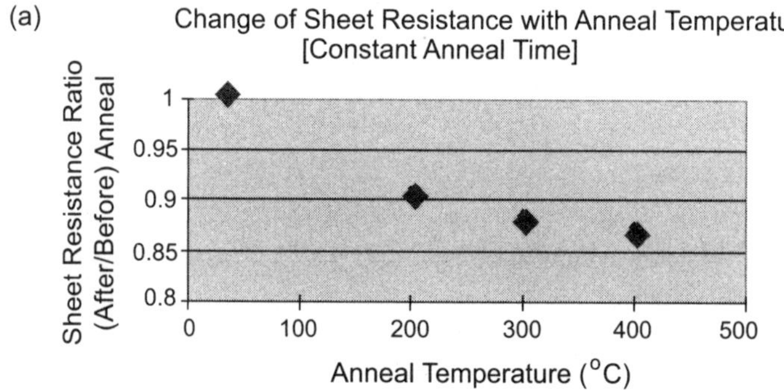

(b)

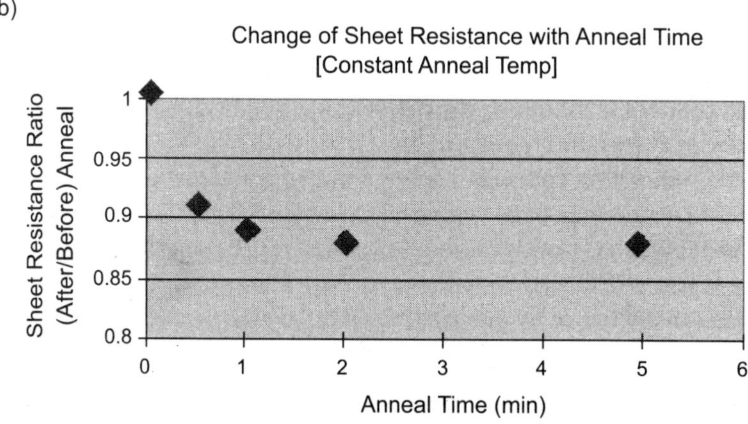

Figure 15.16 Impact on copper sheet resistance of (a) anneal temperature at constant time, and (b) anneal time at constant temperature. (Reproduced with permission from Ref. 8(a).)

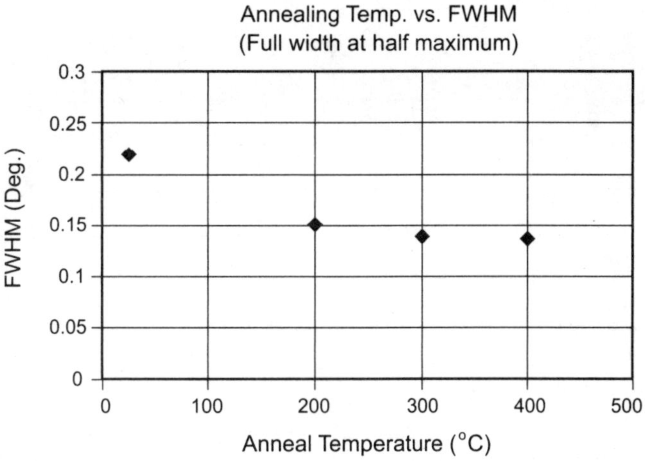

Figure 15.17 Decrease in X-ray diffraction peak width with anneal temperature, indicative of crystal growth in the copper film. (Reproduced with permission from Ref. 8(a).)

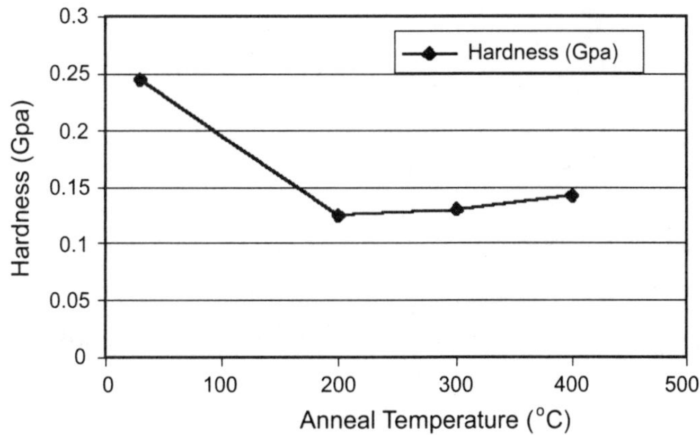

Figure 15.18 Decrease in copper film hardness with anneal temperature. (Reproduced with permission from Ref. 8(a).)

resistance. A post-electrodeposition anneal improved the sheet resistance of the copper film by about 15% compared with the as-deposited copper films. At room temperature, this improvement takes place over several days, which is unacceptable in semiconductor device manufacturing, and a rapid anneal as shown in Figure 15.16b is required after electroplating.

Characterization of the changes in the copper film properties with post-deposition anneal was conducted and the data are shown in Figures 15.17 and 15.18. Figure 15.17 shows the decrease in the width of the Cu(111) peak detected by X-ray diffraction with temperature, indicative of crystal structure growth. Figure 15.18 shows that the hardness of the copper film decreased after anneal, but increased slightly with increasing anneal temperature.

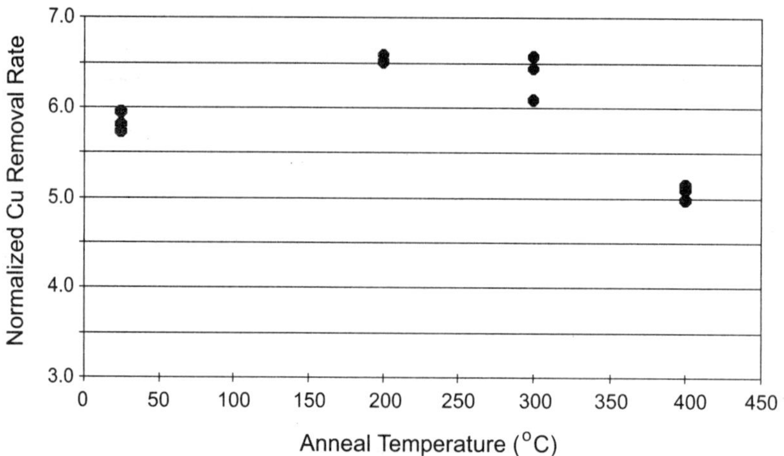

Figure 15.19 Change in CMP removal rate with copper anneal temperature. (Reproduced with permission from Ref. 8(a).)

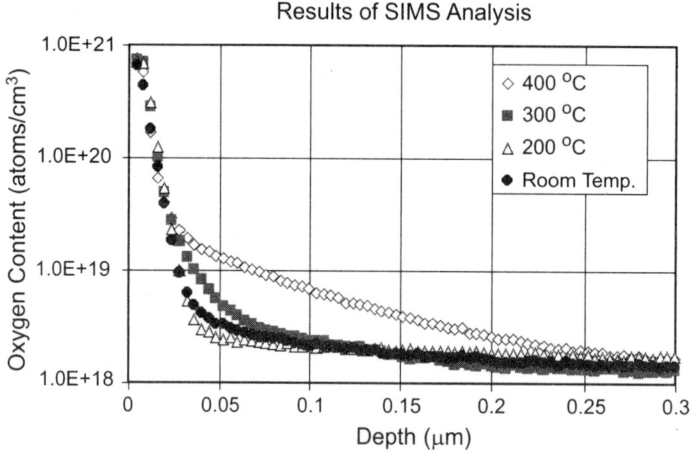

Figure 15.20 SIMS depth profile showing increase in oxygen content and depth with post-copper deposition anneal temperature. (Reproduced with permission from Ref. 8(a).)

To determine the impact of anneal on CMP removal rates, post-electroplated wafers were processed with different anneal temperatures and with a typical copper CMP process. Results are shown in Figure 15.19. The removal rates increased with anneal temperature, as might be expected from the hardness data (Figure 15.18), and then decreased significantly with further increase in copper anneal temperature. Depth profiles of the copper film composition after anneal are shown in Figure 15.20. The results exhibit significant increase in the content and depth of oxygen in the copper films with increasing anneal temperature.

We can explain our observations in the context of the mechanism for metal CMP proposed by Kaufman et al.[3] In this simplistic mechanism, a passivating copper film

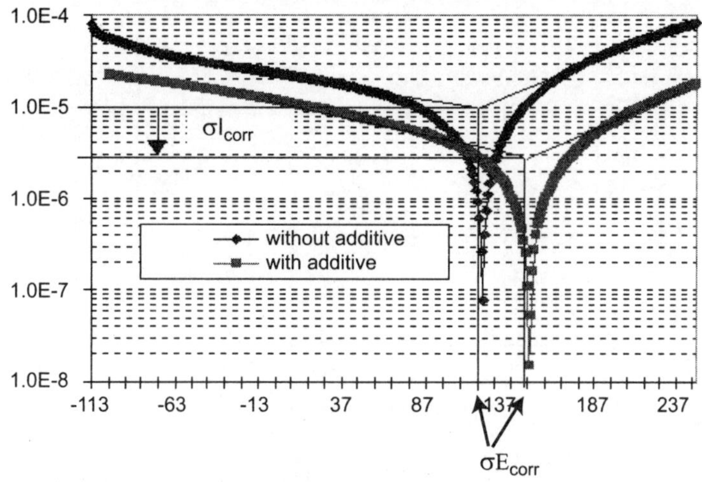

Figure 15.21 Rotating copper disk electrode experiments in CMP slurry to control passivating layer (Rabke et al.[23]) (Reproduced with permission from Ref. 23).

is generated during the CMP process that is easily removed by mechanical action. The existence of such a passivating film during CMP was observed electrochemically (see Figure 15.21), and its control has been demonstrated to be critical to the CMP process.[23] During CMP, the apparent decrease in copper removal rate for higher anneal temperatures is likely due to the generation of a relatively thick copper oxide film at the surface of the copper film. We propose a chemical inhibition of passivating layer formation due to the oxidized copper film, which retards the CMP process until the oxide layer is removed.

Data obtained by monitoring the platen motor current during the CMP process are consistent with the existence of an oxide layer. The platen motor current has been demonstrated to correlate directly to the shear forces during CMP.* In this study, the platen motor current traces obtained for the copper CMP process all showed an increase in current early in the CMP process to a peak in current, followed by a decrease in current as the copper film was processed down into the bulk. Such behavior was consistent with a change in the copper film from the surface compared to the bulk. Further, an increase in both the peak current and the time to reach the peak was observed on wafers processed with increasing anneal time (Figure 15.22), consistent with thicker oxidized copper film at higher anneal temperatures that required longer time to remove by CMP. This study of the interaction between the post-electroplating anneal and the CMP processes have demonstrated the need to incorporate a controlled environment during the post-electroplating anneal process.

Copper electroplating and CMP processes are still under continuous improvement to maximize performance such as increased process rates for throughput and minimum topography for integration.

---

* Platen motor current traces have been used successfully as an endpoint technology for CMP processes that polish one film and stop on another (e.g., STI and metal Damascene).

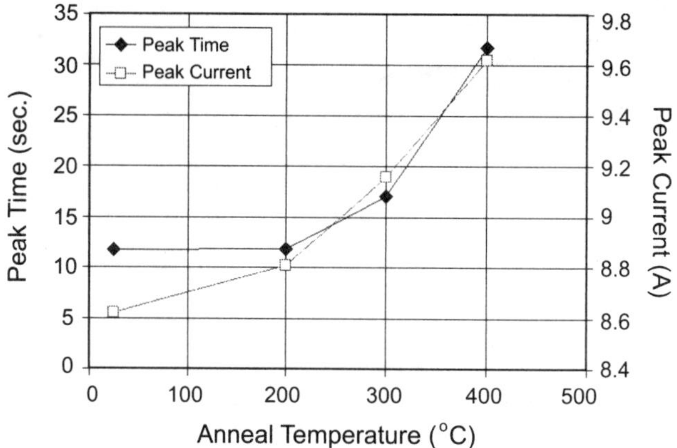

Figure 15.22   Increase in CMP platen motor current peak time and peak current with post-copper deposition anneal temperature. (Reproduced with permission from Ref. 8(a).)

15.3   Advanced applications of CMP

15.3.1   CMP of new interconnect materials

New materials are required in order to meet future performance requirements, further decrease semiconductor device dimensions, and minimize interconnect RC delay. In device interconnects, initial focus has been on decreasing line resistance (R). As we discussed in the previous section, copper ($\rho = 1.67\ \mu\Omega$ cm) has been integrated into semiconductor devices to replace aluminum ($\rho = 2.66\ \mu\Omega$ cm). In addition to copper, we have also conducted research on and demonstrated the feasibility of silver metal ($\rho = 1.59\ \mu\Omega$ cm) integration in interconnects (see below).

Recently, tremendous attention has been given to minimizing the capacitance (C) of the RC delay. Research and development has focused on integration of novel, low dielectric constant ILD materials to replace silicon dioxide ($k_d \sim 4.0$ to 5.0). Extension of silicon dioxide material by fluorine doping has reached its limits ($k_d \sim 3.2$ to 3.9), and dramatic materials changes are being explored to reduce the dielectric constants to <3.0. Several organic and inorganic polymers have been investigated as interconnect dielectric candidates (see Table 15.1).[25] Besides having a low dielectric constant, there are many other requirements for successful integration of these materials (see Table 15.2). One of the most critical factors is the compatibility of the low k material with the CMP process. The materials must be chemically and structurally compatible to prevent both surface and structural bulk damage to the material during CMP. The burden on the CMP process is significant because the hardness and modulus of elasticity for these materials tend to decrease as the dielectric constant is reduced. This becomes even more of an issue as these materials are made porous to extend the dielectric constant to $k_d \leq 2$ in future generations and presents severe challenges to the metal CMP process. The process must operate at lower vertical and shear forces to prevent damage to the surface of the softer low k material. In addition,

Table 15.1 Candidates for interconnect dielectric materials.

Dielectric material	Dielectric constant (k_d)
Silicon dioxide (SiO_2)	4.0–5.0
Fluorosilicate glass (FSG)	3.2–3.9
Polyimides	3.0–3.4
Hydrogen silsesquioxane (HSSQ)	2.9–3.2
"Black Diamond" (SiCOH)	2.7–3.3
Methyl silsesquioxane (MSSQ)	2.6–2.8
Polyarylene ether	2.6–2.8
Benzocyclobutenes	2.6–2.7
Fluorinated polyimides	2.5–2.9
Parylene-F	2.4–2.5
Teflon-AF	2.1
Porous polyarylene ether	1.8–2.2
Porous HSSQ and MSSQ	1.7–2.2
Xerogels (porous silica)	1.2–2.2

Table 15.2 Top ten requirements for low dielectric constant materials.

10. Environmental compliance in deposition process
9. Low moisture absorption
8. Long lifetime
7. Low integration cost
6. Good mechanical properties (hardness, modulus, etc.)
5. Chemically compatible (solvents, metals, etc.)
4. High selectivity to RIE etch
3. Thermal stability to downstream processing (~ 400°C)
2. High adhesion to barrier materials
1. Chemical and structural compatibility with CMP

the metal CMP process must be gentle enough to prevent breaking the thin metal lines and causing shorts, because for the first time the modulus of elasticity for the lines and ILD are significantly different, with the dielectric support material being softer than the embedded metal lines.*

As mentioned previously, in addition to evaluating the integration feasibility of new copper/low *k* we have also evaluated silver metal due to its lower resistance. Figure 15.23 compares the conventional copper (a) to a silver (b) Damascene

---

* It should be noted that with respect to the weaker mechanical properties of low *k* dielectric materials, the CMP process is the most challenging in interconnect integration and development work is proceeding to enable processing these materials. There are also, however, many potential challenges in device packaging (e.g., wire bonding) that may present severe issues with the mechanical strength of these materials, so the development of suitable packaging equipment and procedures should also be explored for these materials.

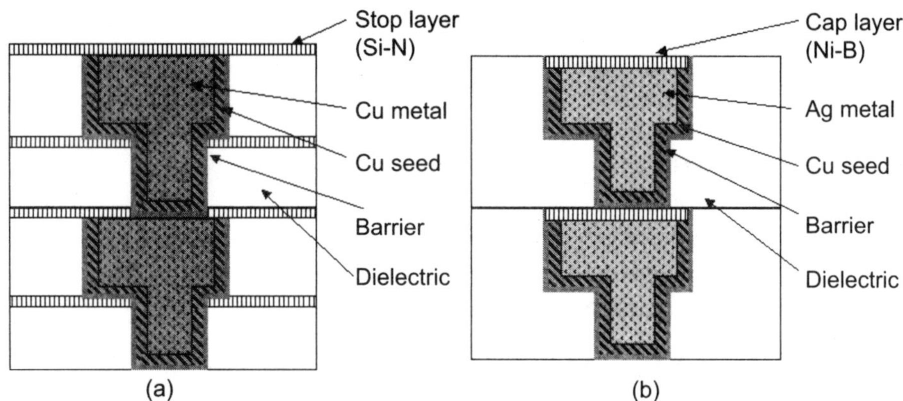

Figure 15.23   Comparison of (a) copper Damascene and (b) possible future silver Damascene integration schemes.

integration scheme. In addition to developing both silver electroplating and silver CMP processes with similar requirements to those previously discussed for Cu/low *k* metallization, we also developed a selective cap layer deposition process in which a barrier film layer was selectively deposited on the silver lines by electroless plating. This integration approach was developed to prevent the diffusion and oxidation of copper during downstream processing steps, while eliminating the silicon nitride etch stop layer.[26] Removing the Si_3N_4 layer from the integration scheme gets rid of its contribution to the effective dielectric constant of the dielectric material and can potentially decrease RC delay by up to 15%.[27]

We have evaluated several cap layer materials as potential selective barriers for both copper and silver. Figure 15.24 shows elemental analysis by atomic emission spectroscopy (AES) at the surface (a) and below the surface (b) of a test structure in which 200 nm of Cu was deposited by PVD followed by 500 nm of silver by electroplating, and finally a Ni–B(3.2%) cap layer was deposited by electroless plating. The capability of the cap material as a barrier was demonstrated with no Ag or Cu detected by AES at the surface after deposition, or after annealing at 400°C for 1 hour. Comparison of depth profiles before and after anneal reveal that the annealing causes copper to migrate through the silver layer to the Ag/Ni–B interface, but does not migrate significantly into the film. The Ni–B material demonstrates good barrier properties for both silver and copper.

15.3.2   *CMP of new materials in other advanced applications*

CMP continues to find novel applications in many aspects of advanced semiconductor processing. In the front end, silicon dioxide is reaching its limits as a gate dielectric material, and new high k dielectric materials are being explored. Coupled with this, metal gates with good compatibility to high k materials have been evaluated for their low gate resistance. One approach to metal gate formation that has recently been reported is a Damascene metal gate.[11] Figure 15.25 shows a transistor with a

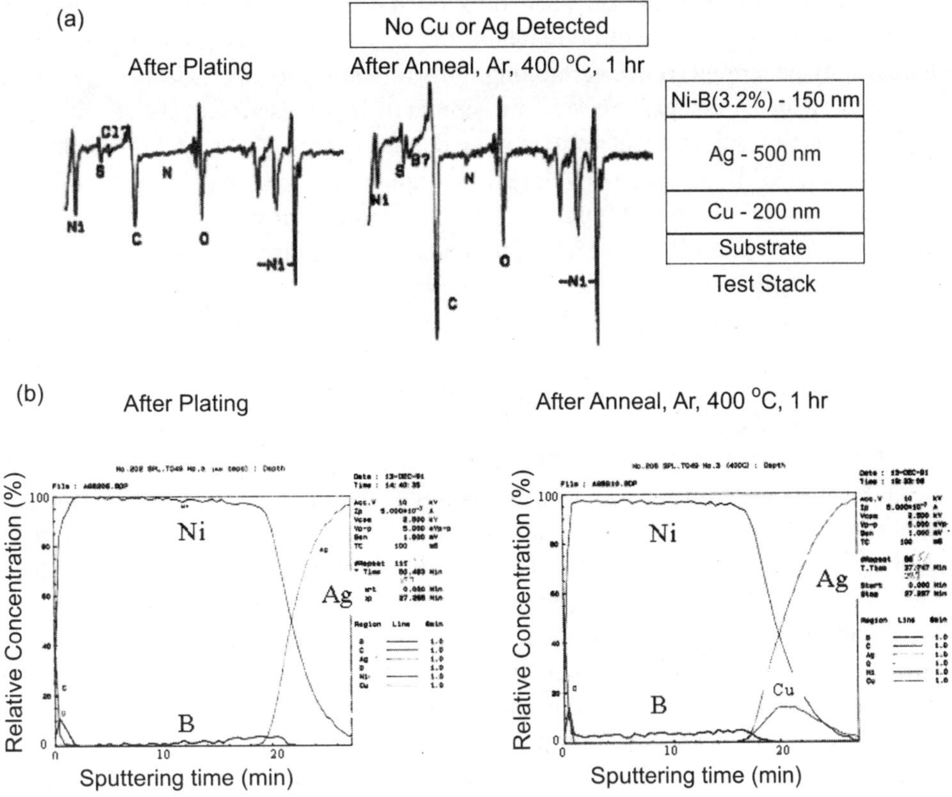

Figure 15.24 (a) Surface and (b) depth profile elemental analysis of Ni–B(3.2%) cap layer by atomic emission spectroscopy before and after a 1 hour anneal at 400°C.

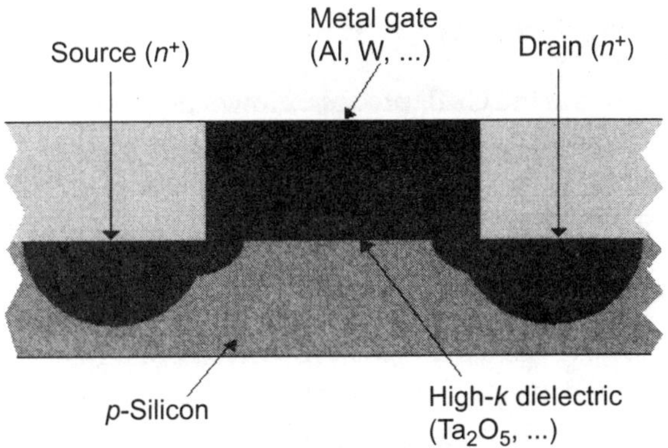

Figure 15.25 Schematic of transistor (MOSFET) with Damascene metal gate formed by CMP over thin high-k gate dielectric material.

metal gate formed by CMP. Significant challenges are presented in the development of a manufacturable, metal gate CMP process due to the combination of requirements for local planarity and global uniformity.

CMP of poly-silicon has been developed for a number of applications, for example, CMP of gate electrodes where the CMP process is used to polish away topography generated from overlapping poly-silicon layers, or CMP for deep trench isolation, with an integration scheme similar to STI discussed above, but using poly-silicon to fill the trenches and using either SiO_2 or Si_3N_4 as a CMP stoplayer.[28]

Polymer CMP has been developed for applications such as photoresist planarization,[29] used to minimize topography due to photolithography depth of focus limitations. CMP of gap fill polymer material has also been proposed to enable via-first dual-Damascene processing in future generation devices.[30] In this application spin-on deposition of the gap fill polymer material is used to fill the via holes and the polymer is planarized by CMP prior to trench patterning. Polymer CMP has similar challenges to those discussed previously for new low k_d materials due to the softness of these materials.

CMP processes for noble metals such as platinum, iridium, and ruthenium have been developed for forming capacitor electrodes in FeRAM devices.[31] In one such application, the CMP process is used to planarize the platinum film, leaving it coplanar with an SiO_2 stop layer.[32] Noble metals present a fundamental challenge to the CMP process. As discussed above, the CMP mechanism is proposed to occur by the combination of surface chemical modification and mechanical removal of the chemically modified surface. Since these metals are inherently chemically resistant, this presents a significant hurdle in the design of appropriate slurry chemistries for noble metal CMP. If chemical modification of the surface is not achieved, then the process amounts to little more than mechanical abrasion of the surface that creates unacceptable levels of defects. To avoid highly reactive, environmentally unacceptable chemistry, that would otherwise be required, carefully chosen complexing agents can be used to reach acceptable removal rates without surface damage.

15.4   Critical peripherals to CMP processing

We have focused so far on the CMP process. However, no discussion on CMP would be complete without discussing some external elements critical to CMP processing. Indeed, early CMP tools were developed strictly for the CMP process so that immediately after the CMP process, the wet wafers were taken to a separate scrubber for cleaning and drying. The wafers were then taken to a separate metrology tool for any post-CMP measurements that were required (film thickness, surface defects, dishing, erosion, etc). As device dimensions decreased, tighter control on the CMP process was required and the cleaning process became more critical to CMP performance demanding improvements to and control of post-CMP cleaning. This led to CMP tools with integrated cleaning modules for "dry-in, dry out" processing,* which allowed for each wafer to be cleaned immediately after CMP processing and sent to onboard metrology for inspection. This not only gave more consistent cleaning, but significantly improved productivity.

---

* The first CMP tool with integrated cleaning was delivered to the market in 1990.

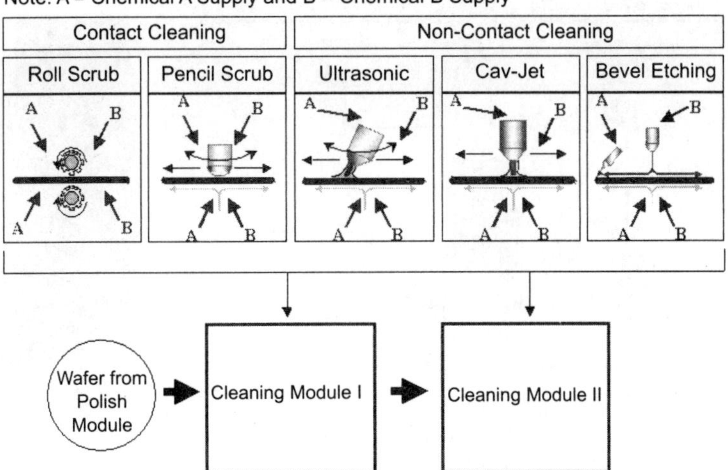

Figure 15.26   Schematic of flexible post-CMP cleaning options.

15.4.1   Post-CMP cleaning

Optimization of the cleaning process and chemistry is required for each CMP application. There are several types of defects that can be introduced from the CMP process such as ionic impurities (e.g., metal ions) from the CMP chemistry or from the film being removed (e.g., copper), organic contaminants from the CMP chemistry (e.g., complexing agents or surfactants), or abrasive particles (e.g., silica, alumina, or ceria). All these surface contaminants can cause device failure and will require removal by post-CMP cleaning in order to maximize device yield. The approaches to cleaning will depend on the contaminants produced by the CMP process, which in turn will depend on the material being removed (note the various CMP applications discussed in preceding sections). As a result, post-CMP cleaning processes must optimize for specific applications, and CMP tools must include the capability to clean after multiple CMP processes. An example of the integrated post-CMP cleaning options is shown in Figure 15.26. In this case there are two cleaning modules with a number of cleaning options of contact and non-contact cleaning available for each module.

It is beyond the scope of this chapter to review all the cleaning options that are available to the numerous CMP applications. The reader is referred to the literature for a more extensive review.[18] We will briefly discuss here a unique approach to post-CMP cleaning that we have developed using electrolytic ionized water.[33] The supply unit and principle of operation are shown in Figure 15.27.* In its simplest form, ultra pure (deionized — DI) water is electrolyzed to create "anodic" and "cathodic" water:

---

* Figure 15.27 is a simplified schematic of the actual unit, which has two separate cells to allow for simultaneous generation with independent control of the anode and cathode.

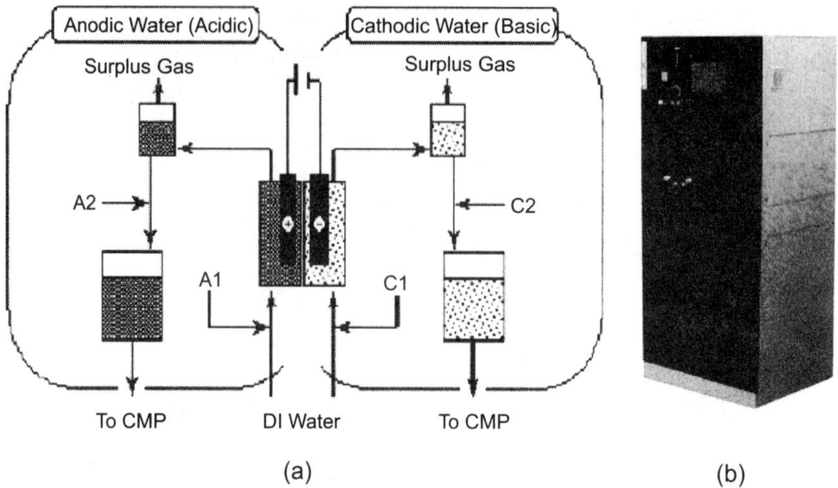

Figure 15.27 Schematic showing (a) principle of electrolytic operation and (b) picture of independent supply unit (dimensions: Footprint = 70×90 cm, Height = 210 cm).

$$\text{Anodic: } 2H_2O \rightarrow 4H^- + O_2 + 4e^- \tag{3}$$

$$\text{Cathodic: } 2H_2O + 2\,e^- \rightarrow 2OH^- + H_2 \tag{4}$$

The system allows for dosing with some proprietary additives before and after the DI water is electrolyzed in both the anodic (A1 and A2) and cathodic (C1 and C2) compartments. Additives injected prior to electrolysis will cause additional reactions to occur at the cathode and anode. This approach allows for tuning of the redox potential of the cleaning chemistry to optimize removal of particulate, ionic, and organic contamination from the wafer surface. It has the potential (no pun intended) to give superior cleaning at lower cost with a more environmentally friendly approach compared to conventional cleaning chemistries. This approach has demonstrated effective cleaning across multiple CMP applications. We present here the application of electrolytic ionized water clean to improve cleaning after poly-silicon and copper CMP.

One of the biggest hurdles to poly-silicon deep trench isolation integration is the defectivity caused by CMP. After a typical poly-silicon CMP process, the wafer surface is contaminated with silica particles, organic and metallic contamination that has proven very difficult to remove by conventional cleaning techniques. Cleaning was evaluated on thermally oxidized Si(100) test wafers on which 100 nm Si_3N_4 was deposited, 450 nm deep trenches were patterned and 1,000 nm of poly-Si was deposited by LP-CVD for the trench filling, and thermally oxidized Si(100) test wafers with blanket films of Si_3N_4, SiO_2 and poly-Si. Polishing and cleaning were carried out with a PVA roll scrub and scanning ultrasonic in cleaning modules I and II, respectively.

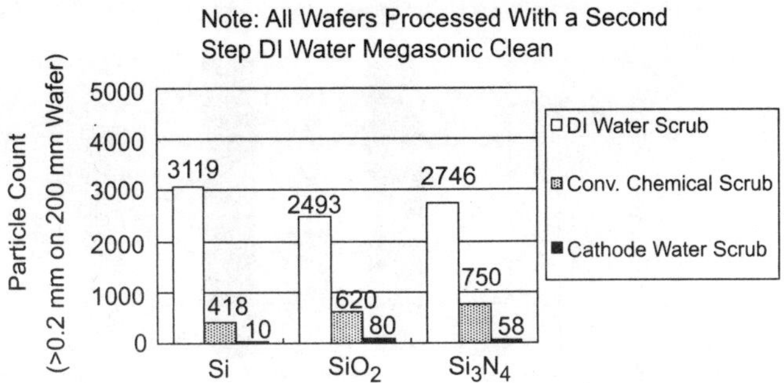

Figure 15.28 Comparison of silica particle counts for two different first module scrub clean chemistries (and DI water control) after the poly-silicon CMP process.

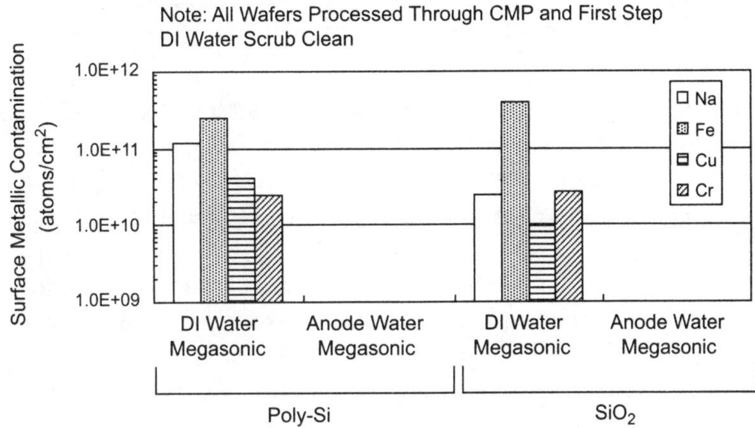

Figure 15.29 Comparison of surface metallic contamination on blanket wafers after poly-silicon CMP process.

Cathodic water can be tuned to be particularly effective at removing post-poly-Si CMP slurry particles. Figure 15.28 shows silica particle counts using a KLA-Tencor SURFSCAN 6420 on 200 mm blanket wafers with the three different film types relevant to poly-Si CMP. The numbers of particles using a conventional chemistry in cleaning module I dramatically decreased the particle counts compared to the DI water control, but the use of cathodic water at −0.50 V (versus Ag/AgCl) showed even greater reduction in post-CMP particle counts. Anodic water was optimized to be particularly effective in removing metallic contamination. Figure 15.29 shows the metallic contamination measured by vapor phase dissolution inductively coupled plasma and time of flight secondary ion mass spectroscopy. In cleaning module II, anodic water at +1.00 V versus Ag/AgCl is delivered through the ultrasonic nozzle and compared to DI water. The anodic water was able to decrease all metallic contaminants below the detection limit (5E8 atoms/cm²). The severe oxidizing capability of anode water also makes it effective for removal of organic contamination from surfactants used in the CMP process.

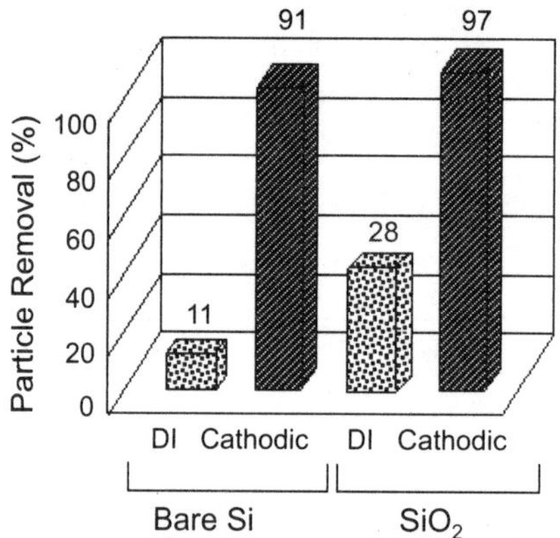

Figure 15.30 Alumina particle removal after a copper CMP process for first module scrub cleans with cathodic and DI water as a percentage removed compared to DI water spin-rinse dry only.

A similar approach using an anodic water scrub after copper CMP has also been shown to reduce metal ions, including copper on SiO_2, to levels below TXRF detection limits ($1E10$ atoms/cm^2). Organic contamination from organic compounds used in the slurry was also removed by anodic water. Further, the copper surface after an anodic water clean was shown by X-ray-photoemission spectroscopy (XPS) to have a thin oxide surface that was much more resistant to further oxidation over time (days) compared to a copper surface cleaned by an oxygenated DI water control. The copper CMP process used alumina instead of silica particles. In Figure 15.30 the percentage removal of alumina particles is shown for a DI water spin-rinse-dry only compared to DI and cathodic water ultrasonic cleans. Appropriately tuned cathodic water exhibits strong alumina particle removal capability compared to DI water on both silicon and silica surfaces. For multiple CMP applications, we have been able to demonstrate that the right combination of anodic and cathodic water cleans can be used effectively to remove post-CMP surface contamination.

15.4.2   Metrology

In addition to integrating the post-CMP cleaning, there has also been an industry push to integrate the post-CMP metrology used to characterize the CMP process (film thickness, surface defects, dishing, erosion, etc). The move to onboard metrology has allowed for greater control of the CMP process because it is possible to get rapid feedback of metrology data from one processed wafer to another wafer about to be processed. One example of such automated process control (APC) is the feedback of onboard film thickness measurements used to adjust CMP process time. Figure 15.31 shows the computer interface screen for a 50 wafer run used to test the closed loop control (CLC) of a SiO_2 CMP process, with 25 blanket wafers run in parallel on each side of the CMP tool. The screen shows post-CMP thickness measurements for all the wafers. Also shown are the upper control limit (UCL) and lower control limit (LCL) of this CMP process in manufacturing. In this example, the first three or four

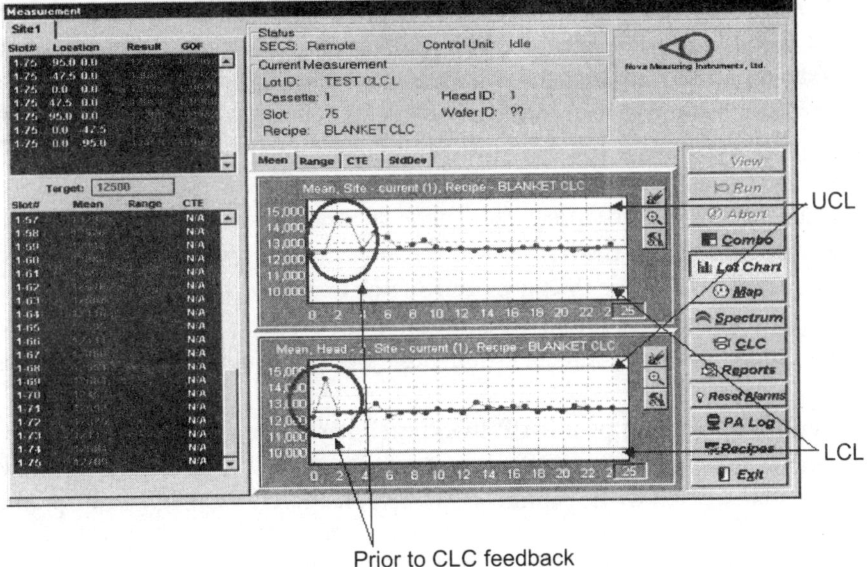

Figure 15.31 Interface screen showing example of closed loop control (CLC) on an Ebara CMP tool using onboard thickness measuring tool from Nova Measuring Instruments, Ltd.

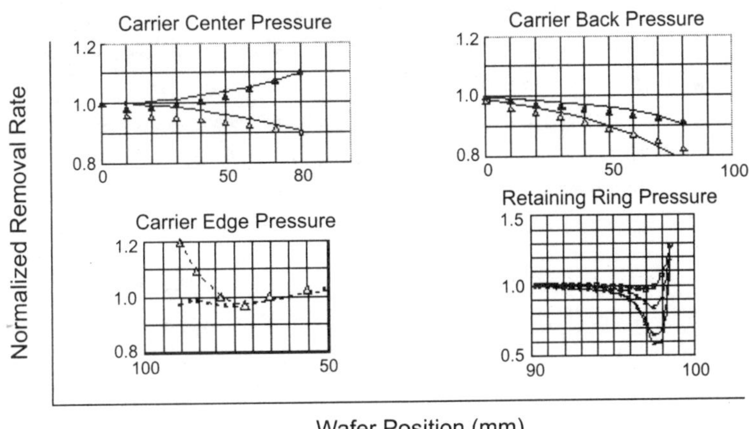

Figure 15.32 FEM analysis curves and experimental data used to determine parameter coefficients.

wafers (depending on the side of the tool) did not receive any CLC feedback since no thickness measurements had been made at the onboard Nova thickness measurement tool. These wafers show the expected post-thickness variation within the control limits. After wafers begin to get CLC feedback, the remaining wafers show much less variation from the post-CMP target thickness.

While onboard metrology allows for rapid feedback by CLC, there is still room to tighten process control even further as can be seen in Figure 15.32, especially in the first few wafers through the tool. CMP tool suppliers have already developed endpoint technologies for several CMP applications. By *in situ* monitoring of the process, an algorithm can send a real-time signal to the tool when to stop processing. We have taken this one step further by combining real-time thickness measurements with endpoint capability to give significant improvements in process control.[34] For example, in the copper Damascene CMP process, we can scan across the wafer during the CMP process to collect data on the thickness profile with time. This allows us to optimize a bulk Cu CMP process for maximum removal rate and planarity that can be stopped at a specific thickness and switch to an interfacial process that can be stopped by a separate endpoint technology. Further, the profile scan of thickness allows the potential for immediate thickness profile feedback to control the CMP process.

In addition to using post CMP thickness measurement data for CLC of CMP polish times, there are many CMP parameters that may be controlled by APC. Table 15.3 shows the top ten CMP parameters to be considered for APC. Since prediction

Table 15.3 Top ten CMP parameters for advanced process control.

10. Post-CMP process time (buff polish, clean etc.)
9. Platen temperature
8. Conditioner pressure
7. Conditioner sweep profile
6. Flow rates of slurry or chemical supply
5. Linear velocity (platen and carrier rotation rates)
4. Carrier retaining ring pressure
3. Carrier profile control pressure (center and edge)
2. Polish pressure
1. Polish time

of the removal profile across the wafer is critical for APC, we have evaluated several parameters with finite element method (FEM) analysis using an adaptation of Preston's equation:

$$R = \mathrm{d}T/\mathrm{d}t = k_0\, k_1\, k_2\, k_3\, k_4\, P\, v \tag{5}$$

where P and v are pressure and linear velocity as defined in Equation 1 above, k_0 is the proportionality coefficient, k_1 is the carrier back pressure coefficient, k_2 is the carrier retaining ring pressure coefficient, k_3 (r) is the carrier center profile pressure coefficient, and k_4 is the carrier edge profile pressure coefficient. Using a particular set of consumables (pad, slurry, etc.), an SiO_2 process was optimized for removal rate and step height reduction. A model was then developed for this process to enable prediction of the removal profile with changes in the four CMP parameters in Equation 5.

Figure 15.32 shows experimental data versus the FEM model used to obtain the coefficients for each of the four parameters. The coefficients were then used to

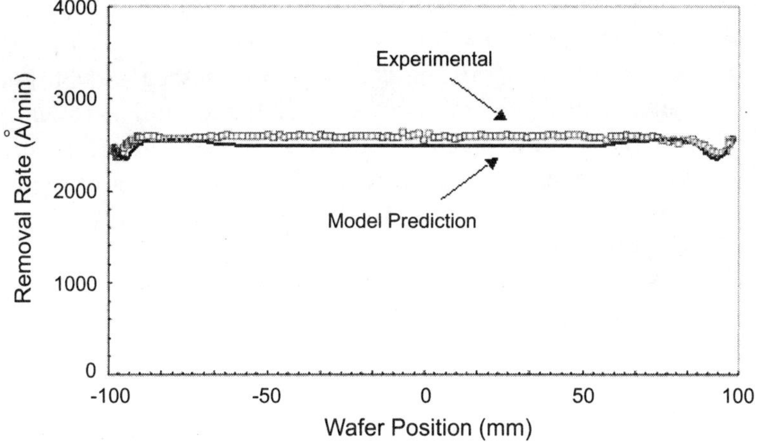

Figure 15.33 Comparison of experimental diameter scan versus model prediction.

predict the removal profile diameter scans as shown in Figure 15.33 for various process conditions with reasonable accuracy. It should be noted that any changes in consumables would require a new FEM analysis to enable predictive capacity. Such analyses are critical to developing robust and accurate APC capability.

15.5   Future of CMP in microelectronics

For CMP to continue enabling future generations of microelectronics, research and development efforts will need to focus on improving process capability as well as compatibility with new materials such as those we have discussed in this chapter. In addition, the CMP process must be made more productive to help reduce the overall cost of ownership with development efforts including CLC, high throughput, and advanced consumables. Such improvements will extend the use of the CMP process over the next five years.

Looking farther down the road, there is a need for significant technological advances that will enable advanced semiconductor device fabrications. New technologies for copper Damascene planarization will be required that avoid damage due to abrasives and contact pressures used in conventional CMP. One proposed approach is a controlled etching called "spin etch planarization."[35] Wet etch chemistries are dispensed onto the wafer surface while the wafer is rotated. Another method proposes the use of localized selective copper electropolishing.[36] The wafer is divided into zones that are locally polished using an electric current. Both of these approaches face serious challenges with pattern sensitivity that will need to be overcome, especially in planarizing wide lines.

A unique approach for polishing that uses electrochemistry together with CMP has also been explored.[37] Recently, we have been developing a copper CMP process based on such a combination of electrochemistry with CMP. The process achieves copper removal rates comparable to conventional CMP, but at extremely low contact pressures and without the use of abrasives. Both of these features are highly desirable for the processing of copper over soft low k materials where high pressures and abrasive particles can cause significant damage (as discussed above). Further, the process has demonstrated capability to planarize both wide and narrow lines. While still in the early stages of development at the time of this writing, the process has the potential to improve both process capability for future generations of semiconductor devices, and overall cost of ownership. We will discuss this advanced technology further in a future publication.

In short, over the next five years, the CMP process will improve capability, expand application, and improve efficiency and productivity, all at a reduced overall cost of ownership.

References

1. Sir Isaac Newton, *Optiks,* Dover Publications, New York, 1652, from 4th Edition, London, 1730.
2. Brown, N.J., *Proc. SPIE,* **306**, 1981, p. 42.
3. Kaufman, F.B., Thompson, D.B., Broadie, R.E., Jaso, M.A. Guthrie, W.L., Pearsons, D.J., and Small, M.B., *J. Electrochem. Soc.,* **138**, 1991, p. 3460.
4. Runnels, S.R., *J. Electr. Mater.,* **25**, 1998, p. 1574 and references therein.
5. Paul, E., *J. Electrochem. Soc.,* **148**(6), 2001, p. G359.
6. Luo, J., *CMP-MIC Proceedings, IMIC,* February 2002, p. 49.
7. (a) Watts, D.K., 4th International Symposium on Chemical Mechanical Polishing, *Center for Advanced Materials Processing,* August 2000, Lake Placid, New York. (b) Ollison, C., Pierce, K. and Stephenson, B., *CMP-MIC Proceedings, IMIC,* February 2002, p. 260.
8. (a) Watts, D.K., Chikamori, Y., Kohama, T., and Kimura, N., *Materials. Research. Society. Symposium Proceedings,* **671**, 2001, p. M3.1.1–8. (b) Contonili, R.J., *J. Electrochem. Soc.,* **141**, 1994, p. 2503. (c) Komanduri, R., *Annals of the CIRP,* **46**, 1997, p. 545.
9. (a) Beyer, K., Guthrie, W., Makarewicz, S., Mendel, E., Patrick, W., Perry, K., Pliskin, W., Rieseman, J., Schiable, P., and Standley, C., U.S. Patent 4,944,836, 1990. (b) Patrick, W.J., Guthrie, W.L., Standley, C.L., and Shiable, P.M., *J. Electrochem. Soc.,* **138**, 1991, p. 1778–1784.
10. (a) Park, Y.B., *J. Electrochem. Soc.,* **148**(10), 2001, p. G572. (b) Gan, T., *J. Electrochem. Soc.,* **148**(3), 2001, p. G159. (c) Boyd, J.M., *J. Electrochem. Soc.,* **143**(11), 1996, p. 3718. (d) Cooperman, S., *J. Electrochem. Soc.,* **142**(9), 1995, p. 3180.
11. (a) Yagishita, A., *IEEE Trans. Electron Devices,* **49**(3), 2002, p. 422. (b) Yagishita, A., *IEEE Trans. Electron Devices,* **48**, 2001, p. 1604.
12. Rutten, M., *Semiconductor International,* September, 1995, p. 123.
13. Venkatesan, S., Gelatos, A., Misra, V., Smith, B., Islam, R., Cope, J., Wilson, B., Tuttle, D., Cardwell, R., and Watts, D. K., IEEE International Electron Devices Meeting, *IEDM Tech. Dig.,* 1997, p. 769.
14. Edelstein, D., Heidenreich, J., Goldblatt, R., Cote, W., Uzoh, C., Lustig, N., Roper, P., McDeVitt, T., Motsiff, W., Simon, A., and Dukovic, IEEE International Electron Devices Meeting, *IEDM Tech. Dig.,* 1997, p. 773.
15. Jeng, S.P., *Mat. Res. Soc. Symp. Proc.,* **337**, 1994, p. 25.
16. Bohr, M., *IEDM Tech. Dig., IEEE,* 1995, p. 241
17. Tomozawa, M., *Solid State Technology,* July 1997, p. 169.
18. Steigerwald, J.M., Muraka, S.P., and Gutmann, R.J., Chemical Mechanical Planarization of Microelectronic Materials, John Wiley & Sons, 1997, and references therein.
19. Preston, F., *J. Soc. Glass Technol.,* **11**, 1927, p. 214.

20. Cook, L.M., *J. Non-Crystal. Solids,* **120**, 1990, p. 152 and references therein.
21. Zhao, B., *CMP-MIC Proceedings, IMIC,* February 1999, p. 13.
22. Watts, D.K., Bajaj, R., Das, S., Farkas, J., Dang, C., Freeman, M., Saravia, J.A., and Gomez, J., U.S. Patent 5,897,375, 1999.
23. Rabke, S., Watts, D.K., Saravia, J., Gomez, J., Dang, C., Islam, R., Klein, J., and Farkas, J., Symposium on Chemical Mechanical Planarization in IC Device Manufacturing, *Proc. Electrochem. Soc.,* **98**(7), 1998, pp. 9–18.
24. Sources: (a) VLSI Data Archives. (b) Dataquest Report, March 2002.
25. Ryan, J.G., Goldblatt, R., McGahay, V., Davis, C., Narayan, C., Hedrick, J., Hay, J., Edelstein, D., Chen, S.T., and Rodbell, K., 4th International Symposium on Chemical Mechanical Polishing, *Center for Advanced Materials Processing,* August 1999, Lake Placid, NY.
26. Itabashi, T., *IEEE Electron Devices Society, International Interconnect Technology Conference,* June 2002, Session 14(3).
27. Saito, T., *IEEE Electron Devices Society, International Interconnect Technology Conference,* June 2001, p. 15.
28. (a) Lin, Z., *Proceedings of the 17th International CMP-MIC Conference,* 2002, p. 138. (b) Abbas, S.A., *IBM Tech. Disclosure Bull.,* **7**, 1997, p. 2754.
29. Keast, C.K., *Proceedings of the 11th International VMIC Conference,* 1994, p. 204.
30. Hu, S., *IEEE Electron Devices Society, International Interconnect Technology Conference,* June 2002, Session 12(3).
31. Beitel, G., *Proceedings of the 16th International CMP-MIC Conference,* March 2001.
32. Wang, H., *Proceedings of the 17th International CMP-MIC Conference,* 2002, p. 333.
33. Miyashita, N., *Mater. Res. Soc. Symp. Proc.,* **613**, 2000, p. E5.3.1.
34. Allen, R., Chen, C., Trikas, T., Lehman, K., Shinagawa, R., Bhaskaran, V., Stephenson, B. and Watts, D.K., International Symposium on Semiconductor Manufacturing, San Jose, CA, October 2001.
35. (a) Levert, J.A., *Electrochemical Society Proceedings,* Oct. 1999, Honolulu, Hawaii, p. 162. (b) Mukherjee, S.P., Levert, J.A., and Debear, D.S., *Mater. Res. Soc. Symp. Proc.,* **613**, 2000, p. E8.10.1. (c) DeBear, D.S., Levert, J.A., and Mukherjee, S.P., *Solid State Technol.,* **43**, 2000, p. 53.
36. (a) Tian, Y., *Proceedings of the 17th International VMIC Conference,* 2000, p. 423. (b) Yih, P.H., Wang, D.H., and Chiao, S.H., 18th International VLSI Multilevel Interconnection Conference, September, 2001, p. 59. (c) Contolini, R.J., Mayer, S.T., Graff, R.T., Tarte, L., and Bernhardt, A.F., *Solid State Technol.,* June, 1997, p. 155–161.
37. Samitsu, Y., *J. Jpn. Soc. Precision Eng.,* **61**(4), 1995, p. 552.

16 Electrochemical deposition equipment

Tom Ritzdorf and Dakin Fulton

16.1 Introduction

Historically, the equipment used for electroplating has consisted of tanks (sometimes made of wood or concrete) containing plating chemistries, pre- and post-treatment chemicals, and rinse tanks arranged in a line or "U" shape. The parts to be electroplated are "racked," or hung on an electrically conductive structure that is used to hold them suspended in the plating bath and also to supply the current for electrolysis. The typical plating line that is used to plate copper onto printed circuit boards, for instance, is a room full of plating and rinse tanks. The automation of these systems consists of using large hoists on linear tracks to move the racks between the various tanks: pretreatment, electroless plating, electroplating, post-treatment, and various rinsing steps associated with the process. While effective for many types of products, this type of system is not sophisticated enough for the semiconductor manufacturing industry, and does not meet the specific requirements of wafer processing.

In electroplating systems used for depositing metals on semiconductor wafers and similar substrates, there are typically requirements for several different types of chambers in an automated processing system. These chambers or "cells" are used for the several different wet processes that must occur in sequence in order to ensure a repeatable, high-quality deposition on all areas of the wafer where metal should be deposited. The electrochemical deposition processes, whether used for metal conductor deposition or for forming an attachment structure, occur in the latter parts of the chip manufacturing process. As such, their yield is especially important in order to avoid scrapping devices that have made it this far through the manufacturing process. In order to ensure the high yields that are required in semiconductor manufacturing, electrochemical deposition systems must be designed with attention to detail, and with robust hardware and processes. In order to produce robust processes, steps such as cleaning and pre-wetting must be used to ensure high yields. Furthermore, automated wafer processing equipment is usually designed to unload dry wafers from a cassette, or front opening unified pod (FOUP), and deliver dry, processed wafers back to the same carrier.

In many processes (especially for packaging applications) more than one metal is required to be deposited on the wafer in a sequential process. Many times (especially in the case of eutectic solder compositions) a copper or nickel stud is desired beneath the solder in order to provide a barrier or sufficient material to react and form intermetallic compounds without consuming the underbump metallurgy (UBM) to the point of causing an adhesion failure. Other examples include the deposition of different metals to provide high conductivity, strength, or mechanical properties; and resistance to oxidation or corrosion. Another example is in the deposition of layers of gold and tin that will later be reflowed to form a eutectic gold–tin solder ball. These processes require the deposition of at least two metals in sequence. In addition, when wafers are coated with a photoresist to be used as a plating template, as is the case with solder bump deposition, a pre-wet step is usually necessary to ensure complete wetting of the features prior to starting the deposition process. Finally, there are requirements to rinse the wafers between process steps and after the final chemical process, and the wafers must be dried before being placed back into the cassette. These requirements necessitate the integration of multiple processing chambers, or cells, in a single piece of automated process equipment. The different types of cells will be reviewed in detail, followed by a discussion of the system level issues involved in the design of automated wafer electrochemical deposition (ECD) processing equipment.

16.2   Electroplating chambers

The most basic system components necessary to perform an electroplating process consist of an electrolyte, a container (or electrochemical cell) to hold the solution, a counter-electrode (anode), and a means of driving current (or supplying voltage) between the electrodes such as a battery or a rectifier. Additionally, most practical systems have a means for producing solution agitation in order to decrease the diffusion boundary layer thickness and increase the limiting current density, and therefore the operating current or deposition rate. In systems used to deposit metal films on wafers for microelectronic applications, it is typical to have closed loop solution flow control, microfiltration, and sensors to monitor various solution properties integrated into the system.

16.2.1   Electroplating chamber design considerations

The particular choice of chemistry to be used may affect the equipment design to some extent but the important issues will be the same for most typical plating solutions. Many factors must be taken into account during the design of a piece of equipment to be used for electroplating of metals on semiconductor wafers. Some of the factors that affect the electroplating reactor design are:

- Thickness uniformity
- Diffusion boundary layer control
- Electric potential and terminal effect
- Alloy composition uniformity
- Edge exclusion zone management
- Electrical contact to the wafer
- Wafer backside contamination requirements

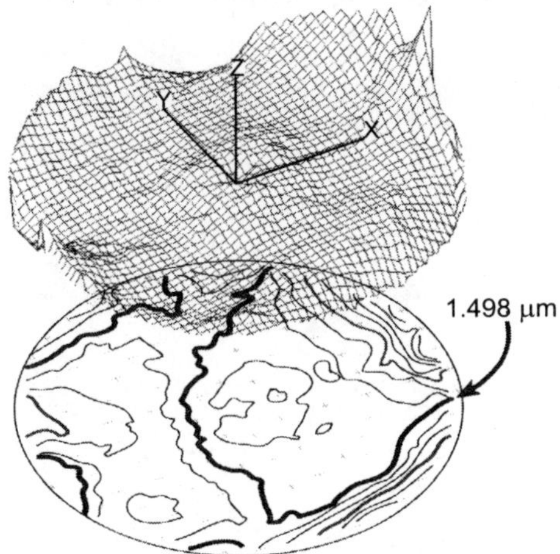

Figure 16.1 Thickness map taken from a sheet resistance measurement of electroplated copper blanket wafer showing a within wafer non-uniformity of 1.77%, 3 sigma with a 1.498 µm average film thickness. The measurement position is 4.5 mm inside the edge of the 300 mm wafer. The total range of all measurements is 720 Å.

- Anode material
- Alpha-particle emission
- Considerations for lead-free solder deposition
- Power supply

The process factors to be considered must begin with the ability to produce structures with the appropriate electrical and reliability characteristics to be useful for their respective applications. As these are usually most strongly dependent on the chemistry chosen and the deposition process used, we will assume that these requirements are met by optimizing the chemistry and process, and move on to those factors that more strongly affect the system design.

16.2.1.1 Thickness uniformity

Uniformity of the deposit thickness is one of the most basic factors that must be taken into consideration during the design of the electroplating reactor. Of course, the uniformity required of the process will be determined by the particular application. The reactor design should be appropriate for meeting the requirements of the application without too much "extra" capability that serves only to add cost to the system. The total uniformity associated with a piece of (single wafer) equipment is the combination of the within wafer uniformity and the wafer to wafer uniformity. The within wafer uniformity (Figure 16.1) is dependent on the current density distribution, the diffusion layer thickness distribution (usually determined by fluid flow), and other means that are used to randomize these effects with respect to the wafer surface. The wafer to wafer uniformity (Figure 16.2) is dependent on the contact design, the reliability of making electrical contact to the wafer, and the method used to determine

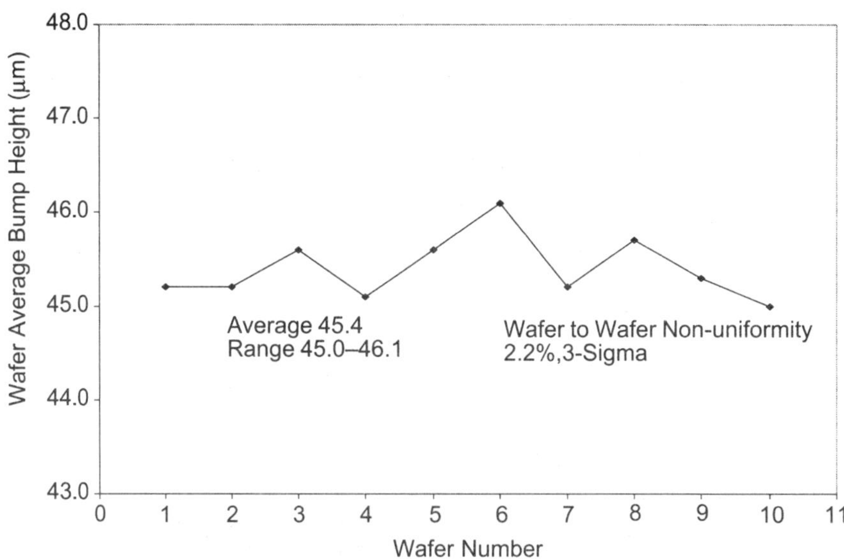

Figure 16.2 Wafer to wafer bump-height thickness variation in a copper stud/solder bump
sequential deposition process. The variation of average bump height from wafer
to wafer is 2.2%, 3 sigma.

the end of the process, as well as the reliable control of parameters such as electrolyte
composition, temperature, and fluid flow. When multiple reactors are used for the
same process step, chamber-to-chamber matching also plays a role in wafer to wafer
uniformity. In most packaging applications, it is the thickness uniformity within each
die that is important, because this affects the robustness of the packaging process.
The distribution, or coplanarity, of the bump thickness determines the reliability of
the connection of the chip to the substrate. This is especially important considering
pattern density effects, as discussed in Section 16.2.1.3, which may contribute to the
within die non-uniformity (Figure 16.3).

The thickness uniformity of the deposit is determined mainly by three parameters: uniformity of the diffusion boundary layer, uniformity of the electric potential
on the surface of the workpiece, and methods used to randomize any irregularities in
these parameters over the surface of the wafer. The plating chemistry will have some
effect on the within wafer thickness uniformity because of the "throwing power"
associated with the particular bath composition and operating parameters, but these
factors must be optimized for their effect on the deposited material and their ability
to fill the particular requirements of the application. The reactor must then be
designed to optimize the uniformity for a given set of chemistry and processing
conditions.

16.2.1.2 *Diffusion boundary layer control*

The diffusion boundary layer has a strong impact on the deposition of metals in an
electroplating system. This boundary layer is generally thought of in terms of the

Figure 16.3 (a) As-deposited electroplated solder bumps on a wafer with a challenging group of patterns of varying densities. (b) Reflowed solder balls showing variations in pattern density within a single die.

metal ions being deposited on the surface but the diffusion of species such as organic additives, grain refiners, or contaminants is also important. The concentration gradient of each of these species across its respective diffusion boundary layer determines the diffusion rate of that species or, in the case of metal ions being deposited, the maximum deposition rate that can be supported.

In a well-designed electroplating system the diffusion boundary layer thickness will actually be determined by the hydrodynamic boundary layer at the cathode (wafer) surface. This ensures that the diffusion layer is controlled by the fluid dynamics of the system and allows the system designer to produce a very uniform diffusion layer thickness. This also provides more control of the diffusion layer thickness for components of the chemistry that are present in very small amounts, such as some additives.

If the rate of metal deposition approaches the maximum diffusion rate for the metal ions being deposited, the system approaches the limiting current density (LCD), at which no further increase in the metal deposition rate can be made without changing the operating conditions (metal concentration, temperature, solution agitation) of the system.[1,2] The practical maximum deposition rate is typically related to the LCD, because the deposit morphology becomes nodular or dendritic as the LCD is approached.[3] Therefore, it is important to pay attention to the LCD of a system when maximizing the deposition rate is the goal. This is especially important in the case of deposition in which a plating template is used for patterned plating. This is the case because, even if the wafer is rotating, the LCD varies from the center to the edge of the wafer due to the effect of convection on the localized diffusion boundary fields at the areas that have been cleared of photoresist. This is in direct contrast to the case of blanket plating (as in copper Damascene interconnect applications) where the LCD is uniform over the surface of a rotating wafer.

16.2.1.3 *Electric potential and the terminal effect*

The current density distribution across the wafer surface is determined by the electrical contact locations and the resistance of the seed layer relative to the resistance within the bath. The ratio of the resistance across the wafer to the resistance in the bath is a measure of the magnitude of the terminal effect.[4] As the wafer size grows to 300 mm (or seed layer thickness is reduced, or plating rate is increased), the terminal effect becomes more pronounced and harder to counteract. The localized high current density near the electrical contact must be compensated for in some manner in order to optimize the deposit thickness uniformity. The resistance of the plating base is constantly changing throughout a blanket film deposition (such as in copper interconnect applications), and so this must be considered to be a dynamic parameter during the deposition process. The best way to compensate for the terminal effect to produce uniform film deposition at the beginning and the end of the process is to have some kind of dynamic control of the current density distribution across the wafer.

The terminal effect has been modeled, and the impact of various parameters on the magnitude of the terminal effect is well-understood.[4–6] Most of the work done on the terminal effect, however, is simply to reduce the magnitude of the effect without considering that the terminal effect is dynamic during the deposition process itself. Some have even advocated optimizing the plating chemistry in order to provide better within wafer uniformity by reducing the terminal effect. The terminal effect is greatest at the beginning of a copper interconnect deposition process, which happens

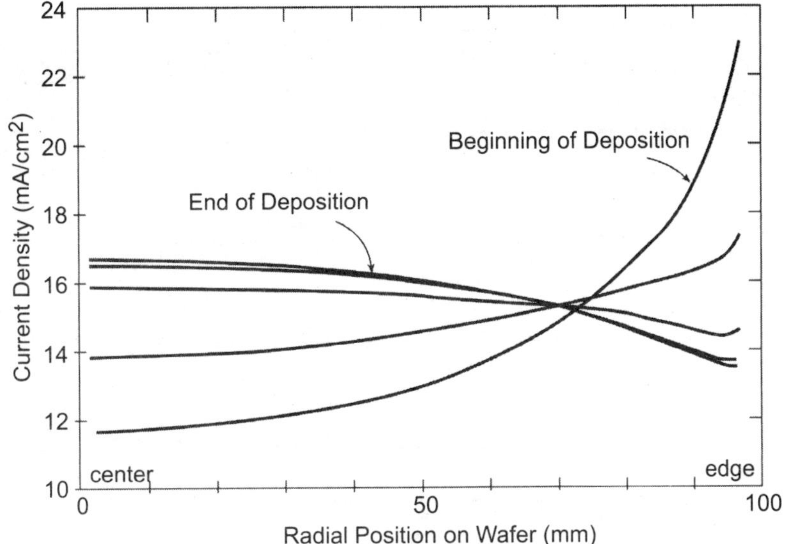

Figure 16.4 Time-dependent current density profiles in a conventional wafer electroplating reactor. The variation of the terminal effect with deposition time causes the decrease of current density at the edge, and increase of current density at the center, as deposition proceeds.

to be when the critical filling of the smallest features on the wafer occurs. The terminal effect can cause a large center-to-edge variation in the current density across the wafer surface, especially at the beginning of the process, as seen in Figure 16.4. In order to produce a uniform film at the desired target thickness, the deposition profile of a conventional reactor (one in which the hardware setup determines the thickness uniformity) must be optimized to complement the edge-thick deposition due to the terminal effect at the beginning of the process. This means that the primary current distribution in the reactor is usually optimized to produce a center-thick profile in the latter parts of the deposition process (Figure 16.4). This provides a uniform film at the target thickness due to the summation of the incremental deposition thickness throughout the process, as represented in Figure 16.5. If the deposition thickness is changed, however, the reactor must be re-optimized or a non-uniform film will result. This non-uniform film will be relatively edge thick if the deposit is thinner than the optimal thickness or edge thin if it is thicker than the optimal thickness (Figure 16.6). Obviously, these variations in film thickness profile (as between metal 1 and metal 6 in a semiconductor device) are far from ideal when considering that this will be the input to a CMP process in the case of copper Damascene interconnects.

One way to get around this problem is by using dynamic cathode current density control. In a system with this type of control, the distribution of the current density across the wafer can be changed during the process in order to compensate for the changing magnitude of the terminal effect. The result is more uniform cathode current density throughout the process (Figure 16.7), which translates to a more

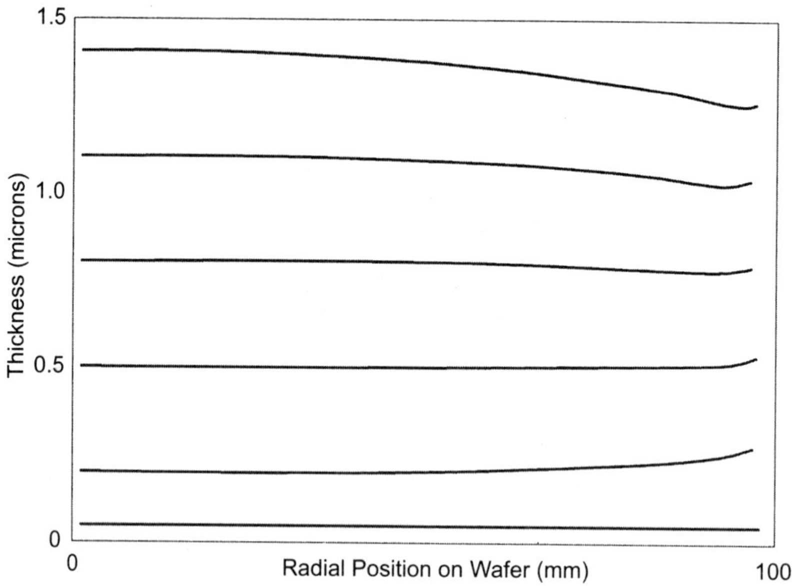

Figure 16.5 Time-dependent film thickness evolution that results from the current density curves in Figure 16.4. Note the edge thick profiles for thin films evolving into edge thin profiles for thicker films.

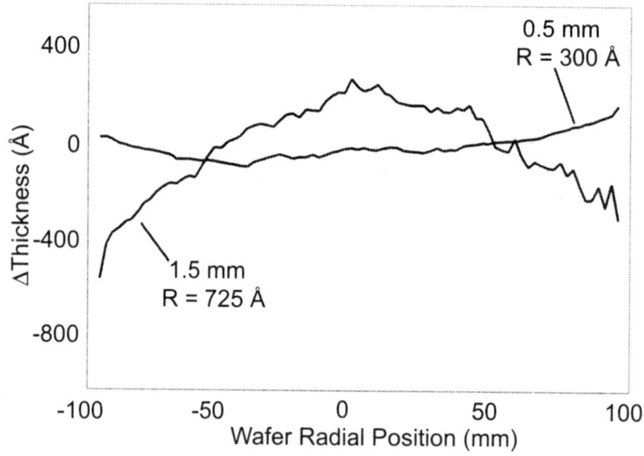

Figure 16.6 Experimental demonstration of the thickness profiles resulting from the models in Figure 16.5. These curves are from a conventional electroplating reactor.

uniform distribution of fill capability, grain size, impurity incorporation, and film morphology across the wafer surface. This type of system can also be used to compensate for variations caused by different product types having different plated areas in patterned plating applications such as solder bump deposition.

The alternatives for compensating for edge effects, whether localized or distributed around the wafer perimeter, include the use of current shielding, current thieving (auxiliary electrodes), or compensating solution agitation effects.

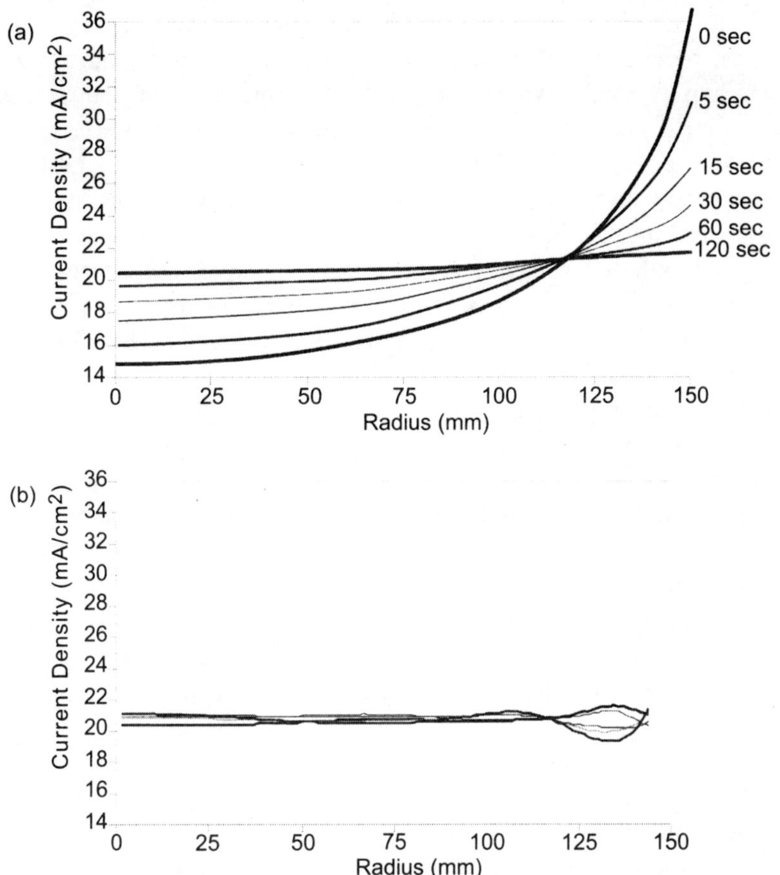

Figure 16.7   Time-dependent current density profiles for electroplating of 1.0 μm of copper on 1,000 Å copper seed layers on 300 mm wafers. (a) Curves from a conventional reactor. (b) Curves from a reactor with dynamic cathode current density control.

Current shielding may be used in wafer electroplating systems to modify the current distribution within the electroplating cell. This is simply the insertion of a non-conducting element within the electrolyte to shape the electric field distribution. The use of these field-shaping elements is well known in the field of electroplating; they can be used to compensate for non-uniform field distribution at the wafer surface. This usually consists of inserting a non-conducting ring inside the reactor relatively close to the wafer surface. Shields may also be used near the anode to modify the anode current distribution. In the extreme case, where shields practically obstruct the line-of-sight current path between the anode and the cathode, they form what may be termed a virtual anode. That is, the open area around the shielding elements through which the current must flow acts as if there were a physical anode in that location, when viewed from the cathode.[7]

Current thieving is the inverse of shielding and can be used in much the same way. It is especially common for reducing edge effects in wafer electroplating systems. A current thief is simply a cathode area that is adjacent to the workpiece that "steals" some of the current that otherwise would have passed through the workpiece. In practical application, it typically consists of a ring of metal that surrounds the wafer and compensates for edge effects caused by the terminal effect, the pattern density effects, or an undesirable primary current distribution. Thieves may be simply connected with the wafer or they may be connected to a separately controlled power supply channel. As in the case of dynamic cathode current density control, the current applied to the thief may be varied during the deposition process. Of course, the use of current thieves necessitates more system maintenance, while introducing some level of inefficiency in the use of electrical power and metal ions. Current thieves are not desirable for use with noble metals, which are hard to remove from the electrode and are expensive.

If all other methods are not sufficient to produce the uniformity that is desired from a particular plating process, it may be necessary to randomize the effects of the remaining non-uniformity at the wafer surface. This may be done simply by rotating the wafer to eliminate any circumferential non-uniformity present in the reactor or by using some more elaborate method of providing pseudo-random motion within the process chamber. This has been done through using orbital motion with rotation[8] as well as through moving louvered shields between the anode and cathode in a complicated motion that leads to a relatively uniform current distribution over the wafer when it is integrated over the time of the deposition process.

Deposit thickness uniformity may also be affected by the solution agitation. This is especially true if the deposition rate is high enough to be impacted by the mass transfer limitations of the system. In this case, changes in the hydrodynamic boundary layer (and therefore the diffusion boundary layer) will have a significant impact on the thickness distribution of the deposit. In some cases, the solution agitation, in the form of flow rate or distribution, wafer rotation, or agitation from paddles, may be adjusted to vary the thickness distribution.

When electroplating a metal in a dielectric template such as a photoresist there may also be a pattern density effect on the uniformity, both at the wafer scale and at a localized feature scale. This effect is due to current crowding near the edges of patterns and is explained in the literature.[9-11] This effect is important to consider during the mask design process. If the pattern cannot be made uniform in the mask design, however, it may be necessary to compensate for pattern density effects through the use of shields, current thieves, or dynamic cathode current density control. It is important to note, however, that these methods are capable of modulating the wafer scale uniformity, but not the die scale uniformity associated with the same effect.

16.2.1.4   *Alloy composition uniformity*

In the case of alloy deposition, the compositional uniformity of the alloy is usually at least as important as the thickness uniformity. The relative fluxes of the ions that make up the alloy and the deposition potential, or current density, control the alloy composition. Therefore, it is critical to control the distribution of these variables across the wafer surface. The concentration of the ions in the solution must be tightly controlled, both in the bulk and at the cathode interface. This is especially true if one or more of the alloy constituents has a low concentration in the solution. The requirement for current density control is essentially the same requirement that is necessary to produce good thickness uniformity.

16.2.1.5 Edge exclusion zone management

It is important to consider the edge of the wafer, or exclusion zone, when designing wafer processing equipment. The condition at the edge of the wafer may not be as well controlled as the device area, and may play a large role in determining what occurs in the deposition process. It is important to consider such things as the edge exclusion of the current-carrying layer, the edge bead removal of the photoresist (if applicable), and the materials that will be exposed to the process.[12]

16.2.1.6 Electrical contact to the wafer

The electrical contact to the wafer will be plated to the wafer if exposed to the plating solution. The simplest way to imagine compensating for this effect is to seal the electrical contact area from the plating solution, thereby eliminating electro-deposition in this area. Unfortunately, any area that is sealed from the solution is area that is lost as yieldable die; and if the contact is isolated from the plating bath, it cannot benefit from the effect of being plated to the substrate as metal is being deposited. This effect improves the electrical contact (lowers the contact resistance) at the onset of deposition, thereby eliminating any potential problems caused by contact resistance. When plating metals that commonly utilize consumable anodes, the electrical contacts may be electrochemically cleaned of deposits in the process chamber.

The choice of a continuous ring contact versus individual contact points is another decision that must be made when designing the deposition equipment. The preferred method may be somewhat determined by the process integration of the wafers that the equipment will be processing.

16.2.1.7 Wafer backside contamination requirements

The contamination of the backside of the wafer with plating solution is an issue that is of concern to many in the semiconductor industry. Whether this is a real or perceived concern is not always clear. In the case of gold deposition on GaAs substrates, though, it is important to manage backside wetting of the wafer in order to avoid depositing gold on the back of the wafer. In any case, it is possible to minimize backside contamination through proper system design. It is also common practice to include backside cleaning chambers in an automated wafer processing tool if low contamination levels are required.

16.2.1.8 *Anode material*

Two classes of anodes may be used in electroplating processes: consumable and inert. The consumable anode provides simple control of the metal concentration in the bath for the price of some attention to the proper operation and dissolution of the anode material. In theory, (at 100% efficiency) the anode and cathode reactions will exactly balance each other and the system will have a perfectly stable metal concentration. In practice, the results may be very close to ideal, but some monitoring of the metal concentration is desirable. The consumable anodes tend to dissolve somewhat even when not being used, which can cause a rise in the metal concentration over extended periods of time. Consumable copper anodes must typically be "burned in" to produce an anode film composed of copper oxides, copper chlorides, copper phosphates, and organic additive materials which provides optimal dissolution characteristics and prevents the anode from polarizing during operation. Also, consumable anodes may contain a small amount of phosphorus or sulfur to help provide optimal dissolution at the expense of adding this element to the bath over time. Any other impurities contained in a consumable anode will also tend to build up in the plating bath over time. It is important, therefore, to consider the anode purity as a possible source of chemical contamination, even though impurities in the plating bath will not necessarily show up in the deposited film (or if they do, not in the same concentration as in the bath) because of the purification effects of electroplating.

An inert anode such as platinized titanium or ruthenium oxide has the advantage of being practically maintenance free as far as the anode is concerned. The trade-off is that the replenishment of the bath chemistry becomes more complicated. In the case of noble metals, inert anodes are the only option available. Metal ions (in the correct stoichiometric ratio) must be replaced by the addition of metal salts or oxides, with the resultant effect of increasing the anion concentration in the bath. Also, the oxidation processes that occur at the inert anode include the evolution of oxygen or the oxidation of other components of the bath, which can lead to problems with the system operation, pH changes, breakdown of organic additives normally contained in these baths, and decreased bath life.

Gas management is an important issue in ECD systems, especially in the cases of electrolytic deposition with inert anodes and electroless deposition. In these cases, relatively large amounts of oxygen or hydrogen can be produced. It is usually important to control these gases in order to prevent them from forming bubbles that interfere with the deposit or from being incorporated into the deposit.

16.2.1.9 *Alpha-particle emission*

In the case of lead–tin solder deposition, it may be very important to limit the amount of alpha-particle radiation that is emitted from the deposit in order to reduce the number of soft errors caused by the interaction of the alpha particles with the semiconductor device. Since the largest contribution to the alpha-particle emission of the material is due to contaminants or the conversion of the ^{210}lead isotope to ^{206}lead through the pathway indicated in Figure 16.8, it is important to have material

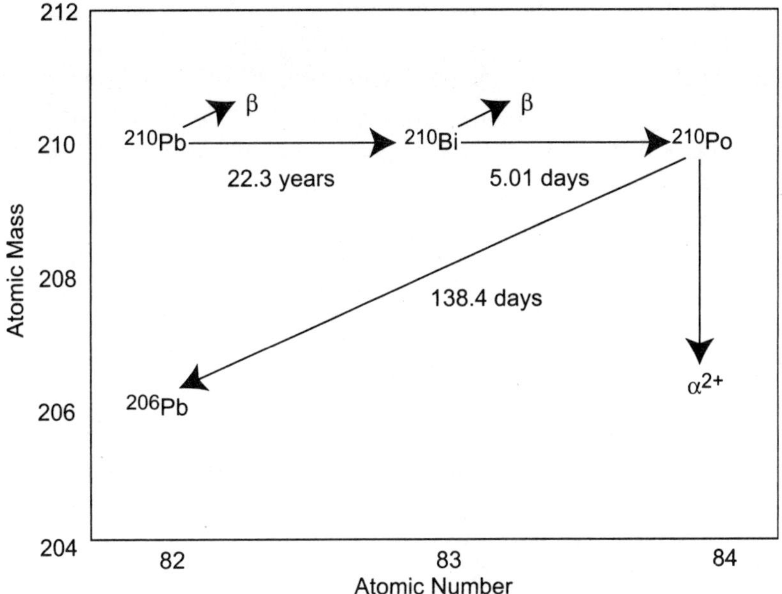

Figure 16.8 ^{210}Pb decay series showing decomposition path from ^{210}Pb to ^{206}Pb with alpha-particle and beta-particle emission.

that is chemically pure, as well as isotopically pure as possible. The source of the lead that is used to produce low alpha-particle emission material is important and people have been very creative in their search for low alpha lead. Since lead that was chemically pure and has been around long enough for most of the ^{210}lead to decay to ^{206}lead has low alpha-particle emissions, lead that was refined long ago is one source of the material. People have found low alpha lead in sunken ships or in material used for roofing houses long ago. There are also some mines (or parts of mines) that contain material that is pure enough to be considered low alpha material, especially with the use of special refining practices.[13]

The measurement of the alpha-particle emission of these anode materials is also a tricky process, because the levels that are being sought by the semiconductor industry are hard to distinguish from the background levels of alpha-particle radiation. Especially during periods of increased sunspot activity it becomes difficult to measure the alpha-particle emission of a material sample. In general, the measurement process is very lengthy (on the order of weeks) and subject to statistical variations in the sample as well as the background.[14]

16.2.1.10   *Considerations for lead-free solder deposition*

The chemistries that are being developed for the deposition of lead-free solder alloys such as tin–copper, tin–silver, and tin–silver–copper must be managed more closely than the typical methane sulfonic acid bath that is used for lead–tin solder deposition. Because of the larger difference in electrochemical potential of the metal ions in these solutions, they tend to exhibit substitution plating on the tin anodes that are used when a consumable anode is desired. Once this happens, the noble metals are depleted from the bath, the scale on the anode may flake off, and the anode may even become passivated. In most plating systems, the only way around this problem is either to use an inert anode or to remove the consumable anode from the chemistry when it is not being used.

Inert anodes cause oxidation reactions other than the complementary dissolution of the metal being deposited at the cathode to occur. This is either the decomposition of water to produce oxygen gas or the oxidation of some component(s) of the bath, which leads to bath instability and short bath life.

16.2.1.11   Power supply

The power supply is an important part of any electrolytic deposition system. The requirements on the power supply are determined in large part by the process conditions to be used for the deposition process. Power supplies capable of direct current, pulsed current (square wave), pulse reverse current (square wave), or arbitrary waveforms, are available from several manufacturers. The cheapest power supplies are only capable of direct current. This may be acceptable for some processes, and may even be used under manual operator control in the case of some wet bench-type systems. For automated electrodeposition equipment, however, several other features are typically required.

In automatic equipment used for electrolytic processing, control of the energy applied during the deposition process is usually microprocessor controlled. The use of a microprocessor to control the power supply provides many features that are required in automated equipment. First of all, the microprocessor accepts commands from the system controller and interprets them to provide commands that the power supply "understands." The microcontroller on the power supply will also usually send status signals or commands back to the system controller so the system knows the status of the wafer being processed. This allows the applied current, voltage, etc. to be collected for each wafer and integrated into the data log of the system. Some sort of coulombmeter, or amp-minute counter, is usually incorporated into the power supply controller as well. This allows a pre-determined set point to be used for the deposition. Because of the relationship of the total charge passed to the number of ions deposited as metal, as expressed in Faraday's law, this is a very effective method of controlling the exact amount of material deposited on each wafer. The coulombmeter may also be used to accumulate a charge counter, or "totalizer," that may be used to determine when some maintenance event should occur.

It may be necessary to have more than a single control setting during a deposition step. This is the case during the deposition of alloys such as solder alloys, where current density determines the alloy composition and where the deposit surface area changes throughout the course of the deposition. This is the case when depositing solder bumps in a "mushroom" configuration, where the metal is deposited to a thickness significantly higher than the resist template used to provide the pattern for the deposit. In such a case multiple current (or voltage) settings may be desired or a programmable current ramp that compensates for the change in plated area through the course of the deposition may be used.

In some cases, a waveform more complex than direct current may be desired. This may be the case for promoting leveling of copper for Damascene copper interconnect formation, for providing smooth gold deposits for electroplated gold interconnects or bumps, or for promoting a specific deposit morphology or composition of a solder alloy. In such situations, switching power supplies capable of providing some range of square-wave waveforms are typically employed. In these systems, there are several parameters that must be set in order to determine a specific wave shape. These parameters typically include on and off times in the forward (cathodic) and reverse (anodic) directions, total time for the forward and reverse wave trains, and power levels (voltage or current) for each of the "on" steps. In addition, the length of time, or total charge passed, during one of these steps must also be specified. Most power supplies today have the option of controlling either the applied voltage (between the anode and cathode terminals) or the current that is driven through the circuit. Finally, the power supply controller may be set up so as to allow a sequence of these steps to be performed under recipe or "macro" control. Current control effectively controls the deposition rate of the metal in the system (given constant chemistry conditions), while the voltage control method would control the reactions occurring and the overpotential used to drive them. Voltage control is susceptible to errors caused by changes in the electrical resistance of the circuit, such as those caused by corrosion, for instance, that will change the voltage applied between the anode and cathode.

Reference electrodes may be used to attempt to utilize voltage control without succumbing to the problem outlined above. Since it is the electrochemical voltage that is important (with reference to the standard hydrogen electrode) rather than the cell voltage *per se*, a reference electrode may theoretically be added to the system to provide a three-electrode cell. The voltage between the reference electrode and the cathode (as the electrode of interest) is then controlled in an attempt to control the deposition voltage itself. This method has some practical limitations in an automated wafer processing system, however. First of all, it is still subject to the problem of changes in the resistance of the wiring of the circuit used to sense the voltage between the cathode and the reference electrode. Second, the choice of a reference electrode is critical in order to provide a reference voltage that is stable over a long period of time, while not subjecting the plating chemistry to contamination concerns from the electrolyte in the reference electrode.

As mentioned above, the microprocessor that controls the power supply also communicates with the system controller on an automated wafer processing system. This communication can range from something as simple as relays to indicate when the supply should be turned on or off, or when it has an alarm, to a more complex computer network communication system utilizing a standard or custom set of commands. Systems are available that use serial communication via RS232 connections or Ethernet communication using a microprocessor and a command set such as the GEM/SECS communication protocol.

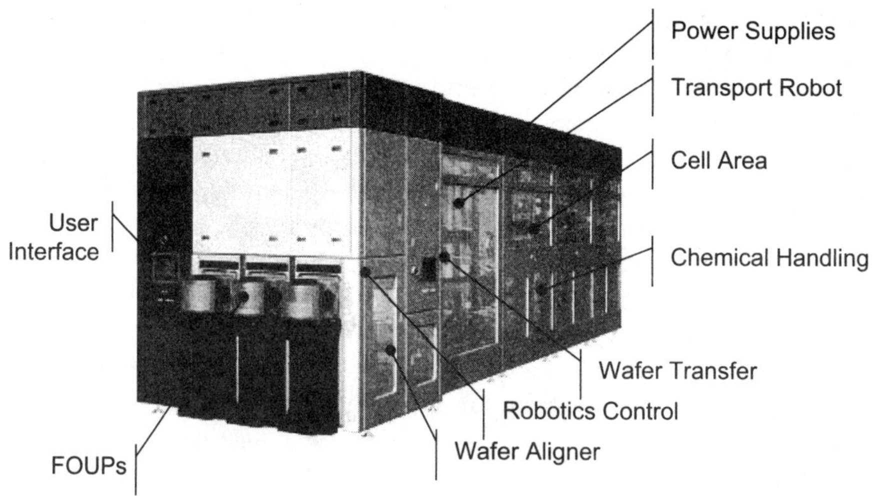

Figure 16.9   A modern wet bench-type electrochemical deposition system used for depositing solder bumps on semiconductor wafers. (Courtesy of Ebara Corp.)

16.2.2   Tank (wet bench) plating equipment

Many of the first systems that were used to plate wafers or other microelectronic substrates were based on a concept similar to a printed circuit board plating line. These systems were based on the automated (or manually operated) wet benches that have been a common fixture of the semiconductor fab for many years. These systems usually required that an operator load one or more wafers into a plating fixture, which was then moved along the series of tanks in a manner similar to a standard plating line. As the rack moved into the plating tank, power would be turned on and the rack would be moved through a series of rinse tanks after the deposition was completed. A modern version of this type of system is depicted in Figure 16.9.

16.2.3   Paddle plating cells

Another type of electroplating equipment that has been used for many years in the thin film recording head industry is the paddle cell (Figure 16.10).[15] In this type of plating "cell" the wafer, or substrate, is loaded (usually manually) into a fixture that supplies current to the substrate and holds the planar surface to be plated in precise relationship to the anode. The fixture is loaded into the plating cell and a paddle is actuated so as to move in a reciprocating fashion back and forth very near the surface of the substrate. The paddle is typically made up of two triangular bars with a slight gap between them. The reciprocating action of the paddle acts to disturb the diffusion boundary layer close to the cathode and aid mixing during the deposition. This type of plating cell has been used very extensively in the deposition of oriented magnetic alloys in the magnetic recording head industry. It is especially useful for magnetically oriented films due to the fact that the fluid agitation is provided without using wafer rotation. The systems are typically homemade, or designed and contracted out for

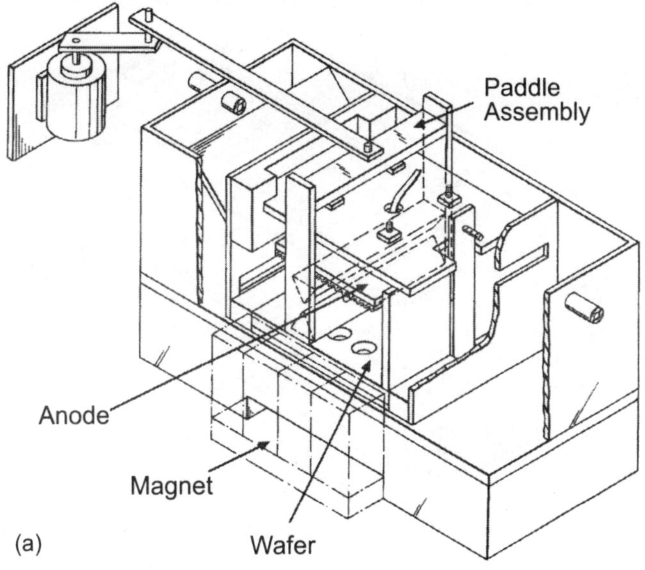

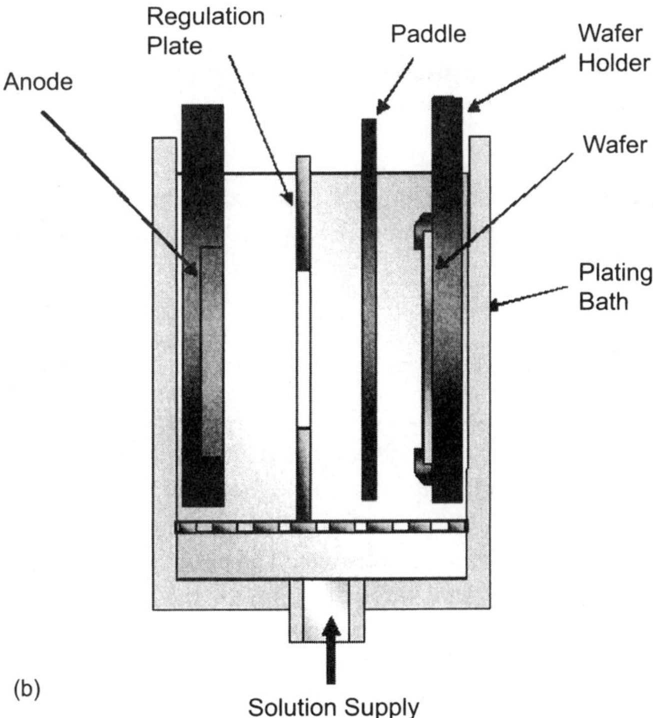

Figure 16.10 (a) Drawing of a horizontally oriented paddle plating cell. (U.S. Patent 4,102,756). (b) Modern paddle cell in a vertical orientation. The anode may be consumable or inert (non-consumable). The paddle reciprocates back and forth in front of the wafer to provide agitation of the solution. The solution flow is vertically oriented past the wafer. (Courtesy of Ebara Corp.)

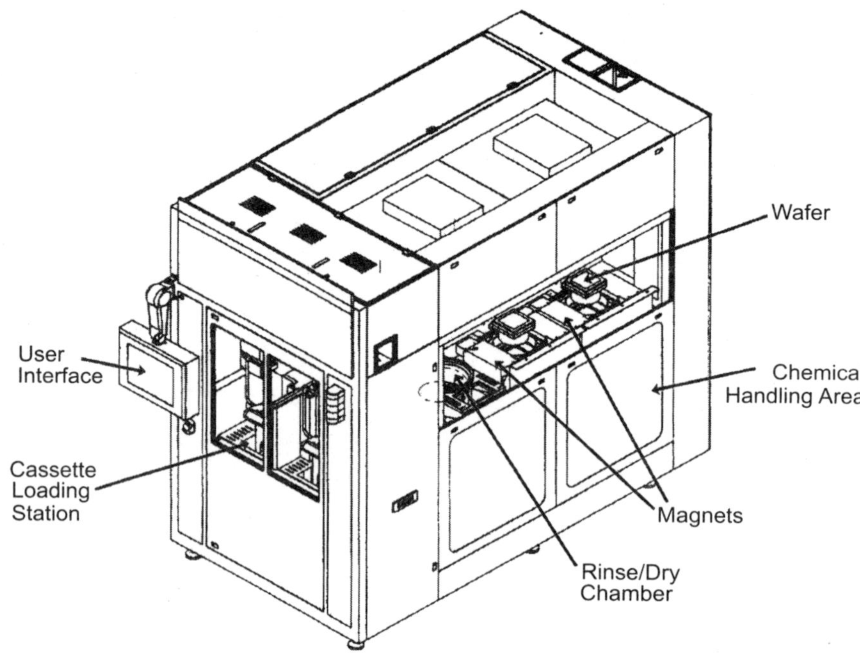

Figure 16.11 Drawing of an automated wafer processing system for plating magnetic alloys. (Courtesy of Semitool, Inc.)

fabrication. Recently, however, these ECD chambers have been incorporated into automated wafer processing equipment (Figure 16.11), which allows more flexibility in their application.

16.2.4   *Fountain plating systems*

The first fully automated semiconductor processing equipment that was used to electrolytically deposit copper for ultra-large-scale integration (ULSI) interconnect applications was a fountain plating system (Figure 16.12), also sometimes referred to as velocity plating.[16] In this type of processing chamber the wafer is held face down over an impinging flow of electrolyte. The anode is located below the wafer immersed in the electrolyte, and the wafer may or may not be rotated during the deposition. Anodes may be either inert or consumable anodes. In this configuration, the impinging flow of electrolyte is used to minimize the boundary layer and provide continuous replenishment of fresh electrolyte to the surface. Most commercially available systems make use of wafer rotation in order to improve the within wafer thickness uniformity. Support chambers are typically utilized to provide pre-cleaning steps and rinse and dry operations after electrodeposition. The electrolytic deposition is typically carried out by multiple chambers in parallel, fed from a single electrolyte source.

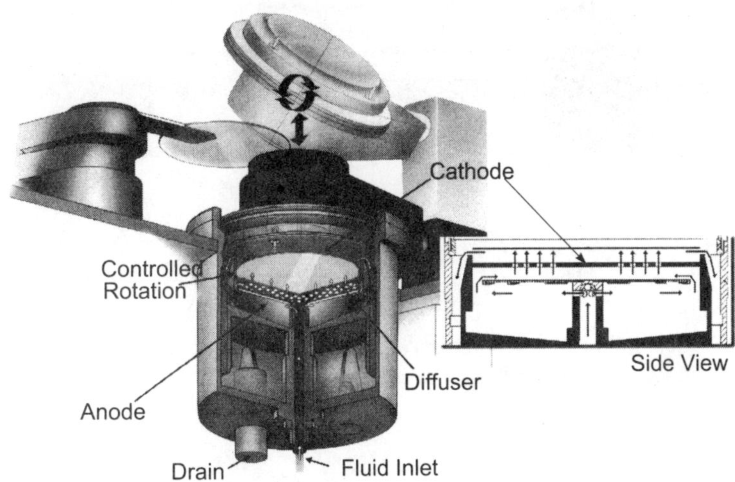

Figure 16.12   Fountain electroplating cell for automated wafer processing. (Courtesy of Semitool, Inc.)

16.3   Electrophoretic photoresist deposition cells

An electrophoretic photoresist is an aqueous emulsion of the organic materials that comprise a photoresist that may be applied to a substrate through the application of a voltage in an electrochemical cell.[17] The electrophoretic photoresist is deposited electrochemically to form a thick (several microns) conformal layer of resist over relatively large (up to hundreds of microns) three-dimensional structures. This process requires an inert counter-electrode and operates at higher potentials than traditional plating processes. Special considerations in reactor design must include the management of evolved gases and prevention of photoresist buildup on exposed hardware. Manual or semi-automated tank configurations have traditionally been used to deposit the resist in the printed wiring board industry, but they have not typically been well characterized for this process. Recently, an automated wafer plating system was developed to address the strict process control requirements of the microelectronics industry (Figure 16.13).

16.4   Electrochemical etching chambers

Electrochemical etching has been used to remove metals in certain wafer fabrication processes. The most common use of this type of process in semiconductor manufacturing is in the removal of gold seed layers in GaAs wafer processing. This is done after electroplated gold deposition for interconnects or bumps and photoresist removal in GaAs semiconductor processing. Any of the ECD chambers mentioned above may be used with the wafer as an anode for the electrochemical etching of various materials, or for electropolishing or anodization processes. There is also

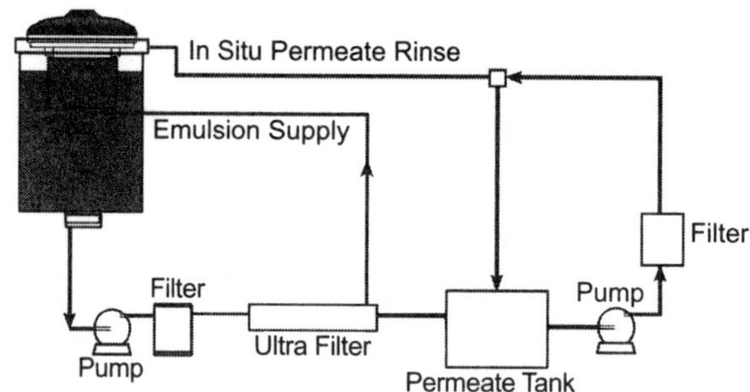

Figure 16.13 Schematic block diagram of a system for the electrophoretic deposition of photoresist. (Courtesy of Semitool, Inc.)

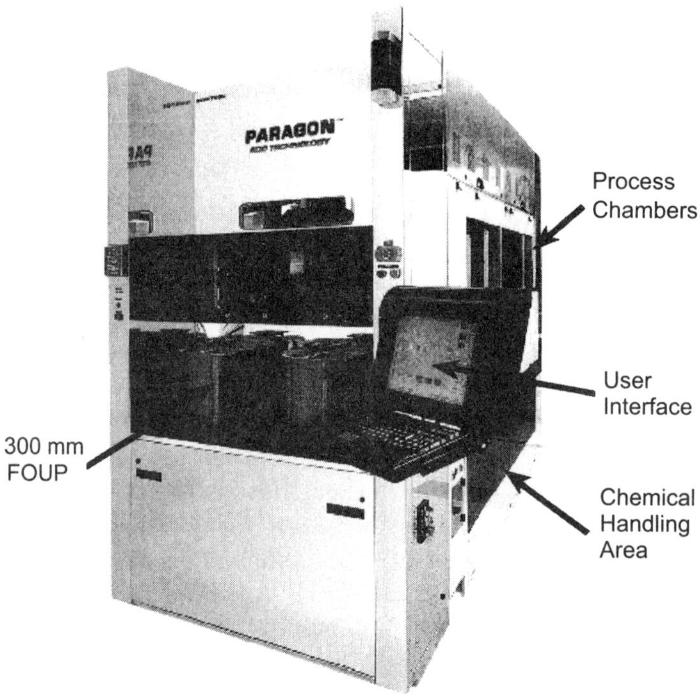

Figure 16.14 A fully automated 300 mm electrochemical deposition tool, which can be used for electroplating, electroless deposition, electrophoretic photoresist deposition, or electrochemical etching. (Courtesy of Semitool, Inc.)

literature describing process cells developed for the electrochemical etching of UBM in solder bump processes.[18–20] The chambers described in these references utilize a bar cathode through which jets of electrolyte impinge on the anodic wafer surface to accomplish dissolution of the metal from the surface. The cathode bar translates across the surface of the wafer to effect complete removal of the UBM material. These processes have also been incorporated into automated wafer processing equipment for high volume manufacturing, as in Figure 16.14.

16.5 Electroless deposition chambers

In the case of electroless deposition, the design requirements for a process chamber are somewhat different than they are for electrolytic deposition. In this form of electrochemical deposition, the electrons required to produce the reductive deposition of metal atoms come from a complementary reaction (or reactions) occurring in the solution. The solution contains the reducible metal ion, or complex, and an oxidizable chemical that functions as a reducing agent. There is no need for an external power supply and these processes are run in wet bench tanks or in immersion cells similar to those used for electrolytic deposition. Because the chemical mixtures used for these processes are inherently unstable, careful control of the solution concentrations, temperature, and flow characteristics are required in order to avoid "plate-out" of the metals onto the equipment. Sometimes the equipment is configured with tanks of etchants such as nitric acid ready in case plate-out does occur.

Since there is no current or voltage control available in electroless plating equipment to control the reactions occurring or the rate of reaction, the process parameters must be carefully controlled to provide stable deposition rates in practical ranges. These processes are usually somewhat slow, and so it is necessary to raise the solution temperature in order to increase the deposition rate. The rate is also dependent on other process parameters such as surface activation, flow rate, reducing agent concentration, etc. so these must be tightly monitored and controlled as well. As the temperature increases, so does the chance of plate-out, and so it is important to provide good thermal mixing of the solution and not to have any high temperature points in the fluid. This means that the heat supply, whether a resistive heater or a heat exchange coil, must be operated so that its temperature is not much higher than the solution temperature at the operating conditions. Because of the high temperature of these systems, it is important to minimize surface area for heat loss and provide insulation to minimize the thermal load. These processes are usually considered to be cheap due to the lack of a power supply, but when the energy requirements for heating and the ancillary equipment to control the chemical constituents and clean deposits from the system are considered, the cost advantages are not so clear.

Pumps and filters are also important considerations in equipment that is used for electroless plating. Pumps must be able to supply adequate flow rate to ensure good chemical and thermal mixing, while not increasing the temperature of the solution locally in the pump. This means that the bearings in centrifugal pumps must be made of compatible materials and be designed to operate without adding thermal energy to the solution. The pump also must not excessively agitate the solution or cavitate to minimize the likelihood of plate-out. The materials used in filters, the pore size, and the replacement schedule are also important in avoiding plate-out. If particulates build up in a filter that is used in a solution that provides autocatalytic deposition, the filter can quickly initiate deposition of metal onto the filter cartridge itself.

16.6   Pre-wet chambers

A pre-wet step is sometimes necessary to ensure complete wetting of features prior to starting the deposition process. This is especially true when wafers are coated with photoresist to be used as a plating template, as is the case with solder bump deposition. The surface energy of the wafer is important in determining the wettability of the wafer. This is one reason that a plasma treatment step to improve the wettability of the photoresist is usually a good idea prior to electroplating with a photoresist template patterned plating process. This is a more important consideration when the feature size is several microns than it is for the submicron features common in copper Damascene processes.

The most common types of equipment used for pre-wetting the wafer are tanks in wet bench systems or spray chambers that make use of de-ionized water and possibly surfactant spray directed at the surface of the wafer. There are interactions between the surface state (hydrophobicity) of the wafer, the pre-wet step, and the surface tension of the deposition bath that will determine the aggressiveness of the pre-wet required. Each of these factors can be adjusted to improve the ability of the plating bath to wet the surface and therefore provide a robust, high yield process.[21]

16.7   Rinse and dry chambers

Finally, there is a requirement for rinsing and drying the wafers after the deposition process, prior to any subsequent process steps. Residual chemicals must be removed from the wafer to prevent corrosion or contamination of other equipment and the wafers must be returned to their carrier in a dry condition. This process usually takes place in some kind of a spin rinse dry chamber that has de-ionized water spray capability and nitrogen purging for drying the wafer.

16.8   Important design factors for wafer ECD equipment: System level considerations

Besides the considerations mentioned above that must be taken into account in the design of an ECD system, there are several other items to be considered at the system level. These items include such things as the choice of materials used for the system level components, both those in contact with the plating solution and those in proximity to the corrosive chemical environment. The equipment used in manufacturing semiconductors is usually automated, which provides several additional challenges for ECD equipment including mechanical automation, control systems, recovering from errors, and managing data.

16.8.1   Chemical compatibility

Typical plating chemistries are very corrosive, and so there is the danger of building equipment that is unreliable over the long term if the equipment manufacturer is not familiar with handling these types of chemicals. It is important to choose materials that will not be degraded by the liquids or the vapors inside the equipment, even over a long period of time. It is also important to locate the various components of the system in a way that minimizes the chance of corrosive liquids coming into contact with mechanical or electrical components of the equipment. In addition, materials in the fluid path that are not very pure will tend to leach impurities into the plating chemistry over time, which will change the performance of the process. The cathode contact is an area of particular concern for chemical compatibility, because it is also exposed to mechanical and electrical stresses, and it must be made of a material that does not promote etching of the seed layer due to galvanic interactions.

Management of the fluid capture area and exhaust stream from the chemical handling area are other important issues that must not be overlooked. Some plating chemistries make use of hazardous chemicals, including cyanides, heavy metals, and carcinogens. It is important that cyanides and acids never be allowed to mix to prevent the formation of poisonous hydrogen cyanide gas. It also may be important to prevent certain acidic and basic chemicals or exhausts from mixing in order to prevent unwanted reactions or salt formation in the exhaust ducts.

16.8.2 Automation

Semiconductor wafer processing equipment is normally automated to provide wafer handling between process stations. Besides the issues mentioned above concerning materials and components in the proximity of the plating chemicals, there are several handling requirements that are unique to the semiconductor industry. Of course, cleanliness of the wafer surfaces is of critical importance to ensure the yield of chips on the wafer. The system must be able to handle wafers cleanly and safely in transporting from the cassette, SMIF pod, or FOUP to the appropriate process chambers and back into the carrier.

Automation is further complicated by the fact that the wafers are being processed in a wet chemical environment. Handling wafers as metal films are being deposited in a wet environment is not a trivial task. Often it is important to seal the edge of the wafer from the solution in order to prevent buildup of metal on the cathode contacts, which bonds the wafer to the contact and complicates removal of the wafer. It is also not trivial to provide seals in an automated system on wafer after wafer without any human intervention. When combined with the normal requirement of semiconductor processing equipment to require very little maintenance, the handling and automation issues become very important, indeed.

16.8.3 Control system

The control system for all the components of a piece of semiconductor processing equipment can be very complex. Because it is important to the semiconductor manufacturer to have high equipment availability, the uptime of the equipment is important and the equipment should be designed in a way that makes it easy and fast to diagnose problems or to perform preventive maintenance. These requirements push the manufacturers of semiconductor processing equipment toward providing modular components and software with expert systems to aid in tracing and solving equipment problems. This also means that all of the control components of the system should be connected to a central controller for monitoring and diagnostics. The controller must also be capable of supporting the myriad of specifications required by the semiconductor industry, including communications, ergonomics, and safety specifications. The control systems in most semiconductor manufacturing equipment today make use of multiple interconnected computers to provide hardware control and user interfaces.

16.8.4   *Error recovery*

It is important to remember that the value of the product that is being produced in semiconductor manufacturing is often more than the value of the equipment being used, even in the case of multi-million dollar processing tools. This puts the equipment design in a perspective that is fairly unique to the semiconductor industry. When an error occurs in the equipment, it is important to be able to recover and save information in a way that prevents misprocessing of the wafers that are in process. It is also necessary for the equipment to be able to communicate with the operator or factory automation system which wafers were processed in what way.

16.8.5   *Data logging*

Semiconductor processing equipment is complex and produces a large amount of data. These data must be managed effectively in order to be useful. A typical piece of semiconductor processing equipment can produce enough data in a day to fill up a large hard disk drive. Some of these data are critical to the performance of the semiconductors being produced and should be saved for tracking and quality control purposes. Other data should be utilized by the equipment to ensure that no errors occurred during processing; then they may be deleted. In any case, any time an error or an unexpected event occurs, that information must be captured in a way that enables the factory to determine what happened to the product and to quickly fix any equipment problems.

16.8.6   *Transition to 300 mm wafers*

The semiconductor industry is constantly moving to larger and larger wafers in the effort to keep up with Moore's law. As the industry moves from 200 to 300 mm diameter wafers, several issues arise that are important for automated electrochemical deposition equipment. The size of 300 mm wafers necessitates that they are handled in sealed boxes, or FOUPs, to maintain cleanliness and provide a standard interface for loading multiple types of equipment. These FOUPs are heavy and large enough that it is difficult for humans to move them around the fab. This means that automated handling systems are utilized and that the equipment interface must be precise enough to interface with the automated handling system. The placement of the equipment inside the fab must also be done with precision to allow integration with the fab transport system.

The transition to 300 mm wafers also increases the magnitude of the terminal effect in electrodeposition processes, as mentioned in Section 16.2.1.3. Some manufacturers have advocated the reduction of the terminal effect through decreasing the conductivity of the electroplating bath used for the process. This approach compromises the properties of the deposit in order to compensate for a reactor issue, however. As more people discover this, they tend to move toward systems that have active cathode current density control which can effectively compensate for the terminal effect. The additional advantage provided by this type of reactor is that it works well with advanced process control (APC), because it can react to information provided by metrology systems. This allows the yield to be increased by eliminating process drift. These systems can also be matched to the input requirements of downstream processes, such as CMP, in order to provide more robust process integration.

16.9   Bath analysis and replenishment systems

As the use of electroplating in semiconductor manufacturing has increased, so has the need for automatic bath control systems. While electroplating has been used for many years in various applications, other applications have not had the stringent requirements for tight bath control that semiconductor electroplating has. The two main drivers for this are the increased value of the product being electroplated and the challenging process requirements for semiconductor plating. The significant economic impact of "scrapping" a wafer because it was processed on an "out-of-control" plating bath points to the obvious need for tightly controlled plating chemistries. Semiconductor plating also tends to have narrower process control requirements than other plating processes. In copper Damascene plating this is due to the need to fill sub-micron features on the wafer surface. In solder bump plating, tight process requirements are needed to maintain consistent and uniform alloy compositions and thickness across the whole wafer surface.

Figures 16.15 and 16.16 show two examples of the importance of controlling tin–lead solder plating baths. Figure 16.15 demonstrates how varying the relative concentration of tin to lead in the plating bath directly affects the composition in the deposited solder bump. The chart shows that as the tin concentration decreases in the plating bath, the tin weight percent in the deposit also decreases. The data were collected over a three-week period on a production eutectic solder bath used for depositing bumps on semiconductor wafers. While Figure 16.15 shows the rather intuitive correlation between the relative concentration of metals in the plating bath and their concentrations in the deposited bumps, Figure 16.16 shows the importance of controlling other bath components. Here the effect of varying one of the organic additives in the eutectic solder bath is explored. In the low current density case, the organic additive is varied almost 100% (range of 5 to 9 units) but that change only impacts the bump composition by a 3% relative change. However, for the same variation of the organic additive at higher current density, a 10% relative change in the bump composition is experienced. Thus, the importance of tightly controlling

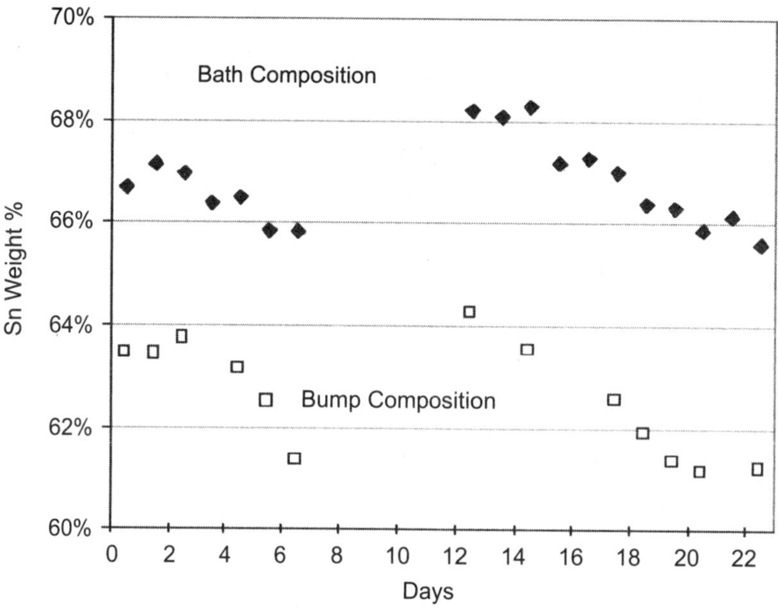

Figure 16.15 Tin composition of eutectic tin–lead solder bumps varies in relation to the bath composition. As tin concentration decreases in the bath, the tin content of the deposited bump decreases.

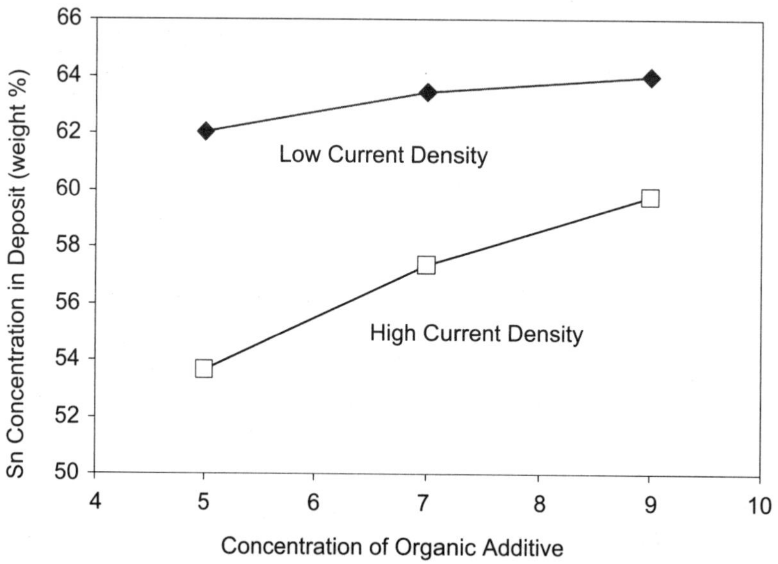

Figure 16.16 Eutectic tin–lead solder bump composition relative to organic additive concentration in the plating bath. The effect of the organic additive concentration in the plating bath on the deposited tin–lead alloy composition is dependent on the deposition current density.

certain components in the bath can vary depending on required process conditions. In this case, if a semiconductor manufacturer wants to run a solder plating process in the higher current density region, the components in the bath must be controlled more closely.

The plating baths used in semiconductor plating typically include both inorganic and organic components. The most important inorganic components are the metal ions that will be deposited during the plating process. In the case of acid baths (typical copper Damascene plating baths) there is also a large amount of inorganic acid. Often there are trace amounts of other inorganic components found in the chemistry that can affect the results of the plating process. The concentration of all these inorganic components can be critical for maintaining the plating process, although they are often relatively stable when compared to organic components. Soluble anodes are normally employed in copper and solder electroplating to help keep the concentration of metal ions constant in the bath. During processing, metal ions in solution are reduced to solid metal at the cathode (wafer surface), while ions enter the solution at the anode due to oxidation of the anode. Since electroplating of gold and other noble metals must utilize an insoluble anode, the gold ions in solution must be maintained through replenishments to the bath. These replenishments can be calculated by determining the mass of gold deposited and replenishing that amount to the plating bath. Unfortunately this does not account for any other upsets to the process that could affect the concentration of the metal in the solution, making it important to analyze the bath on a regular basis to verify its condition.

While the inorganic components are important in the plating process, the organic components tend to play a stronger role in affecting the properties of the deposited metal. However, the organic components also tend to be relatively unstable in the bath environment. Figure 16.17 provides a good example of how organic components can degrade in the plating bath. The chart shows the relative concentration of an organic component in a solder plating bath used in semiconductor production. As seen by the analysis results decreasing over time, the component is steadily consumed. The occasional spikes in the concentration of this component correspond to the relatively infrequent additions of this additive. While this bath is not completely out of control, the control interval, or time between replenishments, is too long to sufficiently control this component.

Along with the trend to automate the wafer processing in semiconductor manufacturing, there has been a strong push to automate the plating bath control. Historically, bath control has consisted of taking samples to be analyzed off-line and manually making replenishments based on those analysis results. The complexity and cost of automation have been the main deterrents to implementing automated systems on plating lines running less expensive products. To improve the economic returns, automatic bath control systems are normally designed to handle several different plating baths at the same time. A full automatic bath control system consists of an automatic analysis system, an automatic replenishment system, and a sophisticated controller. These components can be distributed among several locations or (as in the case of the commercially available equipment shown in Figure 16.18) can all reside in the same cabinet.

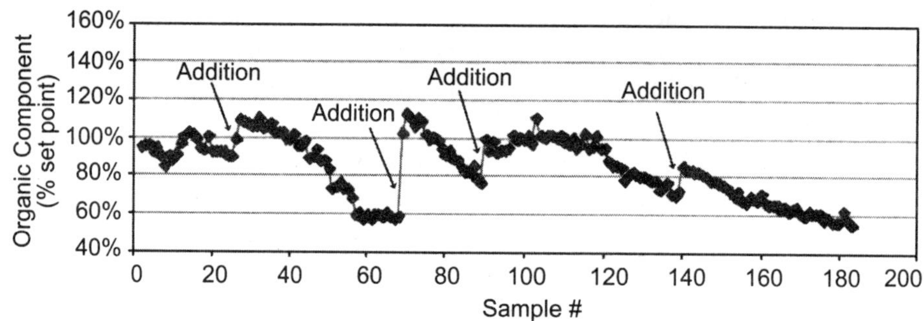

Figure 16.17 Production eutectic solder plating bath data showing the decomposition rate of an organic additive in the bath. Over time this component is consumed, causing its concentration to decrease. Occasional replenishment to the bath raises the concentration of the additive.

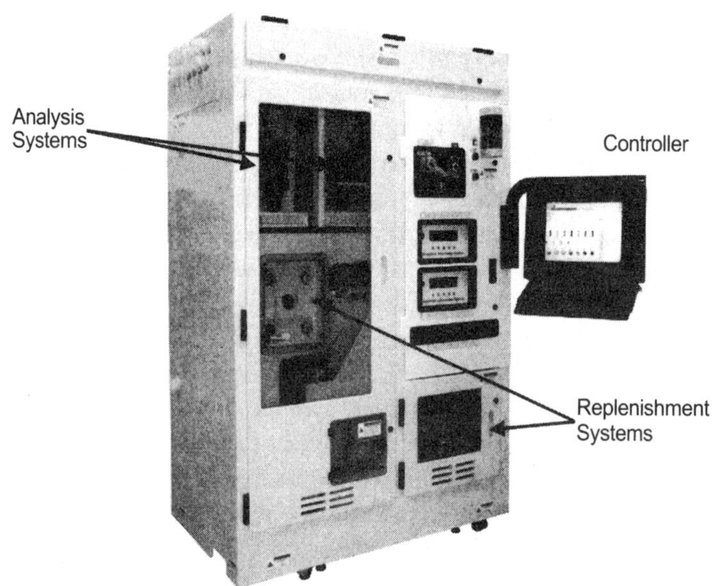

Figure 16.18 Photograph of an automatic chemical management system. (Courtesy of Semitool, Inc.)

The most basic bath control systems have the capability of providing automatic replenishment of one or more of the bath constituents. These systems must be designed with consideration for the bath volume and associated rate of depletion (generation) of the appropriate reactants (by-products), as well as the heightened safety concerns of a semiconductor fab. The frequency of additions from an automatic replenishment unit will be determined by the acceptable range of variation of a component, as well as timing issues that arise as the replenishment events come closer together in time.

Automated bath analysis tends to be somewhat more complicated than the replenishment process. While the number of chemicals to be analyzed usually corresponds to the number of chemicals to be replenished, there is not a single universally applicable method for analysis of all these components. This is further complicated by the fact that most suppliers of organic additives (grain refiners, brighteners, levelers, wetting agents, etc.) will not disclose the chemical composition of their additives, leaving the end users to develop analysis methods based on the empirical responses seen on their analysis equipment.

16.9.1 *Analysis techniques*

The obvious approach to creating an automatic analysis system would be to simply automate the techniques that are currently used for off-line bath analysis. Unfortunately, the needs of an automatic system preclude the use of several laboratory techniques: particularly the need for an automatic system to require very little maintenance. Small size, zero, or very little sample consumption, and low cost are the other requirements which automatic analysis units must meet. Currently most analysis systems have just the opposite characteristics of these requirements. A few of the analysis methods that have been used or considered for automated plating bath analysis include titration (potentiometric, colorimetric, or pH), high performance liquid chromatography (HPLC), ion chromatography, X-ray fluorescence, electro-analytical techniques, spectrophotometry, pH monitoring, conductivity monitoring, oxidation–reduction potential (ORP) monitoring, mass spectrometry, and atomic absorption. While each has its own advantages and disadvantages, only the techniques most commonly used in semiconductor electroplating will be discussed here.

Titration is one of the oldest chemical analysis techniques and in many cases it is the only viable method that meets the requirements for automatic analysis. Simply said, titration consists of reacting the component of interest (reactant) with an analysis reagent (titrant). The consumption of the reactant can be monitored by measuring some other indicator in the system (potential, color, pH, ORP, etc.). Figure 16.19 depicts a titration reaction. The squares represent the amount of the sample component remaining in the analysis cell, the triangles represent the measured response (potential in this case), and the X-axis represents the amount of analysis reagent used. Note how the additions of titrant become smaller as the titration endpoint is approached: a classic technique employed to improve the accuracy of the analysis. The advantages of titration are that it can be automated relatively easily and inexpensively, it is usually very accurate and precise, and it may have relatively short analysis times (<10 minutes). Disadvantages include the need to consume a sample, the need for reagents in the analysis process, and in some cases the relatively long analysis times. Often the benefits of titration outweigh the disadvantages, leading to its use in many automatic analysis units. Figure 16.20 is a photograph of a titration cell used in the automatic chemical analysis system shown in Figure 16.18.

Direct, in-line analysis techniques are preferred for their ability to quickly obtain analysis results without consuming any sample. ORP, conductivity, and pH probes can all be inserted directly into many plating baths to give instantaneous

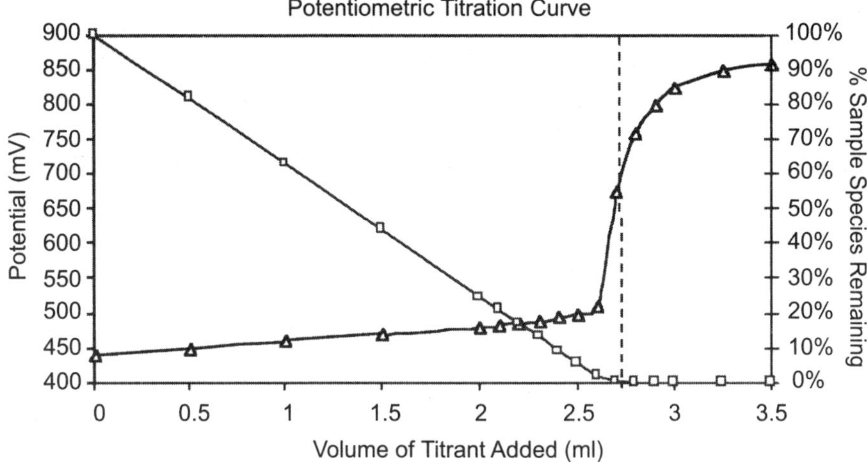

Figure 16.19   An example of a titration analysis curve. As the volume of titrant (reagent) is increased in the analysis cell, the amount of sample species remaining decreases (squares). The potential measured in the cell (triangles) increases sharply at the point where all of the sample species is consumed in the reaction with the titrant. This "equivalence" point is used to calculate amount of the original sample in the analysis cell.

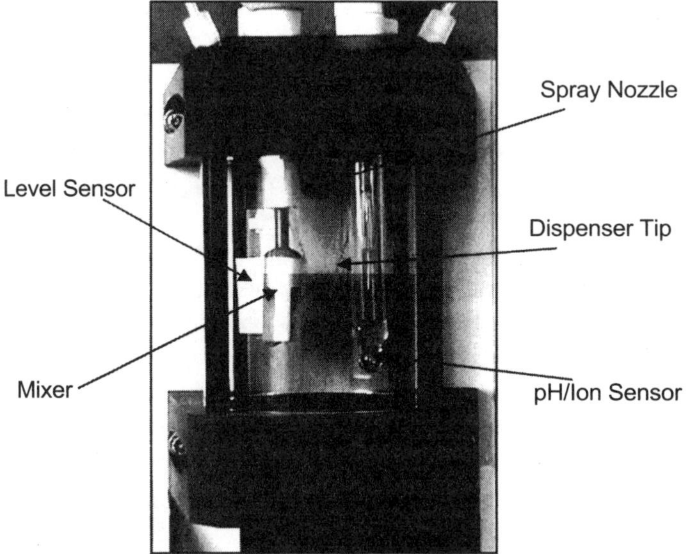

Figure 16.20   Photograph of an auto-titration analysis cell.

analysis results. While pH probes do require regular calibration, the benefits of receiving sample-less, instantaneous bath analysis tend to outweigh the extra maintenance involved. In addition, they can be connected relatively easily to replenishment systems to maintain very tight pH control. ORP and conductivity are

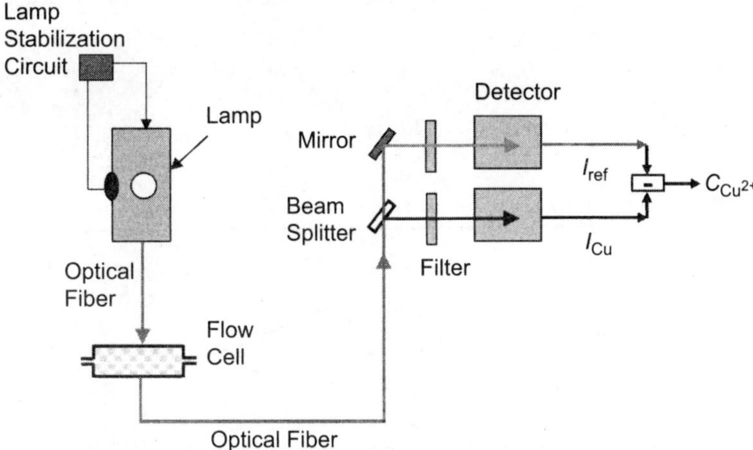

Figure 16.21   Schematic of a photometry hardware configuration for copper ion concentration analysis.

not typically used to control any one component in plating baths, but are used as measurements of the general state of the bath. If the analysis from one of these methods exceeds the normal limits, it is a good indication that one or more components may be out of balance and operator intervention is necessary.

Spectrophotometry is another common in-line analysis technique. This method shines light through a small sample of the chemistry and measures the amount of light received by a detector on the other side (Figure 16.21). The concentration of certain species can be measured by the amount of light they absorb at characteristic wavelengths. Once the key wavelengths are determined, the analysis can be streamlined into measuring the amount of absorption at those specific wavelengths by photometry. Since there are no reagents or sample preparation required, this method can be integrated with the equipment for real-time analysis. Photometry systems are usually very accurate, repeatable, and can be relatively inexpensive. The only disadvantage is that spectrophotometry can be limited by its ability to measure only a few components in any one plating bath.

Electroanalytical techniques — cyclic voltammetric stripping (CVS), pulsed cyclic galvanostatic analysis (PCGA), and others — are attractive methods for measuring the electrochemically active components in plating baths. Essentially, they are based on the same principles being applied in the deposition equipment to deposit metal on the wafers, and thus can closely reflect the actual plating conditions occurring in the plating process. CVS and PCGA are based on changes in electrochemical activity of the bath sample based on the concentrations of the components. These techniques are typically used to measure the concentrations of the organic additives in the bath; however, they can also be used to measure some of the inorganic components.

In CVS analysis, the sample is analyzed by scanning the potential in the cell relative to a reference electrode cathodically and anodically, causing a metal film to be deposited and stripped off the working electrode. These cycles are represented by

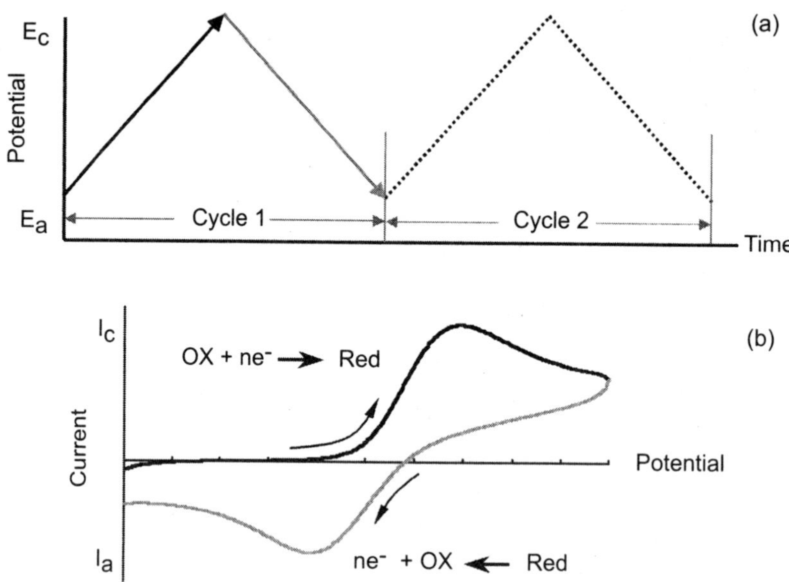

Figure 16.22 The CVS analysis cycle: (a) potential is scanned cathodically and anodically with respect to time, and (b) the induced current is recorded.

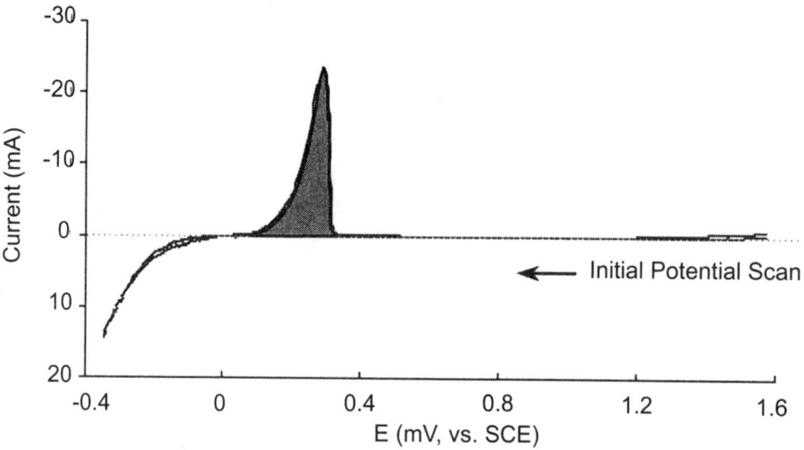

Figure 16.23 The CVS analysis cycle is shown as current versus potential (relative to a standard calomel electrode). The area represented by the shaded region is calculated and used as the analysis metric.

Figure 16.22. Figure 16.22a shows how the potential is varied with respect to time, and Figure 16.22b shows the corresponding current induced in the cell by the varying potential. CVS scans are often shown in the form represented in Figure 16.23. The area represented by the shaded portion corresponds to the amount of charge needed to strip all of the deposited metal from the working electrode. In essence, the basic measurement of a CVS system is a measure of the amount of metal deposited during

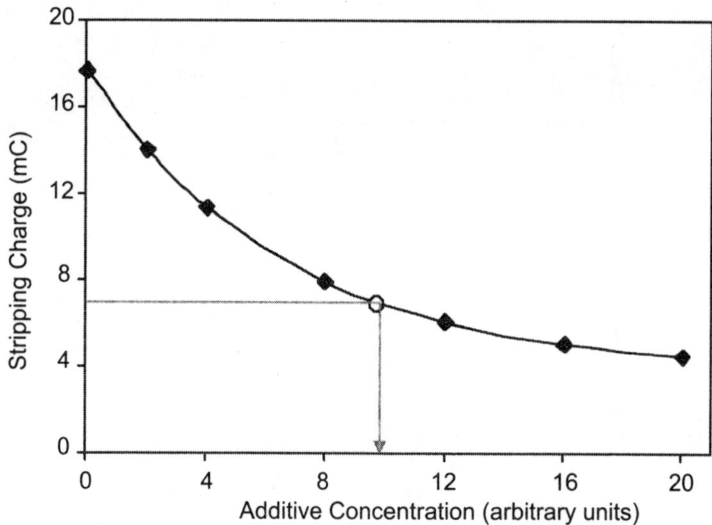

Figure 16.24 The analysis metric obtained in the CVS analysis cycle is used to generate calibration curves for the organic additives. The concentration of the sample can be determined by comparing the sample analysis with the calibration curve.

a cathodic potential scan (which is then removed during the stripping cycle). The amount of metal deposited is directly related to the electrochemical activity of the sample, which is influenced by the concentration of the organic additives. Using the fact that changes in the amount of organic additives in the sample affect the stripping area, a calibration curve can be generated and used to measure the concentration of the additives (Figure 16.24). However, it has been reported that CVS results can be affected by breakdown products of the organic additives.[22]

PCGA differs from CVS in that it applies a current and measures the resultant potential of the working electrode as a function of time. While CVS is commonly used for bath analysis, PCGA is covered by a patent describing chronopotentiometric analysis and thus is not widely available for use.[23] Figure 16.25a illustrates the PCGA cycle: first a nucleation current pulse occurs, and then metal is deposited on the working electrode while the current is held constant. After a short period of equilibration time, the current is reversed to strip the metal off the electrode, and the process is repeated. Figure 16.25b illustrates the corresponding potential in the cell as the analysis cycle is performed. Note that the deposition potential shifts in the cathodic direction (more overpotential) as the suppressor additive concentration increases in the sample. The opposite is true as the accelerator additive concentration increases in the sample. The basic measurement in the PCGA analysis is the cell potential at the end of the plating period, once the deposition has reached steady state.[23] In a method similar to CVS, these data can then be used to create a calibration curve for determining the actual concentration of the organic additives in the sample (Figure 16.26). It has been reported that due to the steady-state nature of the PCGA analysis, it provides a better representation of the conditions occurring at the wafer surface during electrodeposition than CVS techniques.[23,24]

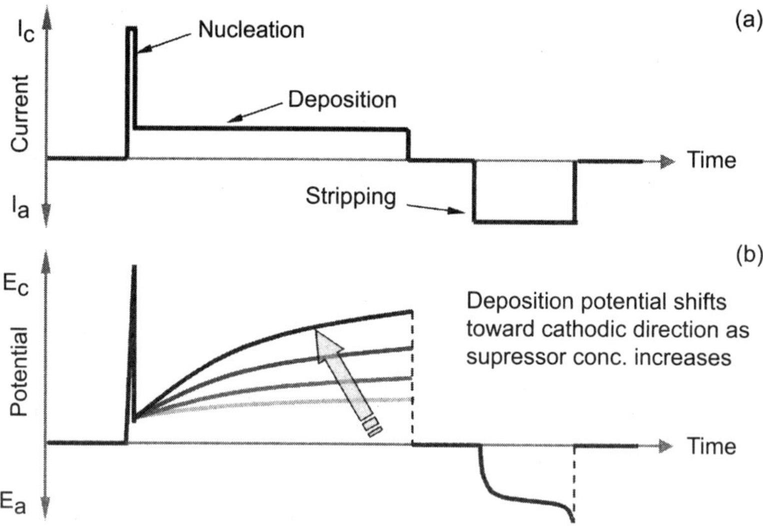

Figure 16.25 The PCGA sequence and response for suppressor organic additive. (a) The current with respect to time in the analysis cell. (b) Induced potential in the cell. Note how the deposition potential shifts as the concentration of organic additives increases in the analysis cell.

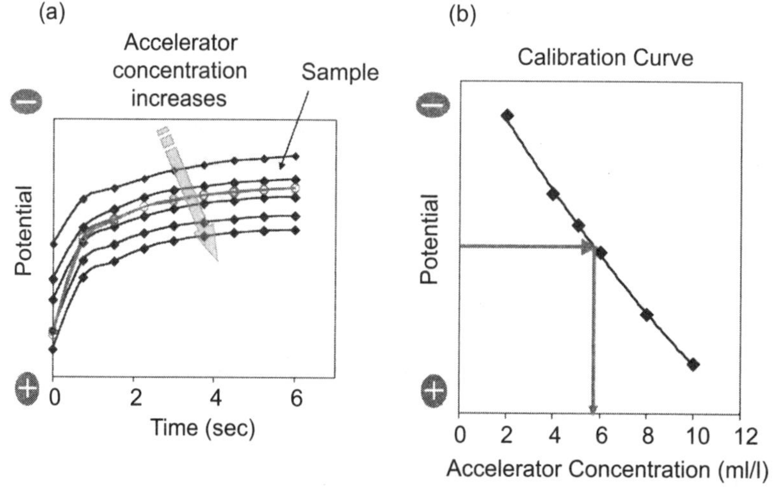

Figure 16.26 (a) Potential during the PCGA analysis deposition step decreases as the concentration of accelerator additive increases in the cell. (b) A calibration curve and subsequent analysis that can be generated from the data gathered in (a).

Chromatographic separation analysis techniques have been used for many years for quantitative analysis of chemical species. HPLC and IC are the two most common methods employed in the analysis of electroplating baths. The drawing in Figure 16.27 illustrates the essential components of a chromatographic system. The pump-driven eluent stream carries the sample into the packed resin bed separation column.

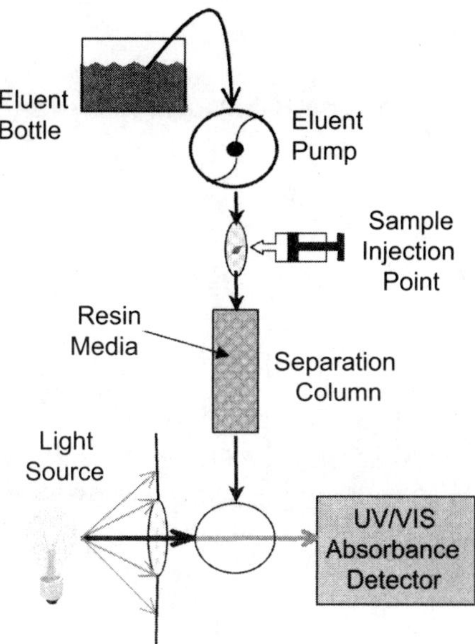

Figure 16.27   Schematic of an HPLC analysis system. (Adapted from Ref. 22.)

Differing affinities to the resin beads cause the components in the sample to remain in the column for differing amounts of time, resulting in their separation into discrete bands before entering the detector. These bands are then detected by either UV/VIS absorption or conductivity. The output from this separation and detection is a chromatogram which is recorded with respect to elapsed time. As with the other techniques already discussed, the concentration of the samples can be determined by comparing their chromatograms with those generated during calibration runs.

The main advantage of HPLC is that virtually all the chemical constituents in a plating bath can be detected, including the inorganic components, the organic additives, the organic additive breakdown products, and other contaminants. The method can be rather difficult to automate, as the sample may need to be diluted by factors up to 10,000:1 before being injected into the analysis unit. In addition, the time required for analysis can be relatively long, as it depends on the residence time of the individual components in the separation column. In light of these limitations, HPLC and IC have been mostly relegated to off-line laboratory analysis tools used to develop and understand the dynamics of plating baths.

16.9.2   *Replenishment techniques*

While analysis is very important in understanding the bath conditions, accurate and dependable replenishment methods are equally important for controlling the chemistry. The ideal case is to have a fully automatic analysis system connected with an automated replenishment system through a controller employing sophisticated control algorithms. However, a few replenishment methods can be employed without having an automatic analysis system.

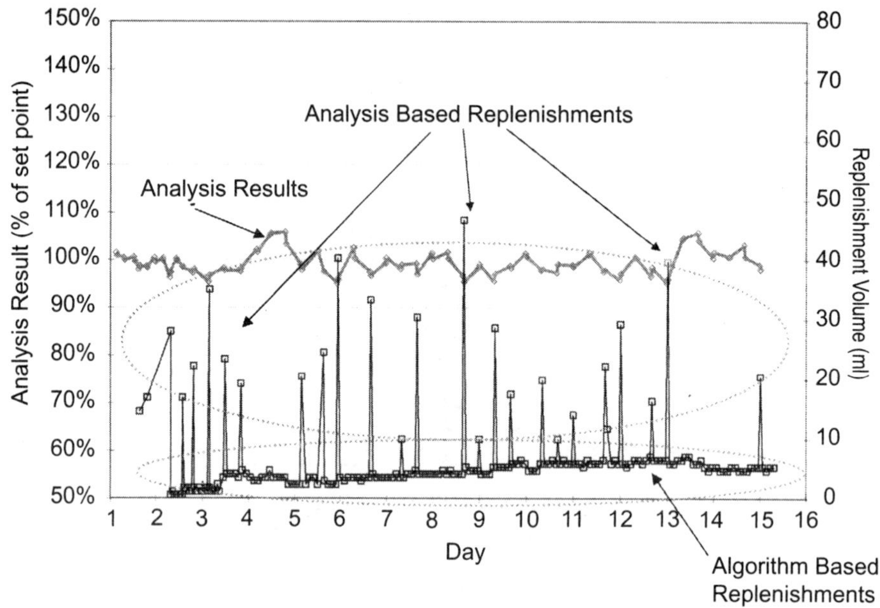

Figure 16.28 Example of automatic bath control: analysis results and replenishments. The concentration of the species is maintained by the algorithm-based replenishments and analysis-based replenishments.

The most basic control scheme is to run a plating bath for some set period of time or usage, and then dispose of it. By employing this method, one can completely avoid the need to analyze or replenish the bath. Unfortunately, the short period of time that the bath would be in the proper concentration range for electroplating for semiconductor applications would require an excessive amount of chemical usage. The cost and time involved in this scheme of control are, in almost all cases, prohibitive.

If one has a means of analyzing the bath components off-line, even by sending samples out to third-party analysis labs, it is possible to use two types of control techniques. First, the off-line analysis results may be compared to the concentration set points and replenishments calculated to bring the components of interest back to the desired levels. Figure 16.28 shows an example of how this would work. In this case, even though there were enough analysis points to tightly control the bath, replenishments were made only occasionally, indicated by the instantaneous concentration increases. After this procedure has occurred several times, the depletion rates of the components can be established. Using this depletion rate, replenishments can be made to the bath based solely on the amount of elapsed time and the amount of usage the bath has received (amount of product plated). This algorithm type of replenishment can be used to maintain the bath concentrations closer to the desired level between each analysis event. This level of control is sufficient for many plating applications, including printed wiring board electroplating. However, it should be noted that control problems could arise if the depletion rate of any component changes and the algorithm is not updated to reflect those changes.

In the ECD processes used for semiconductor products, the stringent requirements for bath control require full closed loop automatic bath control systems. These systems employ the same control methods described above, but on a shorter time scale. In the case where there is instantaneous analysis, components can easily be controlled by comparing the analysis result with the set point and replenishing accordingly. When using HPLC, CVS, or titration, the time between analysis can extend into hours, often making it necessary to use the algorithm control scheme for certain components: particularly the organic bath additives. The advanced control systems used in the automatic bath controller are usually designed to compensate for changes in the critical component depletion rates. This can be seen in Figure 16.28, where the automatic bath control system uses the analysis results to replenish the bath components to the target concentrations and adjust the algorithm parameters at the same time. The advantage of this method is that it allows the system to automatically tune the algorithm parameters based on the results of the analysis. In addition, if there is a large excursion below the set point, the system can do an analysis based (or feedback) replenishment to bring the bath back into line as quickly as possible. Note that the big spikes are the automatic feedback replenishments and the smaller, more consistent replenishments are algorithm-driven events. These systems can even reduce or stop the algorithm replenishments for a particular component when it is over the desired set point and then resume the algorithm replenishment when the concentration has decreased below the set point.

16.10   Waste treatment considerations

Waste streams from the plating process must also be considered when designing processing equipment. There are generally two forms of waste streams from plating equipment that need to be dealt with: rinse water and concentrated plating solutions (spent electrolyte). Regulations governing the discharge of heavy metals such as copper, lead, and tin can vary by country, state, or city; and so a process must be put in place to treat the waste material to the appropriate level before discharge out of the facility.

There are several treatment technologies readily available for metal bearing rinse streams. Ion exchange, electrowinning, and precipitation are the most commonly used methods. Another option is to capture the concentrated and rinse waste and haul it to off-site third-party treatment facilities. Except in the case of small volumes, this approach is typically more expensive than operating a treatment system on-site and the third parties are normally limited to the same technologies (and their limitations) as the on-site systems. Waste streams from other processes in the facility should be considered when designing a waste treatment system. In the copper Damascene process, a significant amount of the metal bearing waste stream comes from the CMP of the deposited metal layer, while another part of the facility may have metal bearing wastewater from the solder bumping process. These waste treatment systems are typically situated in a common area and are configured to handle the waste streams from multiple processes.

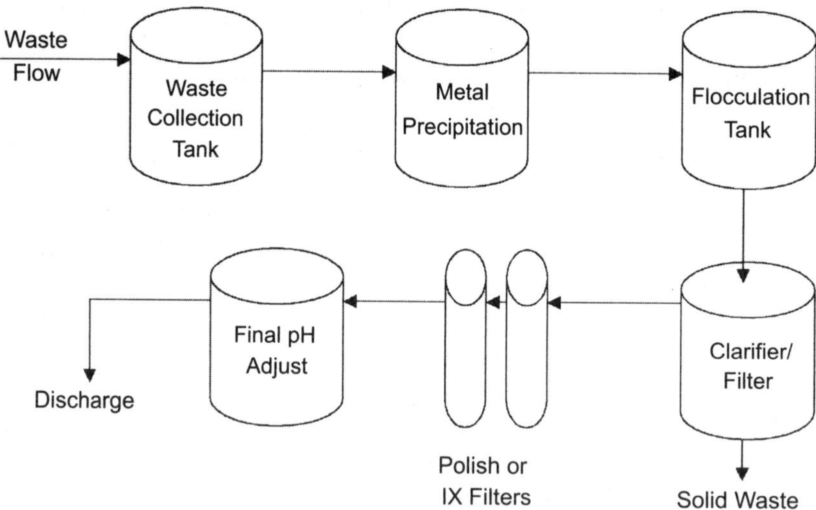

Figure 16.29   A schematic of a wastewater treatment system employing precipitation and ion-exchange technologies.

Treatment of metal ions by precipitation is typically performed by raising the pH of the solution into the basic region (8 to 11) which produces metal hydroxide solid particles. A flocculent is then added and mixed into the solution to agglomerate the fine particles before sending the stream through a clarifier, or filter, where the solids are separated from the water (Figure 16.29). The solids must be collected and disposed of as hazardous waste, while the cleaned water will need a final pH adjustment to bring it back into the neutral range before discharging to the sewer system. Ability to reach low metal ion levels after treatment (<<1 ppm), high flow rates, and relatively low capital costs are advantages of precipitation systems. Larger footprints, higher maintenance, and messy sludge handling are disadvantages associated with these systems. In addition, most of these disadvantages drive a precipitation system to be located in the facilities area of a building, thus requiring a long run of dedicated piping, potentially increasing the cost of treatment.

The electrowinning process uses the same principles that are applied in the electroplating process conducted in plating tools: applied potential or current causes the metal ions to deposit onto the cathode. This leads to one of the main advantages of these systems: the metal can easily be recovered and recycled, making it very attractive for treating precious metal plating baths. Additional advantages include normally lower initial capital costs than other treatment methods, lower maintenance requirements, and smaller footprints. The main disadvantage with this process is its inability to remove metal ions down to the low levels required by regulations. At low concentrations of metal ions in the wastewater, the applied potential in the electro-winning cell must be increased, which tends to lower the efficiency of the process and increase the chance of gas evolution at the electrodes. To overcome this limitation, electrowinning systems are typically paired with ion-exchange units, which can polish the electrowinning effluent down to low metal concentration levels.

Ion-exchange systems are configured to have the metal bearing waste stream flow through a "resin bed" where the metals (cations) attach themselves to functional groups on resin beads. As the metal ion attaches, it replaces a cation (typically hydrogen or sodium) forcing it into solution. When all of the functional groups are full of metal, the resin is "exhausted" and must go through a regeneration procedure to restore it to an active state. The exhaustion point can be checked by monitoring the metal concentration in the resin bed effluent or estimated by measuring the load on the system and extrapolating to the exhaustion point. Ion-exchange systems are known for their ability to remove ions down to low concentrations ($\ll 1$ ppm). In addition, they are relatively easy to maintain, can be configured in small footprints, and may be used to recycle the effluent water from other areas of the facility. Disadvantages include typically higher initial capital costs, the need to regenerate the exhausted resin, the need to dispose of the regenerate waste, and depending on the resin, the need to pre-treat the water to adjust the pH. With an acid resin (one where metal ions are replaced with hydrogen ions) the pH of the liquid will decrease as it passes across the resin bed. Thus, often the pH of the effluent will need to be neutralized before flowing to the drain. In addition, a low pH influent stream to the resin bed can adversely affect the efficiency of the resin because the high concentration of hydrogen ions in solution will not favor additional hydrogen ions coming into solution.

16.11   Summary

In summary, although electrochemical deposition processes have been around for over a hundred years, the requirements being placed on these technologies by their introduction into the semiconductor industry are beyond anything that has been imagined before. The level of effort that has been put into understanding the science behind these processes and the amount of understanding gained in the last decade are very impressive. The analytical resources available to support the electronics manufacturing industry are helping to expand our understanding of other electrochemical applications as well. What is even more impressive, however, is that we are only beginning to tap the possibilities of electrochemical technology in the high technology arena. The next decade should provide even more excitement and uses of electrochemistry that have not yet been imagined as the equipment used for electrochemical deposition continues to evolve.

References

1. F.A. Lowenheim, Modern Electroplating, John Wiley & Sons, New York, 1974, pp. 21–25.
2. M. Paunovic, M. Schlesinger, Fundamentals of Electrochemical Deposition, John Wiley & Sons, New York,1998, pp. 94, 204.
3. B. Kim, M. Funk, T. Ritzdorf, "Effects of Process Conditions on Film Properties of Electrochemically Deposited Sn–Ag–Cu Solder Alloys", Abstracts of the 199th meeting of the ECS, Vol. 2001-1, Abst. 299.

4. K.M. Takahashi, M.E. Gross, "Analysis of Transport Phenomena in Electroplated Copper Filling of Submicron Vias and Trenches", Conf. Proc. ULSI XIV, MRS, 1999, pp. 57–63.

5. O. Lanzi, U. Landau, "Terminal Effect at a Resistive Electrode under Tafel Kinetics", *J. ECS*, **137** (4), 1990, pp. 1139–1143.

6. S. Mehdizadeh, J.O. Dukovic, "Transient Plating-Rate Distribution on Resistive Thin Films with Point Contact Terminals", Extended Abstracts of the 184th Meeting of the ECS, 93-2, *IBM J.*,1997, Abst. 210.

7. R.J. Contolini, A.F. Bernhardt, S.T. Mayer, "Electrochemical Planarization for Multilevel Metallization", *J. ECS*, **141** (9), 1994, pp. 2503–2510.

8. K. Lowery, C. VanHorn, J. Commander, G. Shawhan: "Advancements in Copper Electroplating for Sub Half Micron ULSI Interconnects", Proc. VLSI Multilevel Interconnection Conference, p. 507 (1996).

9. S. Mehdizadeh, J.O. Dukovic, P.C. Andricacos, L.T. Romankiw, H.Y. Cheh, "The Influence of Lithographic Patterning on Current Distribution: A Model for Microfabrication by Electrodeposition", *J. ECS*, **139** (1), 1992, pp. 78–91.

10. S. Mehdizadeh, J. Dukovic, P.C. Andricacos, L.T. Romankiw, H.Y. Cheh, "The Influence of Lithographic Patterning on Current Distribution in Electrodeposition: Experimental Study and Mass-Transfer Effects", *J. ECS*, **140** (12), 1993, pp. 3497–3505.

11. B. DeBecker, A.C. West, "Workpiece, Pattern, and Feature Scale Current Distributions", *J. ECS*, **143** (2), 1996, pp. 486–492.

12. C.R. Simpson, T. Ritzdorf, C. Dundas, "Reducing edge and bevel contamination to help enhance copper process yields", Micro, Oct. 2000, pp. 41–53.

13. D.W. Goosen, "Ultra Low Alpha Lead Manufacture at Teck Cominco's Trail Operations", Peaks in Packaging, 2002.

14. M. Weiser, B. Clark, T. White, "Alpha Particle Issues in Wafer Bumping", Peaks in Packaging, 2002.

15. E. Castellani, J. Powers, L. Romankiw, U.S. Patent 4,102,756 Nickel-Iron (80:20) Alloy Thin Film Electroplating Method and Electrochemical Treatment and Plating Apparatus, 1978.

16. D.C. Luper, B.D. Oberholtzer, W.W. Pcihoda, J. Strautins: "Automatic Plating of Bipolar Integrated Circuits", *Plating Surf. Finish.*, **71**(1), pp. 48–52, 1984.

17. J. Klocke, J. Steeper, "Patterning Three-Dimensional Structures on Wafers with Electrophoretic Photoresist", 2002 Proceedings of the Pan Pacific Microelectronics Symposium, pp. 35–40.

18. M. Datta, R. Shenoy, U.S. Patent 5,486,282, Electroetching process for seed layer removal in electrochemical fabrication of wafers, 1996.

19. M. Datta, D.C. Edelstein, C.E. Uzoh, U.S. Patent 6,103,096, Apparatus and method for the electrochemical etching of a wafer, 2000.

20. M. Datta, R.V. Shenoy, C. Jahnes, P.C. Andriacos, J. Horkans, J.O. Dukovic, L.T. Romankiw, J. Roeder, H. Deligianni, H. Nye, B. Agarwala, H.M. Tong, P. Totta, "Electrochemical Fabrication of Mechanically Robust PbSn C4 Interconnections", *J. ECS*, **142** (11), 1995, pp. 3779–3785.

21. T. Ritzdorf, B. Batz, "Fountain Electroplating of Lead Tin Solder for Semiconductor Flip Chip Applications," International Technical Conference Proceedings of SUR/FIN – 1996, pp. 257–262.

22. T. Taylor, T. Ritzdorf, F. Lindberg, B. Carpenter, M. LeFebvre: "Electrolyte Composition Monitoring for Copper Interconnect Applications", Proceedings 193rd ElectroChemical Society Conference, 1998.

23. L. Graham, T. Ritzdorf, F. Lindberg: "Steady-State Chemical Analysis of Organic Suppressor Additives Used in Copper Plating Baths", Proceedings 196th ElectroChemical Society Conference, Oct. 1999.

24. P. Robertson, Y.V. Tolmachev, D. Fulton: "Galvanostatic Method for Quantification of Organic Suppressor and Accelerator Additives in Acid Copper Plating Baths", Proceedings 199th ElectroChemical Society Conference, March 2001.

25. L.W. Graham, T.C. Taylor, T.L. Ritzdorf, F.A. Lindberg, B.C. Carpenter, U.S. Patent 6,365,033, Methods for Controlling and/or Measuring Additive Concentration in an Electroplating Bath, 2002.

17 Processes and equipment for wet etching and cleaning

Jeffery W. Butterbaugh

17.1 Introduction

Control of contamination is very important during the fabrication of microelectronic components. Great effort is expended to minimize the amount of particles and residues present in the manufacturing environment in an attempt to eliminate the defects that they cause, and improve the yield of functioning components. These efforts include construction of sophisticated clean rooms with ratings of less than one particle greater than 0.5 μm in size passing through 1 ft^3 of air per minute (Class 1, per Federal Standard FED-STD-209D). In addition, the newest manufacturing facilities for 300 mm wafers employ wafer pods which are only opened after being connected to controlled environments built within each piece of process equipment. Use of pods and individual controlled environments drives higher levels of automation and factory control.[1]

Despite these efforts to keep contaminants away from electronic components during fabrication, cleaning of microelectronic components continues to become a larger fraction of the overall manufacturing operation.[2] The fabrication of a typical integrated chip with features sizes of 130 nm on a 300 mm wafer can require up to 100 separate cleaning steps or 25% of the total number of process steps. Cleaning is important in the packaging area as well. Particles can affect the patterning of bumps and leads as well as cause shorts. Residues can cause delamination of insulating layers and debonding of metal leads. Oxidation will also lead to bonding problems and overly resistive contacts to the device. Therefore, cleaning processes and equipment are key components in the successful microelectronics manufacturing operation. It is interesting to note, however, that the Assembly and Packaging section of the 2001 International Technology Roadmap for Semiconductors makes no mention of upcoming challenges in defect reduction, yield improvement, or cleaning in the packaging area.[3] In this chapter the motivation and challenges of cleaning during packaging operations as well as the processes and equipment used for cleaning will be discussed.

Cleaning is essentially an exercise in the selective removal of undesired materials from the surface of the wafer or component while minimizing the removal and damage of the desired materials that make up the electronic device. This exercise is becoming more complicated as the number of different materials used in building and packaging the device increases. The introduction of high dielectric constant materials for transistors and capacitors in the integrated circuit as well as low dielectric constant materials for insulating metal interconnect patterns creates new challenges for cleaning processes. While cleaning equipment and processes used for wafer fabrication have typically been different from those used in packaging operations, the new emphasis on wafer level packaging and bumping has brought those wafer level cleaning processes into the packaging arena. In this chapter we will talk about the challenges of cleaning after soldering operations and also review the cleaning and etching requirements for wafer level and chip level packaging. Further, while new gas-phase and super critical fluid processes are now being developed for cleaning, our focus here will be on wet cleaning processes and the equipment used to perform wet cleaning operations.

In addition to direct wafer and packaging operations, cleaning technology is also required in other aspects of assembly and packaging such as probe card cleaning or stencil mask cleaning.[4] However, the focus here will be on removing contaminants from the device and substrates that make up the final component.

17.2   Typical contaminants

Until recently the main focus of cleaning for microelectronics packaging has been surface preparation for solder mask application and flux residue removal. With the move towards wafer level packaging, cleaning requirements have become more stringent and are similar to the requirements for integrated circuit manufacturing. Flux removal is still required after wafer level solder ball formation. However, additional cleaning steps are required to remove contaminants and prepare surfaces before deposition of underbump metallurgy, to prepare the metal surface before photoresist application, to remove photoresist after solder plating, to remove underbump metallurgy after solder plating, and sometimes to remove solder bump surface oxidation. In addition, with shrinking geometries, it is even more important to remove particulate contamination to prevent patterning errors and electrical shorts.

Typical contaminants that are removed during cleaning operations include surface oxides, organic films, ionic contaminants, solder flux residues, photoresist layers, and residues and particles.

Oxides and organic films will prevent adhesion of deposited metal layers and cause delamination failure during the deposition process or possibly later after the device is put into operation. These surface contaminants will also cause higher resistivity in the electrical connections to the packaged device.

Ionic contaminants arise from many of the materials used in the manufacture and cleaning of the microelectronic device. Chloride, bromide, sodium, and other ionic contaminants can cause electrical leakage and corrosion which may be immediately evident or may not arise until long after the component is in use. Detection and analysis of ionic contaminants by ion chromatography are described by Newton.[5]

Solder flux residues are the source of many problems in microelectronic packaging and a main focus of cleaning efforts. The purpose of the flux is to promote solder flow by providing a clean surface during the solder and reflow process. The acid components added to fluxes to increase the flux activity are what cause corrosion and electrical migration problems if they are not properly removed.[6] Also, in flip-chip operations, flux residues can interfere with the underfill operation causing voids which lead to corrosion and failure.[7] The cleaning process itself can also cause underfill problems if the cleaning chemical is not properly rinsed away.

There are three main types of fluxes currently in use. These are rosin-based, water-soluble, and no-clean fluxes.[8] The term "no-clean" does not necessarily mean that there are no residues after soldering or reflow, but rather that the residues generally do not need to be removed. However, in some cases even the no-clean fluxes require cleaning.[7] Depending on the application, some rosin-based fluxes are left in place and thus offer a no-clean process.[8] The additives which are used to increase the flux activity are the main concern in cleaning operations with rosin-based and water-soluble fluxes. These must be removed or they will cause corrosion and other reliability problems. This is particularly important for water-soluble fluxes.

Photoresist is used to transfer the solder bump and wiring patterns to the surface of the printed circuit board or wafer. In contrast to transistor fabrication, where the photoresists are subjected to plasma etching and ion implantation, the photoresists used in packaging operations are much easier to remove. However, the typical thickness of the photoresist used in packaging is much greater and thus careful attention is required to ensure that all resist is removed and does not redeposit on the substrate surface. Unremoved photoresist residues will interfere with the subsequent etching of the seed layer and underbump metallurgy after the plating operation.

The term "particles" generally refers to contaminants that are embedded or lying on the substrate surface. They are generated by many different sources in the manufacturing environment including humans, clothing, surface abrasion, and equipment flaking. While a large amount of time and money is invested in cleaning particles from the manufacturing environment, they often arise from the etching, deposition, and cleaning equipment used to manufacture the electronic device. Unremoved particles will interfere with patterning and plating operations causing electrical shorts and opens. Depending on the composition of a particle, it can also be a source for ionic contamination or corrosion.

Measurement and characterization of contaminants on the component are necessary to determine what type of cleaning process is required and whether that cleaning process has been effective at removing the contaminant. The cleanliness of circuit boards and assemblies is often measured by soaking or rinsing the object in an aqueous solution and then measuring the contaminants that remain in the solution. Ion chromatography is one method for analyzing the ionic contaminants.[5] This technique gives an average contaminant measurement and may be misleading if contaminants are concentrated in a few small areas where they can cause the most problems. Also, ionic analysis will not detect the presence of organic contaminants. There are several surface analysis techniques commonly used during integrated circuit fabrication at the wafer level that can examine a very localized region. Scanning electron microscopy (SEM) is used to visualize contaminants and also perform some chemical analysis by looking at the X-rays emitted as the electron beam interacts with the surface (EDX). X-ray photoelectron spectroscopy (XPS) is commonly used to identify contaminants on a surfaces as well as auger electron spectroscopy (AES). There are many resources available which describe these and other surface analytical techniques in detail.[9,10] Automated vision systems are also available for wafer level processes that detect and classify defects on the wafer surface and alert the user to contamination problems. Automated vision systems are available from equipment companies such as August Technology and KLA-Tencor.

17.3   Cleaning and etching requirements

As noted earlier, cleaning can be described as the removal of undesired materials from the substrate surface selectively to the substrate. In removing contaminants from a surface, care must be taken to avoid damaging the electronic device that is being cleaned. The term selectivity is often used to describe this ability to remove one material while not removing another. Etching can also be considered a form of cleaning, as its purpose is the selective removal of an undesired material in order to form a pattern or make a device functional.

The requirements of the cleaning or etching operation will depend on the nature of the contaminants being removed and the composition of the underlying substrate. In general, the basic steps of cleaning are (1) delivery of the cleaning agent to the location of the contaminant on the surface of the substrate; (2) reaction; (3) dissolution and/or suspension of the contaminant in the cleaning fluid; and (4) removal of the cleaning agent along with the dissolved or suspended contaminant. The final step involves rinsing as well as drying the substrate.

Parameters that affect the ability of the cleaning fluid to reach the contaminant on the substrate include fluid viscosity, impact pressure, surface tension, and interfacial tension.[11] For post-solder cleaning of circuit board assemblies, component spacing is important, whereas for wafer level packaging operations, spacing of bumps and pads generally does not affect cleaning. One of the biggest factors determining whether a cleaning fluid can effectively reach contaminants on a substrate surface is wettability. Wettability is determined by the critical surface free energy of the substrate and the surface tension of the cleaning fluid. Wetting occurs easily when the surface tension of the cleaning fluid is less than the critical surface free energy. Water, which has a relatively high surface tension of 72 dyn/cm (room temperature) has difficulty wetting most plastic surfaces which have surface energies of 20 to 60 dyn/cm. Organic solvents like ethyl alcohol, with a surface tension of 22 dyn/cm, or trichloroethane, with a surface tension of 25 dyn/cm, do a much better job of wetting surfaces. Glass has a very high surface energy and is easily wetted by most liquids. Often, surfactants are added to water to decrease its surface tension and promote wettability and cleaning effectiveness. Also, heating water lowers its surface tension. The surface tension of water near 100°C is about 60 dyn/cm.

Once the cleaning fluid has reached the contaminant on the substrate, it must dissolve or react with the contaminant to remove it from the surface. The solubility of solid, liquid, and gas materials in solvents has been treated in great detail theoretically and experimentally.[12,13] A common rule of thumb is "like dissolves like." For instance, water, which is a polar solvent, does a good job dissolving ionic contaminants on a surface, whereas hexane, which is a non-polar organic solvent, does a good job at dissolving organic contaminants on a surface. A more rigorous approach compares the solubility parameters of the solute and solvent.[13]

If the contaminant is not readily soluble in the cleaning fluid, chemical reaction may be used to convert the contaminant into a soluble form. The conversion of an insoluble material into a water-soluble material is call saponification. Chemicals used for saponification are discussed in the next section. Mechanical energy can also be used to help remove contaminants. The use of ultrasonic and megasonic energy is discussed in the wet cleaning equipment section.

Finally, the cleaning fluid along with the dissolved or entrained contaminants must be rinsed from the substrate and the substrate dried. Corrosion and other problems can occur if this final step is not carried out correctly. It involves rinsing the chemical and contaminants out of features and away from the surface as well as drying the surface completely to remove the rinsing fluid which is usually water. In some cases the final rinsing step may be accomplished by a volatile organic solvent, such as an alcohol or chlorinated organic compound.

Most importantly, the cleaning process must be carried out with minimal effect on the microelectronic device.

17.4   Cleaning and etching processes

Figure 17.1 shows a simplified sequence for wafer level packaging. Solder bumps are built up on the wafer before it is cut into individual chips. After dicing, or singulation, the chips are flipped and soldered to the circuit board or package. The wet cleaning, etching, and stripping steps are indicated in Figure 17.1.

The purpose of the first cleaning step in wafer level packaging is to prepare the bond pad and top surface of the wafer to receive deposition of the underbump metallurgy. In this step, the surface of the bond pad, typically copper, is cleaned. Oxide is removed from the surface and any other contaminants, such as particles, are removed. A wet cleaning process at this point might involve dilute hydrofluoric acid (about 1% by weight). The dilute HF will serve to remove the oxide, while spraying or agitation will help to remove particle contamination.

After the underbump metal is deposited, the bump plating mask is created by depositing and patterning a photoresist layer. The photoresist is open over the bond pads so that solder bump plating only occurs over the bond pads.

After plating, the photoresist mask must be cleaned or stripped from the surface. This photoresist layer is usually very thick (relative to photoresist masks used in integrated circuit fabrication) at 25 to 100 μm. If the photoresist is positive tone, it is fairly easy to remove with a solvent such as *N*-methyl-pyrrolidinone (NMP). Negative tone resists are cross-linked and are not as easily removed. In many cases the negative resist layer will peel off the surface before it dissolved. This can create problems if these large pieces of undissolved photoresist redeposit on the wafer surface. Special consideration must be given to chemical interaction with the bump metallurgy and underbump metallurgy. Many resist stripping chemicals are

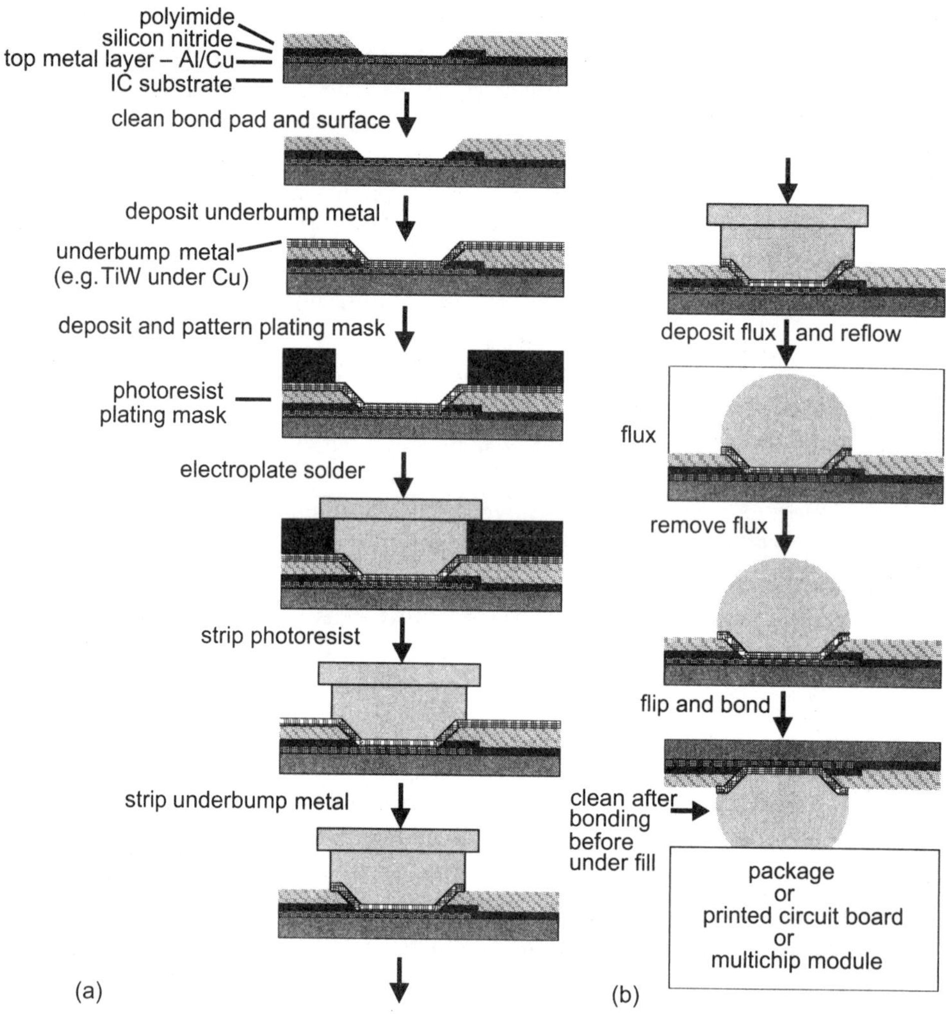

Figure 17.1 Process steps in wafer level packaging (a) through UBM strip and (b) from reflow through flip-chip bonding.

amine-based (contain nitrogen compounds) which have a propensity to attack copper. One might argue that the underbump metallurgy is just going to be removed after the resist strip, so it is not a large concern. However, often the photoresist pattern is defective and must be stripped and reapplied before plating can be carried out. If the underbump metallurgy is attacked during the stripping process it will affect the adhesion and patterning of the new photoresist layer as well as the plating process.

The next step is another cleaning or etching step. The underbump metal must be removed or the solder bumps and bond pads will be shorted to each other. This is a critical process because inadequate etching will cause shorting and over aggressive etching can cause undercutting of the solder bump or etching of the solder bump.

Table 17.1 shows some typical chemicals used to etch various metals used in wafer level packaging. Papers have been published describing optimized mixtures and conditions for specific metal stacks.[14,15] For instance, a chemical sequence for removing an underbump metal layer consisting of Cu over TiW (as shown in Figure 17.1) would use a dilute sulfuric acid (H_2SO_4), or a dilute mixture of H_2SO_4 with hydrogen peroxide (H_2O_2), to remove the copper layer followed by H_2O_2 to remove the TiW layer. Again, selectivity is a key parameter in this step, as it is important not to etch the solder bump or to attack the underlying polyimide layer.

Table 17.1 Chemistries for etching underbump metallurgy.

Metal	*Etching chemistries*
Cu	H_2SO_4, H_2SO_4/H_2O_2, $(NH_4)_2S_2O_8$ (ammonium persulfate), HCl, $CuCl_2/HCl$, $FeCl_3$
Cr	HCl, $HClO_4$ (perchloric acid), $H_8CeN_8O_{18}$ (ceric ammonium nitrate), $H_8CeN_8O_{18}/CH_3COOH$ (acetic acid), HNO_3
Au	KI, HCl/HNO_3 (aqua regia), KCN
Ti	H_2O_2, NH_4OH/H_2O_2, HF, HF/H_2O_2, HF/HNO_3
TiW	H_2O_2, NH_4OH/H_2O_2, HF, HF/H_2O_2, HF/HNO_3
Ni	H_2SO_4, H_2SO_4/H_2O_2, $HNO_3/H_2SO_4/CH_3COOH$
NiV	H_2SO_4, H_2SO_4/H_2O_2, HNO_3, HCl
Al	$H_3PO_4/CH_3COOH/HNO_3$, HCl, H_2SO_4, H_2SO_4/H_2O_2
Pb/Sn	CH_3COOH/H_2O_2,

Now the plated solder is ready to be reflowed into a solder bump. Since the solder will only wet the underbump metal and not the polyimide, it will form a spherical ball when it is heated above its melting point. During the reflow process it is important to prevent metal oxide removal, and so a flux layer is typically deposited over the entire wafer surface. After solder bump reflow the flux layer must be removed. There are many special mixtures available for removing this flux layer and the correct selection depends on the type of flux that is used. Examples of flux removal chemistries for wafer level bump reflow are Kyzen Micronox MX2301 and Arakawa Pine Alpha ST-100SX.

The wafer is diced and the individual bumped chips are now flipped and soldered onto the package, printed circuit board, or multichip module. After mounting the chip into the package or onto the circuit board, either by wire bonding or direct chip attach (flip chip), additional cleaning steps may be necessary. Even when the so-called "no-clean" fluxes are used, cleaning may still be necessary at some level to remove particles and residues and promote good flow of the underfill material.

Common chemicals that have been used in the past for general circuit board cleaning and flux removal include chlorofluorocarbon solvents (CFC) and trichloro-ethylene (TCA).[16] TCA and CFCs were typically used in the vapor form with "vapor degreasing" equipment. In this type of equipment, the cleaning solvent is vaporized and then allowed to condense on the part creating a liquid film and heating the part with the latent heat of vaporization released by condensation. These solvents are effective and relatively safe since they are non-flammable. However, because of growing concern about ozone depletion by these chemicals, their production has been halted and their use has been significantly reduced. New solvents and aqueous-based cleaners are being used in batch and spray equipment, with the use of vapor degreasing equipment sharply declining.

Some of the newer solvents include mixtures of terpenes, which are naturally occurring plant-derived solvents, and alcohols, such as methanol, ethanol, or isopropanol.[17] In order to avoid ignition of these chemicals, which have flash points ranging from 13 to 50°C, they are typically used at room temperature. If they are used at elevated temperature they are often mixed with water or used under an inert atmosphere such as nitrogen or perfluorocarbon blanket.[17] Other new chemicals for vapor degreasing such as hydrofluoroethers (HFE) and *n*-propyl bromide are now being used.[18]

Issues of safety and environmental impact have led to the development of effective aqueous cleaning systems. Water, by itself, is particularly effective at removing ionic contaminants which are the major concerns for cleaning flux residues. However, water, by itself, is ineffective at removing organic contaminants and also has a high surface tension, making it difficult to penetrate fine features and small spaces.[8] Surfactants are added to aqueous cleaning systems to reduce surface tension and promote wetting of the surface being cleaned. On the other hand, higher surface tension assists capillary forces in drawing the cleaning fluid into very small and deep spaces. Therefore, it is desirable to lower the surface tension, but not too much.[8] In order to remove organic or non-ionic contaminants, they must be converted to a water-soluble form. This conversion is called "saponification." This has commonly been done by adding an amine, such as monoethanol amine (MEA) to the solution. Solutions with MEA are basic (pH greater than 11) and convert the insoluble organic contaminates into a water-soluble form. A mixture of alkaline base, such as ammonium hydroxide, with hydrogen peroxide (also referred to as standard clean 1 or SC1) is commonly used in integrated circuit fabrication and is effective at removing organic contaminants. However, this mixture is incompatible with many metals. A thorough treatment of aqueous cleaning can be found in the *Handbook of Aqueous Cleaning Technology for Electronic Assemblies*.[19]

17.5 Cleaning and etching equipment

Equipment used for applying the cleaning chemistry to the electronic components ranges from simple, manual baths to sophisticated, automated, closed spray systems with several variations in between. In this section we will discuss several equipment approaches including vapor degreasing, automated bath or immersion (also called a wet bench, or simply "bench"), batch spray, in-line spray, centrifugal spray, single wafer spray, and centrifugal immersion system. The purpose of the cleaning equipment is to facilitate the basic steps of the cleaning operation which were previously described in Section 17.3. The component or wafer being cleaned must be brought into contact with the cleaning fluid for a time sufficient to solvate the contaminants, then the cleaning fluid must be removed from the component or wafer,

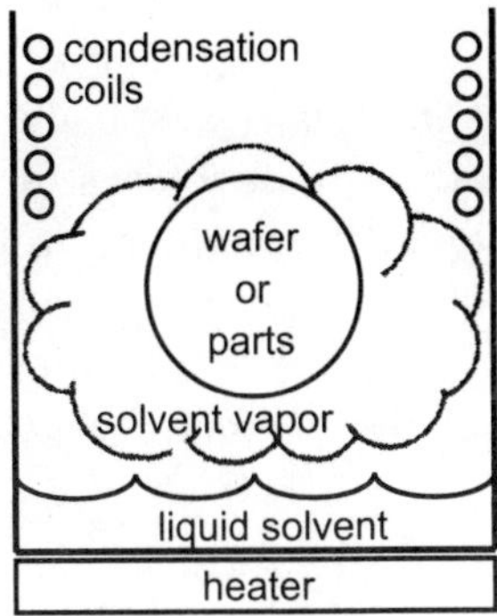

Figure 17.2 Schematic of vapor degreaser system.

and finally rinsing and drying must be carried out. The equipment must also ensure that this operation is carried out in a safe way, and in a way that optimizes the performance of the cleaning operation while minimizing the chemical usage and the cost.

Cleaning of electronic components was probably first carried out in a basin or bath by immersion along with manual, mechanical force, such as a scrub brush. As more sophisticated and more hazardous cleaning fluids were developed, and electronic components became more advanced, this approach was replaced by more automated means to ensure consistency of results, safety of operators, recovery of cleaning fluids, and protection of the components being cleaned.

17.5.1 *Vapor degreaser*

The vapor degreaser was developed along with organic cleaning solvents. The basic premise of the vapor degreaser is to apply the cleaning fluid to the surface of the object being cleaned by condensation of a vapor of the cleaning fluid. The cleaning fluid drains off the component, is collected at the bottom of the apparatus, and then is revaporized and used again. In addition to providing a liquid layer of the solvent on the surface of the substrate, condensation of the vapor also gives off heat, raising the temperature of the substrate and improving cleaning effectives. Figure 17.2 is a simple diagram of the vapor degreaser. Condensation coils in the upper half minimize the amount of vapor that escapes the degreaser, by condensing it and allowing it to flow back into the vaporizer tank in the bottom. With traditional degreasing solvents like 1,1,1 tricholoroethane, the residual fluid would simply evaporate from the substrate as it was removed from the tank. The use of vapor degreasers has declined significantly with the new regulations on Freons and other ozone-depleting substances, but new more "friendly" solvents like HFE or *n*-propyl bromide are now being used.[18]

Vapor degreasers are available from equipment companies such as Branson Ultrasonics, Forward Technology, Sonicor Instrument Corporation, Greco Brothers, and Sonitech.

17.5.2 *Wet bench*

As previously noted, the use of sinks, basins, or baths to immerse the substrate in the cleaning fluid was probably one of the first ways of cleaning. The equipment used for immersing substrates in liquids is commonly referred to as a wet bench. Wet benches are still used extensively for electronics cleaning, making up over half of the cleaning equipment market which reached a total of \$2 billion in 2000. A leading edge, fully automated wet bench for 300 mm wafer cleaning can cost as much as \$3 million.

Stainless steel or fluorinated polymer racks (depending on the type of cleaning chemicals in the bath) are used to insert, withdraw, and support circuit boards and other electronic components being cleaned. For automated wet benches, a robotic arm is used to move the racks in and out of the baths. The racks used to support silicon wafers during processing are called cassettes. Cassettes for processing silicon wafers are highly engineered components able to support up to 50×300 mm wafers at a spacing of 5 mm while preventing the touching of any substrates. Cassettes that are moved from one batch to the next can cause cross contamination between cleaning baths leading to precipitation of salts and shortened bath life. The most advanced wet benches employ a technique referred to as cassette-less processing, meaning that very minimal racks, which are dedicated to a specific batch, are used to support the wafers in that bath. When the wafers are lifted out of the bath, a separate rack is used to transport the wafers to the dedicated rack for the next bath. In addition, silicon wafers are transported through the manufacturing floor in boxes called front opening unified pods (FOUP). These pods have integrated internal racks for supporting the wafers during transportation between process tools.

The fully automated silicon wafer wet bench requires several robots to move the wafers from the FOUP to the bath and back out to the FOUP again. Figure 17.3 is a cut-away view of a fully automated wet bench for 300 mm wafer cleaning available from FSI International. After the FOUP is positioned on the wet bench load port it is moved by a FOUP robot to a storage position in the FOUP buffer. When the system is ready to process the wafers, the FOUP robot moves the FOUP to another position, where it is opened. Another robot extracts the wafers from the FOUP and places them in a batching rack at the 5 mm spacing. A wet bench robot with its own transfer rack removes the wafers from the batching rack and over the first cleaning batch. The wafers are then transferred to the dedicated rack in that bath and lowered into the cleaning solution. The wet bench robot is used to move the wafers from bath to batch and then back out to the batching areas where the cleaned wafers are returned to the FOUP.

Depending on the type of substrates to be cleaned, the volume of a cleaning batch can range from 35 liters, for 300 mm wafer cleaning, to hundreds of liters for large flat panel displays or other electronic components. Chemical usage is a large factor in the cost of wet cleaning, and therefore much attention is paid to minimizing the tank volume and optimizing the amount of time the chemical can be used before it is changed (bath life). Cassette-less fixturing (described above) can reduce cross contamination and extend bath life. Rinsing between cleaning baths can also reduce

Figure 17.3 Cut-away view of a fully automated wet bench for 300 mm wafer processing.

cross contamination, but may introduce water to the next bath, which may or may not be considered a contaminant itself. For cleaning circuit boards and electronic components, baths can be used for several weeks or months before changing. In the manufacturing of integrated circuits on silicon wafers, cleaning baths are changed much more regularly; once per day or even once per shift (8 to 12 hours). In many cases the need to change the bath is determined by the number of substrates that have been processed rather than by time. Recently, real-time diagnostics have been used to monitor cleaning chemical characteristics like pH or water content to determine when the bath needs to be changed.

Cleaning efficiency and bath life are also improved by the use of recirculation pumps. The cleaning fluid is pumped into the bottom of the bath and allowed to overflow into a weir where it is collected and returned to the pump reservoir. Heating and cooling elements are used to control the bath temperature. Recirculation provides more uniform and accurate temperature control of the cleaning liquid.

Ultrasonic (20 to 350 kHz) and megasonic (700 to 1000 kHz) energy is also used to help dislodge particles and residues from the substrate surface.[20] In silicon wafer processing, a quartz tank is often used to efficiently transfer the energy from the megasonic transducer into the cleaning fluid. The megasonic energy then moves through the fluid as pressure waves. The formation of microscopic bubbles, or cavitation, is believed to be a major mechanism for particle removal and cleaning performance with megasonics.[20]

Rinsing is a very important aspect of the cleaning processes. For electronic components and silicon wafers very clean distilled and deionized (DI) water is used for rinsing. In a wet bench, rinsing is often accomplished in a dedicated bath which can be filled and dumped very quickly. The substrates are moved from the cleaning

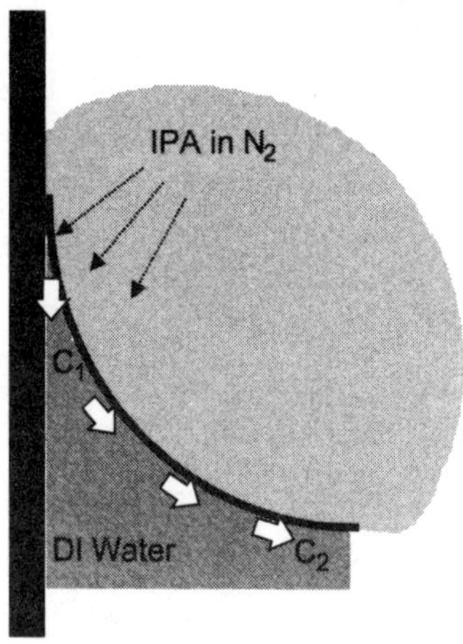

Figure 17.4 Marangoni effect promoting water flow away from the wafer surface. The concentration of IPA in the water meniscus at C_1 is higher than at C_2 giving the region around C_1 a lower surface tension which causes fluid to flow away from the wafer surface.

fluid bath into the rinsing bath which is already full. After a designated soaking time, the water is quickly dumped and refilled. During the soaking time the water is sometimes recirculated or continually replenished from the bottom of the tank with overflow at the top into the collection weir. In semiconductor manufacturing, much more dilute cleaning solutions are now being used which permit fresh chemical baths to be used and dumped for each batch of substrates allowing rinsing to be carried out in the chemical bath. This can significantly reduce the size and cost of an automated wet bench.

Drying is also a critical aspect of the cleaning process. With some of the cleaning solvents, such as HFE, the wafers are dry as they are extracted from the bath and require neither rinsing nor drying. After wafers have been rinsed in DI water, however, they must be dried to remove residual moisture from the surface. Simple methods of drying such as evaporation in air or blowing of heated nitrogen while spinning the substrates can leave residues on the surface. Nonetheless, spin drying is still a very popular method for drying substrates and there are many companies that build equipment for this purpose. Isopropyl alcohol (IPA) has been widely used for final rinsing and drying in electronic manufacturing. In some cases a technique very similar to the vapor degreaser is used to condense IPA on the substrate surface and then extract the substrate slowly so that any residual IPA evaporates leaving a dry surface. More recently, IPA is added to the final rinsing tank during or after extraction of the substrate from the water (either by draining the tank or lifting the substrates). IPA vapor is allowed to condense on the surface of the substrate to displace the water as the substrate and water are separated. Another physical effect, referred to as the Marangoni effect causes a surface tension gradient in the water meniscus at the substrate surface causing the water to flow away from the surface leaving it dry (see Figure 17.4).[21]

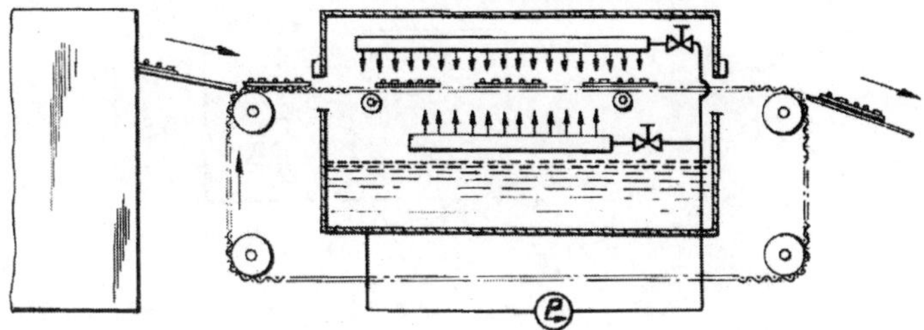

Figure 17.5 Simple in-line spray system disclosed in U.S. patent 3,868,272.

Fully automated wet benches for cleaning and etching silicon substrates are available from several equipment companies including FSI International, Dainippon Screen, Tokyo Electron, SCP Global, and Akrion.

17.5.3 Batch spray

A batch spray system that should be familiar to most people now is the common kitchen dishwasher. A very similar concept is used to clean electronic components in the batch spray system. Circuit boards and other components are loaded into racks and placed in the process chamber which is closed and sealed. Cleaning and rinsing fluids are sprayed over the substrates, collected, and recirculated. Rinsing and drying are also carried out in the same enclosed chamber. Batch spray systems for cleaning microelectronic components are available from equipment companies such as Aqueous Technologies.

17.5.4 In line spray

Spraying of chemicals onto the substrates in conveyorized systems is sometimes used, especially for very large substrates that would require over-sized baths in an immersion system or over-sized chambers in a batch spray system. With an in-line spray system, substrates are moved horizontally through various chambers on a conveyor. A specific cleaning or rinsing fluid is sprayed over the substrate in each chamber as it moves through. The fluid is collected below the substrates and reused within that chamber or moved upstream to a pre-wash chamber. Figure 17.5 is a schematic of a simple in-line spray system disclosed in U.S. Patent number 3,868,272.[22] More advanced systems employing multiple tanks are available from companies such as Aqueous Technology or Speedline Accel.

17.5.5 Centrifugal spray

The second most common cleaning technique (after the wet bench) is centrifugal spray processing. In this technique the substrates are rotated at speeds up to 500 rpm while chemicals are sprayed onto the surface. The chemicals can be collected and

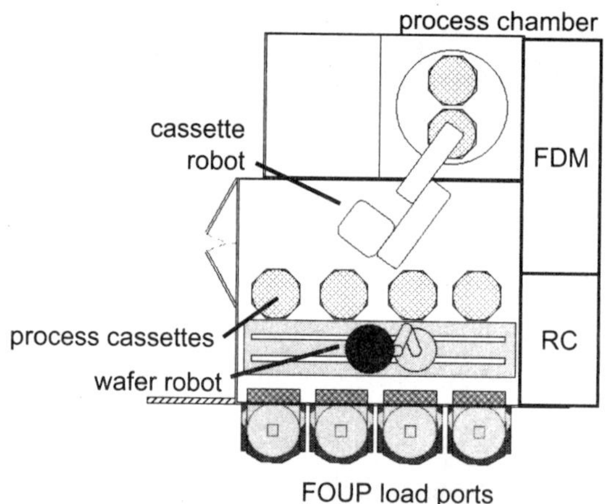

Figure 17.6 Schematic plan view of an automated batch centrifugal spray system for processing 300 mm wafers. FDM=fluid delivery module; RC=robot controller.

recirculated to reduce chemical usage. For critical cleaning of silicon wafers, fresh dilute chemicals are often dispensed and then immediately drained since this technique makes very efficient use of the chemical. With spray processing, fresh chemicals can be blended immediately before dispensing, allowing multiple steps with varying chemicals and concentrations. A cleaning process which may require several separate baths in a wet bench can be done in one continuous process with centrifugal spray.

Figure 17.6 is a plan view schematic of an automated centrifugal spray cleaner for 300 mm silicon wafers available from FSI International. As with the automated wet bench, wafers are extracted from the FOUP and placed into dedicated process cassettes. A robot then places the cassettes onto a turntable in the process chamber. The turntable holds the cassettes in place while they are rotated at high speeds. A lid is closed and sealed over the process chamber which allows the chamber to be completely purged and backfilled with an inert gas such as nitrogen. A single cassette can be placed in the center of the turntable, or multiple cassettes can be placed off center, as shown in the cut-away view of the process chamber in Figure 17.7. In the single cassette system, chemicals and rinse water are sprayed from the side of the chamber. For a multiple cassette system, chemicals and water are sprayed from a center spray post attached to the lid as well as from the side of the chamber. In addition, rinse water and nitrogen are fed through the turntable to help rinsing and drying the process chamber at the end of the process. Drying is accomplished by a final high speed spin while dry nitrogen flows through the chamber. Flow-through heaters are used to heat water which is mixed with chemicals or to heat the chemical mixtures immediately before dispensing.

Some advantages of the centrifugal spray techniques include efficient application of the cleaning fluid to the surface of the substrate, and flexibility in using multiple or repeated chemical sequences without having to move the substrates from bath to bath. In some sense, instead of bringing the substrates to the chemical, the

Figure 17.7 Cut-away view of batch centrifugal spray system process chamber.

chemicals are brought to the substrates. The centrifugal force of spinning the wafers causes a thinning of the liquid diffusion boundary layer at the substrate surface. This provides for more efficient transfer of the cleaning chemicals to the surface and more

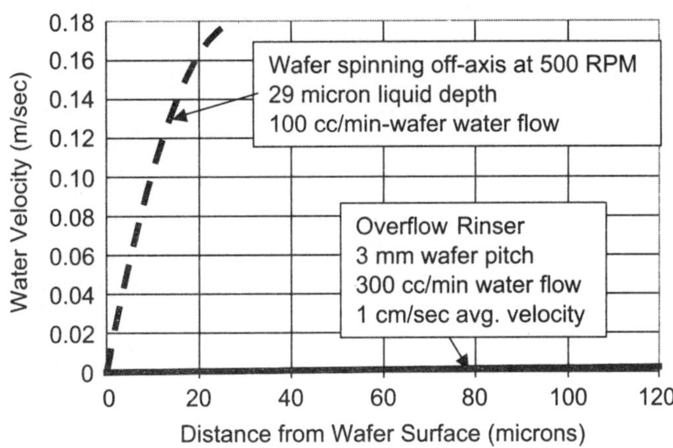

Figure 17.8 Comparison of water velocity near the wafer surface between a batch spray system and a wet bench overflow rinser.

efficient transfer of contaminants from the surface. Figure 17.8 shows a comparison of the fluid velocity at the substrate surface for substrate in an overflow bath and for one spinning at 500 rpm. In the case of the spinning wafer, fresh chemicals are replenished much closer to the surface and have a shorter distance to diffuse while contaminants also have a much shorter distance to diffuse away before being entrained in the convective fluid flow.

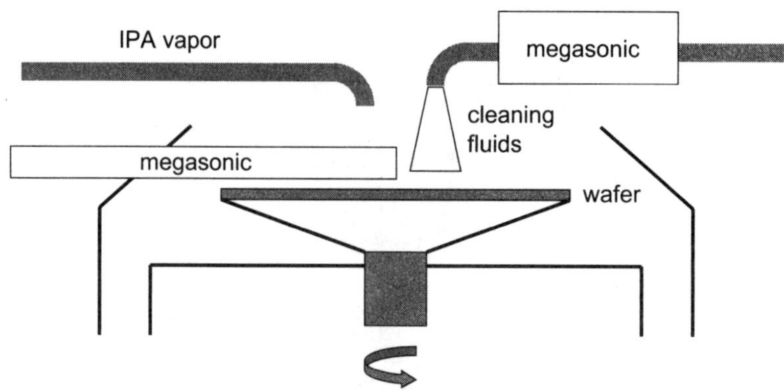

Figure 17.9 Schematic of a single wafer spin spray system.

Many chemical processes can be carried out in chambers constructed of fluorinated polymers. However, when flammable fluids with low flash points (below 60°C) are used, the chamber needs to be constructed with stainless steel.

Centrifugal spray batch systems are available from equipment manufacturers such as FSI International, Semitool, and Tokyo Electron.

A recent variation of the centrifugal spray system is shown in Figure 17.9, where a single substrate is being processed. In this type of system, a single substrate is rotated at high speed while chemicals are dispensed directly onto the spinning surface. Megasonics energy can be applied to the cleaning fluids as they are dispensed or directly to the spinning wafer surface through a quartz rod. Drying can also be assisted by IPA vapor. Depending on the cleaning chemistry being used, this type of system can process a single substrate much faster than a batch system. However, when many substrates must be processed, the throughput and efficiency of chemical usage can be much lower than the batch centrifugal spray system. Single wafer spin systems are available from equipment manufacturers such as SEZ, Dainippon Screen, Tokyo Electron, Solid State Equipment Corporation, and Applied Materials.

17.5.6   *Centrifugal immersion*

Equipment is also available which combines immersion with spinning to impart centrifugal force and more efficient cleaning than immersion alone.[23–25] Equipment currently available allows substrates to be mounted in a cassette or fixture attached to a rotating spindle mounted in the lid of the process chamber. The substrates are then lowered into a bath of the cleaning fluid, and the lid is sealed allowing the chamber to be purged with nitrogen. The substrates are then rotated at speeds of 200 to 500 rpm while immersed in the fluid. The cleaning fluid is drained from the chamber, and the substrates are rinsed by spraying while being rotated in the same way as the centrifugal spray system. Drying is also carried out with nitrogen in the same way as the centrifugal spray processor. This type of equipment is available from Speedline Accel.

References

1. R.N. Castellano, 300 mm is reshaping the factory automation market, *Solid State Technology*, Vol. 45, No. 2, pp. 38, 40, 80 (2002).
2. M. Heyns, P.W. Mertens, J. Ruzyllo, and M.Y.M. Lee. Advanced wet and dry cleaning coming together for next generation, *Solid State Technology*, Vol. 42, No. 3, pp. 37, 38, 40, 42, 44, 47 (1999).
3. *International Technology Roadmap for Semiconductors*, <http://public.itrs.net/>, (accessed 30 June 2002).
4. Single Swipe Makes a Clean Sweep, *Electronic Packaging and Production*, <http://www.e-insite.net/epp/index.asp?layout=article&articleId=CA90860&stt=001>, (accessed 1 July 2002).
5. B. Newton, Tracking ionic contamination: Enhanced troubleshooting helps eliminate electronic failure, *CleanTech*, Vol. 1, No. 5, pp. 38–43 (2001).
6. J.F. Maguire, Metals Joining of Electronic Circuitry, in C.A. Harper and R.M. Sampson (ed.) *Electronic Materials and Processes Handbook, Second Edition*, New York: McGraw-Hill (1994), p. 7.26.
7. M. Todd and M. Bixenman, Improving flip-chip manufacturing through proper cleaning, *Advanced Packaging*, Vol. 10, No. 1, pp. 40–49 (2001).
8. F. Cala, Aqueous Cleaning: The basics & beyond in a world of emerging needs, *Precision Cleaning*, Vol. 5, No. 1, pp. 13–20 (1997).
9. J.C. Vikerman (ed.), *Surface Analysis—The Principal Techniques*, New York: Wiley (1997).
10. H. Bubert and H. Jenett (eds.), *Surface and Thin Film Analysis: A Compendium of Principles, Instrumentation, and Applications*, New York: Wiley (2002).
11. J.F. Maguire, Metals Joining of Electronic Circuitry, in C.A. Harper and R.M. Sampson (eds), *Electronic Materials and Processes Handbook Second Edition*, New York: McGraw-Hill (1994), p. 7.59.
12. A.F.M. Barton, *CRC Handbook of Solubility Parameters and Other Cohesion Parameters Second Edition*, Boca Raton, FL: CRC Press (1990).
13. R.C. Weast (ed.), *CRC Handbook of Chemistry and Physics, 66th edition*, Boca Raton, FL: CRC Press (1986), p. C-676.
14. M. Bernt and G. Solomon, Using wet chemistry for etching under-bump metal, *Solid State Technology*, Vol. 44, 89–92 (2001).
15. L.N. Ramanathan and D. Mitchell, Development of an etchant for selectivity etching TiWN//x in the presence of electroplated 95%Pb–5%Sn solder, IEEE Transactions on Components and Packaging Technologies, Vol. 24, No. 3, pp. 425–430 (2001).
16. C. Cohn and M.T. Shih, Packaging and Interconnection of Integrated Circuits, in C.A. Harper (ed.) *Electronic Packaging and Interconnection Handbook, Third Edition*, New York: McGraw-Hill (2000), p. 7.89.
17. J.F. Maguire, Metals Joining of Electronic Circuitry, in C.A. Harper and R.M. Sampson (ed.) *Electronic Materials and Processes Handbook Second Edition*, New York: McGraw-Hill (1994), p. 7.65.
18. C. Salerno and R. Reynolds, Ghost in the machine: The rumored death and apparent rebirth of the vapor degreaser, *Precision Cleaning*, Vol. 6, No. 11, pp. 24–30 (1998).
19. F.R. Cala and A.E. Winston, *Handbook of Aqueous Cleaning Technology for Electronic Assemblies*, Port Erin: Electrochemical Publications (1996).
20. D.C. Burkman, D. Deal, D.C. Grant, and C.A. Peterson, Aqueous Cleaning Processes, in W. Kern (ed.) *Handbook of Semiconductor Wafer Cleaning Technology*, Park Ridge, IL: Noyes (1993), p. 141.
21. A.F.M. Leenaars, J.A.M. Huethorst, and J.J. van Oekel, Marangoni drying: A new extremely clean drying process, *Langmuir*, Vol. 6, No. 11, pp. 1701–1703 (1990).
22. L. Tardoskegyi, Cleaning of Printed Circuit Boards by Solid and Coherent Jets of Cleaning Fluid, U.S. Patent 3,868,272 (1975).

23. A. Rae, S. Dalton, and M. Bixenman, Cleaning with a centrifugal semi-aqueous process, *Electronic Packaging and Production*, Vol. 41, No. 9, pp. 40–44 (2001).
24. G. Caplinger, Options in flip chip cleaning, *Advanced Packaging*, Vol. 10, No. 11, pp. 39–42 (2001).
25. M.P. McCurdie, Flux residue cleaning for high-lead bumping, *Semiconductor International*, Vol. 23, No. 12, pp. 149, 150, 152, 154, 156 (2000).

Subject Index

Pd metal, 429

Pd-Ni, 426

PEG (polyethylene glycol), 35, 37, 49, 52, 174, 375–377, 379, 380, 386, 407

perimeter attachment package, 201

periodical pulsed reverse current plating, 393, 395

phase change materials (PCM), 284

phase-change films (PCF), 284–286

phenolsulfonic acid (PSA), 230

phosphor-rich layer, 157

photo diode (PD), 305

photolithography, 6, 210, 249, 325, 342, 360, 388, 441, 456

photometry, 499

photoresist, 5, 13, 23, 24, 32, 149, 161, 168, 169, 172, 177, 180, 188,189, 228, 256, 326, 338, 340, 343–351, 354, 356, 456, 470, 474, 478, 487, 512–519

photosensitive dielectric, 325, 337

photosensitive-BCB, 325, 326

physical and mechanical metallurgy, 225

physical vapor deposition (PVD), 27, 29, 55, 63, 62, 63, 73, 80–84, 102–104, 107, 124, 455

pin grid array (PGA), 203, 362, 369

pin-in-hole, 202

planar structure, 23, 339, 337, 339

planarization, 5, 6, 11, 25, 32, 55, 339, 337, 347, 358, 433, 446, 465

plasma ash, 168, 169

plasma etching, 181, 513

plastic ball grid array (PBGA), 22, 250, 251, 253

plastic due in line package (PDIP), 250

plastic land grid array (PLGA), 22, 252, 253

plastic leaded chip carrier (PLCC), 250

plastic package, 16, 21, 147, 249, 250

plastic pin grid array (PPGA), 22, 252, 253

plastic quad flat pac (PQFP), 250

plated through hole (PTH), 4, 16, 21–25, 253, 255, 256, 260, 373, 391–417

platen motor current, 452

plate-out, 65, 489

plating additives, 34, 39, 46, 259, 339

plating bath control, 495

plating bump, 161, 386

plating cell, 180, 257–259, 261, 476, 484, 485, 487

plating coverage, 22, 257, 259

plating rate, 35, 91, 97, 102–104, 180, 230, 231, 255, 257, 259, 351, 395, 396, 399–401, 403, 408, 411, 474

plating technology, 18, 23, 60, 161, 162, 168, 325, 379, 388, 391, 399, 420

Poisson's ratio, 211, 313

polarization, 7, 35, 39, 54, 55, 67, 72, 73, 174–176, 183, 257, 375, 380

polishing ratio, 436, 439

poly silicon, 460, 463, 463

polyalkylene glycol family, 54

polycrystalline materials, 63, 103

polyimide dielectric layers, 300

porous deposit, 67

powdery appearance, 259

PPR, 380

pre-plating etch step, 360

pre-solder, 371, 381, 383, 386

pre-soldering technology, 383

Preston's equation, 11, 435, 463

pre-wet step, 470, 490

printed circuit board (PCB), 3, 4, 14–16, 20–24, 65, 153, 155, 159, 163, 181, 203, 252, 288, 333, 343, 346, 373, 392, 469, 484, 513, 517

printed wire/wiring/wired board (PWB), 198, 200, 202, 204, 206, 207, 210, 250, 312, 313, 315, 396

printing paste, 383

printing process, 383

process window, 53, 271, 272

product mix flexibility, 261

profilometer measurement, 442